Geography of the Physical Environment

The *Geography of the Physical Environment* book series provides a platform for scientific contributions in the field of Physical Geography and its sub-disciplines. It publishes a broad portfolio of scientific books covering case studies, theoretical and applied approaches as well as novel developments and techniques in the field. The scope is not limited to a certain spatial scale and can cover local and regional to continental and global facets. Books with strong regional focus should be well illustrated including significant maps and meaningful figures to be potentially used as field guides and standard references for the respective area.

The series appeals to scientists and students in the field of geography as well as regional scientists, landscape planners, policy makers, and everyone interested in wide-ranging aspects of modern Physical Geography. Peer-reviewed research monographs, edited volumes, advance and under-graduate level textbooks, and conference proceedings covering the major topics in Physical Geography are included in the series. Submissions to the Book Series are also invited on the theme 'The Physical Geography of…', with a relevant subtitle of the author's/editor's choice. Please contact the Publisher for further information and to receive a Book Proposal Form.

More information about this series at http://www.springer.com/series/15117

Ajay Kumar Taloor •
Bahadur Singh Kotlia •
Kireet Kumar
Editors

Water, Cryosphere, and Climate Change in the Himalayas

A Geospatial Approach

Springer

Editors
Ajay Kumar Taloor
Department of Remote
Sensing and GIS
University of Jammu
Jammu, India

Bahadur Singh Kotlia
Centre of Advanced Study in Geology
Kumaun University
Nainital, Uttarakhand, India

Kireet Kumar
Centre for Land and Water
Resource Management
G.B. Pant Institute of Himalayan
Environment and Development
Almora, India

ISSN 2366-8865 ISSN 2366-8873 (electronic)
Geography of the Physical Environment
ISBN 978-3-030-67931-6 ISBN 978-3-030-67932-3 (eBook)
https://doi.org/10.1007/978-3-030-67932-3

Cover image by Sonja Weber, München

This Springer imprint is published by the registered company Springer Nature Switzerland AG
The registered company address is: Gewerbestrasse 11, 6330 Cham, Switzerland

Foreword

It is a well-known fact and an accepted reality that the existing water resources are under constant threat due to various reasons like deforestation, loss of biodiversity, expansion of agriculture and settlement, over-exploitation of natural resources, habitat loss and fragmentation, construction of roads, large dams, unplanned tourism, etc. Though the Himalayan and other mountain chains and the coastal ecosystems have immense ecological, socioeconomic and aesthetic significance to the earth planet as they provide a wide range of ecosystem-based services to the living species, the Himalayan mountains, being young and geotectonically active, remains inherently unstable, fragile and prone to natural disasters. Further, the climate change is likely to impact the Himalayan cryosphere drastically. Therefore, water becomes an issue of concern and a topic of front-line research for scientists, globally.

In order to address these issues and find viable solutions to ensure life on this planet, the space-borne remote sensing, with its ability to provide synoptic and repetitive coverage, has emerged as a powerful tool for the precise assessment and monitoring of existing water resources. I am extremely happy to learn that a very young and vibrant member of the faculty of my institution, Dr. Ajay Kumar Taloor, has taken up the responsibility to edit a book covering topics on water, cryosphere and climate change in the Himalayas.

I am sure that the topics included in this book shall have adequate impact on the readers and that many will find these topics as thought-provoking. I believe that the book is likely to be a valuable asset to the research community on development and planning management for water and climate change in the Himalayas and shall also serve as a useful reference manual for further work in this ever-significant area of research. Wishing the editors all the best for such an academic endeavour.

Jammu, India
March 25, 2021

Prof. Rajni Kant
Dean Faculty of Science
University of Jammu, Jammu

Preface

Water, cryosphere and climate change are linked with each other, each component being vital for human beings. Water is considered as nectar for life on earth and considered as most precious natural resources on Earth surface. Out of 70% of water on earth, only 3% of drinking water is available, 2% potable and 2 % ice. We restore only 6% of rain water and our lives have disconnected from water. Population explosion and urbanization everywhere have created water problems in recent years. Today, half a billion people in the world face severe water scarcity all year round. It is affecting every continent and was listed in 2019 by the World Economic Forum as one of the largest global risks over the next decade. A serious concern is that water is expected to become increasingly insufficient considerably due to the climate change. Further, the blasting, bombing, explosions and testing of nuclear weapons sufficiently add to the loss of atmospheric environment and a number of such human exercises are indeed the fundamental reason for polluting environment as a whole. The impoverished environment causes health hazards to the life on planet earth and that is likely to continue in future. Therefore, there is a need to address a framework for the assessment of water budget, extraction of water from cryosphere under the climate change and overall impacts on civilization, social set-up and health. In the absence of appropriate water usage, management of water, proper augmentation of water, methods of reusing water, water-related social culture and reverence for water, any kind of development is unthinkable.

The cryosphere, especially in the Polar Regions, is very sensitive to changes in global climate. Therefore, understanding how and why the cryosphere changes over time in response to both natural and human influences is important for forecasting future scenario on our planet. Researchers exercise snow and ice as climate indicators by monitoring trends in the cryosphere over time. Examples of such indicators include how many square kilometres of the ocean are covered by Arctic Sea Ice (sea ice extent), the balance between snow accumulation and melting in glaciers (glacial mass balance), and the amount of land covered by snow. The key issues currently being addressed by most researchers are (1) mechanisms of various kinds of glaciers with respect to climate change and the scale-conversion in water resources assessments, (2) water modelling and heat exchanges between frozen soil and undergrowth and (3) parameterization of physical practices in cryosphere as well as coupling with climate models. Climate change is also likely to impact the Himalayan cryosphere drastically. Space-borne remote sensing with its ability to provide synoptic and repetitive coverage has emerged as a powerful tool for assessment and monitoring of these resources. Researchers in this region are extensively studying the nature and consequences of water resources over the last several decades.

The need of the hour is considerable size of outlay on education, skill development and awareness at grass root level to fight against the threats of climatic change, pressure on future water accessibility and overall environmental breakdown. Advanced technology is equally important to maintain a healthy and productive global biosphere, particularly continental and oceanic biomass. Although nature had its own cycle throughout the geological history, humans have misbalanced it by fast and unplanned urbanization, industrialization and agricultural intensification causing soil, water, air and sound pollutions. Important is that we reduce our overall consumption of resources, recycle what we can and whenever we can, and reuse all the resources and materials that we possibly can.

This book presents the scientific foundation for integrated water, cryosphere and glacier resources and resulting climate change in the light of depletion of natural resources and human sources of environmental deterioration. It contains 20 papers, contributed be eminent scholars and researchers focussing mainly on geospatial approach to assess the future scenario of water–cryosphere–climate change relationship in various regions. Among some extremely significant contributions, Tiwari et al. have discussed application of geospatial techniques for monitoring the cryospheric elements of glacier system in Indian Himalayan Region, while Siva Shankar et al. have forecasted snow melt runoff with input from remote sensing technique. By using remote sensing techniques, Thakur et al. have tried to map snow cover and glaciers. Mir et al. have presented their work on the Ladakh Himalaya, whereas Singh et al. have analysed snow dynamics in Beas Basin by employing Combined Terra-Aqua MODIS Improved Snow Product and In-Situ Data during twenty-first century. Rawat et al. have assessed snow cover and land surface temperature of Mana Basin using MODIS satellite data. Taloor et al. have tried to monitor the seasonal ground water fluctuations using GRACE satellite technology. Kannaujiya et al. have made an assessment of groundwater storage using effective downscaling GRACE Data in some of the water-stressed regions of India. Furthermore, Prakasam et al. have delineated the groundwater potential recharge zones by applying the remote sensing and GIS techniques.

We firmly believe that the book chapters will be extremely useful as a reference material to those interested in geospatial approaches, modelling and planning management for water and climate change particularly in the Himalayas.

The editors are thankful to the contributing authors from various universities/research institutions. We also thank Springer Nature for accepting our proposal to publish this book and also express our gratitude to

Ms. Carmen Spelbos, Project Coordinator, Book Production Springer Nature, Ms. Juliana Pitanguy, Publishing Editor and Mr. Ambrose Berkumans for their assistance and coordination throughout the entire publication process.

Ajay Kumar Taloor
University of Jammu, Jammu, India
e-mail: ajaytaloor@gmail.com

Bahadur Singh Kotlia
Kumaun University, Nainital, India
e-mail: Bahadur.kotlia@gmail.com

Kireet Kumar
G.B. Pant National Institute of Himalayan
Environment and Sustainable Development
Kosi Katarmal, Almora, India
e-mail: kireet@gbpihed.nic.in

Contents

About the Editors

Dr. Ajay Kumar Taloor has obtained his Doctorate in Remote Sensing and GIS applications in hydrogeology from University of Jammu, NAAC accredited A^{+} University of India. Thirteen years of research experience in the applications in geospatial technology for natural resources management for land and water resources. He has excelled twice with best paper presentation award in India. Being an expert of remote sensing applications in remote sensing and GIS application in water science, Cryosphere and climate change, tectonic and quaternary geomorphology, he is working on two major research projects on using space-based inputs for glacier mapping and climate change in Himalayas. He has also published many articles in tectonic and quaternary geomorphology in the recent years. He has high scientific temper and strong HR relations in science world, with high professional and managerial skills.

He has edited many volumes in the top-rated journals in the Elsevier and Springer publishers, member of Editorial Board of the Quaternary Science Advances and reviewer of the many top-rated international journals in science world

Dr. Bahadur Singh Kotlia is Research Scientist (Professor) at Centre of Advanced Study, Department of Geology, Kumaun University, Nainital. He received his Ph. D. degree in Geology from Panjab University, Chandigarh. He is a recipient of Alexander von Humboldt Fellowship. He received his Post-Doctoral degree from University of Bonn, West Germany, Institute of Palaeontology. He has executed various projects in the field of Earth Sciences. He has published over 80 research papers in international journals and 50 research papers in national journals

Er. Kireet Kumar is Scientist-G at G.B. Pant National Institute of Himalayan Environment, Kosi-Katarmal, Almora. He received his M.Tech. degree in Environmental Engineering from I.I.T. Kanpur, Uttar Pradesh. His areas of specialization are environmental engineering, hydrology and water resource management. He is a Fellow of Leadership for Environment and Development (LEAD) International, UK. He has executed various projects in the field of glaciology and hydrology. He has published several research papers in peer-reviewed journals

Application of Geospatial Techniques for Monitoring the Cryospheric Elements of Glacier System in Indian Himalayan Region (IHR)

1

Ashutosh Tiwari, Kireet Kumar, Manoj Patley, and Jyoti Sharma

Abstract

The Himalayan region consists of a number of glaciers, many of which are yet to be explored. Glaciers are the important cryosphere feature and are the source of major rivers of the Indian Himalayan Region (IHR) that drains the plains of the Indian subcontinent down the stream mostly throughout the year. The accessibility issues due to remoteness of location, high elevation and risk involved in the study of glaciers through field survey have not motivated much to explore the unexplored regions of IHR in context to glacier studies. However, the recent advancement in the field of the remote sensing and geospatial technology has emerged as a promising tool for glacier study. The Remote Sensing (RS) data products ranging from multispectral satellite imagery including optical and Infrared EMR data; techniques of active remote sensing such as radar datasets; high spatial resolution Digital Elevation/Surface Models (DEM/DSMs), etc. have made it possible to study the dynamics of the glacier over the period of time without the manual field survey. Studies conducted by scientists, researchers and experts using the geospatial technique have proved the results to be very promising. During the ablation season, the glacier dynamics becomes important to analyse as it reflects overall the precipitation pattern and trend of temperature during the season. The chapter is about to discuss the application of geospatial technique using the RS data and in situ observations to study the parameters of glaciers such as mass balance, retreat rate, velocity, snow cover and ELA over the mid-Himalayan glacier named as Chipa located near the Baling village of Darma valley in Dhauliganga Basin of Pithoragarh district, Uttarakhand, India. Additionally, it has been tried to carry out the study using mostly the freely available RS datasets which were also validated using in situ observations.

A. Tiwari (✉) · K. Kumar
Centre for Land and Water Resource Management, G.B. Pant, National Institute of Himalayan Environment & Sustainable Development, Almora, India
e-mail: ashutosh.gbpnihesd@gov.in

K. Kumar
e-mail: kireetkumar@yahoo.com

M. Patley · J. Sharma
G.B. Pant, National Institute of Himalayan Environment & Sustainable Development, Almora, India
e-mail: mpatley7@gmail.com

J. Sharma
e-mail: jyotisharma54478@gmail.com

A. K. Taloor et al. (eds.), *Water, Cryosphere, and Climate Change in the Himalayas*, Geography of the Physical Environment, https://doi.org/10.1007/978-3-030-67932-3_1

Keyword

Ablation season · Climate change · Cryosphere · Geospatial technology · Glacier dynamics

1.1 Introduction

The Himalayas are the youngest mountain chain of the world and the highest snow peaks of the world lie in this region. The Himalayas hold the largest snow and ice cover area in the world outside the polar regions. The Himalayan region is referred to as the third pole (Schild 2008) and the water tower of Asia (Xu et al. 2009; Taloor et al. 2019; Singh et al. 2020; Sood et al. 2020). The Himalayan range encompasses about 15,000 glaciers, which store about 12000 km^3 of freshwater (Fourth assessment of climate change). The greater and middle Himalayas also known as the Himadri and Himachal respectively are the home to numerous glaciers, most of which yet could not be studied to date due to the issues of accessibility, connectivity and vast remoteness. However, it has become really essential to study the cryosphere elements of the Indian Himalayan region which is the source of major rivers of the subcontinent including the Ganga, the Yamuna and other important rivers which ultimately feed to larger river and population lives in the downstream region. Himalayan glaciers supply meltwater for ~800 million people, including for agricultural, domestic, and hydropower use (Pritchard 2017; Singh et al. 2017; Khan et al. 2020; Haque et al. 2020; Sarkar et al. 2020 Kumar et al. 2020) and affect their lives and livelihoods. Not limited only to humans, the water stored in the form of ice in the glacier also caters to the demands of wildlife, forests, irrigation, etc. The cryospheric elements are very sensitive to the climatic parameters and their dynamics study may indicate the trends of climate change. The snowfed mountains of IHR receive precipitation mainly in the form of snowfall throughout the year but mainly during the monsoon season. Over the successive snowfall, the snow layers are formed which under the pressure of top layers is converted to ice masses. Under the action of gravity, these masses slowly move down the valley and these moving ice masses along the valley are termed as glaciers.

The point of termination of glaciers is known as snout after which the valley is mainly drained further by water resulted from the melting ice. The location and ice mass thickness of the snout mainly depends on the climatic variability such as temperature pattern, precipitation trends over the accumulation zone. The two commonly used terms to understand the behaviour of glaciers are advancing and retreating. But over the year if there are consecutive trends of either advancing or retreating, it may be the indicator to interpret the changing climate scenario of the region.

1.2 Some Major Elements of Glacier Study

1.2.1 Snout Monitoring and Velocity Measurement

The elevation of snout of a glacier may vary and may be located at an elevation as low as 3470 m (Chipa Glacier, Uttarakhand) and as high as 5000 m for Khangri Glacier (Bisht et al. 2018) (Fig. 1.1a), Tawang Valley of Arunachal Pradesh (in situ observation under the combined study by ISRO—SAC Ahmedabad and G.B. pant National Institute of Himalayan Environment and Sustainable Development under the project "Integrated Study of Himalayan Cryosphere") . The top of the snout (Glacier terminus) may be monitored using the high-resolution satellite data or in situ observation through the kinematic DGPS survey along the width of the snout to ensure the accuracy at sub-centimetre level. However, GPS observation may induce an accuracy bias up to 5 m.

The velocity of the glacier for a particular location depends on the thickness of ice masses of the glacier. The more the thickness is the more will be the velocity due to larger mass (F = m*a

Fig. 1.1 **a** Showing the snout of glacier and **b** showing the glacial lake (*Source* Authors)

or w = m*g), where 'F' is force, 'm' is the mass, 'a' and 'g' are the acceleration due to gravity and 'w' is weight. Keeping the 'g' constant force (gravitational) is directly proportional to mass.

In situ observation requires the installation of stakes which must be drilled inside the ice masses nearly 1 m to 2 m. The location of these stakes is required to be monitored temporally preferably using DGPS and over a period of time the velocity may be calculated. On other hand, the freely available tools facilitate the velocity measurement using the SAR (Synthetic Aperture Radar) data in combination with the digital elevation model over a period of time.

1.2.2 Glacial Lake

The melting water or precipitation runoff along the glacier surface sometimes accumulates the water over the glacier surface or down the snout dammed by the moraines and features the glacial lakes (Fig. 1.1b). The area of these lakes can vary from small to large depending on the topography of the surrounding region. Moreover, the volume of water stored exerts heavy pressure on the boundaries, especially along the dammed moraines. When the potential energy of the stored water exceeds the tolerance capacity of the moraine dam, the lake is burst suddenly and results in Glacial Lake Outburst Flood (GLOF). GLOF results in havoc downstream and is responsible for the disaster of flash floods in the down valleys. Kedarnath disaster of 2013 was mainly due to the outburst of a glacial lake of Chorabari glacier.

1.2.3 Mass Balance

The term mass balance means net gain and loss of ice from the glacier system for a given period of time. A glacier system thus for any point of time may be thought of as the product of how much mass it receives and how much it loses by melting. If the gain in ice mass is more than the loss due to melting, the glacier will advance and on the contrary, it will recede. If gain and loss are equal the glacier is said to be in the equilibrium state. The differencing between gaining and reducing mass denotes mass balance which can be measured either on a seasonal basis or annual. Snow pits and probing methods are used to directly measure accumulation for the glacier whereas stakes are installed to measure the surface ablation.

Geodetic measurements of glacier thickness change incorporate two dominant terms, surface mass balance and the vertical component of ice flow, necessitating consideration of flux divergence. Conservation of mass at a point on the surface of a glacier can be stated as

$$\dot{h} = \frac{\dot{\mathrm{b}}}{\rho} \nabla \cdot \vec{\mathrm{Q}}$$

where $\dot{\mathrm{h}}$ is the rate of thickness change, $\dot{\mathrm{b}}$ is the specific surface mass balance rate, ρ is density,

$\nabla.\vec{Q}$ is a flux divergence term (Rasmussen and Krimmel 1999; Cuffey and Paterson 2010).

Stakes are installed into the glacier surface and its co-ordinates including the elevation are regularly acquired preferable using the DGPS to measure the rate of vertical thinning in the ablation zone. Whereas the pits in the accumulation zone give the rate of thickening (accumulation) in the accumulation zone of the glacier.

Geospatial method can also be employed for the mass balance studies where the DEM datasets of two different time periods are analysed for the change in surface elevation values within the glacier boundary. The change in volume can be interpreted for net gain or loss of the ice from the Glacier system over a period of time. The detailed approach is explained later in the chapter.

1.2.4 Accumulation Zone

The accumulation zone of the glacier receives the precipitation in the form of snowfall. It is interesting to note that precipitation in the form of water is regarded as a loss and does not contribute to the accumulation. It can contribute to the accumulation only if the water is percolated and collected down the ice masses or crevasses and refreezes due to lower temperature.

1.2.5 Ablation Zone

The ablation zone of the glacier is responsible for losing the mass of the system in the form of melting. The glacier system losses the mass in the form of meltwater, surface meltwater runoff, sublimation, separation of chunks of ice in case of greater slopes, etc.

1.2.6 Accumulation Area Ratio (AAR)

Accumulation Area Ratio (AAR) is associated with the positioning of ELA at the end of ablation season of the year, above the ELA the accumulation zone of the glacier lies. The ratio between the area of accumulation zone and the total area of the glacier surface is known as AAR. To measure the contribution of accumulated area over the total glacier area, AAR is calculated as

$$AAR = \frac{S_A}{S_T}$$

where S_A refers to the surface accumulation area and S_T is whole surface area of the glacier.

1.2.7 Equilibrium Line Altitude (ELA)

Equilibrium Line Altitude (ELA) is the lowest elevation where the net of accumulation and ablation is zero, above ELA there is accumulation throughout the year and below this elevation ablation is more than the accumulation. ELA can be estimated using the temporal datasets throughout the year. It may be noted that the frequency of RS based observation may be increased especially during the ablation season. The worst-case scenario depicting the least snow cover for the season is to be considered for ELA estimation. Moreover, the multispectral dataset mainly having the bands of green and SWIR (Short wave Infrared) may be utilized to derive the Normalized Difference Snow Index (NDSI) that can be thresholded to derive the snow cover mapping over the satellite imagery. The better the resolution the better would be the accuracy of the estimation.

1.2.8 Debris-Covered and Debris-Free Glaciers

The glacier system may be debris-covered and debris free (Fig. 1.2a). A Debris-Covered Glacier (DCGs) is a glacier where part of the ablation zone has a continuous cover of supraglacial debris across its full width (Kirkbride 2011). However, it may not be necessary that all the DCGs have continuous cover across the full width but it may be stated that debris is present over the large portion of width in a continuous manner. Approximately, 23% of all glaciers

Fig. 1.2 **a** Showing the debris-covered glacier **b** showing debris-free glacier (*Source* Authors)

across the Himalaya Karakoram range are debris-covered glacier (Scherler et al. 2011). This debris may be supplied to the glacier in one or more ways through avalanching, rockfalls, vertical thinning of the glacier, glacier movement cause scrapping of valley walls and small landslides onto the glacier surface (Bisht et al. 2020). Debris over the glacier plays very complex role in the dynamics of glacier especially with respect to water melt. This debris insulates the glacier surface beneath it and mostly it has been observed that the debris-covered glacier responds in case of negative mass balance over a period of time in form of thinning rather than the retreat of the snout. Such a type of glacier retreat is common in the glacier which is normally located on lower elevations. The Kangri glacier of Arunachal Pradesh which is located between 4900 m and 6300 m is debris-free glacier (Fig. 1.2b).

The meltwater of the glacier system near the snout may be studied further to calculate its chemical composition to understand the mineralogy of the underlying geological features.

1.3 Methodology

A variety of remote sensing data products are freely and commercially available that may be utilized to study the above-mentioned components of glacier system. The optical remote sensing data includes the spectral resolution of visible spectrum in the range of 0.45 μm–0.69 μm at varying spatial resolution 30 m (Landsat series datasets) 10 m (sentinel series datasets), LISS IV 5.8 m, etc. It is important to highlight that the Short Wave Infrared (SWIR) band may be utilized (sentinel band 12 at 1610 nm). The SWIR may be customized with the Green band to yield the normalized difference snow index (NDSI).

Active remote sensing datasets such as SAR are also available that may be utilized in the study of glacier system. Normally in the active remote sensing technique, the sensor illuminates the objects of interest using its own source of EMR (Microwave) and receiver. The advantage is that such EMR has higher wavelengths than the optical EMR which facilitates higher penetration power. The SAR data such as RISAT, Sentinel 1 (C Band), therefore, can give the surface feature even in presence of clouds and also can yield subsurface information up to some extent.

It is very important to understand the topography of the glacier system in order to better understand the elevation profile, slope, aspect, etc. Using the stereo pairs of the optical data such as the Cartosat-1 having 2.5 m spatial resolution, the Digital Elevation Model (DEM) of the area can be generated at 10 m (Carto DEM). Other freely available DEM are ASTER and SRTM (Shuttle Radar Topography Mission for 2000) at 30 m and ALOS-PALSAR at 12.5 m for 2008.

1.3.1 Preparation of Base Map

Toposheets survey records the historical blueprint of topography at mainly three different scales that is 25 k scale, 50 k and 250 K. Such survey sheets may be utilized to demarcate the glacier boundary and the contour line which may be utilized to digitally generate the DEM profile. This information can be used to draw the crucial findings for the trend analysis of the glacier system over the period of time. Survey of India records the information of Chipa glacier for 1961, 1979 and 1986. The base map has been prepared using the SOI toposheet of 1961 at 1:50000 scale.

1.3.2 Image Preprocessing

The satellite data should be free from any inherent errors that are caused due to sensor alignment with respect to earth surface, relative motion between earth and the satellite, atmospheric noises and mainly due to two-dimensional representations against actual three-dimensional existences. Therefore, the process of orthorectification, image correction and enhancement is adopted which helps in removing the effects of tilt inherent in the imagery and the relief (terrain effect) of the earth surface (Fig. 1.3). For orthorectification mainly two inputs are involved.

- An imagery having accurate sensor geometry.
- A Digital Elevation Model such as SRTM, Carto DEM, ALOS-PALSAR, etc. with lesser Root Mean Square (RMS) error. The better DEM resolution provides more accurate orthorectification.

The resulted orthorectified image yields the planimetrically corrected image having a constant scale in which the features are represented in 'true geometry'. Orthorectification becomes very essential when the terrain is undulating as that of mountains (IHR). Therefore, for the study of the cryosphere in IHR using the RS data, orthorectified/orthocorrected images yield much better calculations for distance, angles and areas.

In the raw satellite dataset, the image might have radiometric, geometric and atmospheric errors. These errors are required to be addressed through the process known as image correction.

Fig. 1.3 Impact of topography in the image (*Source* Authors)

(a) **Radiometric error**: The satellite data is in raster format where the sensors store the Digital number (DN) values against the reflected brightness from the source. But sometimes all the grids of the sensors may not store the DN value due to which the image rendered is inaccurate by the sensors. Few of its pixels are either heavily illuminated (maximum DN value) or is completely black (Zero DN value). The errors are raised due to the failure of particular grids of sensors as shown in Fig. 1.4a.

To correct the imagery from radiometric errors, radiometric corrections are incorporated where the DNs are converted to radiance value which is further converted to the Top of Atmospheric (TOA) reflectance (requires additional information such as Solar zenith angle and exoatmospheric irradiance). Further, using the TOA, surface reflectance values may be calculated. TOA reflectance is given by

$$\rho\lambda = \frac{M_\rho Q_{cal} + A_\rho}{\cos(\theta_{SZ})}$$

where $\rho\lambda$ is TOA reflectance; M_ρ band-specific multiplicative rescaling factor from the metadata; Q_{cal} is quantized and calibrated standard product DN values; A_ρ is band-specific additive rescaling factor from the metadata; θ_{SZ} is local solar zenith angle. The corrected image may be thus obtained as shown in Fig. 1.4b.

(b) **Geometric error**: Satellite imagery suffers from the error due to the relative position of location between satellite sensor and the earth surface, speed of satellite, unstable sensor platform, irregular topographic feature and viewing path of sensors. These errors are termed as geometric errors. These errors affect the actual shape of the objects of interest as captured in the imagery and the edges are skewed and the angles are distorted. Any mathematical calculation such as area and the perimeter is not accurate. This also impacts the location of pixel coordinates (x, y) over the imagery against the real-world position.

Geometric correction is performed to address the image to render them error free or to minimize the effect of error. These errors are corrected using the technique of spatial adjustment where ground control points are utilized to co-register and geometrically adjust the features of the imagery.

(c) **Atmospheric error**: The atmosphere layers have the impact of refraction, scattering, absorption, etc. Also, there are small particles, air pollutants and aerosols that are suspended along the layers of atmosphere. These atmospheric noises make the imagery hazy based on the degree of concentration of these agents of atmospheric errors. This reduces the visibility of earth surface by the sensor and the true reflectance value of the object of interest is not recorded shown in Fig. 1.4c.

Atmospheric corrections may be performed where the true reflectance value of the object is predetermined and a correction factor is required to be adopted which normalizes the complete

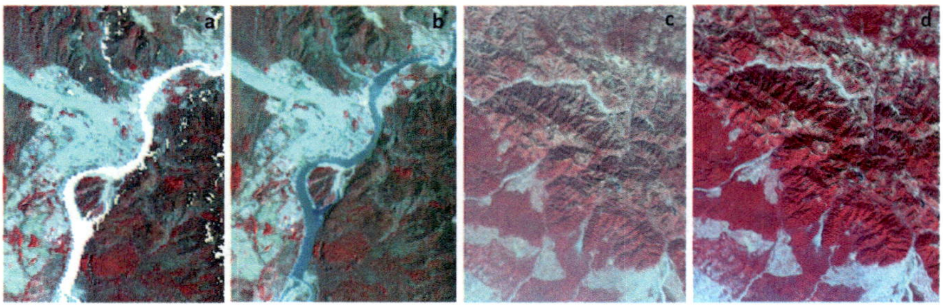

Fig. 1.4 Image errors and corrections **a** Radiometric error (Source LISS II) **b** Radiometric error corrected **c** Atmospheric error (*Source* LISS IV) **d** Atmospheric error Corrected (*Source*LISS IV)

image to nearly the true reflectance value. The corrected image is highlighted in Fig. 1.4d.

In order to make the imagery sharper for visual interpretation, image enhancement methods are applied over the satellite imagery such as contrast stretching, level slicing, histogram equalization, etc. All these methods make the input RS data fit, optimum, and precise for visual interpretation and mathematical calculations.

1.3.3 Image Classification

Image classification is one of the most important geospatial operations in which the input satellite data is classified based on the training samples of different features to be provided to the algorithm (supervised classification), defining the number of different spectral classes in which the imagery is to be synthesized followed by the assignment to most appropriate feature class (unsupervised classification) or segmenting the pixels into the objects that are the set of homogeneously similar pixels which is again assigned to its most appropriate class (object-based classification).

Most suitable from the above-mentioned techniques may be adopted to extract the snow cover area out of the glacier region. As mentioned earlier, indexes like NDSI, NDVI (vegetation index) may be used effectively to extract out the snow and vegetation feature, respectively, from the imagery.

1.3.4 Accuracy Assessment

Observations drawn using the RS data through the geospatial technique is required to be passed through the accuracy assessment. It is accomplished either by comparing the finding with the in situ observation or through comparison of findings over the pilot area where the exact feature distribution is known. Normally, the overall accuracy of any RS based estimation should be equal or better the eighty-five percent to be acceptable. Kappa coefficient may also be calculated which signifies any classification system as how much error has been avoided by the classified image.

1.4 Case Study of Chipa Glacier

1.4.1 Study Area

Chipa Glacier has been chosen as the study area (Fig. 1.5) for the case study which is a part of the Himalayan system located in Dhauliganga basin —Pithoragarh district, Uttarakhand. Chipa glacier area lies between 30°8′53.027″N to 30°11′ 42.595″N latitudes and 80°27′ 43.065″E to 80° 32′14.838″E longitudes. It covers an area of 14.29 km^2 on 3458 m to 5600 m higher elevation. It is an east flowing valley type glacier which is fully covered with debris. The upper surface of the glacier valley walls is fully covered by lush vegetation which is also growing on the glacier surface. This area lies above the Main Central Thrust (MCT) and below the Trans-Himadri Fault (THF) comprises bedrocks of Gneiss and Garnet mica schist. The entire region is dominantly made up of N-E gently dipping rocks of Gneiss which is exposed all along the road cuts and around the village (Baling).

1.4.2 Snout Monitoring

The in situ observation was made during the end of ablation season for snout location acquisition for the year 2017 and 2018. The DGPS survey in rapid static and kinematic mode performs the acquisition of co-ordinates to ensure sub-centimeter accuracy. Leica SR 520 and GX 1220 GPS receiver and AT502 antenna were used for this purpose. The static station is calibrated for 72 h pre-acquisition of the point of interest in kinematic mode. Snout of the glacier was traversed across the width to acquire the co-ordinates. The different location of the snout acquired for the year 2017 and 2018 is mentioned in Table 1.1. These geo co-ordinates were mapped in GIS environment as depicted in Fig. 1.6 to assess the temporal variation in the location of snout during the start of ablation season of respective years.

The DGPS outperforms the accuracy of the handheld GPS device. The handheld GPS normally records the location (latitude, longitude)

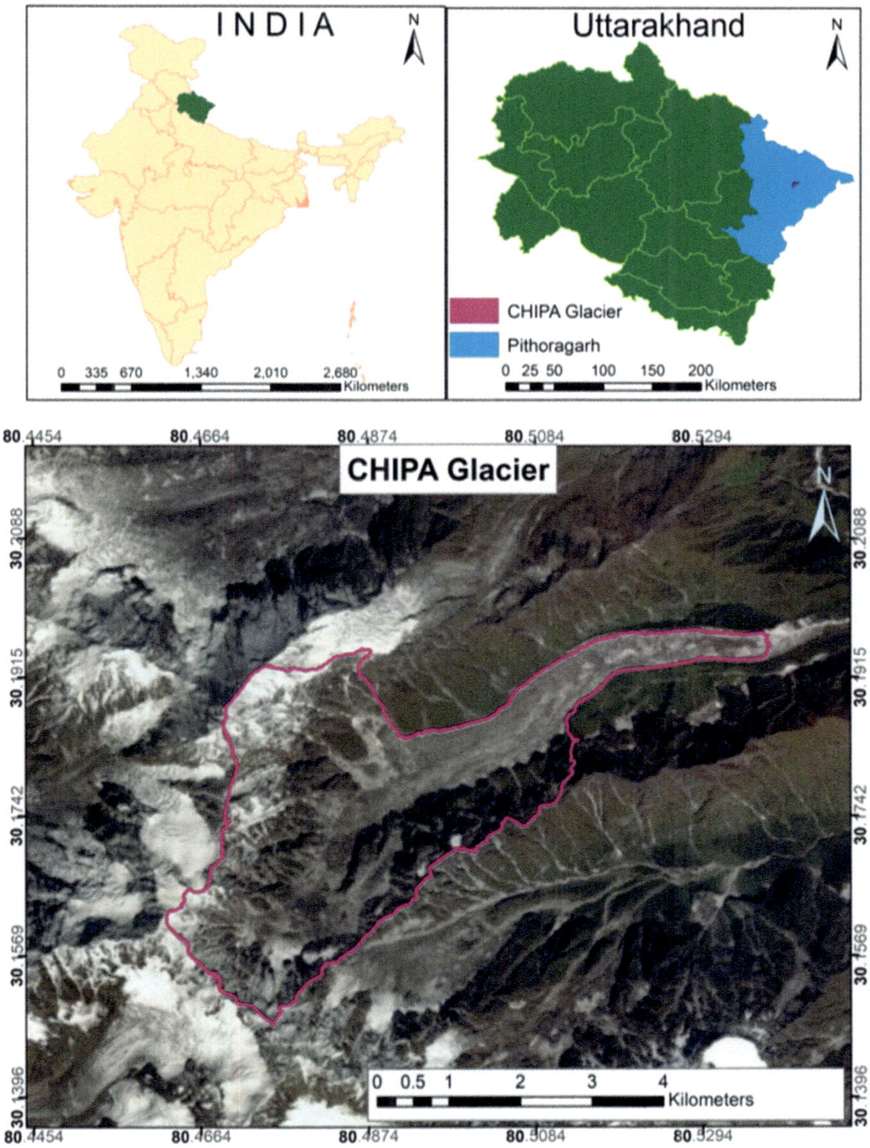

Fig. 1.5 Location map of the Chipa Glacier (*Source* Google image)

within 5 m accuracy. It has been shown that to better understand (Fig. 1.6) the possibility of actual location of the point of interest, a buffer can be drawn around the point with the radius of the accuracy of the device (In this case 5 m case has been shown). The temporal variation may be recorded and mapped to derive monitor the snout location over the period of time. This trend defines the retreating, advancing or equilibrium trend of the glacier. The snout location can also be monitored directly using the high-resolution satellite imagery preferable better than 1 m accuracy. Several web map service (WMS) can be accessed from the open series map services such as Bhuvan, Google Earth, etc. It would be essential to highlight that the time series datasets

Table 1.1 Chipa Glacier snout points collected by GPS during 2017 and 2018

S. No.	Latitude	Longitude	Height (m)	Year
1	30° 11′ 46.6132″ N	80° 32′ 15.2392″ E	3483	2017
2	30° 11′ 46.2984″ N	80° 32′ 15.1908″ E	3779	2017
3	30° 11′ 46.1004″ N	80° 32′ 15.1008″ E	3485	2017
4	30° 11′ 46.050″ N	80° 32′ 15.2016″ E	3481	2017
5	30° 11′ 45.600″ N	80° 32′ 15.2988″ E	3477	2017
6	30° 11′ 45.3012″ N	80° 32′ 15.6012″ E	3480	2017
7	30°11′ 45.7872″ N	80° 32′ 14.3988″ E	3458	2018
8	30°11′ 45.4203″ N	80° 32′ 14.1021″ E	3467	2018
9	30°11′ 46.2084″ N	80° 32′ 14.1023″ E	3463	2018
10	30°11′ 46.0671″ N	80° 32′ 14.2008″ E	3466	2018
11	30°11′ 45.1712″ N	80° 32′ 14.8206″ E	3465	2018
12	30°11′ 46.8456″ N	80° 32′ 14.3016″ E	3476	2018

Source Authors

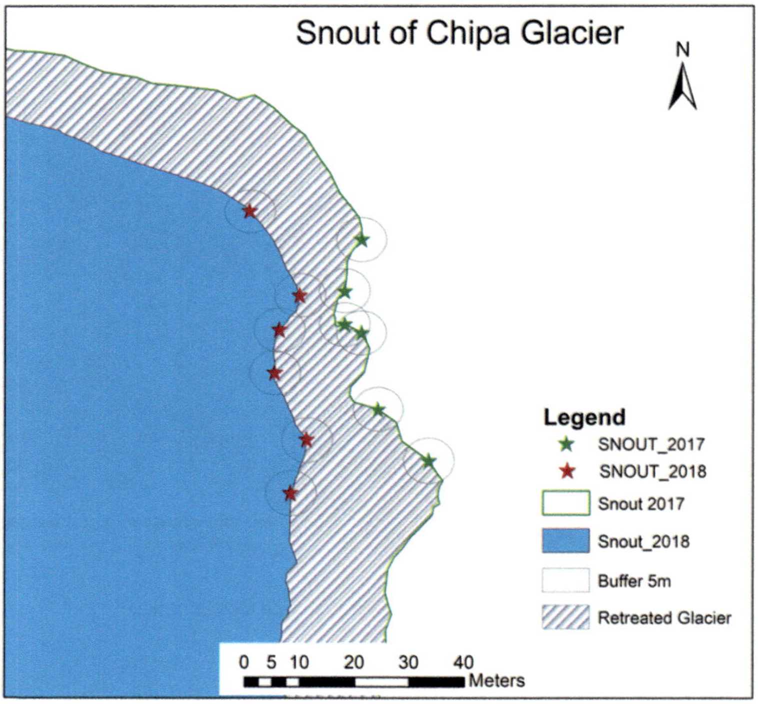

Fig. 1.6 Snout retreat between 2017 and 2018 (*Source* Authors)

should be co-registered with one another, if possible, to avoid geo-referencing pixel errors. The in situ observations were plotted on the Planet Scope imagery at 3 m resolution and also on WMS platform. It was observed that the snout location as retrieved from the imagery was precisely matching with the in situ observations.

1.4.3 Velocity Measurement

The accumulation zone of the glacier receives the snow in the form of precipitation. These snow during successive snowfall marks the layer and is stratified which under pressure is converted to ice masses. The ice masses move down the glacier under the impact of gravity. The velocity of the glacier movement down the valley is required to be studied. The velocity for a particular point of time and location can give the idea of mass balance. When the glacier is thickened, the mass in more and the velocity is measured to be higher and in case of thinning of the glacier mass, the velocity also decreases. The in situ measurement involves stakes installation in the gridded manner. The stakes are drilled down up to 1.5 m from the glacier surface. During the ablation season, the glacier mass is lost both horizontally and vertically. Each stake is marked for its unique identity and its surface measurement for location and elevation is acquired using the DGPS. These stakes are temporally measured from start to end of ablation season to assess the velocity and the rate of thinning of the glacier. The velocity involves both displacement and direction. The temporal locations of the stakes may be plotted on the GIS platform to assess the trend of movement of the glacier to measure the displacement and direction for the given time. It has been also demonstrated that the velocity of the glacier can also be monitored using the geospatial technique. Freely available tools such SNAP (Sentinel Application Platform) can be used to derive the velocity over the whole glacier. The tools consume the Sentinel 1 C band SAR data as input for two time periods and suitable digital elevation model. The velocity map is generated as the output where the value of velocity can be retrieved over any location along the glacier.

1.4.4 Mass Balance Estimation

Mass balance of the glacier describes the overall loss or gain in the glacier system in form of ice. The mass balance study requires the demarcation of the glacier boundary. Toposheet prepared by the survey of India in 1961 at 1:50000 scale has been utilized to demarcate the glacier boundary and its contours to prepare the base map. The contours are used to generate the Triangulated Irregular Network (TIN) of the glacier region. The TIN can be further used to create the DEM for the year 1961. It is to be carefully noted that the projection for the system should be used as SoI'if the toposheet survey is conducted prior to 1984.

The fishnet utility creates the grids of 75 m which is clipped along the boundary of the glacier as shown in Fig. 1.7. These grids are further utilized to extract the mean elevation of DEM using the 'Zonal statistics' feature over the GIS platform. Each grid (including the partial grids along the edges) is calculated for its area which is further multiplied with mean elevation about the grid to calculate the volume. The method is repeated for the 1961 Survey of India (SOI) and 2008 (ALOS-PALSAR) data. Here the mass balance estimation has been estimated using the DEM differencing approach. The time series DEM data SRTM at 30 m for the year 2000 and ALOS-PALSAR at 12.5 m for the year 2008 is utilized on GIS platform. It is important to establish the Root Mean Square Error (RMSE) inherent in the DEM data. This can be calculated by comparing the precise ground observation of control point (object whose location and elevation has been constant for a very long period of time) with the predicted value of the DEM. The formula for the RMS estimation is mentioned below:

$$\mathrm{RMSI} = \sqrt{\frac{\sum_{i=1}^{n}(\mathrm{Oi} - \mathrm{Pi})}{n}}$$

where O_i = Observed value through DGPS Survey

P_i = Predicted Value from ALOS-PALSAR DEM

It was observed that the RMS error of ALOS-PALSAR is much lesser than the SRTM and falls below 3.5 m. The comparison table of the predicted versus observed value is given in Table 1.2.

$$\mathrm{RMS} = \sqrt{\frac{(3663.55 - 3659)^2 + (3658.48 - 3661)^2 + (3609.42 - 3606)^2 + (3612.59 - 3610)^2}{4}}$$

$$= \sqrt{\frac{20.7025 + 6.3504 + 11.6964 + 6.7081}{4}}$$

RMSE = 3.3711 m

A total loss in volume from 1961 to 2008 was differenced out to be (0.275 ± 0.017) km^3. It can

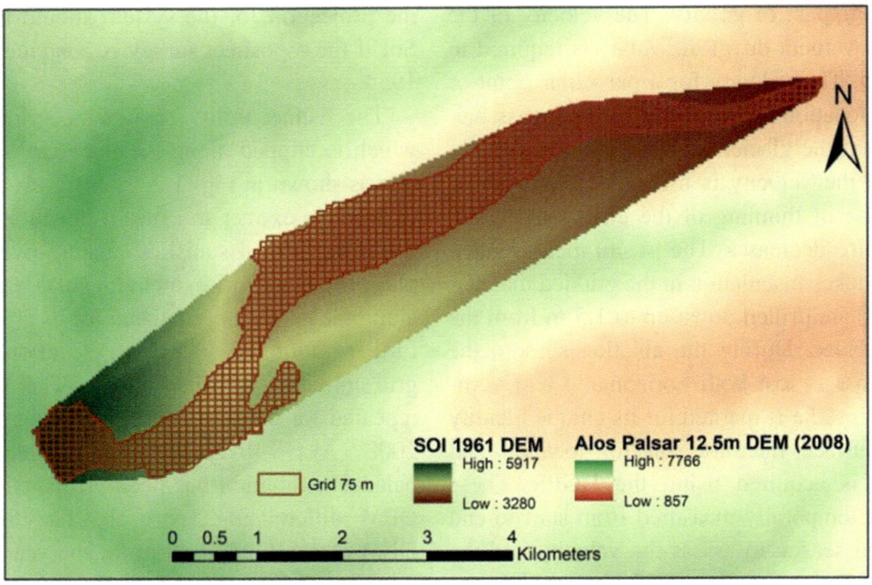

Fig. 1.7 Grid creation (75 m) along SOI 1961 (*Source* Authors)

Table 1.2 Observed and predicted value for RMS calculation for ALOS-PALSAR DEM

S. No.	Latitude (N)	Longitude (E)	DGPS Elevation (m)	DEM Elevation (m)
1	30° 11′ 59.38993″	80° 31′ 30.27837″	3663.55	3659
2	30° 11′ 59.424″	80° 31′ 30.6516″	3658.48	3661
3	30° 11′ 57.1776″	80° 31′ 30.3204″	3609.42	3606
4	30° 11′ 57.1092″	80° 31′ 32.1060″	3612.59	3610

be envisaged that the mass balance calculated using DEM differencing approach from 1961 to 2008 shows loss of mass of the glacier region. Also, the snout monitoring using the time series data of high-resolution satellite data also confirms the retreating trend of the glacier.

1.4.5 Snow Cover Area Mapping and Estimating the Snow Line (ELA) at the End of Ablation Season

Generally, snow has dynamic characteristics and depends on weather and climatic variables. It is textural and physical property changes continuously. The glacier system gains the mass from the precipitation in the form of snow. The accumulation zone of the glaciers deposits the snow in layers which under compression from successive layers is compressed to ice. The study of snow is therefore very crucial as in the glacier system the mass is added by the snow.

Remote sensing high-resolution temporal imageries preferably with Green and Short wave infrared (SWIR) band helps is the precise demarcation of snow and snow cover area analysis. Green and SWIR can be indexed using the Normalized Difference Snow Index (NDSI). To extract the snow cover area, temporal imagery of Sentinel 2B for Band 3(Green) and Band 11 (SWIR) at 10 m and 20 m spatial resolution respectively has been used to calculate Normalized Difference Snow Index (NDSI) which is given by

$$\text{NDSI} = \frac{\text{Green} - \text{SWIR}}{\text{Green} + \text{SWIR}}$$

Figure 1.8 shows the snow cover area mapping in the glacier region during different time periods for the year 2018 and is tabulated in Table 1.3. The intersection of all the snow cover areas gives the area of permanent snow cover throughout the year. The lowest elevation where the snow is found throughout the year gives the snow line or the Equilibrium Line Altitude (ELA). Fig. 1.9 graphically shows the variations of snow cover during different months of the year 2018.

Basically, snow line is the boundary between snow-free and snow-covered surface. The ELA is a theoretical snowline at which the glacier mass balance is zero. Snow Line Altitude (SLA) at the end of melting season is generally regarded as the ELA (Pandey et al. 2013). It may be noted that the elevation of ELA varies at different glaciers due to the relative positioning from the tropics. Annual ELA study helps in understanding the trend of mass balance of the glacier. If the elevation of snow line is decreased, the mass balance increases and if it goes up, the mass balance is assumed to be decreased.

1.5 Results and Discussions

1.5.1 Snout Monitoring

The DGPS inputs of the snout locations were plotted on GIS platform for the end of ablation season for the year 2017 and 2018. It was measured that the mean rate of retreat of snout was 5.89 m in one year. The temporal high-resolution data set of Planet Scope at 3 m for the years 2017 and 2018 were downloaded from https://www.

Fig. 1.8 Snow cover mapping for the year 2018 and intersection for Chipa Glacier (*Source* Sentinel 2B data)

Table 1.3 Fluctuation of snow cover area of Chipa glacier (*Source* Authors)

S. No.	Data acquisition date	Snow cover (km^2)	Snow cover deviation from mean	Accumulation/ablation period
1	07.01.2018	13.19	0.38	Accumulation
2	21.02.2018	13.79	0.98	
3	18.03.2018	13.92	1.11	
4	22.04.2018	12.93	0.12	
5	12.05.2018	12.43	−0.38	Ablation
6	24.10.2018	10	−2.81	
7	08.11.2018	12.75	−0.06	
8	28.12.2018	13.41	0.6	Accumulation

Source USGS

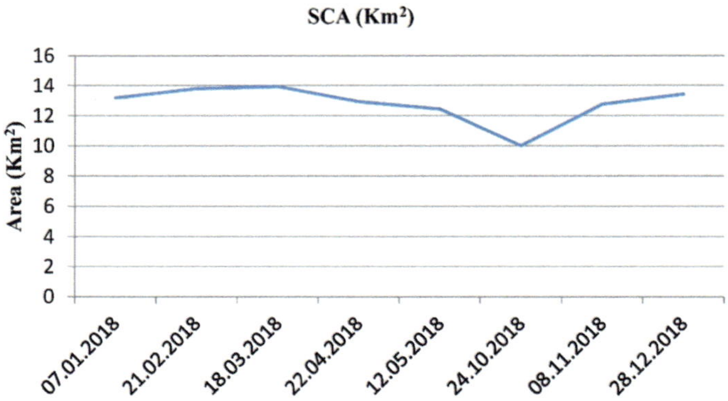

Fig. 1.9 SCA variation Graph of Chipa Glacier for the year 2018 (*Source* Authors)

planet.com/ and also the high-resolution satellite dataset of 0.5 m had been utilized using the WMS over the suitable GIS tool for demarcating the snout boundary, it was observed that the average retreat rate of snout was 6.1 m/year from 2014 to 2018 (Inherent geo-referencing error between imageries over Google Earth engine has been ignored).

1.5.2 Velocity Measurements

Velocity was measured on ground using the installation of network of stakes over the ablation zone of glacier surface. DGPS positions acquired during the beginning (14 June 2018) and end (05 November 2018) of ablation season. The geo coordinates were plotted on the suitable GIS platform to derive the displacement for calculating the velocity as shown in Fig. 1.10 Average displacement of the stakes was calculated to be 5.09 m in 145 days. Therefore, the velocity was calculated to be ~0.035 m/day. 1.5.3. Mass balance.

The mass balance can be estimated using the DEM differencing approach by the given formula:

$$B = \rho \delta V$$

And, $\delta V = (A_1Z_1 - A_2Z_2)$

Where δV is change in volume in (m^3); A_1 and Z_1 is the area of glacier boundary in(m^2) and mean elevation* in (m) respectively derived from DEM prepared using contours at 40 m demarcated from 1961 SoI toposheet at 1:50000; A_2 and Z_2 is the area of glacier boundary in (m^2) and mean elevation* in (m)derived from ALOS-PALSAR DEM at 12.5 m for the year 2008.

(*Mean elevation has been calculated over each grid of the fishnet along the glacier boundary).

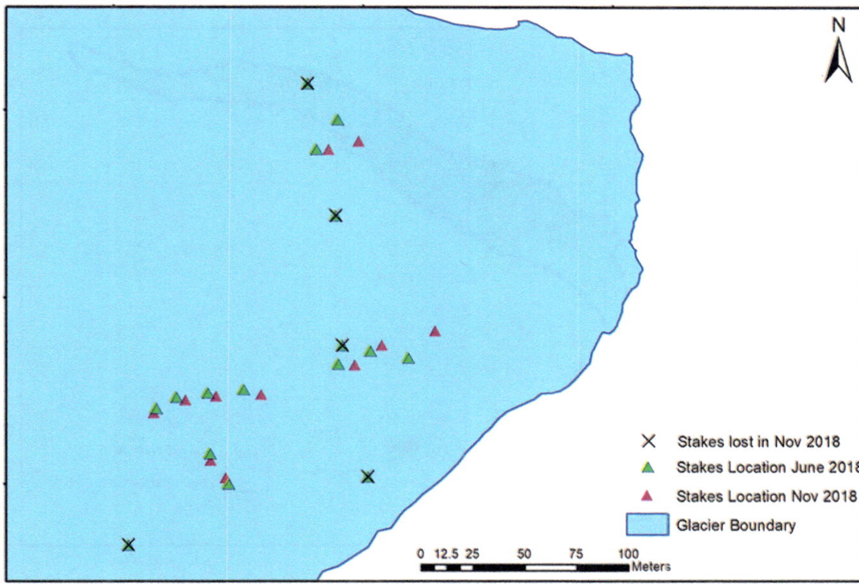

Fig. 1.10 Temporal variation in stakes (*Source* Authors)

B is the mass balance in Kg; ρ = ice density in Kgm^{-3}.

Density varies in the ablation and accumulation zone of the glacier. The uncertain density value due to the density assumption is explored by either setting $\rho = 900\ kgm^{-3}$ for the entire glacier, or 900 kgm^{-3} in the ablation area, and 500 to 600 kgm^{-3} in the firn zone for converting volume changes to mass changes (Huss 2013). In our study, density of ice ρ has been assumed to be 900 kg m^{-3} $\delta V = (0.275 \pm 0.017)*10^9 m^3$as calculated between 1961 and 2008.

Mass balance (B) = $(247.5 \pm 15.3)*10^9$ kg

Mass balance value reveals the loss of $(247.5 \pm 15.3)*10^9$ kg of ice mass between 1961 and 2008 (forty-seven years).

1.5.3 Snow Cover Area Mapping and Accumulation Area Ratio (AAR)

The snow cover mapping done for accumulation and ablation zone carried out using the temporal sentinel 2B datasets reveals:

SCA_{max} on 18th March 2018 = 13.93 km^2
SCA_{min} on 24th October 2018 = 10.0 km^2
Total glacier area = 14.17 km^2
$AAR_{max} = 0.98$
$AAR_{min} = 0.71$

1.5.4 Equilibrium Line Altitude (ELA)

ELA has been estimated using the intersection of temporal datasets of Sentinel 2B at 10 m for the year 2018. The ELA was found at 3990 m. Determination of snow line on Chipa glacier accomplished using a number of spatial data monitoring. We found that the snow line lies on a 3990 m elevation above to which permanent snow cover lies throughout the year (Fig. 1.11).

1.6 Conclusions

The study of Chipa glacier reveals the trends of the cryosphere in the Indian Himalayan Region (IHR). The cryosphere is highly sensitive to climatic parameters such as increase/decrease in temperature or precipitation patterns. Even a small change in climate may have a severe impact on such system. The Chipa glacier clearly demonstrates the retreating pattern through its

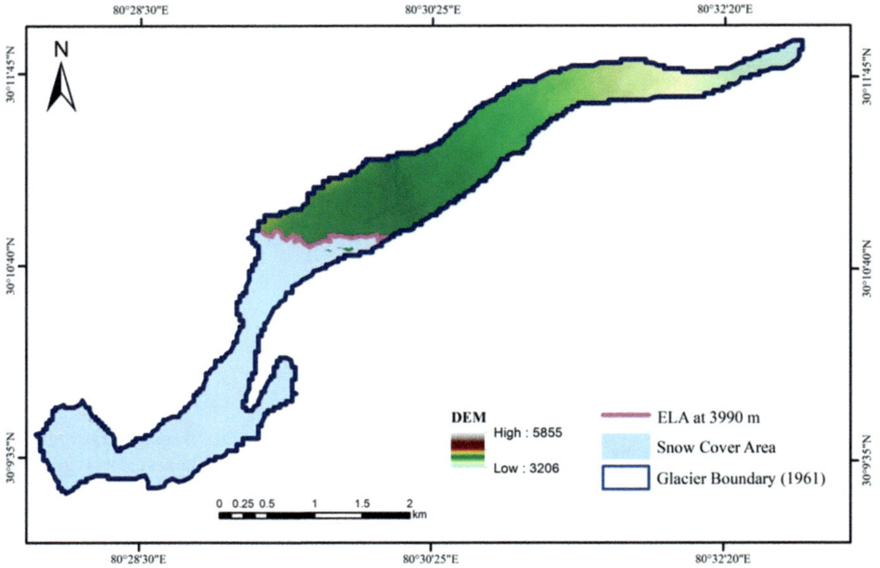

Fig. 1.11 ELA of Chipa glacier for the year 2018 (*Source* Authors)

snout and mass balance dynamics. Snout monitoring revealed the retreat of snout at an average rate of approximately ~6 m per year. The mass balance of the glacier is declining at an average rate of $5.27*10^9$ kg per year. The studies on other glacier systems of IHR are also suggesting similar trends of the decline of the cryosphere of the Himalayas. It is now evident that the Himalayan region is being hit by the changing climate at an alarming rate.

Acknowledgements The authors extend their high acknowledgement to the Director of GB Pant National Institute of Himalayan Environment and Sustainable Development for his support and facilities of the institute to carry out the study. All due acknowledgements to 'Cryospheric Sciences Division, ISRO SAC Ahmedabad' for the project "Integrated studies on Himalayan Cryosphere". The contribution of the research and field staff is also acknowledged for their dedication and support.

References

Bisht H, Arya PC, Kumar K (2018) Hydro-chemical analysis and ionic flux of meltwater runoff from Khangri Glacier, West Kameng, Arunachal Himalaya, India. Environ Earth Sci 77:1–16. https://doi.org/10.1007/s12665-018-7779-6

Bisht H, Kotlia BS, Kumar K, Joshi LM, Sah SK, Kukreti M (2020) Estimation of the recession rate of Gangotri glacier, Garhwal Himlaya (India) through kinematic GPS survey and satellite data. Environ Earth Sci 13(6):1–12. https://doi.org/10.1007/s12665-020-090780

Cuffey KM, Paterson WSB (2010) The physics of glaciers, 4th edn. Butterworth-Heinemann, Oxford

Haque S, Kannaujiya S, Taloor AK, Keshri D, Bhunia RK, Ray PKC, Chauhan P (2020) Identification of groundwater resource zone in the active tectonic region of Himalaya through earth observatory techniques. Groundwater Sust Develop 10:P100337. https://doi.org/10.1016/j.gsd.2020.100337

Huss M (2013) Density assumptions for converting geodetic glacier volume change to mass change. The Cryosph 7:877–887

Khan A, Govil H, Taloor AK, Kumar G (2020) Identification of artificial groundwater recharge sites in parts of Yamuna river basin India based on remote sensing and geographical information system. Groundwater Sust Develop 11: P100415. https://doi.org/10.1016/j.gsd.2020.100415

Kirkbride M (2011) Debris-covered glaciers. Encylopedia of Snow, Ice and Glaciers: Encyclopedia of Earth Series, 180–182. https://doi.org/10.1007/978-90-481-2642-2_622

Kumar D, Singh AK, Taloor AK, Singh DS (2020) Recessional pattern of Thelu and Swetvarn glaciers between 1968 and 2019, Bhagirathi basin, Garhwal Himalaya, India. Quat Int. https://doi.org/10.1016/j.quaint.2020.05.017

Pandey P, Kulkarni AV, Venkataraman G (2013) Remote sensing study of snowline altitude at the end of melting season, Chandra-Bhaga basin, Himachal Pradesh, 1980–2007. Geocarto Int 28(4):311–322. https://doi.org/10.1080/10106049.2012.70533

Pritchard (2017) Asia's glaciers are a regionally important buffer against drought. Nature 545:169–174

Rasmussen LA, Krimmel RM (1999) Using vertical aerial photography to estimate mass balance at a point. Geografiska Annaler: Series A, Phy Geog 81(4):725–733

Sarkar T, Kannaujiya S, Taloor AK, Ray PKC, Chauhan P (2020) Integrated study of GRACE data derived interannual groundwater storage variability over water stressed Indian regions. Groundwater Sust Develop 10: P100376. https://doi.org/10.1016/j.gsd.2020.100376

Scherler D, Bookhagen B, Strecker MR (2011) Spatially variable response of Himalayan glaciers to climate change affected by debris cover. Nature Geosci 4 (3):156–159. https://doi.org/10.1038/ngeo1068

Schild A (2008) ICIMOD's position on climate change and mountain systems. Mt Res Dev 28:328–331

Singh AK, Jasrotia AS, Taloor AK, Kotlia BS, Kumar V, Roy S, Ray PKC, Singh KK, Singh AK, Sharma AK (2017) Estimation of quantitative measures of total water storage variation from GRACE and GLDAS-NOAH satellites using geospatial technology. Quat Int 444:191–200. https://doi.org/10.1016/j.quaint.2017.04.014

Singh S, Sood V, Taloor AK, Prashar S, Kaur R (2020) Qualitative and quantitative analysis of topographically derived CVA algorithms using MODIS and Landsat-8 data over Western Himalayas. Quat Int, India. https://doi.org/10.1016/j.quaint.2020.04.048

Sood V, Gusain HS, Gupta S, Taloor AK, Singh S (2020) Detection of snow/ice cover changes using subpixel-based change detection approach over Chhota-Shigri glacier. Quat Int, Western Himalaya, India. https://doi.org/10.1016/j.quaint.2020.05.016

Taloor AK, Kotlia BS, Jasrotia AS, Kumar A, Alam A, Ali S, Kouser B, Garg PK, Kumar R, Singh AK, Singh B (2019) Tectono-climatic influence on landscape changes in the glaciated Durung Drung basin, Zanskar Himalaya, India: a geospatial approach. Quat Int. 507:262–273

Xu J, Grumbine RE, Shrestha A, Eriksson M, Yang, Wang YUN, Wilkes A (2009) The melting Himalayas: cascading effects of climate change on water, biodiversity, and livelihoods. ConserBio 23(3):520–530

Mr. Ashutosh Tiwari, a Scientist by profession, is working in G.B. Pant National Institute of Himalayan Environment (GBPNIHE), an autonomous institute of Ministry of Environment, Forest and Climate Change, Government of India. Additionally, he is a Ph.D. scholar in the discipline of Geo-engineering, Andhra University and ISRO. He holds M. Tech. degree in GIS and Remote Sensing from National Institute of Technology, Allahabad. Earlier, he received his B. Tech. in Computer Science Engineering. He has more than 8 years of extensive experience in research and development in the field of Geospatial Technologies and Remote Sensing in diverse organizations under ministries and industries including scientific, space, IT, PSU and academia. Earlier he has served in various organizations such as Tata Consultancy Services, NRSC-ISRO, NIT—Nagpur and Kolkata Port Trust in different roles pertaining to application of GIS, remote sensing and computer science. He has been the contributor of policy documents of NITI Aayog for two out of five working group reports on 'Contributing to Sustainable Development in Indian Himalaya Region'. His areas of interest include the field of Himalayan Cryosphere, urban studies, ground and surface water hydrology, water scarcity mapping, image classification through rulesets, big data analytics, mathematical and GIS modelling, cartography, algorithms for image analysis, database, etc.

Er. Kireet Kumar is Scientist-G at G.B. Pant National Institute of Himalayan Environment, Kosi-Katarmal, Almora. He received his M.Tech. degree in Environmental Engineering from I.I.T. Kanpur, Uttar Pradesh. His areas of specialization are environmental engineering, hydrology and water resource management. He is a Fellow of Leadership for Environment and Development (LEAD) International, UK. He has executed various projects in the field of glaciology and hydrology. He has published several research papers in peer-reviewed journals

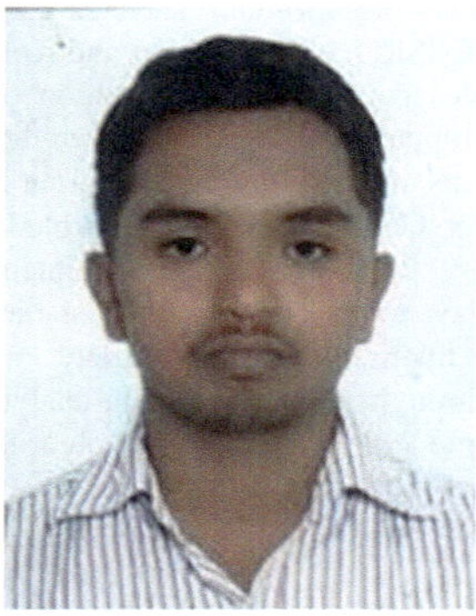

Mr. Manoj Patley received his M. Tech. in Geoinformatics from Bharathidasan University Tiruchirappalli, India, in 2017 and M. Phil in Geography from Department of Regional Planning and Economic Growth, Barkatullah University, Bhopal in 2015. He completed his Master's thesis from Polar Remote Sensing Division, NCPOR, Goa in Glacier mapping. He joined Centre for Land and Water Resource Management (CLWRM), G. B. Pant National Institute of Himalayan Environment Almora, Uttarakhand India in 2018 as a Junior Research Fellow and worked on Himalayan Cryosphere project funded by Space Application Centre (SAC), ISRO, Ahmedabad. He participated in High altitude Glacier expeditions to Himalaya (Uttarakhand and Arunachal Pradesh). His research interest includes high-resolution remote sensing, glacier mapping, Himalayan landscape and urban studies

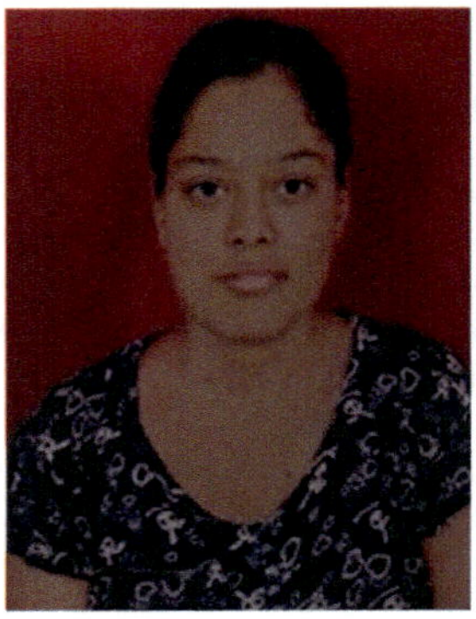

Jyoti Sharma is currently working as Research Fellow at Haryana Space Application Center (HARSAC), Node Gurugram, Haryana. She has completed her post-graduation in Remote Sensing and GIS from Centre of Excellence for NRDMS in Uttarakhand, Department of Geography, Kumaun University, SSJ Campus Almora. She has worked on Himalayan Cryosphere project at G.B. Pant National Institute of Himalayan Environment, Almora, Uttarakhand India in 2018 as a Junior Project Fellow and worked on Himalayan Cryosphere project funded by Space Application Centre (SAC) ISRO, Ahmedabad. She participated in high altitude glacier expeditions to Himalaya (Uttarakhand). She is profoundly interested in RS and GIS applications including advance techniques, cryospheric studies, agricultural mapping, etc.

2 Snowmelt Runoff Forecasting in Himalayan Basins Using Remote Sensing Inputs

E. Siva Sankar, B. Simhadri Rao, and K. Abdul Hakeem

Abstract

In Himalayas, snow, ice and glaciers play an important role in the hydrology of the region for the growth and sustainable development of the inhabitants. The snowmelt runoff, occurring mostly during April, May and June months, constitutes a substantial part of the water resources of the major perennial rivers of Northern and Eastern India, namely the Indus, Ganga, Brahmaputra and their tributaries. In the summer months, the snowmelt runoff is of vital importance for hydroelectric power generation, irrigation and water supply in the northern states of India. All major multipurpose projects like Bhakra depend heavily on snowmelt runoff inflows especially during spring months. Therefore, correct and timely information about the snow cover conditions in the watersheds and on the volume of snowmelt runoff likely to occur is of great importance to the water resources managers. A study has been carried out for Central Water Commission for forecasting seasonal snowmelt runoff for few selected basins located in the Himalayas, to estimate the snowmelt runoff using an energy balance approach. The energy balance approach involves computation of energy flux components which exchange between snow pack and the atmosphere. Major components have been estimated using satellite data derived inputs making suitable assumptions. The satellite data derived inputs include snow cover area, snow albedo, snow emissivity, land surface temperature, digital elevation model, land use, and land cover. The field data such as observed discharge and rainfall have been used. Independent seasonal snow melt runoff models were developed for Chenab, Beas, Sutlej, Yamuna, Bhagirathi and Alaknanda basins, to estimate snowmelt runoff during 01 April–30th June period. The developed model has been calibrated and validated using historic data (2006–2012). In the forecast situation, snow cover area as of 30th March is only available by 1st April. In the absence of other input data sets, the current snow cover area as a primary input in conjunction with other data inputs derived from historic years, were used for estimating snowmelt volume. The runoff from glacier melt and rainfall is suitably computed considering historic averages. The aggregate of estimated snowmelt runoff, rainfall–runoff, glacier runoff, and base flow is considered as

E. Siva Sankar (✉) · B. Simhadri Rao · K. Abdul Hakeem
National Remote Sensing Centre, ISRO, Hyderabad, India
e-mail: sivasankar_e@nrsc.gov.in

B. Simhadri Rao
e-mail: simhadrirao_b@nrsc.gov.in

K. Abdul Hakeem
e-mail: abdulhakeem_k@nrsc.gov.in

A. K. Taloor et al. (eds.), *Water, Cryosphere, and Climate Change in the Himalayas*, Geography of the Physical Environment, https://doi.org/10.1007/978-3-030-67932-3_2

the total runoff. The accuracy of forecasts was better than 90% in most of the basins except during 2013. There was unprecedented rainfall and unusual snowfall occurred in the central Himalayas during June 16–18, 2013. In view of this, substantial additional snow cover was observed and resulted in more discharge due to which there is a large deviation of the forecast with respect to observed discharge.

Keywords

Snowmelt runoff · Satellite data · Remote sensing · Himalayan rivers

2.1 Introduction

Water in its frozen state accounts for more than 80% of the total fresh water on the Earth and is the largest contributor to rivers and groundwater over major portions of the middle and high latitudes (Dozier 1989; Kumar et al. 2020; Singh et al. 2020). Snow, ice and glaciers play important interactive roles in the Earth's radiation balance, because snow has a higher albedo than any other natural surface. Over 30% of the Earth's land surface is seasonally covered by snow, and 10% is permanently covered by glaciers. Understanding of global and regional climate and assessment of water resources require that we monitor the temporal and spatial variability of the snow cover over land areas, from the scales of small drainage basins to continents. Much of the uncertainty and sensitivity in the global hydrological cycle lies in these reservoirs of frozen water, and the melting of mountain glaciers during the last half century appears to account for much of the corresponding rise in sea level (Meier 1984; Taloor et al. 2019; Sood et al. 2020).

In the case of Himalayas also, snow, ice and glaciers play an important role in the hydrology of the region. The snowmelt runoff, occurring mostly during April, May and June months, constitutes a substantial part of the water resources of the major perennial rivers of Northern and Eastern India, namely the Indus, Ganga, Brahmaputra and their tributaries. In summer months, the snowmelt runoff is of vital importance for hydroelectric power generation, irrigation and water supply in the states of Himachal Pradesh, Haryana, Punjab, Jammu & Kashmir, Delhi, Uttar Pradesh, Bihar, West Bengal, etc. Important multipurpose projects like Bhakra depend heavily on snowmelt runoff inflows. Therefore, correct and timely information about the snow cover conditions in the watersheds and on the volume of snowmelt runoff likely to occur is of great importance to the water resources managers.

Since the year 2000, National Remote Sensing Centre (NRSC) provided seasonal forecasts of snowmelt runoff inflows into Bhakra reservoir to Bhakra Beas management Board (BBMB), in the first week of April every year. Snowmelt runoff was estimated based on the hypothesis of depletion analysis, i.e. the thick snowpack having high snow water equivalent depletes later and slower compared to a thin snowpack having low snow water equivalent which depletes early and faster. With the availability of medium resolution daily satellite images on snow albedo and surface temperature from MODIS and similar sensors, it gave an opportunity to characterize the snowmelt runoff process based on the surface energy balance approach. A study was carried out by the National Remote Sensing Centre, (NRSC) of ISRO during 2009–2014 period to develop snowmelt runoff model and provide snowmelt runoff forecast during the summer months for few selected basins in the Himalayas. This chapter briefly discusses the study area, energy balance principles, methodology, input data preparation, model development and forecast results.

2.2 Study Area

The study area comprised of five river basins in the western and central Himalayan region, namely Chenab, Beas, Sutlej, Yamuna and Ganga. Under the Ganga river system, Bhagirathi and Alaknanda rivers were considered for the present study. Considering the current objectives, study area was limited up to snow bound regions where snowmelt runoff is dominant. The basin outlets are selected considering the location of discharge sites and the surface runoff is unobstructed with any storage structures upstream of these outlets. The following are the outlet locations for which the model was set up and the forecast was provided. The study area location is shown in Fig. 2.1. The drainage area up to the basin outlet for each basin is given in Table 2.1.

1. Chenab up to Premnagar,
2. Beas up to Bhuntar,
3. Sutlej up to Bhakra,
4. Yamuna up to Tajewala,
5. Ganga basin
   (a) Bhagirathi up to Uttarkashi, and
   (b) Alaknanda up to Rudraprayag.

2.3 Methodology

The snowmelt runoff is modelled either by a lumped approach using degree day index or by energy balance approach. In view of the availability of satellite data based products on Snow Cover Area (SCA), Glacier Cover Area (GCA), Land Surface Temperature (LST) and emissivity, Snow Albedo, Land Cover, etc., the present study was proposed to use energy balance based approach for modelling the snowmelt runoff

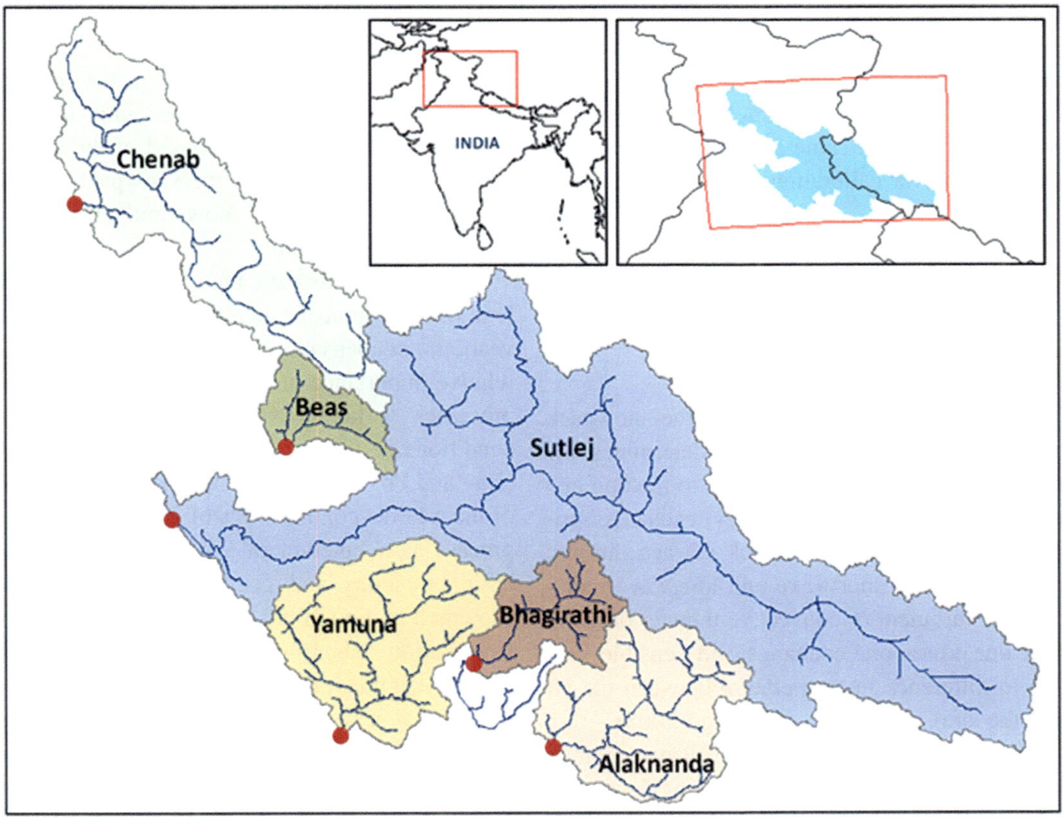

Fig. 2.1 Location map of the study area (*Source* NRSC 2014)

Table 2.1 Drainage area for each basin

Basin	Drainage area (km^2)
Chenab up to Premnagar	17,273
Beas up to Bhuntar	3,160
Sutlej up to Bhakra	51,451
Yamuna up to Hatnikund/Tejawala	11,323
Bhagirathi up to Uttarkashi	4,527
Alaknanda up to Rudraprayag	10,201

Source NRSC (2014)

process for forecasting the snowmelt runoff. The runoff at the basin outlet comprises snowmelt, glacier melt, runoff due to rainfall and baseflow components. The present runoff model addresses all these four components. The approach followed in estimating each of these components is explained in the following sections.

The total runoff (Q) during snowmelt season (Apr-May-Jun months) is the sum of snowmelt runoff (Q_s), glacier melt runoff (Q_g), rainfall–runoff (Q_r) and base flow (Q_b). The runoff measured in the field at the outlet point represents total runoff and the same is compared with estimated runoff for calibration and validation.

$$Q = Q_s + Q_g + Q_r + Q_b. \quad (2.1)$$

2.3.1 Snowmelt Runoff

The exchange of energy between the snowpack and its environment ultimately determines the rate of snowpack water losses due to melting and evaporation/sublimation. Energy exchange primarily occurs at the snowpack surface through the exchange of shortwave and longwave radiation and turbulent or convective transfer of latent heat due to vapour exchange and sensible heat due to difference in temperature between the air and the snow (DeWalle and Rango 2008).

The sources of energy that cause snowmelt include both shortwave (Q_{sn}) and longwave (Q_{ln}) net radiation, convection from the air (sensible energy, Q_h), vapour condensation (latent energy, Q_e), and conduction from the ground (Q_g), as well as the energy contained in rainfall (Q_p). These fluxes are usually measured as energy per time per unit area of snow. The energy budget equation that describes the energy available for snowmelt is given in Eq. 2.2 below (U.S. Army Corps of Engineers 1998). The total energy available for snowmelt is Q_m.

$$Q_m = Q_{sn} + Q_{ln} + Q_h + Qe + Q_g + Q_e - \Delta Q_i \quad (2.2)$$

where ΔQ_i is the rate of change in the internal energy stored in the snow per unit area of snowpack. This term is composed of the energy to melt the ice portion of the snowpack, freeze the liquid water in the snow, and change the temperature of the snow.

Evaluating snowmelt theoretically is a problem of heat transfer involving radiation, convection, condensation, and conduction. The relative importance of each of these heat transfer processes is highly variable, depending upon conditions of weather and the local environment. Gray and Prowse (1992) tabulate selected results of the relative contributions of each heat transfer process as a function of site environment. The basic equations and coefficients that describe snowmelt at a point have been derived primarily from various laboratory and field experiments.

The summation of all sources of energy (Eq. 2.2) represents the total amount of energy available for melting the snowpack (Q_m). The amount of snowmelt (M) at a point may be expressed as Eq. 2.3.

$$M = Q_m / L\rho_{wB} \quad (2.3)$$

where

M = snowmelt, m of water equivalent,

Q_m = algebraic sum of all heat components, $kJ/m^{2,}$

B = thermal quality of the snow (e.g. ratio of heat required to melt a unit weight of the snow to that of ice at 0 °C),

L = latent heat of fusion of ice, 334.9 kJ/kg,

ρ_w = density of water, $kg/m^{3.}$

A melting snowpack consists of a mixture of snow (ice) and a small quantity of free (liquid) water trapped in the interstices between the snow grains. The relative proportion of a snowpack that consists of ice determines the thermal quality (B) of the snowpack. A snowpack that contains no free water has a thermal quality of 1.0. However, after melt has begun, there is some free water held within the snow matrix, yielding a thermal quality of less than 1.0. Using the Eq. 2.3 above, the snowmelt is calculated spatially with remote sensing based inputs from various sources (Fig. 2.2). The basin boundaries were delineated using ASTER digital elevation model and Advanced Wide Field Sensor (AWiFS) sensor data on-board Resourcesat satellite, considering the basin outlet points for which the runoff forecast is planned. The AWiFS satellite data has been used to map the land use and land cover of the study area. The land cover is categorized primarily into open areas, evergreen forest, deciduous forest and water bodies.

The snow cover present in a basin can be mapped using appropriate satellite data. Due to the high probability of the presence of cloud cover spatially or temporally, it is difficult to acquire suitable satellite data of predefined time periods. Therefore, it was decided to use 8-day time composite snow cover products available on MODIS website. The 8-day time composites are for specific Julian 8-day periods commencing

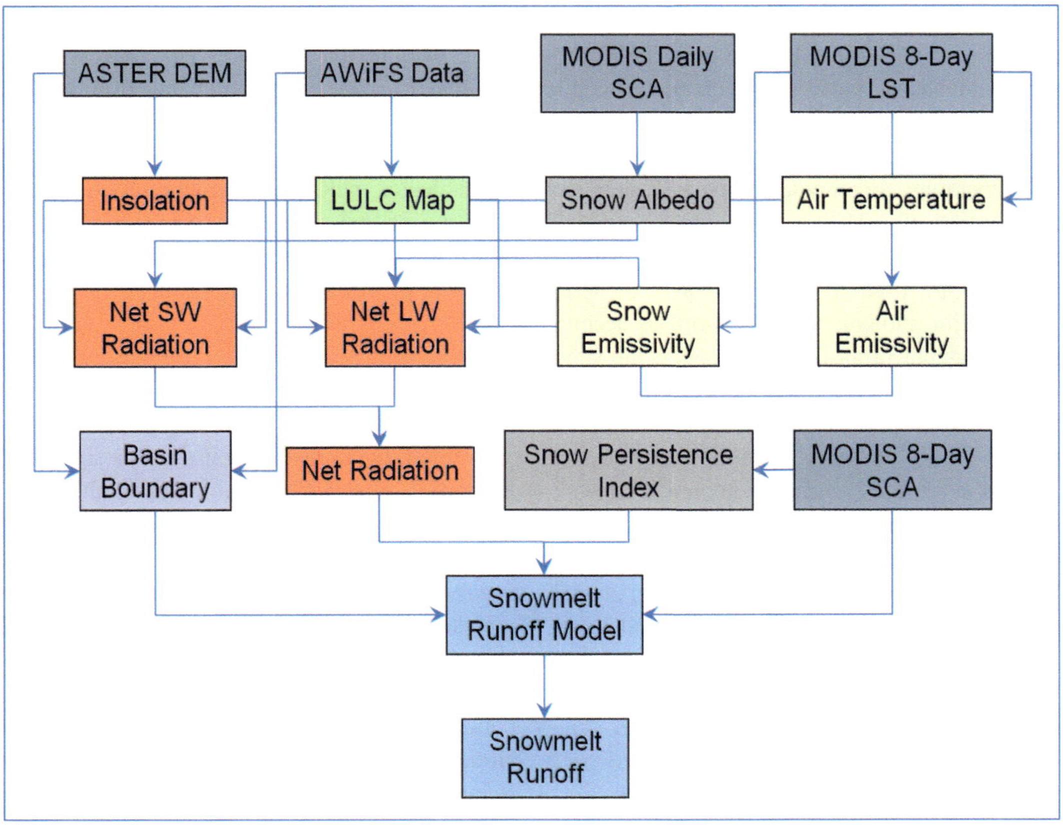

Fig. 2.2 Flowchart of methodology for snowmelt runoff modelling (*Source* NRSC 2014)

from 01 January of a year. The data from 30 March to 30 June are downloaded and used for several years (from 2002 to 2014) in the model setup.

The total net energy flux is primarily contributed by incoming solar radiation and outgoing longwave radiation. The incoming solar radiation is a function of the location of a place (latitude, longitude), elevation, slope, Julian day and time. The longwave radiation is a function of surface temperature, air temperature and emissivity. The incoming solar radiation is computed for each 8-day time period using tools available in ArcGIS s/w and digital elevation model. The total insolation is computed for the 91-day period from 01 April to 30 June. The insolation is corrected for atmospheric transmittance/scattering and cloud absorption/scattering. The net shortwave radiation is computed as the difference of incoming solar radiation corrected for interception due to land cover and outgoing shortwave radiation computed using snow albedo.

For the purpose of estimating longwave radiation, the Land Surface Temperature (LST) product available as 8-day time composite on the MODIS website for free download, are used. The snow emissivity is derived from LST data. The air temperature is computed using LST data and DEM empirically. The air emissivity is computed using air temperature. The energy balance equations are used to compute the net energy available for snowmelt. In addition, to account for snow depth in an indirect manner, the snow cover depletion concept is used. The snow cover depletion concept states that a thicker snowpack depletes slower and later and a thinner snowpack depletes faster and earlier. Based on this concept, to account for snow depth in an indirect manner, snow persistence index is computed for each year depending on the snow residency period at each pixel. The index is scaled between 0.1 and 1 with 1 representing a pixel containing snow for a full 91-day period and 0.1 representing a pixel containing the snow for a minimum period. This persistence index is used to qualify the snowpack in runoff estimation.

The snowmelt runoff (Q_s) in a basin is computed using the total energy available for melt during that 8-day time period and snow persistence index. The snowmelt runoff for all the 12 time periods of 8 days between 01 April and 30 June are computed individually with corresponding data inputs and then integrated to arrive at seasonal snowmelt runoff.

2.3.2 Glacier Melt Runoff

The glacier melt runoff during the summer months is a significant component of total runoff. It depends on the extent of glacier area within a basin, the level of exposure of the glaciers and duration of exposure during the snowmelt season. The level of exposure of glaciers in the summer months depends on the presence of seasonal snow cover. The probability of glacier melt and its magnitude is more when the seasonal snow cover is less. The glacier melt runoff increases as the snowmelt season progresses. For the purpose of estimation of glacier melt runoff, it is necessary to map the glaciers in each basin using satellite data of the September–October months. The glacier map prepared from satellite data is used to estimate glacier melt volume during summer period considering the net energy available for melt. The glacier melt component (Q_g) during the 3 months period is computed with energy available for melt and the level of exposure of glacier in each basin.

2.3.3 Rainfall–Runoff

The runoff due to rainfall constitutes a significant proportion of the total runoff during the snowmelt season in the Himalayan region. The rainfall in a region varies spatially and temporally and needs to be measured in a systematic and well representative manner. The accuracy of runoff estimation due to rainfall depends on the number of rain gauge stations available and their distribution within the basin. Generally, in the Himalayan mountainous region where elevation is more than 4500 m only snowfall occurs and rainfall does not occur. Hence, in this study, areas with an elevation below 4500 m are considered as areas contributing to runoff from rainfall. Mean rainfall of a basin is estimated depending on the number of rain gauge stations available within a basin. In basins where only one rain gauge station is available the rainfall at this station is assumed to be representative of the entire basin. In basins where multiple rain gauge stations are available Thiessen polygon method is used to estimate the mean rainfall of the basin. The rainfall–runoff is estimated using a suitable runoff coefficient. In the present study, the scope is restricted to runoff due to rainfall during the snowmelt period, i.e. April–May–June months. An appropriate runoff coefficient is assumed for each basin which varies temporally during the season. Basin level runoff (Q_r) from rainfall is computed by multiplying the average rainfall with rainfall contribution area and runoff coefficient as given below.

$$Q_x = C_r * \text{mean rainfall} * \text{rain - fed area}, \quad (2.4)$$

where

Q_r—runoff due to rainfall,
C_r—coefficient of runoff.

2.3.4 Base Flow

Base flow as defined by Hall (1968) is the portion of flow that comes from groundwater or other delayed sources. During snowmelt season, the base flow is a part of total runoff. When the snow melts, a significant portion of it percolates in the underlying ground and joins the stream within the basin, sometimes with a time delay. In the initial part of snowmelt season, depending on the soil conditions, the snowmelt water percolates until the ground is saturated before it becomes direct runoff. The base flow varies marginally depending on the snow cover conditions prevailing in the basin. The base flow during winter months is generally constant. The base flow during the summer months may be similar to that of the winter months or it can vary marginally. In the present study, the base flow (Q_b) is empirically estimated as a function of seasonal snow cover available in each basin.

$$Q_b = C_b * \text{snow cover area} \quad (2.5)$$

where

Q_b—runoff due to base flow,
C_b—coefficient of runoff.

2.3.5 Seasonal Runoff Forecast

Seasonal snowmelt runoff forecast for the summer period (April–May–June) needs to be provided in the first week of April for a given year. In the forecast situation, snow cover area as of 30th March is available by 1st April for that year. Other input data sets such as snow albedo, LST, snow emissivity, etc., are not available on the day of forecast. Hence, the current snow cover area as primary input, in conjunction with other data inputs of historic years is used for estimating snowmelt volume. Present year snow cover area and other data inputs of each historic year were modelled to arrive at snowmelt volume. The snowmelt volumes thus computed are averaged and considered as the present year snowmelt runoff. The runoff from glacier melt and rainfall are suitably computed considering historic averages. The aggregate of estimated runoff from snowmelt, glacier melt, rainfall and base flow was given as total seasonal runoff forecast.

2.4 Data Preparation

The energy balance based snowmelt runoff model requires several data inputs to compute each component of energy. These energy components are estimated using satellite data derived information along with appropriate assumptions as applicable, based on the review of literature in the Himalayan context. In addition to this, remote sensing based data inputs were also used in this study to estimate different model inputs. The procedures followed in the preparation of these input datasets are discussed in this section.

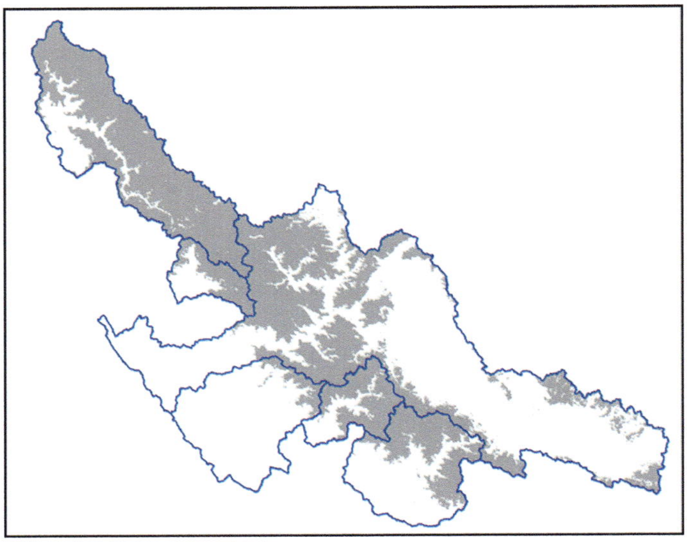

Fig. 2.3 Snow cover image of the study area *Source* NRSC (2014)

2.4.1 Snow Cover Area

The Moderate Resolution Imaging Spectroradiometer (MODIS) Snow Cover (SC) 8-Day L3 Global 500 m SIN Grid V005 product (MOD10A2) was downloaded from the REVERB website (Source: http://reverb.echo.nasa.gov/reverb/) for the period from 2002 to 2014 during the snowmelt season (April–May–June). The MODIS snow cover data are based on a snow mapping algorithm that employs a Normalized Difference Snow Index (NDSI) and other criteria tests. Each year a total of twelve 8-day products were downloaded with the starting Julian day of 89, 97, 105, 113, 121, 129, 137, 145, 153, 161, 169 and 177. The study area is covered in two tiles, namely, h24v05 and h25v05. The MODIS data tiles are mosaiced, re-projected to the Lambert Conformal Conic (LCC) projection and then clipped for the study area. The snow cover data is edited for cloud cover using time compositing technique. A sample snow cover image is shown in Fig. 2.3.

2.4.2 Seasonal Snow Cover Persistence Index

The 8-Day MODIS snow cover images for the snowmelt period (April–June) were used for the generation of snow cover persistence index map. The map is generated by adding all the 8-day snow cover products in a snow season and normalizing the sum to one. This snow cover persistence index map (Fig. 2.4) shows the trend in snow cover depletion in a season and used as one of the inputs in the model as a proxy for snow depth.

2.4.3 Land Surface Temperature

The MODIS Land Surface Temperature (LST)/Emissivity 8-Day L3 Global 1 km SIN Grid V005 product (MOD11A2) was downloaded from the REVERB website (Source: http://reverb.echo.nasa.gov/reverb/) for the period from 2002 to 2014 during the snowmelt season (April–May–June). This product is comprised of day time and night time LSTs, quality assessment, observation times, view angles, bits of clear sky days and nights, and emissivity estimated in bands 31 and 32 from land cover types. Layer 1 of this product contains 8-Day daytime 1 km grid land surface temperature in degree Kelvin. This layer was used for preparing the land surface temperature image. Each year a total of twelve 8-day products were downloaded similar to snow cover images. The digital values of each product was converted into LST image

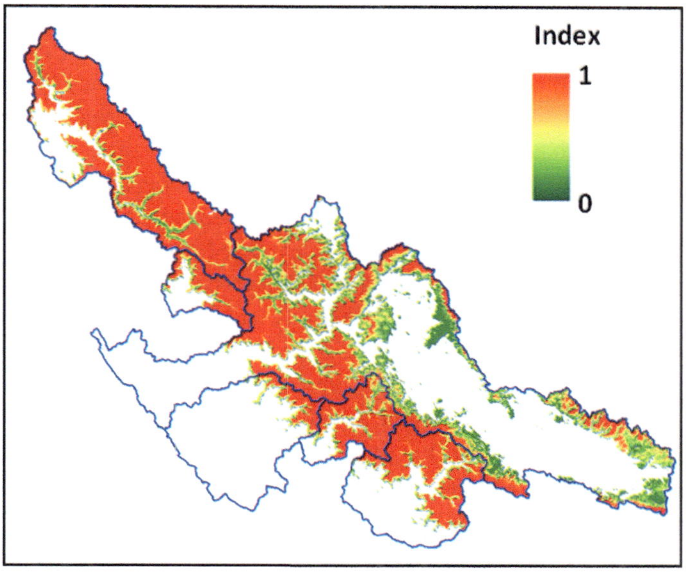

Fig. 2.4 Snow persistence index map of the study area (*Source* NRSC 2014)

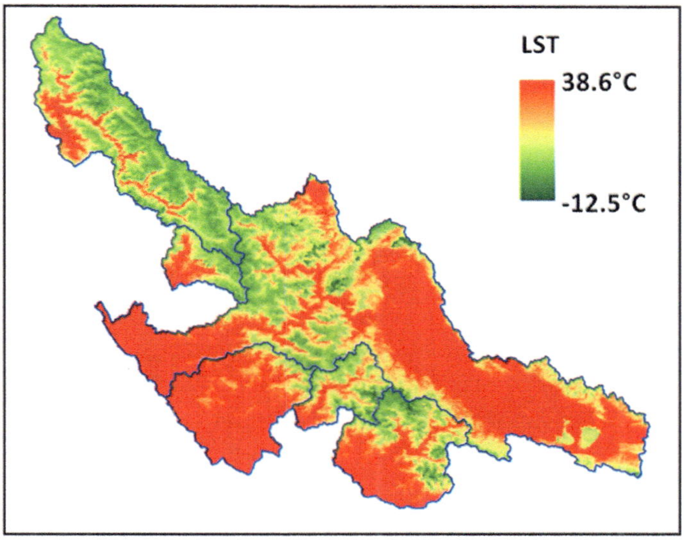

Fig. 2.5 Land surface temperature image of the study area (*Source* NRSC 2014)

in degree Celsius (°C), after applying the multiplication factor (0.02). The LST data is edited for cloud cover using time compositing technique. A sample LST image (Fig. 2.5).

2.4.4 Emissivity

The emissivity of a material (usually written ε or e) is the relative ability of its surface to emit energy by radiation. It is the ratio of energy radiated by a particular material to energy radiated by a black body at the same temperature. The air emissivity is calculated based on an empirical equation.

The MODIS MOD11A2 product has emissivity estimated in bands 31 (10.780–11.280 μm) and 32 (11.770–12.270 μm) for different land cover types. The average emissivity of these two bands was computed after multiplying with the scale factor (0.002) and added with a constant (0.49) to prepare snow emissivity images.

2.4.5 Snow Albedo

The MODIS Snow Cover Daily L3 Global 500 m SIN Grid V005 (MOD10A1) (Source: http://reverb.echo.nasa.gov/reverb/) contains snow cover, snow albedo, fractional snow cover and Quality Assessment (QA) data. The second layer containing snow albedo is extracted for each day (with values between 0 to 100 per cent). The snow albedo data for the study area was prepared similar to snow cover and LST. It was edited for cloud cover using time compositing technique. The daily snow albedo images are then time-composed to prepare 8-day snow albedo image. The variation in snow albedo for different time periods (Fig. 2.6).

2.4.6 Incoming Solar Radiation

The sources of energy that cause snowmelt include both shortwave and longwave net radiation. Incoming solar radiation (insolation) received from the sun is the primary energy source for this net radiation. With landscape scales, topography is a major factor that determines the spatial variability of insolation. Variation in elevation, orientation (slope and aspect), and shadows cast by topographic features all affect the amount of insolation received at different locations. This variability also changes with time of day and time of year, and in turn contributes to the variability of microclimate. The solar radiation analysis tool, in the ArcGIS Spatial Analyst extension, enables to map and analyse the effects of the sun over a geographic area for specific time periods. It accounts for atmospheric effects, site latitude and elevation, steepness (slope) and compass direction (aspect), daily and seasonal shifts of the sun angle, and effects of shadows cast by surrounding topography. The solar radiation analysis tools calculate insolation across a landscape or for specific locations, based on methods from the hemispherical viewshed algorithm. The output has units of watt hours per square meter (WH/m^2).

Generally, direct radiation is the largest component of total radiation, and diffuse

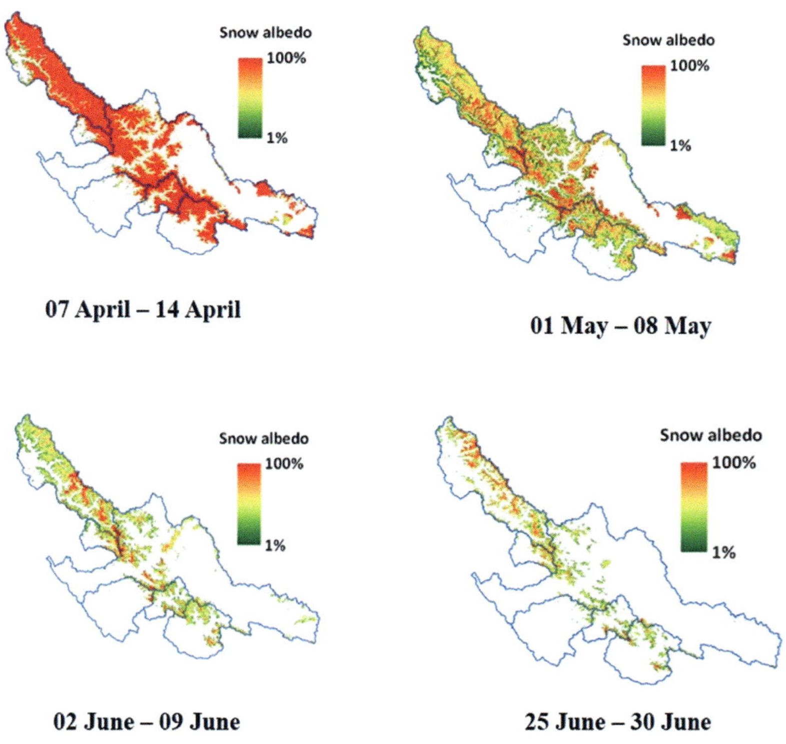

Fig. 2.6 Snow albedo image of the study area (*Source* NRSC 2014)

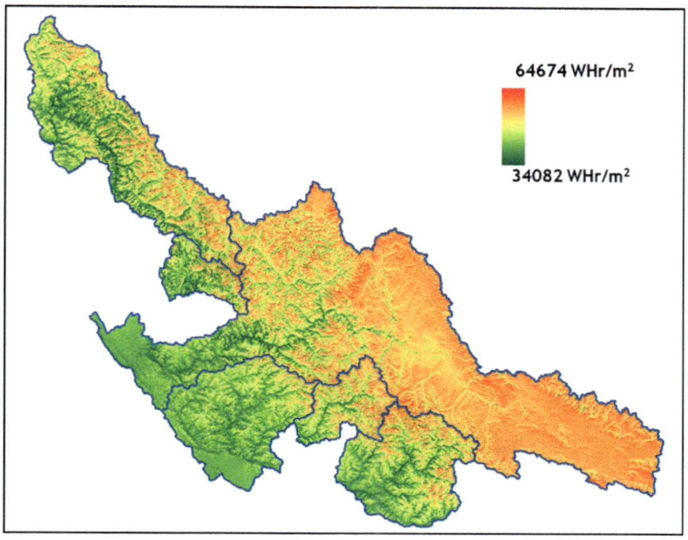

Fig. 2.7 Sample Incoming Solar Radiation Image (*Source* NRSC 2014)

radiation is the second largest component. Reflected radiation generally constitutes only a small proportion of total radiation, except for locations surrounded by highly reflective surfaces such as snow cover. The solar radiation tools in ArcGIS Spatial Analyst do not include reflected radiation in the calculation of total radiation. Therefore, the total radiation is calculated as the sum of the direct and diffuse radiation. Since radiation can be greatly affected by topography and surface features, a key component of the calculation algorithm requires the generation of an upward-looking hemispherical viewshed for every location in the digital elevation model (DEM). In the present study, the ASTER DEM has been used to generate the surface insolation images without considering atmospheric transmission and land cover effects (Fig. 2.7).

2.4.7 Land Cover Map

The incoming solar radiation reaching the land surface will vary depending on the land cover. Similarly, the outgoing longwave radiation emitted by the snow surface is influenced by the overlying vegetation. To account for these land cover interceptions, land cover map is necessary. The AWiFS satellite data was used to prepare the land cover map (Fig. 2.8) with classes such as evergreen forest, deciduous forest and water.

2.4.8 Glacier Map

For computing the runoff due to glacier melt, the glacier cover area is essential. The cloud-free satellite images from the AWiFS sensor during the least snow cover period (generally October) is used for visual interpretation of glaciers in all the basins. The presence of cloud cover can hamper the discriminability of snow and clouds. The SWIR band was used under partial cloud cover conditions for the interpretation of glaciers. The glacier boundaries were digitized on-screen using visual interpretation techniques. The digitized vector layer of glaciers for each basin is converted into raster images and basin wise glacial mask image has been created. The sample glacier map for the Sutlej basin is shown in Fig. 2.9.

2.5 Model Implementation

2.5.1 Snowmelt Runoff Component

The incoming surface solar radiation is estimated as mentioned in the previous sections using

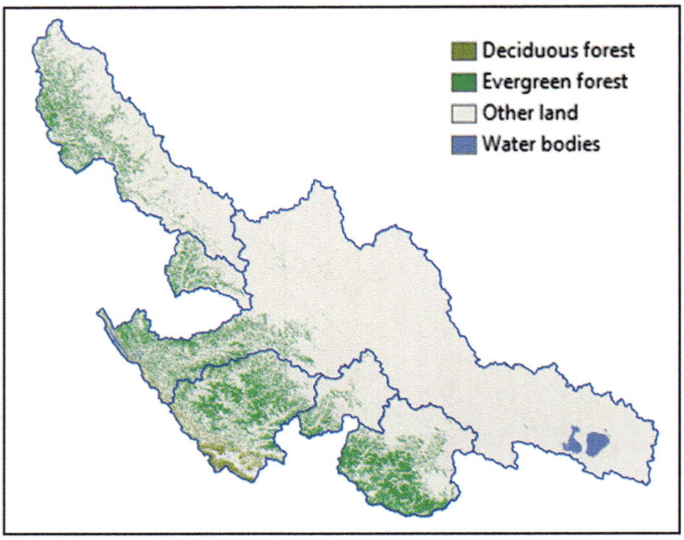

Fig. 2.8 Land cover image of the study area (*Source* NRSC 2014)

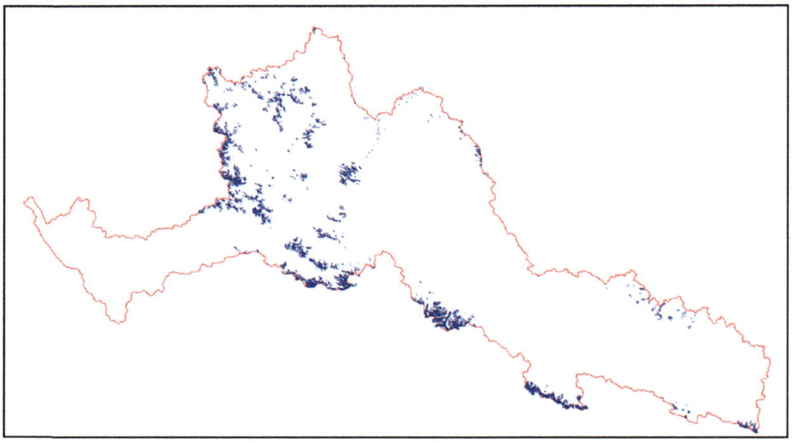

Fig. 2.9 Glaciers in Sutlej basin (*Source* NRSC 2014)

ArcGIS software. The incoming surface solar radiation needs to be corrected for atmospheric transmittance factor and cloud cover absorption which may vary spatially and temporally. The correction for atmospheric transmittance and cloud cover absorption was adopted using available literature. The net flux density of shortwave radiation generally represents the major source of energy for snowmelt. The incoming solar radiation after correcting for atmospheric transmittance and land cover effects is used for estimating net shortwave radiation. Net shortwave radiation (Q_{ns}) represents the sum of incoming and outgoing flux densities of shortwave radiant energy according to

$$Q_{ns} = K\downarrow - K\uparrow \quad (2.6)$$

where

K = shortwave radiation,

$\downarrow\uparrow$ = incoming and outgoing radiation flux densities, respectively.

The net shortwave radiation (Q_{ns}) is calculated based on the following equation:

$$Qns = (1 - \alpha)I_i \quad (2.7)$$

where

α = albedo,

I_i = incident solar radiation after atmospheric correction.

The albedo images and incoming solar radiation images generated earlier are used to prepare shortwave radiation images.

Net longwave radiation exchange (Q_{nl}) for an unobstructed, horizontal snowpack surface can be defined as the sum of incoming (L↓) and outgoing (L↑) longwave radiation flux densities as

$$Qnl = L\downarrow - L\uparrow \quad (2.8)$$

Incoming longwave radiation is estimated using Stefan–Boltzmann law as

$$L\downarrow = \sigma\varepsilon T^{4}{}_{a} \quad (2.9)$$

where

σ = Stefan–Boltzmann constant = 5.67 × 10 − 8 Wm^{-2} K^{-4},

ε = emissivity of snowpack surface,

T_a = air temperature, K.

Outgoing longwave radiation can be computed using the Stefan–Boltzmann equation derived to describe emission from a perfect or blackbody radiator as

$$L\uparrow = \sigma\varepsilon T_s^4 + (1-\varepsilon)L\downarrow \quad (2.10)$$

where

σ = Stefan–Boltzmann constant = 5.67 × 10 − 8 Wm^{-2} K^{-4},

ε = emissivity of snowpack surface,

T_s = surface temperature of the snowpack, K.

The emissivity and temperature images generated earlier are used to prepare net longwave radiation images (Fig. 2.3).

The algebraic sum (Q_m) of all heat components is the energy available for the snowpack to melt. The net radiational energy computed earlier is assumed to be 60% of the total energy. The other components of energy in Eq. (2.2) are assumed to be remaining 40% based on the available literature. Accordingly, the algebraic sum (Q_m) is computed by dividing the total radiational energy by 0.6.

Once the sum of energy available for melt is obtained, the mass flux density of meltwater (M) was computed using Eq. 2.3. Then the snowmelt in terms of depth units is computed for all the pixels using the 8-day energy images for each 8-day period. This is converted into a spatial snowmelt volume image for the area under snow cover by taking into account the size of each pixel (56 m) and the corresponding 8-day snow cover image. The spatial snowmelt computed is cumulated for the snowmelt period of April, May, and June each year by adding the snowmelt volume for all the 8-day time periods. This is integrated with snow cover persistence image (index of snow depth) to generate snowmelt volume for the 3 months period in each basin.

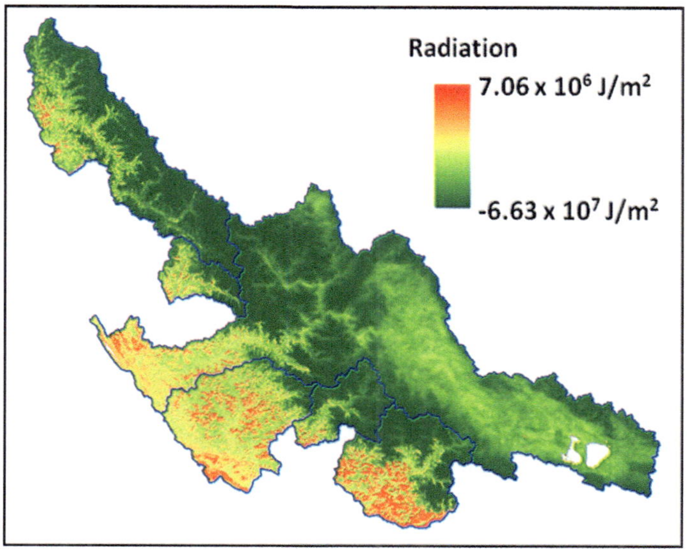

Fig. 2.10 Net longwave radiation sample image (*Source* NRSC 2014)

2.5.2 Glacier Melt Runoff Component

The net radiation energy at the glacier areas is estimated following similar steps as in the case of snowmelt runoff for all the 8-day time periods during the April–May–June months for all the years 2002–2014. The magnitude of glacier melt depends on the level of exposure of glaciers and increases as the snowmelt season advances. The energy balance based glacier melt runoff volume is estimated by adopting suitable progressively increasing runoff coefficients during the snowmelt season. The glacier melt runoff cannot be validated in isolation as there are no such suitable field discharge measurements. It cannot be separated from snowmelt runoff in the field. It is difficult to estimate the level of glacier exposure during the snowmelt season. Therefore, the glacier melt runoff coefficients are estimated considering the prior knowledge of the basins.

2.5.3 Rainfall–Runoff Component

Runoff due to rainfall can be estimated if well-distributed rain gauges are present in the basin. However, in the present study area, very few (sometimes only one) rain gauges are present for which the data is available. In the Himalayas, it is generally accepted that rainfall occurs in areas having an elevation of below 4500 m. In the present study, the scope is restricted to runoff due to rainfall during snowmelt period, i.e. April–May–June months. Therefore, for estimating the runoff from rainfall during the snowmelt period, it is assumed that the basin area below 4500 m elevation is the contributing area. Accordingly, for each basin rainfall contribution area is computed. The average rainfall from the rainfall stations during the melt period is computed based on the Thiessen polygon. An appropriate runoff coefficient is assumed for each basin which varies temporally during the season. Basin-level runoff from rainfall is computed by multiplying the average rainfall with rainfall contribution area and runoff coefficient.

In the case of the Sutlej basin, rainfall is measured at Rampur, Suni, Kasol, Kahu, Berthin and Bhakra. A varying runoff coefficient for rainfall component is adopted month-wise during April to June months. The rainfall and corresponding runoff for April–May–June months are shown in Fig. 2.11.

2.5.4 Base Flow Component

Base flow for each basin is computed separately as a function of snow cover area existing as on 30 March. In the case of the Sutlej basin, the base flow is estimated by dividing into 3 sub-basins and integrated to account for spatial variability.

2.5.5 Calibration and Validation

The runoff due to snowmelt, glacier melt, rainfall and base flow computed earlier were integrated. The total runoff volume in the 3 months period is the sum of all the components snowmelt runoff, glacier melt, runoff due to rainfall and the base flow.

$$Q = Q_s + Q_g + Q_x + Q_b \quad (2.11)$$

The total runoff was computed for all the years. The estimated runoff is calibrated with observed runoff using 4 years (2002, 2003, 2004 and 2005) data. The calibrated model is validated with the remaining data (2006–2012). The results of calibration and validation along with forecast results for the Sutlej basin are given below.

In the Sutlej basin, the snowmelt runoff component and base flow components were computed for each sub-basin. Figure 2.12 shows the snowmelt volume for the 3 months forecast period in the year 2012. Figure 2.13 shows the snowmelt runoff component, glacier melt runoff component, base flow component and rainfall–runoff component. Table 2.2 and Fig. 2.14 show the comparison of the observed runoff and estimated runoff at Bhakra in the Sutlej basin.

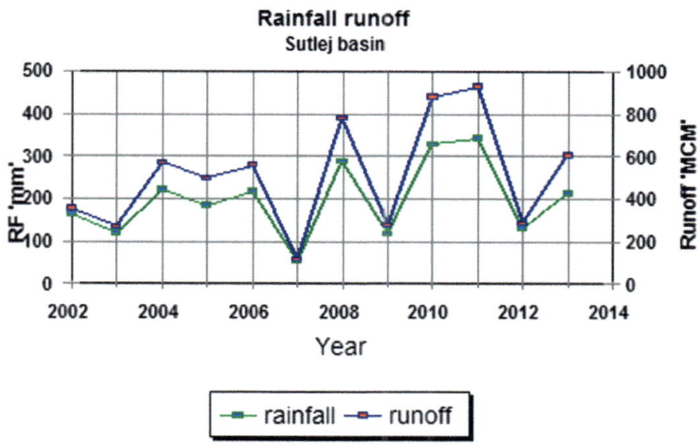

Fig. 2.11 Rainfall–runoff in Sutlej basin (*Source* NRSC 2014)

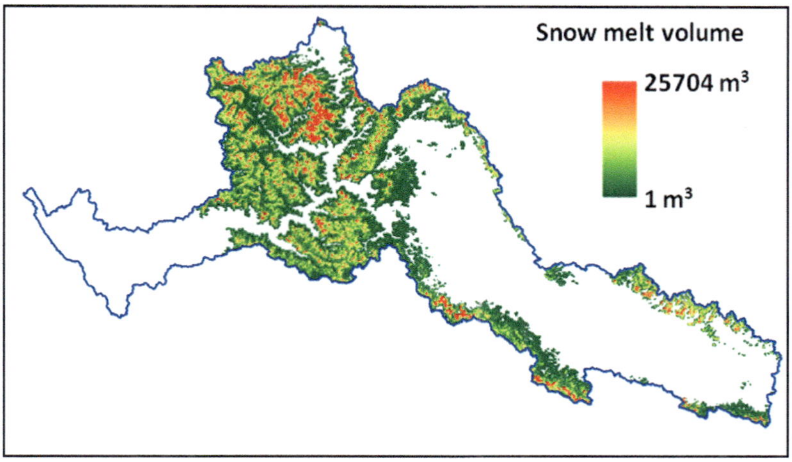

Fig. 2.12 Snowmelt volume during 2012 season in Sutlej basin (*Source* NRSC 2014)

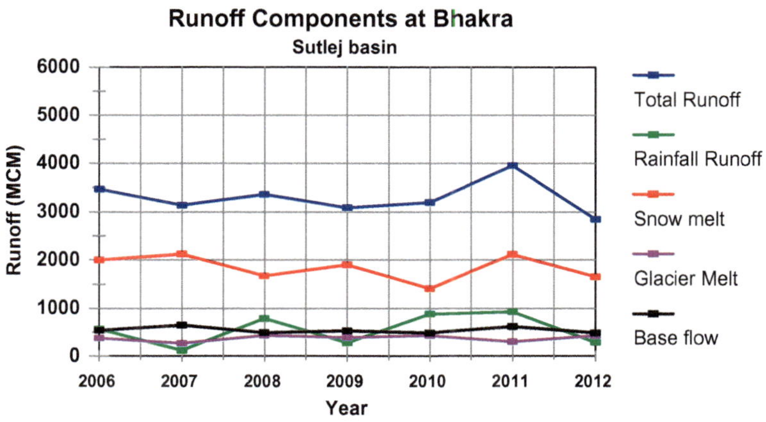

Fig. 2.13 Runoff components in Sutlej basin (*Source* NRSC 2014)

Table 2.2 Comparison of estimated runoff and observed runoff in Sutlej basin

Year	Observed runoff (MCM)	Estimated runoff (MCM)	Deviation (%)
2006	3820	3461	9
2007	3056	3130	−2
2008	4148	3354	19
2009	2730	3077	−13
2010	3186	3191	−0
2011	5389	3949	27
2012	2637	2841	−8

Source NRSC (2014)

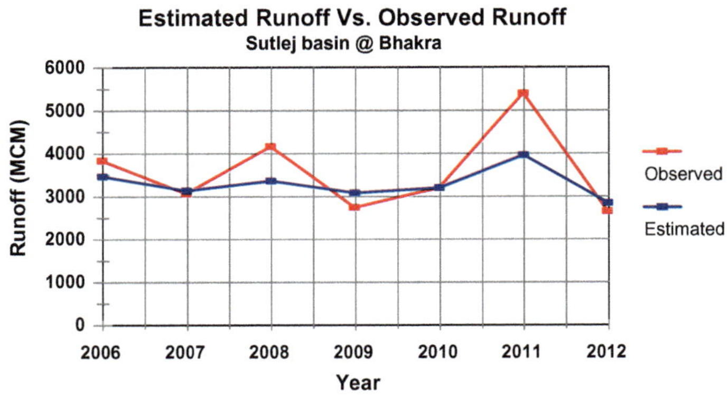

Fig. 2.14 Comparison of estimated runoff and observed runoff in Sutlej basin (*Source* NRSC 2014)

Table 2.3 Comparison of forecasted seasonal runoff with observed runoff

Year	Forecasted runoff (MCM)	Observed runoff (MCM)	% Deviation
2012	2800	2637	6.2
2013	3700	5173	−28.4
2014	3750	3425	9.5

Source NRSC (2014)

2.6 Runoff Forecasting Results

The seasonal runoff forecasts provided for the Sutlej basin for the snowmelt period during 2012, 2013 and 2014 were compared with observed discharges and the deviations are computed. The year-wise estimated runoff and actual observed runoff are compared and the details are given in Table 2.3. The accuracy of forecasts was better than 90% except during 2013. There was unprecedented rainfall and unusual snowfall occurred in the central Himalayas during 16–18 June 2013. In view of unusual snowfall, there was substantial additional snow cover was observed from satellite images as depicted in Fig. 2.15 (before the snowfall event) and Fig. 2.16 (after the snowfall event). This resulted in substantially high discharges in most of the basins of the study area, including the Sutlej basin.

2.7 Conclusions

In the present study, snowmelt runoff was estimated using the energy balance approach. Energy balance approach involves several energy input components. Major components have been estimated using satellite data derived inputs making suitable assumptions. The satellite data derived inputs include snow cover area, snow

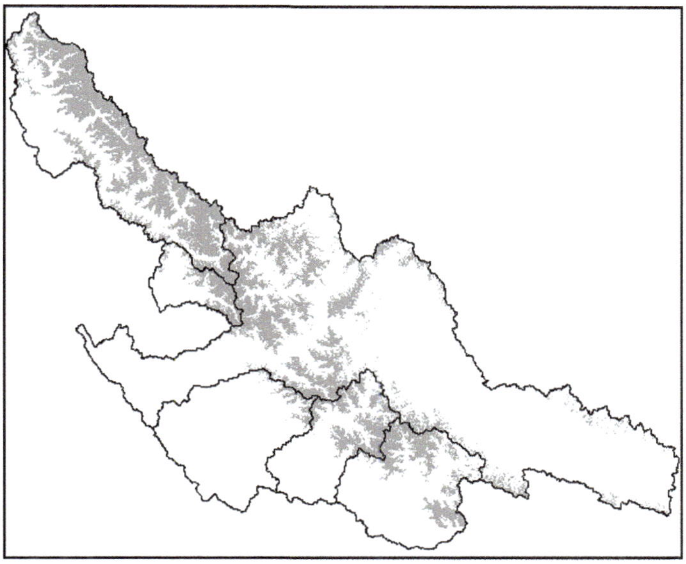

Fig. 2.15 Snow cover image as of 02-June-2013 (*Source* NRSC 2014)

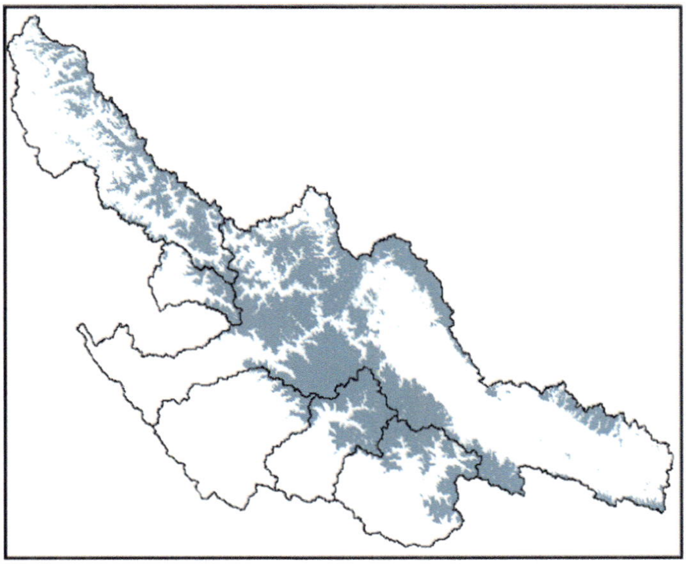

Fig. 2.16 Snow cover image as of 18-June-2013 (*Source* NRSC 2014)

albedo, snow emissivity, land surface temperature, digital elevation model, land use and land cover. The field data such as observed discharge and rainfall have been used. In the forecast situation, the current snow cover area as a primary input in conjunction with other data inputs of historic years was used for estimating snowmelt volume. The runoff from glacier melt and rainfall are suitably assumed considering historic averages. The aggregate of estimated snowmelt runoff, rainfall–runoff, glacier runoff and base flow is considered as the present runoff forecast.

In the case of Sutlej basin, the average snow cover area is varying from 49% to 11.5% of the basin area from beginning to the end of the season. Glacier area is about 1655 km^2 (3.2%) of the basin area. The rain-fed area as estimated from DEM is about 10100 km^2. Only about 78% of the basin area is above 4,000 m altitude. Most of the basin area (58%) that is in the Tibet region is relatively flat and contains a thinner snowpack. The lower Sutlej region and Spiti region contribute most of the snowmelt runoff. Therefore, in spite of the large basin area, the snowmelt runoff contribution is less from the Tibet region.

Acknowledgements The authors would like to thank the Deputy Director (RSAA) and the Director of National Remote Sensing Centre (NRSC), ISRO for their guidance and encouragement during the study. The authors would also like to thank the Central Water Commission of the Ministry of Water Resources, River Development & Ganga Rejuvenation for providing the financial grant for this research work and also the field data for calibration and validation.

References

DeWalle DR, Rango A (2008) Principles of snow hydrology. Cambridge University Press

Dozier J (1989) Estimation of properties of alpine snow from the Landsat Thematic Mapper. Adv Space Res 9:207–215. https://doi.org/10.1016/0273-1177(89)90487-0

Gray DM, Prowse TD (1992) Snow and floating ice in handbook of hydrology. In: Maidment DR (ed), McGraw-Hill, Inc, New York: pp 7.1–7.58

Hall FR (1968) Base flow recessions–a review. Water Resour Res 4(5):973–983

Meier MF (1984) Contribution of small glaciers to global sea level. Science 226:1418–1421

NRSC Technical Report (2014) Snow Melt Runoff Modelling and Forecasting in Himalayan Basins using Remote Sensing Inputs. National Remote Sensing Centre, Hyderabad

US Army Corps of Engineers (1998) Engineer Manual 1110-2-1406 on Runoff From Snowmelt, Engineering and Design, Department of the Army, Washington, DC 20314-1000, 31 March 1998

Er. E. Siva Sankar is a Civil Engineer graduated from Osmania University and has a Master's degree in Water Resources from Osmania University. He has 30 years of experience in remote sensing applications to natural resources management. He is heading Water Infrastructure Division in National Remote Sensing Centre, ISRO, Hyderabad. His contributions in the domain of snow and glacier hydrology are worth mentioning. He implemented energy balance approach to modelling snow melt runoff using remote sensing technologies. His interests are in water resources monitoring and management, infrastructure security, hydropower generation and spatial data analytics. He published many papers in conferences and journals

Er. B. Simhadri Rao received B.E. in Civil Engineering from Madras University. He obtained his M. Tech. In Water Resources Engineering from IIT, Madras. He worked as an Engineer in Water And Power Consultancy Services (WAPCOS) Ltd. During 1996–1998. He joined National Remote Sensing Centre/ISRO in 1998 and currently working as Scientist/Engineer 'SG' in Water Resources Group. He has more than two decades of experience in utilizing remote sensing and GIS in water resources and hydrology. His research interests are geospatial technology applications in snow hydrology, glacial lakes and glaciers, GLOF modelling, water resources assessment, monitoring, hydrological modelling, hydraulic modelling, flood hydrology. He has published 18 research papers in journals/conferences/seminars. He is recipient of ISRO Team Excellence award in 2012 (AIBP monitoring project). He is life member of three societies (IWRS, ISRS, ISSE)

Sh. K. Abdul Hakeem is a Civil Engineer and obtained his Master's in Irrigation Water Management from Centre for Water Resources, Anna University, India in 1994. He joined National Remote Sensing Centre as Scientist/Engineer 'SC' and currently he is heading the Water Resources Informatics Division. Mr. Abdul Hakeem has more than 25 years of experience in the field of geospatial technology applications for natural resources mapping, monitoring and management. He has expertise in handling more than 30 projects in the domain of water resources, agriculture and infrastructure, with diversified roles. He has significantly contributed in research and development of new techniques/applications in the field of water resources management using geospatial technologies. He has expertise in the areas of hydrological modelling, decentralized planning, water informatics and cryosphere studies. He has more than 35 publications to his credit and received three ISRO team excellence awards. He has visited Tunis in Tunisia and Kathmandu in Nepal, representing ISRO. Currently, his interests are in actual evapotranspiration estimation and hydrological drought assessment using satellite data, as Deputy Project Director under National Hydrology Project

3 Climate Change and Its Impacts with Special Reference to India

Omkar Verma

Abstract

Increases in anthropogenic emissions of greenhouse gases to the atmosphere are considered to be the major driving force behind current climate change. Although climate change is not a new phenomenon, the impact of these emissions on the earth's climate and environment has serious implications for human occupation. The geological record of the earth provides numerous examples that climate has not been uniform through its history, and has been significantly altered from time to time. It also indicates that events of past climate change have severely impacted the earth's environment and caused widespread destruction of ecosystems. Understanding climate change is therefore of great concern to human lives because it is expected to have wide-ranging effects on the future sustainability of Mother Earth. It poses a serious threat to India—the largest agricultural nation in the South Asian region. India has a population of over 1.2 billion that makes it one of the most vulnerable regions in the world. Climate scientists estimate an increase of 2.1° to 2.6 °C in temperature by 2050 and around 3.3° to 3.8 °C by 2080 over the Indian region that is already experiencing climate change. Further change is predicted to have adverse impacts on natural resources such as freshwater supply, Himalayan glaciers and rivers, agriculture, biodiversity and human health. Change is expected to result in increases in the frequencies of extreme weather events, including increased precipitation, droughts, rising sea level and the submergence of low-lying coastal areas, floods and cyclones. Climate induced-immigration from neighbouring nations may also put additional strain on its resources. Consequently, an understanding of climate change and its potential impact on natural resources, both in general and in the Indian region in particular, is important because it will affect the lives of millions of people. This article presents an overview of climate change research, the potential impacts of climate change on natural resources (with reference to India) and the possibilities of mitigation.

Keywords

Climate change · Natural resources · Extreme weather events · India

O. Verma (✉)
Geology Discipline Group, School of Sciences,
Indira Gandhi National Open University,
New Delhi 110068, India
e-mail: omkarverma@ignou.ac.in

A. K. Taloor et al. (eds.), *Water, Cryosphere, and Climate Change in the Himalayas*,
Geography of the Physical Environment, https://doi.org/10.1007/978-3-030-67932-3_3

3.1 Introduction

In the last few decades, there have been shocking scientific and media reports on the undoubted fact that the earth's average temperature is increasing at an alarming rate (McMichael 2003; Le Treut et al. 2007; IPCC Report 2019; Hegerl et al. 2019; Sarkar et al. 2020). It has been estimated that global temperatures have increased by around 0.74 °C since the advent of the industrial era (Ritter 2009; Ogurtsov et al. 2013; Henderson et al. 2020). Several lines of evidence link this rise to an increase in the concentrations of greenhouse gases in the earth's lower atmosphere. Among these gases, carbon dioxide is one of the major contributors and currently, concentrations of this are reaching 380 ppm (parts per million) (Ruddiman 2008; UNDP Report 2008; Bhandari 2018). An increase in the mean temperature of the troposphere has caused global warming that, in turn, is forcing the climate to change as it has already happened many times in the geological past. Climate change is a reality and there is a strong consensus in the science community (e.g., Intergovernmental Panel on Climate Change (IPCC)) that the global climate is changing rapidly (IPCC Report 2019; Hegerl et al. 2019; Henderson et al. 2020). Using current climate change trends, the IPCC, in its third assessment report (IPCC Report 2001), projected that temperatures could rise globally by 1.4° to 5.8 °C in the next 100 years. It is also concluded that the increasing concentrations of greenhouse gases, resulting from human activity such as the burning of fossil fuels and deforestation, have been responsible for most of the observed temperature increases since the middle of the twentieth century (Kumar et al. 2020; Singh et al. 2020; Taloor et al. 2020). This communication presents an overview of the history of climate change research, climate systems and palaeoclimate, together with potential impacts of climate change on natural resources in general and on India in particular and on mitigation strategies.

3.2 Climate and Greenhouse Effect

Climate is generally defined as "average weather". Weather represents the state of temperature, precipitation, humidity, wind speed, cloud cover and other physical phenomena of an area of the lower atmosphere over hours or days. The climate system is one of the most complex and dynamic systems of the earth. It comprises numerous elements, including inbound solar radiation, atmosphere, lithosphere, biosphere, cryosphere and hydrosphere. It also encompasses the rotation and revolution of the earth around the Sun. Indeed, the variability of temperature and precipitation and their averages over periods ranging from months to centuries (a typical period is ~30 years) describes the climate (Rohli and Vega 2008). Thus, the climate system develops over time and is controlled both by its own internal deviations and by variations in external factors such as solar radiation, volcanic eruptions and anthropogenic atmospheric changes (Le Treutet al. 2007). Energy from the Sun is the basic power behind this system and without solar energy the earth would be a gloomy and frigid planet with no life. The atmosphere is a substantial component of the climate system. The atmosphere is basically a gaseous envelope surrounding the earth, comprising nitrogen, oxygen, carbon dioxide, ozone, methane and various inert gases, together with water vapour and particulate matter. It provides variation within internal components (temperature, precipitation, wind, cloud cover, reflection and absorption of solar energy) of the system and necessary gases for the survival of life (Le Treutet al. 2007). It also acts as a filter to incoming solar radiation, shielding the surface from the more destructive effects of ultraviolet radiation and preventing overheating.

The atmosphere is divided into distinct layers: the troposphere, stratosphere, mesosphere and thermosphere (Fig. 3.1). Its effective thickness is estimated to be between 16 and 29 thousand kilometres from sea level, but there is no definite outer edge, it gradually becomes thinner until it merges into space. More than 80% of atmospheric gases are held by gravity within 20 kilometres of the earth's surface. The surface of the earth receives energy from solar radiation in the form of very short wavelengths, predominantly in the visible or near-visible part of the electromagnetic spectrum. The earth reflects nearly 30% of incoming radiation and absorbs the remaining 70% that warms the land, atmosphere and oceans (Le Treut et al. 2007). As a result, the earth emits terrestrial radiation into the atmosphere in the form of longer wavelengths, primarily in the infrared portion of the electromagnetic spectrum. Up to 90% of outgoing terrestrial (thermal) radiation emitted by the land and oceans is absorbed by the atmosphere. However, part of this is subsequently radiated back to the earth's surface, warming the lower troposphere and surface, mainly because of the presence of greenhouse gases, notably carbon dioxide and water vapour (Salby 1992). This process, by which energy is recycled in the lower atmosphere to warm the earth's surface, is referred to as the "Greenhouse Effect", and is an essential component of the earth's climate. Virtually, all weather and climate-related phenomena occur within the lower atmosphere, mostly in the troposphere, but partly in the stratosphere (Fig. 3.1). Hence, this region is critical to life in the biosphere. Since the formation of the earth, around 4.6 billion years (Ba) ago, greenhouse gases have played a crucial role in the origins, evolution, extinction, diversification, migration and adaptation of life forms under different climatic conditions.

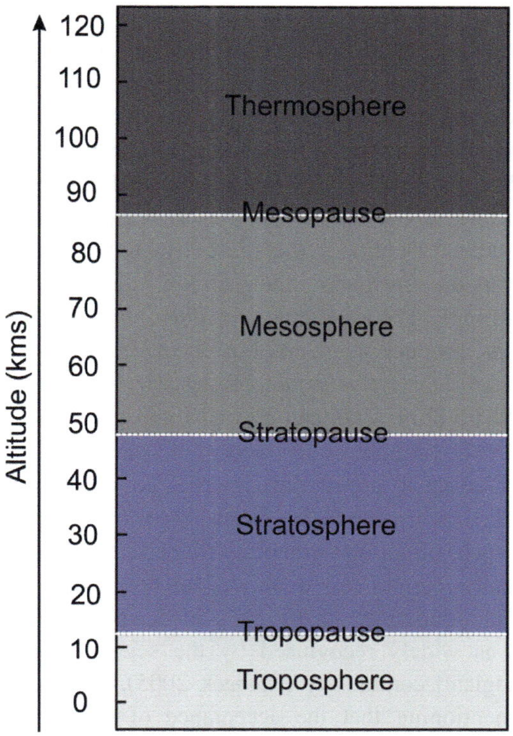

Fig. 3.1 Vertical layering of the atmosphere. Tropopause, stratopause and mesopause are the boundaries between troposphere-stratosphere, stratosphere-mesosphere and mesosphere-thermosphere, respectively (modified after Salby 1992)

3.3 A Concise History of Climate Change Research

The beginning of the climate change research goes back to eighteenth century, when a French political philosopher and jurist, Charles-Louis de Secondat, Baron de La Brède et de Montesquieu (1689–1755) put forward the views that the physical environment substantially influences the human nature, their societies and the political systems created by it (Kriesel 1968; Bhandari 2018). While examining the writings of Montesquieu, Kriesel (1968) noted that Montesquieu outlined his statement of "The Spirit of Laws", in which, he summarised the impacts of the physical environment and non-physical societal attributes (customs, manners, morals, traditions and technologies) on the growth of complex bio-physical, politico-economic and socio-cultural systems. In fact, climate change science was a part of academic and philosophical domains much earlier as many ancient Greeks and other scholars suspected in the sixteenth and seventeenth centuries that humans had the capacity to alter temperature and rainfall patterns by clearing forests, irrigating deserts, plowing agriculture fields and by animal over-grazing (Fleming 2005; Bhandari 2018; https://www.history.com). In 1824, Joseph Fourier—a French physicist—described the earth's natural "greenhouse effect" and stated that energy received by the earth from solar radiation must be balanced by energy returning back to the space after emitting by the earth's surface. Further, he argued that some portion of the returning energy must be absorbed by the atmosphere, which keeps the earth warm. Later, in 1861, John Tyndall, an Irish physicist, while advancing the work of Joseph Fourier ahead, identified the gases including water vapour, methane and carbon dioxide that strongly absorb radiation emitted by the earth and produce a greenhouse effect. In the meantime, Jean Louis Rodolphe Agassiz, a renowned Swiss-born American biologist and geologist, in 1837, proposed the theory of an "Ice Age" referring to a past time interval, in which large glaciers covered the North Pole, Europe and much of North America (Riebeek 2005). Initially, people cast doubts about his theory, but, later, during the 1870s, the evidence of an ice age was widely recognised by the scientific (geological) community (Riebeek 2005). It is worth mentioning that the acceptance of an ice age theory triggered many questions: what had caused the ice age?, why and when it had ended?, could it come again?, will climate change again? and others. Following the acceptance of the theory of ice age and the then separate concept of the greenhouse effect, Svante Arrhenius, a Swedish chemist, in 1896, envisioned that the ice age could have been caused by a decrease in the concentration of greenhouse gases, particularly carbon dioxide, that would reduce the temperature during the ice age (Weart 2003; Fleming 2005; Riebeek 2005; Bhandari 2018). Additionally, he proposed that industrial emissions could increase the temperature of the earth in the coming centuries. In fact, Arrhenius' work helped to explain historical climatic variations and also laid the foundation for the advancement of the theory of the greenhouse effect and climate change (Weart 2003; Bhandari 2018). Later, in 1938, Guy Stewart Calendar, a British engineer, based on data collected from weather stations, suggested that land surface temperatures had increased in the last preceding 50 years due to anthropogenic activities. He also demonstrated that concentrations of carbon dioxide also had increased in the same time interval, and finally, he established a link between them, which is now referred to as the "Calendar effect" (Hawkins and Jones 2013; Bhandari 2018).

It is evident that climate change research has a long history and its origin is associated with palaeoclimate studies of ice ages. However, it took a long time to be verified because a majority of workers considered that humanity could not alter the climate system. However, in the 1960s, evidence of anthropogenic interventions in the natural climate system was becoming increasingly obvious (Riebeek 2005; Singh et al. 2017; Bhandari 2018). The modern development of climate change research took place in the 1950s, when scientists around the globe agreed not only that the content of carbon dioxide in the atmosphere had been increasing, but also that the average global temperature might further increase by some degrees Celsius prior to the end of the twenty-first century (Weart 2003). The modern era of climate change research actually started between the 1950s and 1960s. In 1955, based on carbon-14 isotope investigations, Hans Suess found that the carbon dioxide released from burning fossil fuels was not instantly absorbed by oceanic waters (Baxter and Walton 1970). Following this, in 1958, Charles David Keeling started systematically measuring the concentration of atmospheric carbon dioxide at Mauna Loa observatory in Hawaii (USA). Within 4 years of this project, he presented unequivocal results showing that the level of carbon dioxide in the atmosphere was increasing (Weart 2003; Black 2013; Hawkins and Jones 2013). The graph showing data year by year with the increase of atmospheric carbon dioxide is described as the "Keeling Curve".

Global interest in the environment has grown since the early 1970s, when the first United Nations (UN) Conference on Human Environment (also known as the Stockholm Conference) was organised in Sweden in 1972. International environmental issues (whaling, chemical pollution and atomic bomb testing) were the main focuses of this conference that gave birth to international environmental politics (Black 2013; Bell 2014). Indeed, climate change was not formally part of the conference, but the meeting is nevertheless considered as its beginning, leading to the establishment of the United Nations Environment Programme (Black 2013; Bell 2014). Subsequently, in 1979, World Meteorological Organisation (WMO) organised the first World Climate Conference in Geneva. In this, experts around the globe enunciated, for the first time, the major consequences of climate change on the earth, which later became the subject of political talks at both national and international levels (e.g., Fleming 2005). In 1988, the UN Assembly (by the acceptance of Resolution No. 43/53) considered climate change to be a problem common to the whole of mankind. In 1988, the UN and WMO, with the support of the United Nations Environment Programme, established Intergovernmental Panel on Climate Change (IPCC) as an independent scientific organisation (Fleming 2005; Black 2013; Bell 2014; https://www.ipcc.ch/). The main functions of the IPCC are to document past climate (also known as palaeoclimate) changes, the current climate and its environmental, social and economic impacts, and to project future changes as well as proposals for mitigating climate change effects. The IPCC has continued this work since its inception and has released a series of assessments and supplemental reports consisting of summaries of scientific work on climate change within every 5 to 6 years. So far, the IPCC has published five assessment reports: the First Assessment Report in 1990, followed by 1995, 2001, 2007 and 2013–2014. The Sixth Assessment Report is due to be issued in 2022 (https://www.ipcc.ch/reports/).

Apart from the IPCC, the introduction to the Montreal Protocol, an international treaty signed in 1987 that came into force in 1989, is a global agreement protecting the ozone layer by phasing out the production and consumption of ozone-depleting substances has also contributed to climate change research (https://www.unenvironment.org/ozonaction/). In 1992, the United Nations Framework Convention on Climate Change (UNFCCC) was adopted at the UN Conference on Environment and Development held in Rio de Janeiro, popularly known as the Earth Summit. This is also an international agreement, and its main objective is "to stabilise greenhouse gas concentrations in the atmosphere at a level that will prevent dangerous human interference with the climate system, in a time frame, which allows ecosystems to adapt naturally and enables sustainable development" (Leggett 2020). The UNFCCC has two subsidiary treaties: the Kyoto Protocol and the Paris Agreement. The Kyoto Protocol was introduced in 1997 and its primary aim is to limit and reduce the greenhouse gas emissions of industrialised/developed countries in accordance with agreed individual targets (https://unfccc.int/kyoto_protocol). The Paris Agreement, adopted in 2015 by almost all countries, aimed at reducing climate change and speeding up and strengthening the responses and investments required for a sustainable low carbon future. Its main focus is to support the world response to the threat of climate change (Leggett 2020). Both the Montreal Protocol and the UNFCCC added new dimensions to climate change research.

3.4 Impacts of Climate Change

The IPCC has established that climate change is a serious concern and its impacts on biotic and abiotic resources are heterogeneous, varying in scale from region to region and latitude to latitude (IPCC Report 2007, 2019). Climate models indicate that global temperatures have increased by several degrees during the twentieth century. The increase in temperature has altered the global and regional climate zones and ecosystems of earth, and thus, has affected the environmental quality and livelihoods of populations and the economies of various countries. The effects of global warming and climate change are manifold (Fig. 3.2), including the melting of glaciers and sea-ice, leading to a rising sea level and flooding in low-lying areas, and an increase in the frequency and severity of extreme weather events. These have the effects of shifting vegetation zones, with the loss of wildlife species, collapsing dynamics of socio-economic structures, reduction in crop yields and instability in food supplies, the rapid spread of insect transmitted

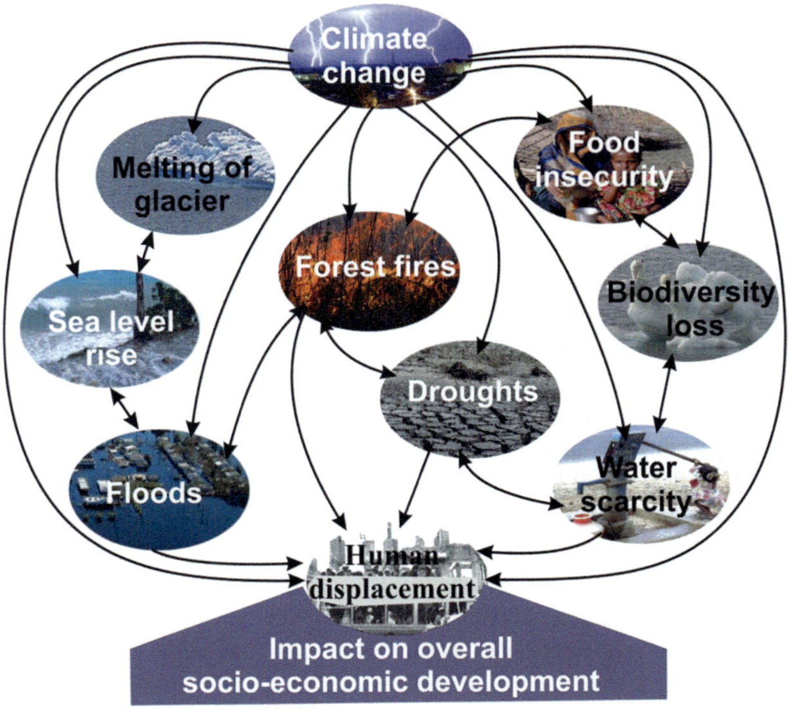

Fig. 3.2 Impacts of climate change (modified after IPCC Report 2019; Verma 2019)

diseases and an increase in the intensity of severe weather with a concurrent increase in related natural hazards. The damage resulting from climate change imposes a heavy toll on society and environment. For instance, floods along the Yangtze River (China) in 1998 caused nearly 4,000 deaths and a financial loss of $30 million (Yin and Li 2001). In the same year, extreme weather conditions in Florida (USA) led to drought and widespread wildfires, causing a loss of 483,000 acres of forest and 356 buildings, with an estimated economic loss of US $276 million (Vellinga and van Verseveld 2000). In August 2010, an abnormal cloudburst hit the Union Territory of Ladakh (part of the former Jammu and Kashmir State) that lies in a rain shadow area. Nearly 600 people were reported missing, with 130 confirmed deaths and nearly 300 injured (India Today 2010). During 1900–1995, the frequency and intensity of droughts are reported to have increased in many regions of the Asian and African continents (Dai et al. 1998). The recent Australian bushfire season (2019–2020) caused the loss of millions of acres of forest, destroying more than 5,900 buildings, and killed 34 persons (Phillips and Nogrady 2020). Many millions of animals were also burned (Phillips and Nogrady 2020). The incidence of these extreme events at almost the same time raises concerns about their relationship to climate change (Verma 2019; Phillips and Nogrady 2020). In the following section, some of the adverse impacts of climate change are briefly discussed.

3.4.1 Impacts on Himalayan Glaciers

Nearly 10% of the earth's surface is covered by ice, and particularly by glaciers. Glaciers are produced by the accumulation of ice above the snowline under cold climate conditions. At present, the earth has permanent major ice covers in the Arctic, Greenland and Antarctica. However, snow and ice also occur in various mountain regions of the globe, notably the Rockies (North America), the Andes (South America), the Alps (Europe), the Himalaya (Asia) and Mount Kilimanjaro (Africa). It is important to note that although many are small, snow and ice covers are enormously important, providing a source of freshwater to millions of people. The most noticeable impact of global warming is the melting of polar ice caps and the progressive retreat of mountain glaciers of the Alps, Mount Kilimanjaro and the Himalayas (e.g., Rasul and Molden 2019; Krishnan et al. 2020). The Himalayas have the largest concentration of glaciers outside Polar Regions and are an important freshwater source in Asia; hence, the range is referred to as the "Water Tower of Asia" (WWF Report 2005; Mukherji et al. 2015). Glaciers supply melt waters to nine Asian river systems including three Indian rivers, the Indus, Ganga and Brahmaputra. Himalayan glaciers supply fresh water to river systems throughout the year and sustain life for almost one-third of humanity on the planet. Climate change has been impacting the glacial ecosystem very rapidly. Research has shown that Himalayan glaciers have been melting at alarming rates (Anthwal et al. 2006; Bisht et al. 2019; Krishnan et al. 2020). In the last two decades, nearly 60% of Himalayan glaciers have retreated, with climate change identified as the causative driver (e.g., Miller et al. 2012; Sood et al. 2020). The Chaturangi and Onglaktang-Rathong glaciers in the Garhwal and Sikkim Himalaya have each retreated about 46 m, at 1 and 1.2 km per year, respectively. The Kolhani and Machoi glaciers of the Kashmir Himalaya have retreated at rates of 16 m and 8.1 m per year, respectively (Bisht et al. 2019). Lastly, the famous Gangotri glacier in the Garhwal Himalaya has been retreating at an average rate of about 23 m per year (WWF Report 2005). Melting of the Himalayan glaciers, due to increasing temperature, is enhancing the number and size of glacial lakes but also increasing the risk of glacial lake outburst floods. Initially, melt waters will increase runoff through the major river systems to the ocean, but this is likely to be followed by dry spells once the glaciers disappear. It is suggested that some major rivers of the Himalaya will become ephemeral, with a monsoonal behaviour like those in peninsular India (e.g., Haque et al. 2020).

The reservoirs of ice and snow in Himalayan glaciers are a major source of freshwater in South Asia and, if the observed trends of rising temperature continue, they may vanish entirely (WWF Report 2005). As they decline, the area is likely to see an increased risk of sudden failure of large structures (dams) and an increase in the frequency of natural hazards (glacial lake outburst floods) as the volume of melt water increases (e.g., Krishnan et al. 2020). They will also impact supplies of freshwater for drinking, irrigation and industry, reducing the potential for hydropower. Eventually, their reduction will have major effects on basic human indices due to the lack of water. Glaciers are sensitive to climate change, and in the Himalayan region provide an excellent opportunity to assess the impact of global climate change (Mukherji et al. 2015).

3.4.2 Impacts on the Sea Level

There is strong evidence that the global sea level has increased in the recent past. According to the IPCC (IPCC Report 2007) during the last century, there has been a rise in sea level of 0.3 to 0.8 m. In the IPCC special report on emission scenarios; it is projected that sea level will increase by 0.22 to 0.44 m above the current sea level at the end of this century (2090–2099). The melting of land-based ice and snow cover, due to global warming and the thermal expansion of the oceans, are regarded as the major reasons for the observed global rise in sea level (Mimura 2013). The rise is expected to lead to the submergence of low-lying and deltaic regions, with the loss of coastal ecosystems and wetlands, and the intrusion of saline water in coastal aquifers. Satellite data have shown that, since 1992, sea-level rise has been the greatest in the western Pacific and eastern Indian oceans (Krishnan et al. 2020). The additional rise predicted will pose serious threats to humanity. In the Indian subcontinent, the projected rise in sea level will lead to varying submergence of the coastline, with the most severe impacts along the most important rice-growing coastal agricultural lands of Myanmar, Bangladesh, Sri Lanka, India and Pakistan (Krishnan et al. 2020). It will place highly populated cities like Dhaka (Bangladesh),

Visakhapatnam, Mumbai, Kochi, Mangalore (India) and Karachi (Pakistan) at high risk. Over 70 million people in Bangladesh, 22 million in Vietnam and 6 million in Lower Egypt would be at risk due to the expected rise (UNDP Report 2008).

3.4.3 Impacts on Agricultural Productivity and Food Security

India is primarily an agricultural country, in which a wide variety of crops ranging from Kharif to Rabi are grown. The Kharif crops account for nearly 78% of cereal production and Rabi about 72% of food production in India. The Indian economy is wholly dependent upon agricultural productivity as it alone contributes more than 27% of the gross domestic product. India is the second largest populated country on the globe and in the near future, high agricultural productivity will be required to meet the food demands of a growing population. Agricultural productivity in India depends entirely on climate and land resources, specifically soil and water. At present, the Himalayan and peninsular river systems and the monsoon system fulfil all of the water requirements for irrigation, which, in turn, supports high agricultural productivity (Krishnan et al. 2020). According to the 1996 IPCC report, by 2070 the mean temperature in India is projected to increase by 0.4° to 2.0 °C in lands growing Kharif and 1.1° to 4.5 °C for those relying on Rabi (IPCC Report 1996). It is also predicted that the timing of the onset of the monsoon system may change and that the frequencies of droughts and floods may increase (Mujumdar et al. 2020). It is notable that in the past few years, regions of the country have frequently experienced extreme events including floods, droughts, unusual heavy rainfall, landslides and cyclones (Seneviratne et al. 2012). Himalayan glaciers, one of the major sources of rivers in India, have been retreating in recent decades, and this retreat may change the volume of water supplied. It is clear from these facts that in India the impacts of climate change on agriculture are significant.

Agriculture is itself a major contributor to climate change as it adds nearly one-third of emissions of greenhouse gases (nitrous oxide and methane) into the atmosphere. Such changes can have both beneficial and detrimental impacts on crop productivity, increasing productivity in some areas and reducing it in others. Cereals such as rice and wheat are the major food grains for consumption in India and are sensitive to high temperatures. It is projected that their yields would decrease with a rise in temperature, but increase with a rise in precipitation. An increase in temperature would also reduce the yields of fruit crops such as apples and berries that require a long winter chilling period. It is projected that agricultural output in developing countries may decline by 20% due to climate change and by 2080 yields in these areas could decrease by 15% on average (Fisher et al. 2005). Overall, an increase in temperature will have serious effects not only on the nutritional content and quality of cereals, pulses, fruits and vegetables but also on the qualities of cotton, tea, coffee, aromatic and medicinal plants (Battist and Naylor 2009).

India also has the largest population of livestock, with animals used as milk producers, nutrient recyclers and as a source of food. An increase in temperature would likely lower reproduction rates and make conditions more favourable for disease, ultimately reducing production (Rojas-Downing et al. 2017). Climate change is also likely to reduce food availability, because it adversely affects the basic components of food production: soil, water and biodiversity. The decrease of agricultural production due to climate change, and rising world prices of many food items in the past decade, have put over 1 billion people in what is regarded as food insecurity. Within the next decades, it will increase the risks of hunger and malnutrition on an unprecedented scale.

3.4.4 Impacts on Biodiversity

Biodiversity is simply the variety of all life forms on the planet earth. It is usually measured as the number and variability of genes, species and communities in both space and time. The basic component of biodiversity is the ecosystem that represents the interactions between biotic and abiotic elements of the earth. Changes in temperature, sea level and rainfall along with increasing frequencies of extreme weather events due to global warming will have a direct impact on ecosystems (Miller 2007). It is evident that biodiversity is under the control of climate, and changes in climate may alter its configuration and the productivity of ecosystems. As a consequence of climate change, all forms of biodiversity (animals, plants and micro-organisms) will be put under pressure. It will increase the risks of changes in the composition and location of ecosystems, including species extinctions, and reduce social and economic benefits in response to the challenge of adapting to a warmer climate (Miller 2007; Ruddiman 2008). Plant diversity is particularly significant; on the one hand, plants act not only to sequester carbon dioxide and bring its concentration down, but they also provide fuel, food, fodder, timber and medicinal and aromatic plants. The fourth assessment report of the IPCC (IPCC Report 2007) predicted that many mid-latitudinal areas will experience decreased rainfall and an increased risk of droughts, which would allow forest or wildfires to occur on a larger scale. This will release stored organic carbon into the atmosphere (a positive feedback to global warming), and will also reduce the overall forest cover and sequestration of carbon dioxide. Tropical wet and warm rain forests may disappear and cool temperate vegetation would become warm and temperate. Global warming may in the future lead to a shift of lower altitude forests to higher altitudes. Forest types/species unable to disperse or migrate fast enough, or have no place to go, as in the Arctic, alpine and coastal species and island endemics, will decline or face extinction (e.g., Aitken et al. 2008).

Tropical insects play an important role in the health of tropical habitats, as they provide vital ecological services, breaking down organic materials and pollinating flowers to provide fruits and nuts. Many insect species are sensitive to temperature and are only able to tolerate a narrow range of variation; an average rise of 1° to 2 °C could kill many. Tropical insects are expected to be among the first terrestrial species to become extinct as a consequence of global warming. Polar bears and Emperor penguins are the next likely casualties (Miller 2007). Polar bears use sea-ice as a platform for hunting, and global warming is reducing floating ice in the Arctic and other regions, placing them under food insecurity. Some plant and animal species will adapt to a warmer climate, but others will not do so well. It is projected that a rise of 3 °C in temperature may result in 20–30% of land species facing extinction. Overall, global warming will reduce biodiversity—a major pillar of sustainability of the earth.

3.4.5 Impacts on Human Health

Variations in climate are likely to impact the basic requirements of health, including drinking water, air, food and shelter. A warmer climate may bring either positive or negative effects on human health (McMichael et al. 2003; Singh and Dhiman 2012; Sheridan and Allen 2015). However, the intensity of these impacts will depend upon the location and rate of temperature increase. Localised benefits of global warming may occur in some regions, including lower rainfall in wet regions, higher rainfall in dry regions and consequent increases in food production, hence, fewer deaths related to food insecurity (Miller 2007). Elsewhere, regions may experience extreme weather events (heavy rainfall, droughts, landslides, floods and forest fires), water scarcity, loss of food production and immigration that may make them more vulnerable to vector- and water-borne, cardiovascular, respiratory and diarrhoeal diseases, together with diseases transmitted through insects and malnutrition (Miller 2007; Costello et al. 2009; Singh and Dhiman 2012). Rising temperatures have two opposing impacts on humans; higher temperatures in winter tend to reduce deaths from cold, whereas extreme temperatures in summer increase deaths from heat exhaustion.

3.5 Climate Change: The Indian Scenario

India is a developing nation and its rapid growth is leading to a progressive increase in greenhouse gas emissions (Krishnan et al. 2020). In 2004, it was the world's sixth highest emitter of greenhouse gases, after the United States (22%), China (14%), the European Union (13%), Russia (6%) and Japan (5%) (Miller 2007). Its contribution to net global greenhouse emissions has been reported at 4%. Agriculture, biomass burning, burning fossil fuels and deforestation have been considered the main factors for observed emissions. Carbon dioxide emissions in India have increased from about 20 million metric tonnes of carbon in 1950 to more than 175 million metric tonnes of carbon in 1988, increasing at a rate of nearly 5.6% per year (Roy and Prasad 1991). It is projected that India will become the world's third biggest emitter of carbon dioxide by 2030 (NIC Report 2009). The BASIC countries, comprising Brazil, South Africa, India and China, are predicted to together become the largest emitters of greenhouse gases, with China's emissions expected to surpass those of the United States in the next 20–30 years.

India is home for more than 1.2 billion people. And, as it is one of the major emitters of greenhouse gas, thus it is also one of the most vulnerable nations to face the impacts of predicted climate change. The country is already witnessing changes in climate that are impacting natural resources, bringing changes in the timings of monsoons and precipitation, and increasing the frequencies of extreme weather events (Fig. 3.3). Current climate changes, along with rapid population growth, industrialisation, urbanisation and economic growth are all putting pressure on the ecological and socio-economic development of the country. The growing industrial sector is continuously adding more to the production of greenhouse gases; and consequently, there will be a further increase in temperature. Under A2 (740 ppm CO_2, medium-high emissions with priority given to economic issues) and B2 (575 ppm CO_2, medium-low emissions with priority given to environmental issues) scenarios of the IPCC, it is projected that the average surface temperature will likely to be increased by 3–4 °C at the end of the twenty-first century.

The projected increases in temperature will likely result in maximum warming over northern India (Fig. 3.3). There will be an increase in day- and night-time temperatures, but nights will warm more than days and higher latitudes will warm more than lower (Shukla et al. 2003). It is likely that an increase in surface temperatures will push the snowline higher and increase the risk of floods in northern India during wet seasons (Singh 1998). The whole country will experience heavy precipitation; extreme precipitation will increase on the west coast and in western central India, but will probably decrease in Punjab, Rajasthan and Tamil Nadu (Groisman and Kovyneva 1989). Low-lying coastal areas of eastern India will experience greater increases in the frequencies of tropical cyclones and flooding than the equivalent west coast (Fig. 3.3; NIO Report 1988). A significant rise in sea level will submerge many coastal areas of the country, particularly the Gulf of Kutch and the coast of West Bengal, putting the lives of over 7 million people at risk (e.g., Krishnan et al. 2020). The Himalayan river systems (fed by glaciers) will first experience a brief increase in flows, but this will be followed by a decline as glaciers melt. As a result, the country will face water stress and water insecurity for drinking, irrigation and industrial purposes. It is predicted that the Indian region will continue to suffer substantial water stress, with increased effects in the southern part of the peninsula.

Future climate change will pose a serious impact on Indian forests (Fig. 3.3). It is projected that there will be a loss of area under any given forest cover and one type of forest would be

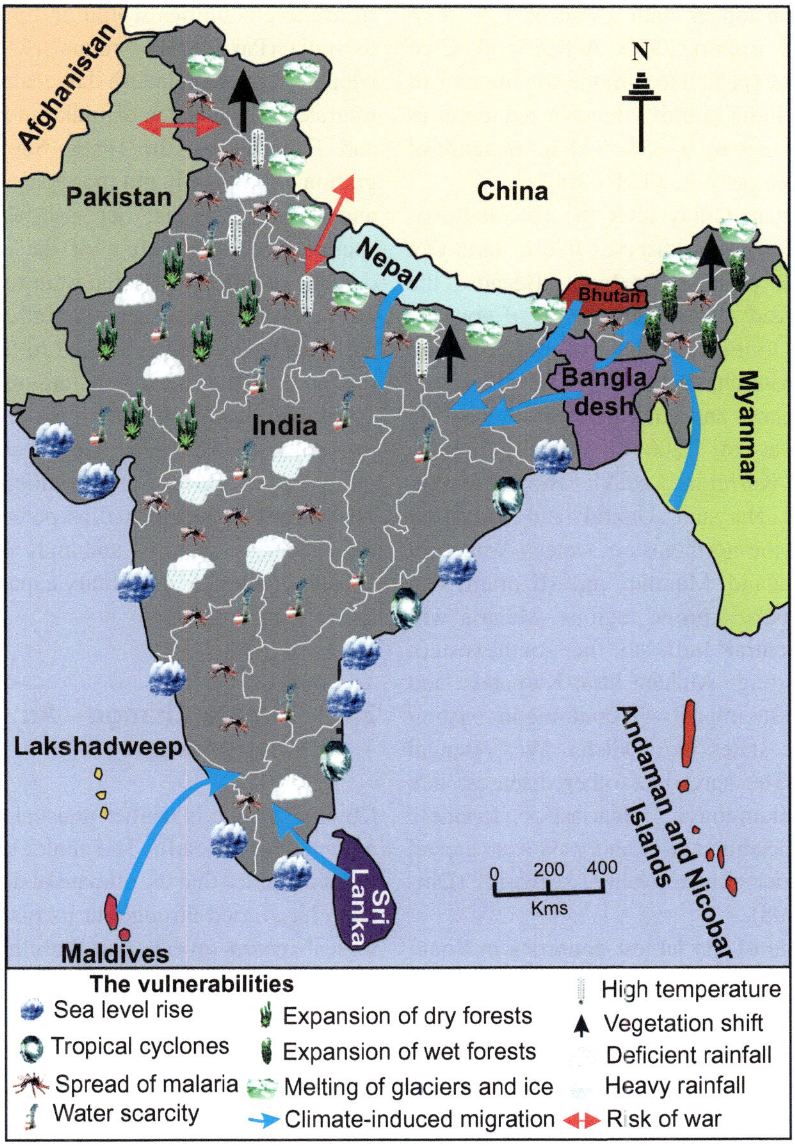

Fig. 3.3 Map of India showing possible impacts of climate change (based on various sources: NIO Report 1988; TERI 2002; Shukla et al. 2003; WWF Report 2005; Anthwal et al. 2006; Ravindranath et al. 2006; Dhiman et al. 2008; Pai 2008; NIC Report 2009; Miller et al. 2012; Singh and Dhiman 2012; Mukherji et al. 2015; Bisht et al. 2019; Krishnan et al. 2020)

replaced by another by 2085. The wetter forests will extend in the north-eastern region and drier forests in the north-western region whereas, warm mixed forests will change into temperate conifer forests. The area under existing tropical evergreen forests will likely increase due to shifts of the tropical deciduous and tropical semi-deciduous forests. However, the Western Ghats evergreen, semi-evergreen and mangrove forests will experience little or no change (Ravindranath et al. 2006).

Global warming will produce more negative impacts on agricultural production. It will reduce the yield of both wheat and rice, but the production of wheat will be more severely affected. The projected drop in wheat production is estimated to be 4–5 million tonnes with a rise of 1 °C temperature (GOI Report 2004). A rise of 3 °C of temperature is expected to shift the sowing time of many crops. India could witness a reduction in farm net revenues by 9–25% with an increase of 2–3.5 °C in temperature (TERI 2002).

Climate change may alter the spatial distribution of vector-borne diseases like malaria that are already a problem in India. Because the country is already facing high levels of poverty, with only a limited capacity of public health systems, climate change will expose millions of Indians to malaria and other diseases. According to Dhiman et al. (2008), northern regions (Jammu and Kashmir, Ladakh, Himachal Pradesh, Punjab, Haryana, Uttarakhand and Uttar Pradesh) and the north-eastern states (Arunachal Pradesh, Nagaland, Manipur and Mizoram) will all become malaria-prone regions. Malaria will shift from central India to the south-western coastal states (e.g., Maharashtra, Karnataka and Kerala) while its impact will continue in existing malaria-prone states like Odisha, West Bengal and Assam. The spread of other diseases, like dengue, chikungunya, filariasis, Japanese encephalitis, leishmaniasis and kala-azar are all suggested under climate change scenarios (Dhiman et al. 2008).

India is one of the largest countries in South Asia sharing international boundaries with Pakistan, Sri Lanka, Bangladesh, Myanmar, Bhutan, Nepal, Tibet and China. High population growth, less adaptive capacity, high poverty, unique and valuable ecosystems, rich mineral wealth, vast productive agricultural regions, various water treaties, long international boundaries and conflicts with some neighbouring countries, all make India more vulnerable in facing challenges of climate change, not only at local and national levels, but also regionally. Within a few decades, India is expected to experience large-scale cross-border immigration. Rising sea level will submerge the Maldives, Lakshadweep, parts of Sri Lanka, low-lying areas of Bangladesh and Myanmar. Melting glaciers and heavy precipitation will pose a serious threat of flash and glacial lake outburst floods in Bangladesh, Nepal, Bhutan and Tibet. As a result, millions of people from these countries would be forced to migrate to India (Pai 2008). Around 12 to 17 million people from Bangladesh have already illegally migrated to the states of India, mostly in Assam and Tripura since the 1950s. Such mass immigration will not only put more pressure on natural and other resources, but also alter the entire socio-economic structure of the country. Agricultural productivity in Pakistan mainly depends on the Indus river waters. As India and Pakistan are already in disputes related to the distribution of waters of the Indus river, any reduction in the supply of water due to climate change may trigger wars between these countries. Mass immigration and existing conflicts with neighbours must be recognised as potential threats to the country's security and may force India to expand diplomatic or military capacities to tackle these issues.

3.6 Climate Change—An Old Phenomenon

Climate change is neither unusual or a new phenomenon to the earth. The geological record provides evidence that the climate of the living planet earth has varied throughout its history. The geological record reveals that the climate has been altered by changes in solar intensity, volcanic eruptions, lithospheric plate motions, weathering reactions, fluctuations of greenhouse gases, changes in oceanic circulation, cyclic variations in the earth's orbit, meteorite impacts and biological evolution. Geologists broadly define the past climate of the earth in terms of non-glacial and glacial periods. During the Precambrian eon (4.6 Ba to 540 Ma [million years] ago), the earth's climate was warm and concentrations of greenhouse gases like carbon dioxide, methane and water vapour were very high in the early oxygen-free atmosphere. The concentration of carbon dioxide was more than 20 times its current levels and methane above 1000 ppm (Berner 1994). Millions of years after the formation of the earth, as the temperature decreased, water vapour in the early atmosphere produced rain. As a consequence, the earth was provided with basic necessities such as soil, free water and air for the origination of life. Around 3.5 Ba ago, early forms of life like cyanobacteria made their first appearance. These used the sun as a source of energy to make their food, but released oxygen as a by-product of photosynthesis. Around 600 Ma ago, enough oxygen was present in the atmosphere for the development of multi-cellular organisms (Canfield et al. 2007). During the Phanerozoic eon (540 Ma to the present), the concentration of carbon dioxide fluctuated, decreasing from perhaps 6000 ppm to reach its current levels (280 ppm pre-industrial level and 380 ppm after industrialisation). The carbon cycle helped to shape the Phanerozoic climate with changes reflected in species radiations of multi-cellular organisms including plants (Beerling and Berner 2005).

Palaeoclimatic data indicate three broad climatic stages in the geological history of the earth: (a) Early Archaean (3850–3200 Ma ago), a non-glacial stage; (b) Late Archaean–Middle Riphaean (3200–1200 Ma ago), a stage with episodic glaciation; and (c) Riphaean–Recent (1200–0 Ma ago), a stage with frequent periodic glaciations. These long-term changes may have played an important role in shaping the planet's ecosystem, determining where and which life forms would survive, contributing to the destruction of ecosystems and the extinction of many species. Five great mass extinctions are recognised in the fossil record (Racki 2019). About 543 Ma ago (the Precambrian-Cambrian boundary), circa 76% of the dominant early fauna and flora vanished in this first great mass extinction. At 254 Ma ago (the Permian/Triassic boundary), the earth experienced a mass extinction resulting in the elimination of about 90% of marine and some 70% of land species. Trilobites, some corals, placoderms and pelycosaurs were the marine and terrestrial animals that did not survive beyond this boundary. The most famous extinction reflected in the demise of the dinosaurs, which took place around 66 Ma ago at the Cretaceous/Palaeogene boundary. About 60% of all species disappeared, making it the second largest mass extinction event in geological history. Two other extinctions occurred at 444 Ma (Late Ordovician) and 359 Ma (Late Devonian) ago resulting in the disappearance of many forms.

3.7 Mitigation of Climate Change

Critics of climate change have argued that the current rate of temperature increase is not unusual, because the earth is a single non-equilibrium dynamic system billions of years old, and it is not easy to understand its climate system. They argue that variation in climate has occurred on all time scales and has been continuous. The dynamic nature of the earth, and the non-equilibrium state of components of the climate system, makes it difficult to assess any cumulative human impact on climate (Freitas 2002). Miller (2007) argued that climate change is a more serious challenge to humanity than the threat of terrorism. The reason being that Mother Earth has only one atmosphere and this is not possible to separate greenhouse gas emissions based on the country of origin. The emissions of one country will cause climate change in another, because the wind systems of the planetary atmosphere transport emissions and their effects worldwide. The poor of the developing countries have the lowest capacity to respond to climate change and will probably be the first victims of change. In the future, the whole of humanity will face the risks of global warming and thereafter, the future generation who did not create this problem will have to be ready to face the consequences of climate change.

However, global warming, from whatever cause, is more likely to produce big losses than benefits. This is a global problem and needs reaction at all levels. There are two basic ways to deal with it; one is mitigation and the other adaptation. Mitigation involves the reduction of greenhouse gas emissions by shifting from fossil fuels to sustainable non-carbon energy sources, doing re-afforestation, taking poverty reduction measures and by slowing population growth. Adaptation involves adjustments in terms of behaviour or economic structures that limit the damage caused by climate change.

3.8 Conclusions

Climate change research has a long history and its beginning can be traced from the sixteenth century, when ancient Greeks acknowledged the capacity of humans to alter the climate system. However, it accelerated in the second half of the twentieth century when human influence on the climate system was more widely accepted. Climate change is a real issue facing humanity today. There is now a broad consensus that anthropogenic activities, notably burning fossil fuels, are increasing the concentration of carbon dioxide in the atmosphere, and this in turn is increasing temperature and is the main reason for current climate change. Climate change has serious impacts on natural resources (biotic and abiotic resources) that are heterogeneous in nature and vary from country to country and latitude to latitude. India is already touched by the impacts of climate change. The changes facing the country are impacting natural resources and also generating and increasing frequencies of extreme weather events. These will eventually put the cultural and socio-economic growth of the country under significant pressure.

Geological and palaeoclimatic studies show that climate change is not a new phenomenon. The earth has already witnessed numerous episodes of past climate change of glacial and interglacial periods, some of which caused widespread destruction of biomass. Five great mass extinctions and several minor extinctions are recorded in earth history. Some of these may have been triggered or exacerbated by climate change and played an important part in deciding the survival of life forms. The threat of climate change on humanity and natural resources is real and requires strategies including mitigation and adaptation to deal with it. As we are living in and sharing a single atmosphere, collective urgent actions are required by all nations including both developing and developed to mitigate climate change. No nation, acting independently, can win the war against global warming, but there is no doubt that the responsibility for mitigation must be shared.

Acknowledgements The author is thankful to Dr. Ajay Kumar Taloor for inviting him to contribute this chapter in the present volume. Special thanks go to Prof. C. J. R. Braithwaite, Dr. S. K. Parcha and Dr. S. Pant for their advice and fruitful discussions that have greatly improved the manuscript.

References

Anthwal A, Joshi V, Sharma A, Anthwal A (2006) Retreat of Himalayan glaciers—indicator of climate change. Nat Sci 4(4):53–59

Aitken SN, Yeaman S, Holliday JA, Wang T, Curtis-McLane S (2008) Adaptation, migration or extirpation: climate change outcomes for tree populations. Evol Appl 1(1):95–111

Battist DS, Naylor RL (2009) Historical warnings of future food insecurity with unprecedented seasonal heat. Sci 323(5911):240–244

Baxter MS, Walton A (1970) A theoretical approach to the Suess effect. Proc Roy Soc Lond Ser A Math Phys Sci 318(1533):213–230

Beerling DJ, Berner RA (2005) Feedbacks and the coevolution of plants and atmospheric CO_2. PNAS 102:1302–1305

Bell A (2014) A very short history of climate change research. https://roadtoparis.info/2014/09/05/history-climate-change-research/

Berner R (1994) Geocarb II: a revised model of atmospheric CO_2 over Phanerozoic time. Am J Sci 294:56–91

Bisht H, Rani M, Kumar K, Sah S, Arya PC (2019) Retreating rate of Chaturangi glacier, Garhwal Himalaya, India derived from kinematic GPS survey and satellite data. Curr Sci 1169(2):304–311

Bhandari MP (2018) Climate change science: a historical outline. Adv Agr Environ Sci 1(1):5–12

Black R (2013) A brief history of climate change. https://www.bbc.com/news/science-environment-15874560

Canfield DE, Poulton SW, Narbonne GM (2007) Late-Neoproterozoic deep-ocean oxygenation and the rise of animal life. Sci 315:92–95

Costello A, Abbas M, Allen A, Ball S, Bell S, Bellamy R, Friel S, Groce N, Johnson A, Kett M, Lee M, Levy C, Maslin M, McCoy D, McGuire B, Montgomery H, Napier D, Pagel C, Patel J, de Oliveira JAP, Redclift N, Rees H, Rogger D, Scott J, Stephenson J, Twigg J, Wolff J, Patterson C (2009) Managing the health effects of climate change. Lancet 373:1693–1733

Dai A, Trenberth KE, Karl TR (1998) Global variations in droughts and wet spells: 1900-1995. Geophys Res Lett 25(17):3367–3370

Dhiman RC, Pahwa S, Dash AP (2008) Climate change and malaria in India: interplay between temperatures and mosquitoes. Regional Health Forum 12:27–31

Fisher G, Shah M, Tubiello FN, van Velhuizen H (2005) Socio-economic and climate change impacts on agriculture: an integrated assessment, 1990–2080. Philos Trans R Soc Lond B Biol Sci 360(1463):2067–2083
Fleming JR (2005) Historical perspectives on climate change. Oxford University Press, Oxford and New York
de Freitas CR (2002) Are observed changes in the concentration of carbon dioxide in the atmosphere really dangerous? Bull Can Pet Geol 50:297–327
India Today (2010) Cloudbursts trigger flash floods in Ladakh; 130 die, 600 missing. https://www.indiatoday.in/india/north/story/cloudbursts-trigger-flash-floods-in-ladakh-130-die-600-missing-79973-2010-08-0
IPCC (Intergovernmental Panel on Climate Change) Report (1996) Climate Change 1995: Impacts, Adaptations and Mitigation of Climate Change: Scientific-Technical Analysis, Report of the Working Group II of the Intergovernmental Panel on Climate Change. Cambridge University Press, Cambridge, UK and New York
IPCC Report (2001) Climate Change—The Scientific Basis. Contribution Report of the Intergovernmental Panel on Climate Change. Cambridge University Press, Cambridge, UK and New York
Report IPCC (2007) Climate change 2007: the physical science basis. Cambridge University Press, Cambridge, UK and New York
IPCC Report (2019) IPCC Special Report on Climate Change, Desertification, Land Degradation, Sustainable Land Management, Food Security, and Greenhouse gas fluxes in Terrestrial Ecosystems. https://www.ipcc.ch/site/assets/uploads/2019/08/Fullreport-1.pdf
GOI (Government of India) Report (2004) India's initial national communication to the UN framework convention on climate change. Ministry of the Environment and Forest, Government of India, New Delhi. https://www.in.undp.org/content/india/en/home/library/environment_energy/india_s_initial_nationalcommunicationtotheunitednationsframework.html
Groisman PA, Kovyneva NP (1989) Preliminary estimates of climate change on Indian subcontinent during the global warming of limited scale. Mausam 40:73–78
Haque S, Kannaujiya S, Taloor AK, Keshri D, Bhunia RK, Ray PKC, Chauhan P (2020) Identification of groundwater resource zone in the active tectonic region of Himalaya through earth observatory techniques. Groundwater for Sustainable Development. p 100337. https://doi.org/10.1016/j.gsd.2020.100337
Hawkins E, Jones PD (2013) On increasing global temperatures: 75 years after Callendar. Q J Roy Meteor Soc 139(677):1961–1963
Hegerl GC, Brönnimann S, Cowan T, Friedman AR, Hawkins E, Iles C, Müller W, Schurer A, Undorf S (2019) Causes of climate change over the historical record. Environ Res Lett 14:123006. https://doi.org/10.1088/1748-9326/ab4557
Henderson RM, Reinert SA, Oseguera M (2020) Climate change in 2020: implications for business. Harvard Business School Publishing, Boston
Kriesel MK (1968) Montesquieu: possibilistic political geographer. Ann Assoc Am Geogr 58(3):557–574
Krishnan R, Sanjay J, Gnanaseelan C, Mujumdar M, Kulkarni A (2020) Assessment of climate change over the Indian region. Springer, Singapore
Kumar D, Singh AK, Taloor AK, Singh DS (2020) Recessional pattern of Thelu and Swetvarn glaciers between 1968 and 2019. Quaternary International, Bhagirathi basin, Garhwal Himalaya, India. https://doi.org/10.1016/j.quaint.2020.05.017
Leggett JA (2020) The United Nations Framework Convention on Climate Change, the Kyoto Protocol, and the Paris Agreement: A summary. Congressional Research Service. https://fas.org/sgp/crs/misc/R46204.pdf
Le Treut H, Somerville R, Cubasch U, Ding Y, Mauritzen C, MokssitA, Peterson T, Prather M (2007) Historical overview of climate change. In: Solomon S, Qin D, Manning M, Chen Z, Marquis M, Averyt KB, Tignor, M, Miller HL (eds), Climate Change 2007: The Physical Science Basis. Contribution of Working Group I to the Fourth Assessment Report of the Intergovernmental Panel on Climate Change. Cambridge University Press, Cambridge, United Kingdom and New York, pp 94–127. https://www.ipcc.ch/site/assets/uploads/2018/03/ar4-wg1-chapter1.pdf
McMichael AJ, Campbell-Lendrum DH, Corvalán CF, Ebi KL, Githeko AK, Scheraga JD, Woodward A (2003) Climate change and human health: risks and responses. World Health Organization, Switzerland, pp 1–322
Miller GT (2007) People and environment. Cengage Learning India Private Limited, New Delhi
Miller JD, Immerzeel WW, Rees G (2012) Climate change impacts on glacier hydrology and river discharge in the Hindu Kush–Himalayas: a synthesis of the scientific basis. Mt Res Dev 32(4):461–467
Mimura N (2013) Sea-level rise caused by climate change and its implications for society. Proc Jpn Acad Ser B Phys Biol Sci 89(7):281–301
Mukherji A, Molden D, Nepal S, Rasul G, Wagnon P (2015) Himalayan waters at the crossroads: issues and challenges. Int J Water Resour Dev 31(2):151–160
Mujumdar M, Bhaskar P, Ramarao MVS, Goswami M, Borgaonkar H, Chakraborty S, Ram S, Mishra V, Rajeevan M, Niyogi D (2020) Droughts and floods. In: Krishan R, Sanjay J, Gnanaseelan C, Mujumdar M, Kulkarni A, Chakraborty S (eds) Assessment of climate change over the indian region. Springer, Singapore, pp 141–177
NIC (National Intelligence Council) Report (2009) India: The impact of climate change to 2030. A commissioned research report. The National Intelligence Council, No.03D. pp 1–49. https://www.dni.gov/files/documents/climate2030_india.pdf

NIO (National Institute of Oceanography) (1988) Report on the workshop on sea level rise due to greenhouse effect: implications for India. National Institute of Oceanography, Goa, India

Ogurtsov M, Lindholm M, Jalkanen R (2013) Global warming—scientific facts, problems and possible scenarios. In: Tarhule A (ed) climate variability—regional and thematic patterns, Intech Open. https://doi.org/10.5772/56077

Pai N (2008) Climate Change and national security: preparing India for new conflict scenarios. Indian Natl Interest Policy Brief 1:1–9

Phillips N, Nogrady B (2020) The climate link to Australia's fires. Nat 577:610–612

Racki G (2019) Big 5 mass extinctions. In: Encyclopedia of Geology, 2nd edition. Elsevier Inc. https://doi.org/10.1016/B978-0-12-409548-9.12028-7

Rasul G, Molden D (2019) The global social and economic consequences of mountain cryospheric change. Front Environ Sci 21:1–18. https://doi.org/10.3389/fenvs.2019.00091

Ravindranath NH, Joshi NV, Sukumar R, Saxena A (2006) Impact of climate change on forests in India. Curr Sci 90:354–361

Riebeek H (2005) Paleoclimatology: Introduction. https://earthobservatory.nasa.gov/features/Paleoclimatology/paleoclimatology_intro.php

Ritter SK (2009) Global warming and climate change. Chem Eng News 87(51):11–21

Rohli RV, Vega AJ (2008) Climatology. Jones and Bartlett Publishers, Sudbury

Rojas-Downing MM, Nejadhashemi AP, Harrigan T, Woznicki SA (2017) Climate change and livestock: Impacts, adaptation, and mitigation. Clim Risk Manag 16:145–163

Roy SN, Prasad K (1991) Climate change and global scientific response. Vayu Mandal 21(3–4):69–82

Ruddiman WF (2008) Earth's climate—past and future. WH Freeman and Company, New York, USA

Salby ML (1992) The atmosphere. In: Trenberth KE (ed) Climate System Modeling. Cambridge University Press, pp 53–115

Sarkar T, Kannaujiya S, Taloor AK, Ray PK, Chauhan P (2020) Integrated study of GRACE data derived interannual groundwater storage variability over water stressed Indian regions. Groundwater Sustain Develop, P100376. https://doi.org/10.1016/j.gsd.2020.100376

Seneviratne SI, Nicholls N, Easterling D, Goodess CM, Kanae S, Kossin J, Luo Y, Marengo J, McInnes K, Rahimi M, Reichstein M, Sorteberg A, Vera C, Zhang X (2012) Changes in climate extremes and their impacts on the natural physical environment. In: Field CB, Barros V, Stocker TF, Qin D, Dokken DJ, Ebi KL, Mastrandrea MD, Mach KJ, Plattner G-K, Allen SK, Tignor M, Midgley PM (eds) Managing the risks of extreme events and disasters to advance climate change adaptation: a special report of working groups i and ii of the intergovernmental panel on climate change (IPCC). Cambridge University Press, Cambridge, pp 109–230

Sheridan SC, Allen MJ (2015) Changes in the frequency and intensity of extreme temperature events and human health concerns. Curr Clim Change Res 1:155–165

Shukla PR, Sharma KS, Ravindranath NH, Bhattacharya S (2003) Climate change and india vulnerability assessment and adaptation. Universities Press, Hyderabad

Singh AK, Jasrotia AS, Taloor AK, Kotlia BS, Kumar V, Roy S, Ray PKC, Singh KK, Singh AK, Sharma AK (2017) Estimation of quantitative measures of total water storage variation from GRACE and GLDAS-NOAH satellites using geospatial technology. Quatern Int 444:191–200

Singh PK, Dhiman RC (2012) Climate change and human health: Indian context. J Vector Borne Dis 49:55–60

Singh S (1998) Effect of global warming on the stream flow of high altitude Spiti River. Enhydrology of High Mountain Areas. International Centre for Integrated Mountain Development, Kathmandu, pp 103–114

Singh S, Sood V, Taloor AK, Prashar S, Kaur R (2020) Qualitative and quantitative analysis of topographically derived CVA algorithms using MODIS and Landsat-8 data over Western Himalayas. Quaternary International, India. https://doi.org/10.1016/j.quaint.2020.04.048

Sood V, Gusain HS, Gupta S, Taloor AK, Singh S (2020) Detection of snow/ice cover changes using subpixel-based change detection approach over Chhota-Shigri glacier. Quaternary International, Western Himalaya, India. https://doi.org/10.1016/j.quaint.2020.05.016

Taloor AK, Kumar V, Singh VK, Singh AK, Kale RV, Sharma R, Khajuria V, Raina G, Kouser B, Chowdhary NH(2020). Land Use Land Cover Dynamics Using Remote Sensing and GIS Techniques in Western Doon Valley, Uttarakhand, India. InGeoecology of Landscape Dynamics 2020 (pp. 37–51). Springer, Singapore. https://doi.org/10.1007/978-981-15-2097-6_4

TERI (Tata Energy Research Institute) (2002) TERI Energy Data Directory and Yearbook 2001/2002. TERI Press, New Delhi

UNDP (United Nations Development Programme) Report (2007) Fighting climate change: human solidarity in a divided world. In: Human Development Report 2007/2008. Human Development Report. Palgrave Macmillan, London, pp 1–18. https://link.springer.com/chapter/10.1057/9780230598508_1

Vellinga P, van Verseveld WJ (2000) Climate change and extreme weather events. World Wide Fund for Nature, Switzerland. http://awsassets.panda.org/downloads/xweather.pdf

Verma O (2009) Extreme Climate Events. In: Ramanan VV (ed) Climate change indicators. Indian Gandhi National Open University, New Delhi, pp 54–66

WWF (World Wildlife Fund) Report (2005) An Overview of glaciers, glacier retreat, and subsequent impacts in Nepal, India and China. WWF Nepal

Programme, pp 1–70. https://assets.panda.org/downloads/himalayaglaciersreport2005.pdf

Weart SR (2003) The Discovery of Global Warming. Harvard University Press, Cambridge. https://www.history.com, https://www.history.com/topics/natural-disasters-and-environment/history-of-climate-change, https://www.unenvironment.org/ozonaction/. https://unfccc.int/kyoto_protocol

Yin H, Li C (2001) Human impact on floods and flood disasters on the Yangtze River. Geomorphology 41(2–3):105–109

Dr. Omkar Verma is a Senior Assistant Professor of Geology at the Indira Gandhi National Open University, New Delhi, India. He received his M.Sc. and Ph.D. degrees in Geology from University of Jammu, Jammu and Kashmir, India. He is the recipient of Senior Research Fellowship and Research Associateship of the Council of Scientific and Industrial Research, New Delhi. He led several expeditions to Cretaceous (145 to 66 million years ago) deposits of western, central, southern and southeastern India in search of small vertebrates that have lived in the shadow of dinosaurs. He has published more than 25 research articles in peer-reviewed national and international journals. He is the life member of the Geological Society of India, Indian Science Congress Association, Palaeontological Society of India, Indian Geological Congress and Gondwana Geological Society. His research focuses on palaeobiodiversity, biotic evolution, biogeography, palaeoecology and palaeoclimate with reference to northward drift of the Indian plate

4 Snow, Glacier, and Glacier Lake Mapping and Monitoring Using Remote Sensing Data

Praveen K. Thakur, Vaibhav Garg, Bhaskar R. Nikam, S. P. Aggarwal, Suruchi Aggarwal, and Dhanendra Singh

Abstract

Nowadays, remote sensing (RS) technology under the space programs of the world's leading agencies provides multispectral optical and microwave data for natural resource mapping of the Earth's surface. Snow and glacier mapping are critical for the accurate assessment of water resource availability on the Earth's cryosphere and to quantify the impact of climate change on these cryosphere components. At present, the availability of multispectral remote sensing data is a major source for studying the snow and glaciers from space. The RS approach provides data visualization, interpretation, and assessment of the time series scenario of both snow cover change and glacier dynamics. The mapping of snow is comparatively easier than glacier mapping. Most commonly, the normalized difference snow index (NDSI) band ratio technique has been adopted for snow cover mapping and monitoring. There are various remote sensing-derived snow cover area products available. To overcome the limitations of optical data in terms of cloud presence, the microwave RS data is highly useful for snow cover mapping and snow physical parameters retrieval. However, for snow physical parameters such as snow water equivalent, the scatterometer data incorporating the synthetic aperture radar (SAR) data was used. Recently, semi-automated algorithms are used for glacier mapping separating clean ice and debris cover area. For the Indian Himalayan region, glacier mapping was primarily based on Survey of India (SoI) and Geological Survey of India (GSI) topographic maps ranged on the scale 1:250,000 to 1: 50,000. The innovation of novel geospatial approaches for glacier mapping like manual delineation, band ratios, image segmentation, and classification is based on the multispectral/panchromatic data. However, mapping clean-ice and debris cover of the glacier still has some limitations for automated methods. Automated approaches for glacier mapping dealt with the comparison of different datasets, measurement of glacier change in length, volume, and snout positions by multi-temporal satellite imagery, and

P. K. Thakur (✉) · V. Garg · B. R. Nikam · S. P. Aggarwal · S. Aggarwal · D. Singh
Water Resources Department, Indian Institute of Remote Sensing (IIRS), Dehradun, India
e-mail: praveen@iirs.gov.in

V. Garg
e-mail: vaibhav@iirs.gov.in

B. R. Nikam
e-mail: bhaskarnikam@iirs.gov.in

S. P. Aggarwal
e-mail: spaggarwal@iirs.gov.in

S. Aggarwal
e-mail: suruchigeog@gmail.com

D. Singh
e-mail: dsingh.iirs@gmail.com

A. K. Taloor et al. (eds.), *Water, Cryosphere, and Climate Change in the Himalayas*,
Geography of the Physical Environment, https://doi.org/10.1007/978-3-030-67932-3_4

digital elevation model. Further, the high-resolution RS data has been extensively used for glacier lake mapping and monitoring. The regular study of glacier lakes is critical for any possible glacier lake outburst flood. In this way, the critical glacier lakes may be identified. It is concluded that geospatial technology plays a significant role in cryosphere studies and analysis from a climate change perspective.

Keyword

Snow · Glacier and glacier lakes · Remote sensing · Automatic methods · GIS

4.1 Introduction

The snowfall, which is one of the solid forms of precipitation, usually occurs when the air temperature drops below zero (0) degree Celsius for most parts of the world. This snowfall occurs in the Himalayas mainly during the winters and reaches its maximum coverage in March–April of a year. The deep Himalayan valleys and mountain tops have been receiving this snow over thousands of years and lead to the formation of Himalayan glaciers due to the process of snow accumulation and metamorphism. The age of glacier ice in Antarctica and other parts can be a few thousand to a few million years old. The seasonal snow line elevation (SLE) of the Himalayas fluctuates approximately between 1800 m as a minimum and 5500 m as maximum, with a few hundred meters of variations on an annual basis, lower SLE during good snowfall season and high SLE during low snowfall season. The trend of snowfall and subsequent snow cover and Snow Water Equivalent (SWE) varies with latitude as well as longitude and altitude for the Himalayas. Similarly, the glaciers, "*Which is a large, perennial accumulation of crystalline ice, snow, rock, sediment, and often liquid water that originates on land and moves downslope under the influence of its weight and gravity*" (USGS-2019), are found in high altitude Himalaya, mainly above 3800 m altitude. The high altitude Himalayan and Tibetan plateaus are places where a large number of lakes are found. Some of these lakes have glacier origin, i.e., they are formed due to the various morphological changes, due to the deglaciation and seismic process or due to the typical topographical and geomorphological setting of the area (NRSC 2015; Aggarwal et al. 2017). Many of these lakes are found at the elevation of 3500–5100, with a few exceptions in East Nepal near the Everest area, where elevation can reach up to 5500 m. Some artificial lakes are also formed due to man-made dams and water storage/diversion structures, but most of these lakes are located below 4000 m (NRSC 2015).

The traditional ground-based snow, glacier, and glacier lake mapping and survey methods are too expensive, labor-intensive, and require network and maintenance of field instruments (Dhir 1951). Satellite-based remote sensing is an efficient and reliable observation system for the systematic, large-area mapping and monitoring of snow, glaciers, and high altitude glacier lakes (Rango and Salomonson 1975; Ramamoorthi and Subba Rao 1981; Dozier 1989; Jain et al. 2001; Kulkarni and Rathore 2003; Thakur et al. 2013, 2017). This chapter gives an overview of basic remote sensing data, characteristics, and its interaction with snow, glacier, and glacier lakes. The basic mapping indices, tools, data products, and available operational maps of snow, glacier, and glacier lakes are provided.

4.2 Snow

Snow is generally porous and permeable aggregates of ice grains and its growth began at a temperature below 0 °C (Bader 1961; Hall et al. 2005). Snow crystals may occur in a variety of shapes, such as flat hexagonal or six-sided (Hall et al. 2005) (Fig. 4.1a). Snow forms an integral part of the global hydrologic cycle and spring snowmelt contributes significantly to total stream flow in many high altitude regions around the world. For the case of the Himalayas, the changes in the snowmelt from seasonal snowpack and ice melt from permanent glaciated ice impact the annual water distribution in its headwater rivers. In the South Asian region, a major amount of water is received from the melting of snow and glaciers of the rivers Indus, Ganges, and Brahmaputra (IGB) (Singh and Kumar 1997). Satellite-based RS applications enhanced our capabilities to study the snow cover and glacier mapping (Dozier 1989; Kulkarni 1991; Martinec and Rango 1981; Prasad and Roy 2005). The large-scale effect of snow water equivalent and grain size of snow on the visible and near-infrared (NIR) portion of EM energy was studied by Choudhury and Chang 1981; Dozier et al. 1981; Warren 1982). Figure 4.1 a shows the microscopic view of snow showing the classic dendritic snow crystal, and Fig. 4.1b shows the effect of grain size/age of snow on its spectral response; note the increase in spectral reflectance with increasing grain size or age of snow.

The standard False-Color Composite (FCC), which uses the band combination of Red→ NIR: Green→Red and Blue→Green band of multi-spectral remote sensing data, fails to differentiate the snow and clouds, as both have high and similar spectral reflectance visible to the NIR region of electromagnetic (EM) energy (Fig. 4.2a.3, a.2). This can be solved with the help of the middle Infrared (MIR) portion of EM energy. As shown in Fig. 4.2a.3, the snow/ice has a strong absorption band near 1.44 to 1.52 μm, whereas clouds still reflect higher in this region of EM energy. Therefore, this portion of the EM spectrum is used in the IRS Satellite (IRS 1C/1D, IRS-P6 (Resourcesat-1), Landsat, and MODIS sensors onboard Terra and Aqua. Table 4.1 gives the specification of these sensors relative to snow/ice studies.

The response of optical and synthetic aperture radars (SAR) on different types of snow cover is shown in the top and bottom panels of Fig. 4.2a, b. As is clear from Fig. 4.2a, the higher the density or grain size of snow, the lesser the reflectance in the visible and NIR part of the EM spectrum. Similarly, for SAR-based backscatter from the snowpack, the dry snow is virtually transparent to the snowpack, with the main backscatter from the snow-ground interface. For the wet snow, backscatter comes from the air-snow interface dominates, as there is very little penetration in wet snow (Thakur et al. 2012).

The visual appearance of snow, glacier, and glacier lakes as seen in SAR and optical images are shown in Fig. 4.3a, b, respectively.

4.2.1 Spectral Characteristics for Snow/Ice (H_2O and CO_2) and Clouds

The main spectral characteristic curves for the water ice and dry ice (solid CO_2) in the Visible-NIR range (VNIR) are given in Fig. 4.4a, b (Thakur et al. 2006; Cull et al. 2010). Over the entire range of the wavelength, dry ice reflects high, but displays a prominent absorption around

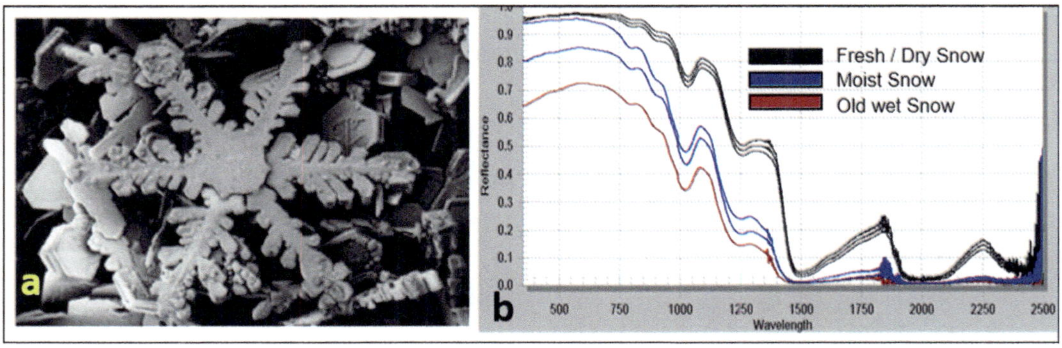

Fig. 4.1 **a** Microscopic view of a fresh Snow crystal and **b** Spectral Reflectance Characteristics of Snow (*Source* SASE Chandigarh)

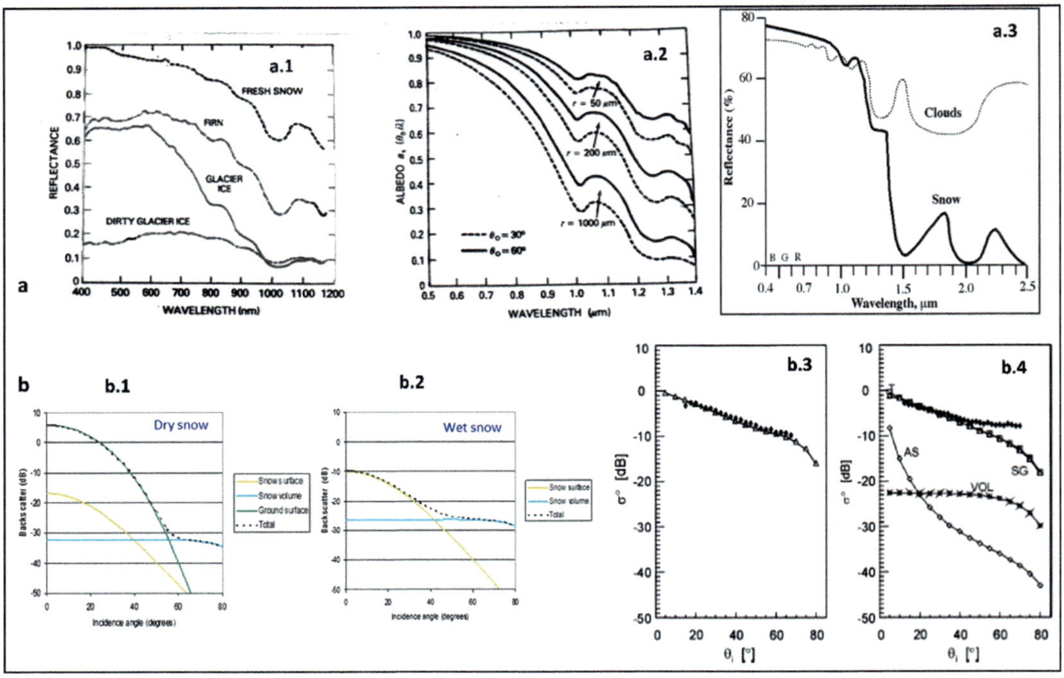

Fig. 4.2 **a** Basic spectra of snow as seen in optical RS data; **a.1** Reflectance curves for snow and glacier ice (Source: Hall and Martinec 1985); **a.2** Reflectance curves for snow at various grain sizes and Sun elevation angles (Source: Hall and Martinec 1985); **a.3** Reflectance curves for snow and cloud (Avery and Berlin 1992). **b** Backscatter response of snow in SAR images (Source: Nagler and Rott 2000); **b.1** and **b2** C-band Backscatter for dry and wet snow; **b.3** and **b.4** Total C-band backscatter from snowpack and its sub-components at air-snow, snow-volume, and snow-ground interfaces (*Source* Nagler and Rott 2000)

Table 4.1 Satellite sensors for snow/ice studies

Satellite/sensor	Spectral band (SWIR) μm	Spatial resolution (m)	Swath (km)	Year of launch
IRS-1C/1D—LiSS-III	B5 : 1.55–1.70	63.6 to 70.5	133–148	1995 and 1997
IRS-P6—LiSS-III	B5 : 1.55–1.70	23.5	141	2003
IRS-P6—AWiFS	B5 : 1.55–1.70	56	740	2003
IRS-R2. 2A—AWiFS (10 bit) LiSS-III	B5 : 1.55–1.70, B5 : 1.55–1.70	56 23.5	740 140	2011, 2016
Landsat TM and ETM$^+$	B5: 1.55–1.75 B7 : 2.064–2.345	30.0	185	1982, 1984, 1999
Landsat 8, OLI	B6: 1.56–165, B7: 2.107–2.294	30	185	2013- cont.
Terra/Aqua-MODIS	B7 : 2.10–2.15	500	2330	1999 and 2002
Sentinel-2A, 2B-MSI	B11 : 1.610, B12 : 2.19	20	290	2014—cont.

Source (WRD-IIRS)

2 μm (Cull et al. 2010). Water ice and snow have high reflectance at visible and NIR wavelengths, but their reflectance decreases after one μm, with substantial absorption near 1.5, 2.0, and 2.5 μm. In the case of the liquid water, it absorbs most of the electromagnetic spectrum (EMS) energy in

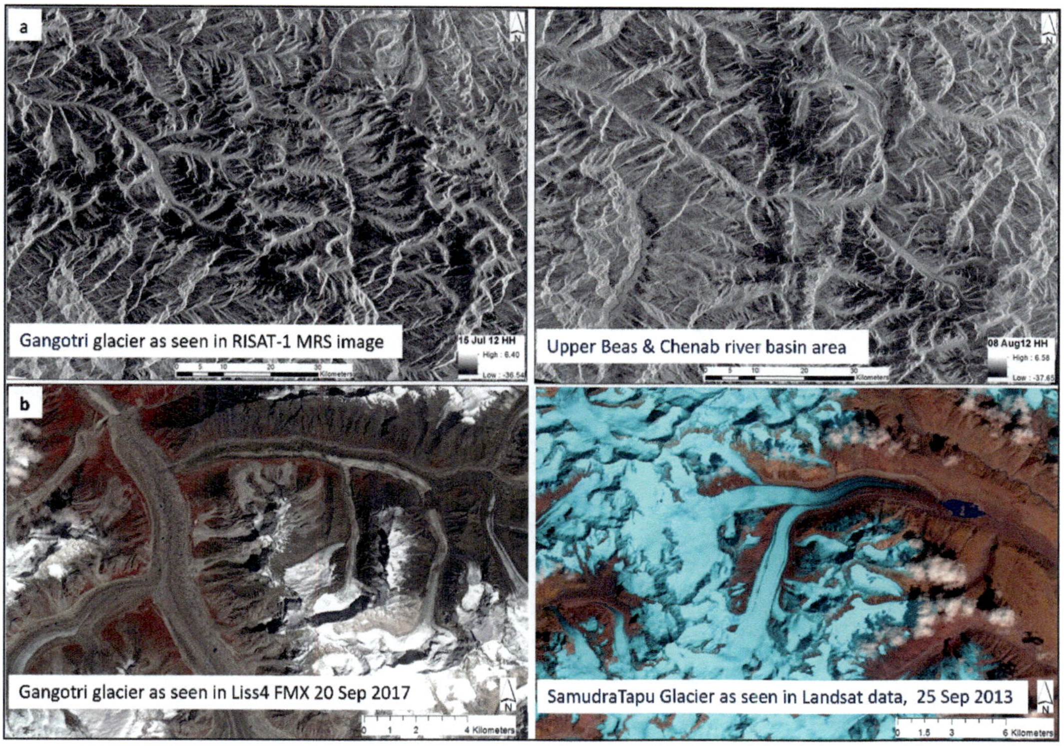

Fig. 4.3 Snow, glaciers, and Glacier Lake, as seen in **a** SAR (RISAT-1 MRS HH) data and **b** Optical images (Satellites Images: RISAT-1, Resourcesat-2 LISS-IV FMX, and Landsat 8, OLI) (*Source* WRD-IIRS)

NIR and beyond that, and only reflects slightly in the visible region of EMS (mainly in blue and green regions).

The Normalized Difference Snow Index (NDSI) is the most common spectral indices, which is using the SWIR and Green for finding the snow and non-snow areas (Hall et al. 1998).

$$\text{NDSI} = (\text{Green} - \text{SWIR})/(\text{Green} + \text{SWIR})$$

The range of NDSI is +1 to −1, with high values for fresh snow (>0.50) and less for old deposited glacier snow/ice (0.35 to 0.45), with the value of 0.4 usually taken as the threshold for snow and non-snow. Note that the mountain

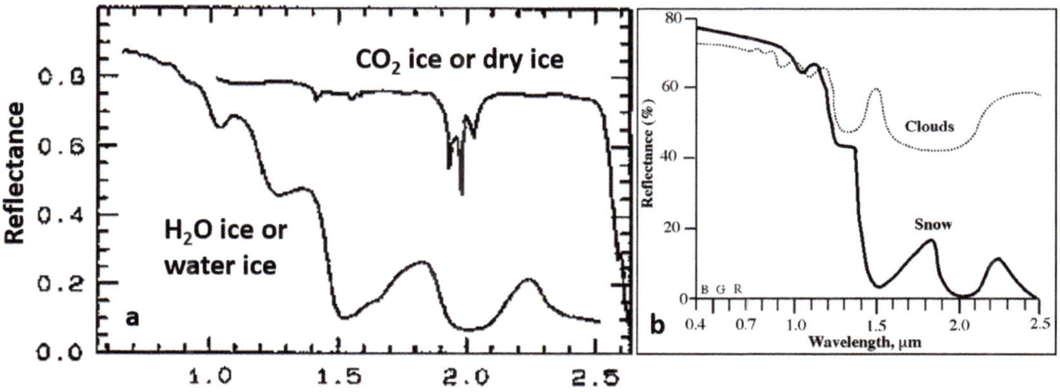

Fig. 4.4 **a** Spectral characteristics of H_2O and CO_2 Ice (*Source* https://www.ssec.wisc.edu/sose/pirs/pirs_m2_dataquality.html). **b** Spectral characteristics of snow and clouds (Avery and Berlin 1992)

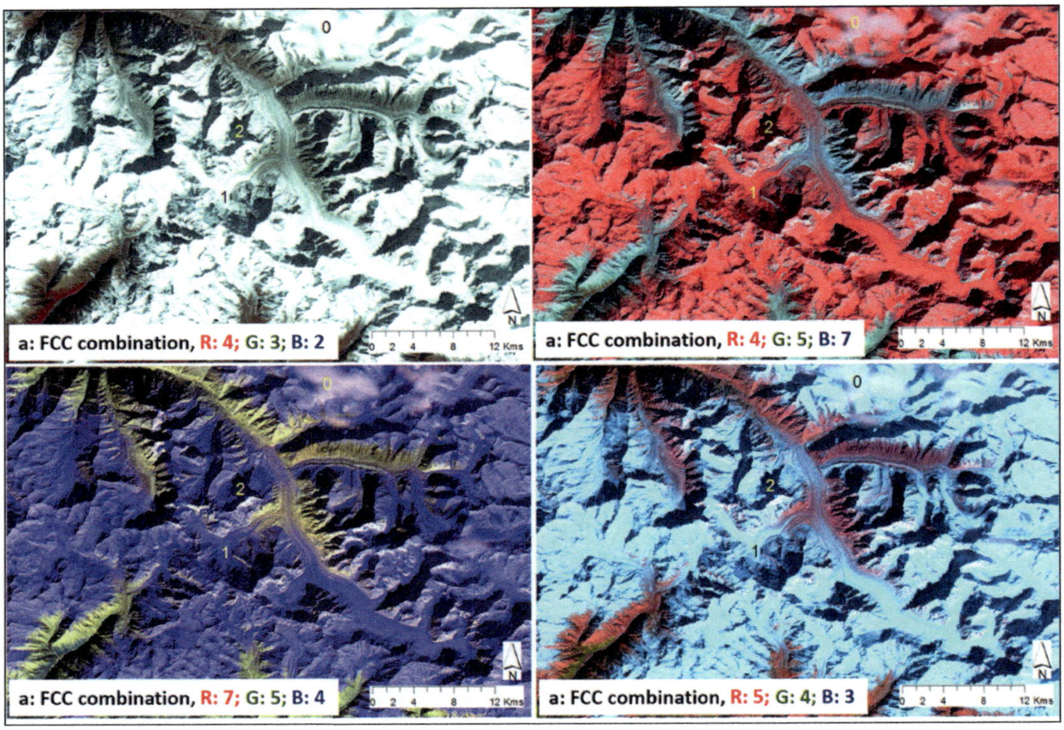

Fig. 4.5 Snow and clouds are seen in various Landsat 7 band combinations (Nov 25, 2000) for Gangotri Glacier in Garhwal Himalaya, Uttarakhand, India. 0: Cloud; 1: Snow; 2: Mountain Shadow. The ETM + band number 2 is Green, 3 is Red, 4 in NIR, 5 and 7 are in the SWIR region (*Source* WRD-IIRS)

shadow and clouds create a problem in identifying the snow/ice areas (Fig. 4.5). The problem of the shadow is solved by band rationing techniques such as NDSI (Fig. 4.6), and cloud discrimination is done using SWIR, NIR, and Red band-based color composites. This enhanced difference is mainly due to the fact that band ratioed images show the slope variations in the spectral reflectance curves of selected bands, regardless of their absolute reflectance values observed (Lillesand et al. 2015). The NDSI image of the Gangotri glacier; the bright areas show the snow/ice area (Fig. 4.6).

The operational snow cover product for the Indian region is available from Bhuvan, National Remote Sensing Centre, NRSC Hyderabad at a 15-day interval with 3 min resolution (Thakur et al. 2017) (Fig. 4.7). Apart from this, the global snow cover area, SCA, product from MODIS is used by various organizations around the world, including in India, to estimate the daily to 8-daily SCA at 500 m resolution. An example of such an SCA product (Nikam et al. 2017) is given in Fig. 4.8, for the northwest Himalaya.

In addition to the SCA products, change in SWE has been estimated using Indian Ku-band-based SCATSAT-1 level 4 sigma0 data (Oza et al. 2019). The ΔSWE maps (Fig. 4.9) were generated using a change detection technique, applied on 15-day temporal SCATSAT-1 backscatter data, to show the scattering properties of the snowpack (Yueh et al. 2008; Takala et al. 2011; Xiong et al. 2014).

4.2.2 Snow Mapping Using Passive Microwave Remote Sensing

The presence of snow over the underlying surface results in a decrease in the emitted microwave portion which is similar to that of the grain size of snow (Chang et al. 1987). This reduction in the emitted portion of wavelength is governed by the

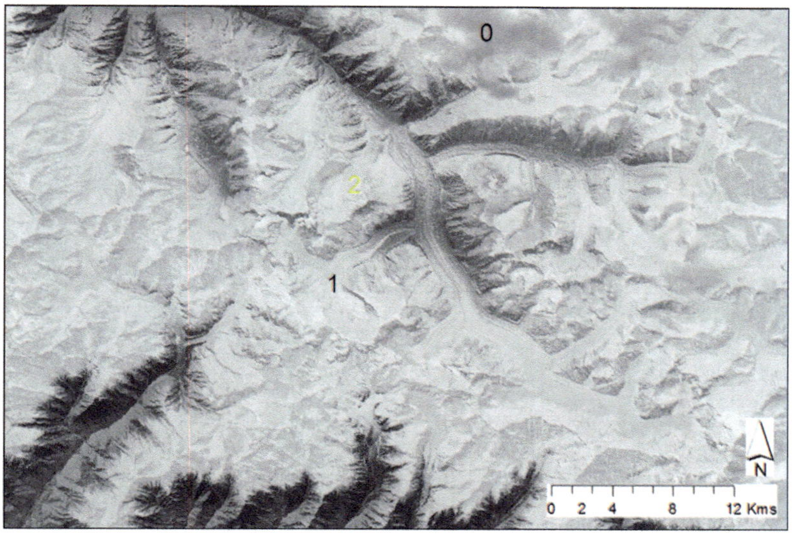

Fig. 4.6 Normalized Difference Snow Index (NDSI) image showing snow extent for the part of upper Bhagirathi basin and Gangotri Glacier area. Note that the shadow part has been removed and NDSI shows high value under the shadow as well, and the same high NDSI is shown for the clouds (*Source* WRD-IIRS)

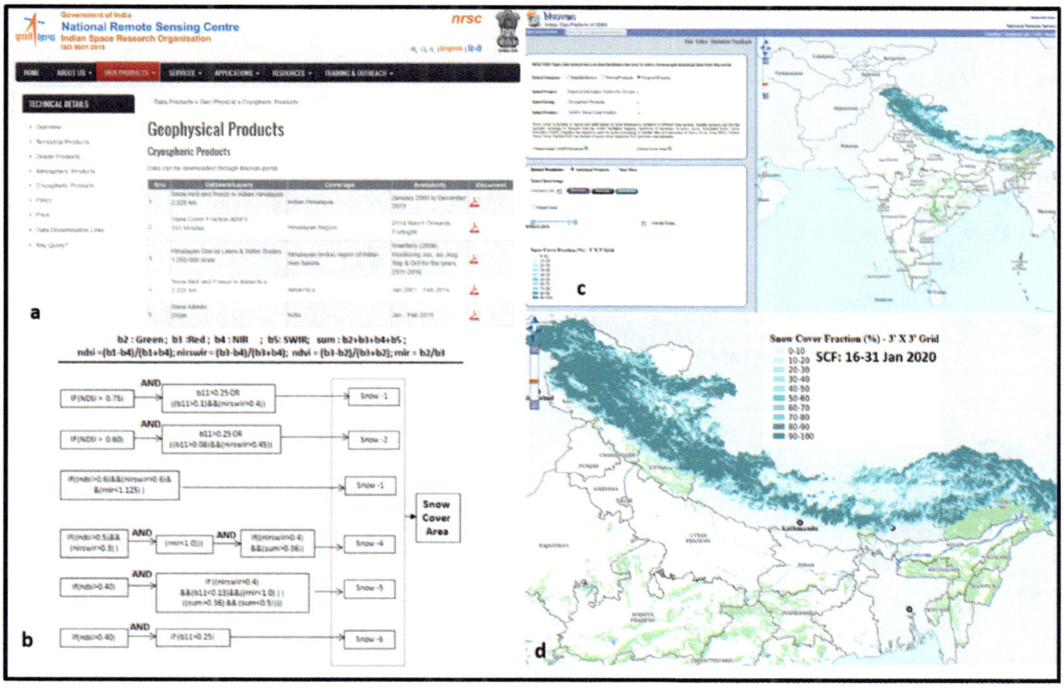

Fig. 4.7 **a–d** Fractional snow cover product from NRSC, Bhuvan-NRSC and its methodology (*Source* NRSC-ISRO)

snow cover area and snowpack mass of snowpack and area of snow cover, i.e., large amount of snow covers with thick snow depth (König et al. 2001). In this case, vertically polarized data have shown improved results in the mapping of thin or shallow SCA mainly due to the better sensitivity of snow volume (Thakur et al. 2017). However, horizontally polarized are usually used to map

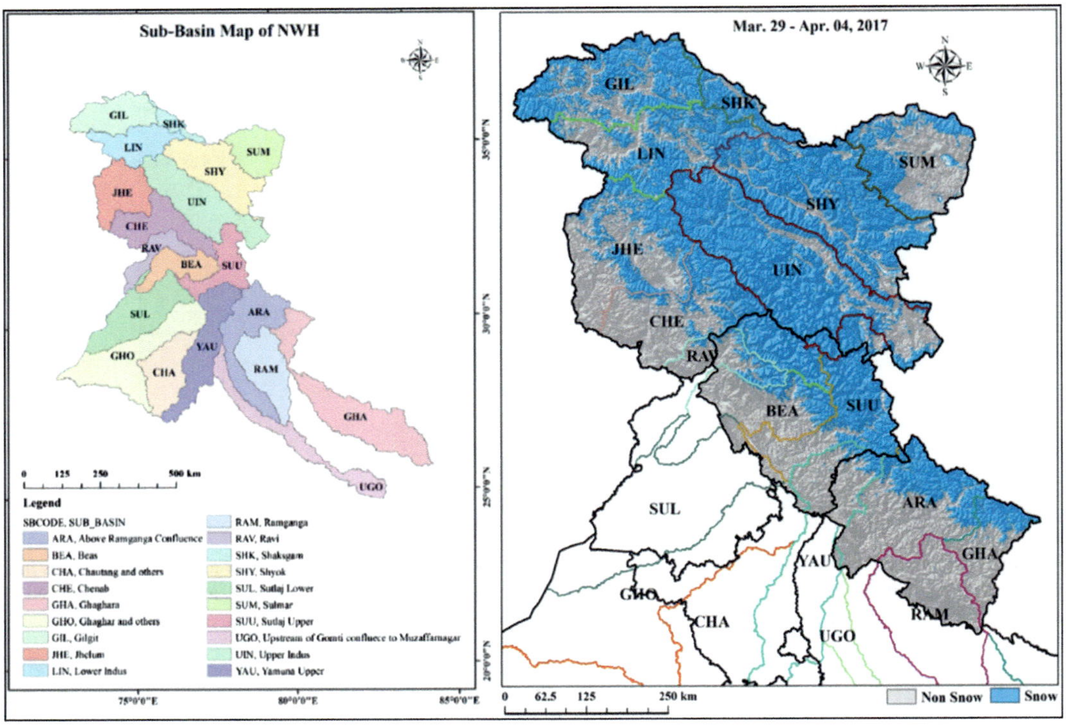

Fig. 4.8 MODIS based 8-day SCA for the early spring 2017 in northwest Himalaya (*Source* WRD-IIRS)

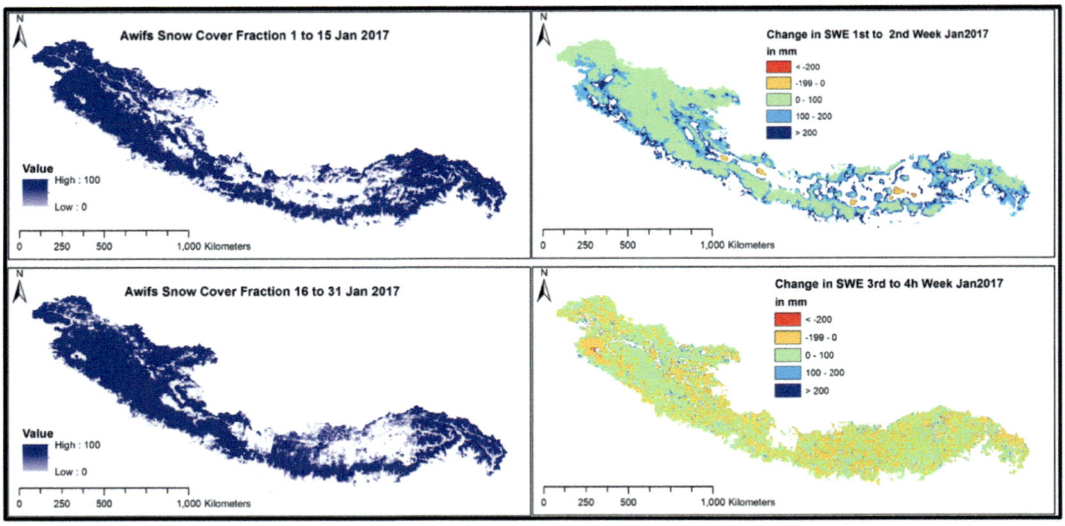

Fig. 4.9 AWiFS and SCATSAT-1-based fractional snow cover and ΔSWE maps for the Himalayas (*Source* WRD-IIRS)

SCA, mainly due to the mixed signals, which are caused by the inclusion of dry soil signals which underlie below the shallow snowpack (Amlien 2008). The initial results and overview of the passive microwave applications for snow research were provided by Foster et al. (1984). The mapping global snow cover has been done using passive microwave RS data and methods

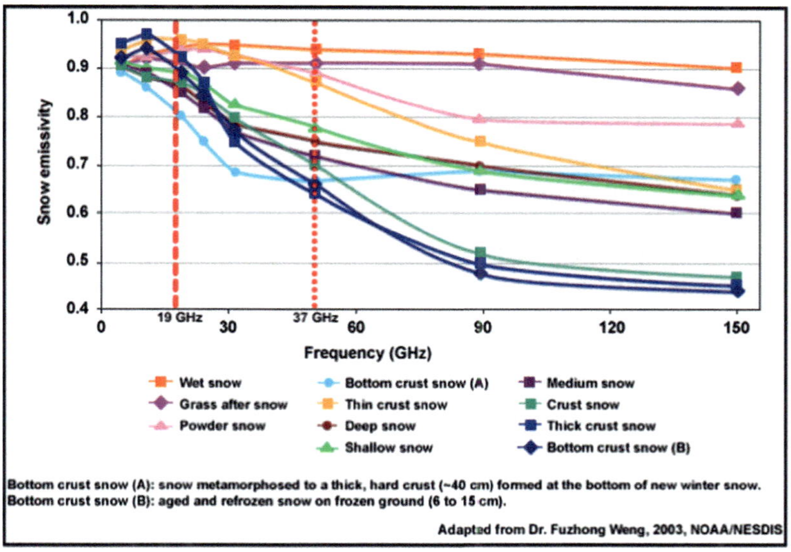

Fig. 4.10 Passive microwave emissivity spectra for different snow types at 50 deg. viewing angle (*Source* https://www.meted.ucar.edu/satmet/microwave_topics/land_ocean_v2/print.php)

from the early 1970 s to the 1990 s using NIMBUS' Scanning Multichannel Microwave Radiometer (SMMR) and Defense Meteorological Satellite Program (DMSP), Special Sensor Microwave/Imager (SSM/I) sensors (Foster et al. 1984; Grody 1991; Chang et al. 1992; Chen et al. 2001). The various types of snow conditions in the microwave emissivity spectra at different wavelengths are shown in Fig. 4.10.

The snow cover and depth parameters were studied by Saraf et al. (1999) in the Indian Himalayan region whereas, NIMBUS 7's SMMR sensor study has been made by Thakur et al. (2017) and the modified algorithm study related to snow cover was studied by Chang et al. 1(992) whereas by using AMSR-E data semi-empirical relation between brightness temperature and ground data was derived by Singh and Mishra (2006) and microwave daul polarized ground-water radiometer study was carried at Dhundi Manali, India, by Singh et al. (2007).

4.2.3 Global Snow Products Based on the Multi-sensor Approach

The global data on various types of cryospheric study can be obtained from the National Snow and Ice Data Centre (NSIDC) website (https://nsidc.org/), as well as from the ESA GlobSnow project website. Since 2014, NSIDC has released a multi-sensor, multi-resolution snow cover product for the Northern Hemisphere. The data at the National Ice Center's Interactive Multisensor Snow and Ice Mapping System (IMS) is available since February 1997 to the present. Satellite imageries, in situ observations along with various other data products were utilized for the generation of this database. The data is available at 03 different resolutions, i.e., 1, 4, and 24 km in both ASCII text and GeoTIFF formats. Canadian Meteorological Centre (CMC) has released the Northern Hemisphere (https://nsidc.org/data/g02156). Similarly, a Northern Hemisphere snow depth was created using Sentinel-1 SAR data at 1-km spatial resolution at a weekly interval (Lievens et al. 2019).

4.3 Glaciers

Glaciers are moving mass of snow and ice, which are generally present above the snow line (Cogley et al. 2011). The glaciers across the globe have varying thickness from a few hundred to more than a thousand meters, which may further change due to global warming significantly. Around 70% of world's non-polar glaciers are located in the Himalayas between countries namely India, Nepal, Pakistan, Bangladesh, Afghanistan, and Bhutan, supporting millions of lives. The runoff from Himalayan snow and glaciated regions feeds three major rivers in the world, i.e., the IGB. In the Himalayan cryosphere, glaciers are the most important constituents of the natural system and very sensitive to climate dynamics. The accurate monitoring and mapping of glaciers are of vital significance for the water resources as well as mankind in the Himalayan region. As glacier change in time, are the key for climate indication and the outlines of glaciers are mandatory for such type of mapping and monitoring, over the planet Earth (Thakur et al. 2017). The Randolph Glacier Inventory (RGI) is globally available in the current version (RGI Version 6.0: released July 28, 2017). At the starting of snow and glacier studies with optical satellite data, like Landsat False Color Composite (FCC) from Multi-Spectral Scanner (MSS) and later the Thematic Mapper (TM) were used to delineate the glacier boundaries and its different zones like accumulation and ablation area, also classify ice and snow facies in an FCC.

4.3.1 Manual Delineation

The first method in the sequence for glacier mapping was applied, especially to the Landsat MSS data for delineating the glacier boundary combined with an FCC (Fig. 4.11). This method takes excessive time and required well approach of image interpretation skills. This method, based on the subjectivity of the goal, includes human errors in recognition of glacier terrain and associated features on the glacier surface and surroundings. The results also may vary from person to person by the interpretation of satellite imagery, especially in the debris and lateral

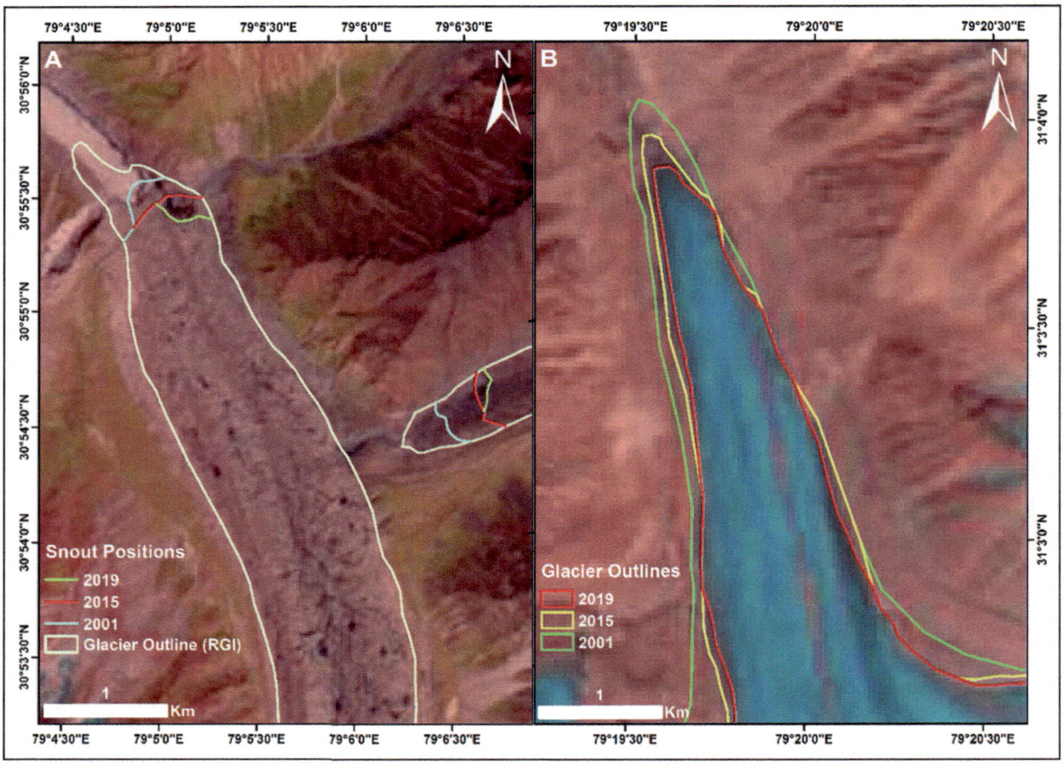

Fig. 4.11 On-screen manually digitized glacier snout positions and boundaries (Sentinel-2A FCC image, acquired on September 19, 2019). **a** Snout positions of Gangotri glacier systems in different years and **b** A clean-ice glacier in the Alaknanda valley. (*Source* WRD-IIRS)

moraine side. The obtained results by manual digitization depend on the accuracy of skills and individual's expertise in image interpretation with scene characteristics. However, this method was proved most robust and is likely to be provided with more precise results other than automatic methods. It is highly recommended for the classification of glacier surface separating clean ice, debris cover, rock glacier area, proglacial and supraglacial lakes, and debris cover from each other. The accuracy of the method also depends on the meteorological conditions (seasonal snow, cloud cover, and shadow) of the imagery obtained.

The manual method of glacier mapping is often required to correct automated mapping results (Rott and Markl 1989; Hall et al. 1992; Williams et al. 1997; Paul 2002; Andreassen et al. 2008; Bhambri and Bolch 2009). Also, panchromatic data (air photos) and Corona satellite images (Bishop et al. 1998; Leonard and Fountain 2003; Rabatel et al. 2008; Racoviteanu et al. 2009; Bhambri and Bolch 2009; Bhambri et al. 2011) and those without SWIR bands (MSS, QuickBird, ALOS AVNIR, IKONOS, and others) are applied for manual digitization in glacier mapping. In studies regarding the Indian context, almost all the glacier inventories have been prepared by manual delineation based on satellite imageries, from coarse-resolution FCC (MSS and LISS 1) to the high-resolution satellite data (LISS IV and PAN). The Survey of India (SoI) topographic maps ranged 1:250000 to 1: 50000 provide a base for historical data in support of remote sensing methods to prepare glacier inventories (Raina and Srivastava, 2008; Bhambri and Bolch 2009). However, these base maps lie some cartographic errors in their studies due to the lack of high skills and more sophisticated surveying instruments.

4.3.2 False-Color Composites

The FCCs are the combination of different bands that provide differences in reflectance of surface features of the landscape. Landsat 7 ETM + (Enhanced Thematic Mapper Plus) 5, 4, 3 RGB (Red: band 5; Green: band 4; Blue: band 3), Landsat 8 OLI (Operational Land Imager) 5, 4, 3 (NIR: band 5; Red: band 4; Green: band 3), and Sentinel-2 MSI (Multi-Spectral Instrument) 8, 4, 3 composite bands reflected snow and ice clearly and differentiated rocks, debris, vegetation, and clouds due to the FCC color differences. The FCC composites also benefitted from manual digitization and automatic classification for glacier mapping (Fig. 4.12). They work very precisely to extract clean-ice and snow over the glacier surface (Hall et al. 1988;Rott 1994; Bayr et al. 1994; Paul et al. 2002, 2016; Pellikka and Rees 2009).

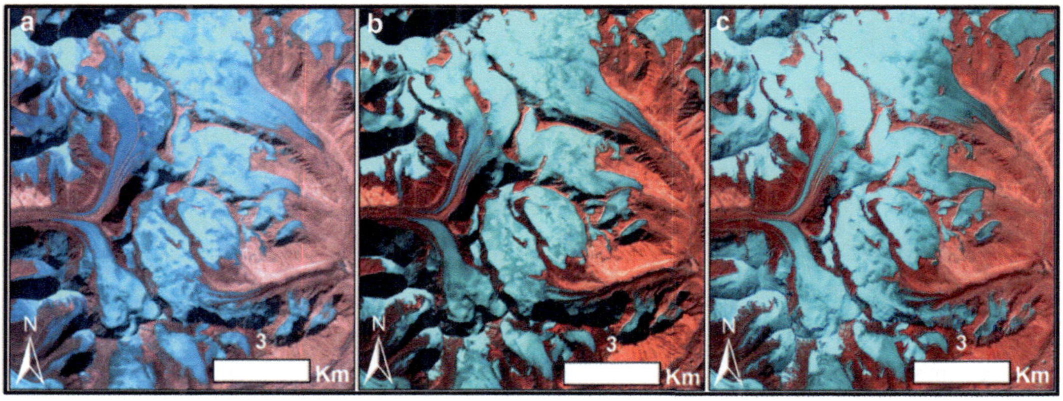

Fig. 4.12 FCC combinations for glacier mapping: **a** ETM + : 5-4-3 dated Oct. 8, 2000; **b** OLI: 5-4-3, dated Oct. 29, 2013 **c** MSI: 8-4-3 dated Sep. 19, 2018 for the Chaturangi and surrounding glaciers in Garhwal Himalaya, Uttarakhand, India (*Source* WRD-IIRS)

4.3.3 Automated Glacier Mapping by Segmentation of Ratio Images

Automated glacier mapping involves digital image processing techniques by applying simple band ratio/mathematics and classification. The automated method is based on the reflectance of the ice and snow, and the fact is that snow and ice have a strong reflectance in the visible and NIR regions whereas, the shortwave infrared (SWIR) region has lower reflectance in the spectrum. Bayr et al. (1994) and Rott (1994) applied and proposed threshold values of ratio image for TM band 4 to TM band 5 (NIR/SWIR) and TM band 3 to TM band 5 (RED/SWIR) ratios to extract glacier ice area. Paul (2000) and Bhambri et al. (2011) evaluated this method and proposed that the TM 4 to TM 5 ratio is the more appropriate and accurate for clean-ice glacier mapping whereas, for shadow regions, TM 3/TM 5 ratio images produced better results than TM 4/TM 5 ratio. The average reflectance of the glacier basin was calculated for 3 years to assess the area of each glacier. The band ratio RED/SWIR performed better than others (NIR/SWIR) in the dark and shadow areas, and also over the thin debris layer (Paul 2000, 2002; Paul and Kääb 2005; Andreassen et al. 2008; Racoviteanu et al. 2009; Bhambri and Bolch 2009). Recently, a large number of glacier inventories have been prepared with many new approaches and methods (Haeberli et al. 2000; Paul 2002; Campbell 2005; Bolch and Kamp 2006; Kulkarni et al. 2007; Andreassen et al. 2008; Kaushik et al. 2019; Taloor et al. 2019; Singh et al. 2020). An example of mapping glaciers using band ratio methods is shown in Figs. 4.13 and 4.14.

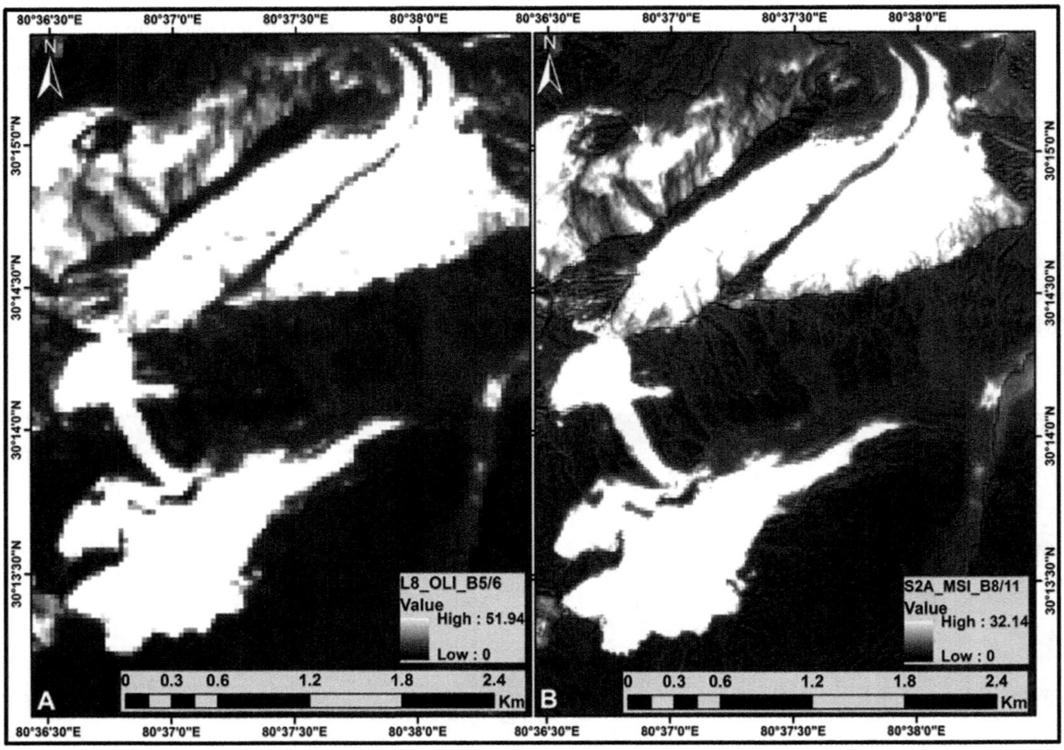

Fig. 4.13 NIR/SWIR band ratios for glacier mapping: **a** Landsat OLI (band 5/band 6), **b** Sentinel 2A MSI (band 8/band 11) (*Source* WRD-IIRS)

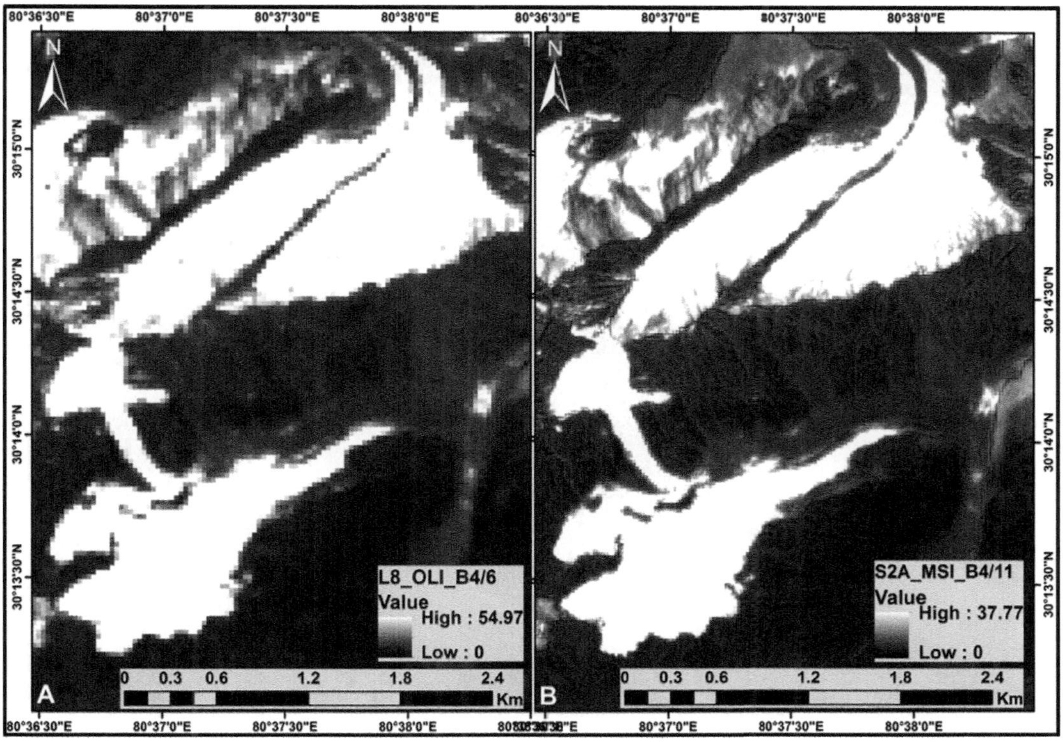

Fig. 4.14 Red/SWIR band ratios for glacier mapping: **a** Landsat OLI (band 4/band 6), **b** Sentinel 2A MSI (band 4/band 11) (*Source* WRD-IIRS)

4.3.4 Normalized Difference Snow Index (NDSI) Method

Due to cloud cover, it is not easy to differentiate between snow and cloud although at 1.6 mm, snow cover absorbs energy from the Sun, so it looks darker to the clouds which provides an accurate difference between clouds and snow cover. The RS-based observations used 0.66 and 1.6 mm for snow presence, therefore, the band ratio technique named as normalized difference snow index (NDSI) has widely been used for snow cover mapping which is the normalized difference between the reflectance of visible (Green) and Shortwave Infrared (SWIR) wavelength bands and is quite sensitive for the snow and not to bare ice (Hall et al. 1998, 1995). The low reflectance of snow and, on that contrary, the high reflectance of clouds in the SWIR region allow a good difference between snow cover and area covered by clouds. This technique has been successfully applied by Racoviteanu et al. 2008 for the mapping of glaciers in Cordillera Blanca. However, this automatic method failed to map the debris over the glacier surface due to the similarity in the spectral signature of rocky terrain. For the precision of results by the NDSI method, Sidjak and Wheate 1999 got the best output by the combination of principal component and the TM-4/TM-5 ratio.

For the Indian context of glacier mapping in the Indian Himalayas, Racoviteanu et al. (2008), Gupta et al. (2005), Sood et al. (2020a, b) applied the NDSI method for the mapping of glaciers using the NDSI technique over satellite data. By these results, it is confirmed that these methods are applicable for clean-ice glacier/area mapping/extraction and unsuitable for the debris cover mapping.

The results obtained from NDSI and segmentation of the ratio image are explained as (**A**) Band ratio, and NDSI methods are not capable of differentiating debris-covered ice due to their similar reflectance between debris-covered

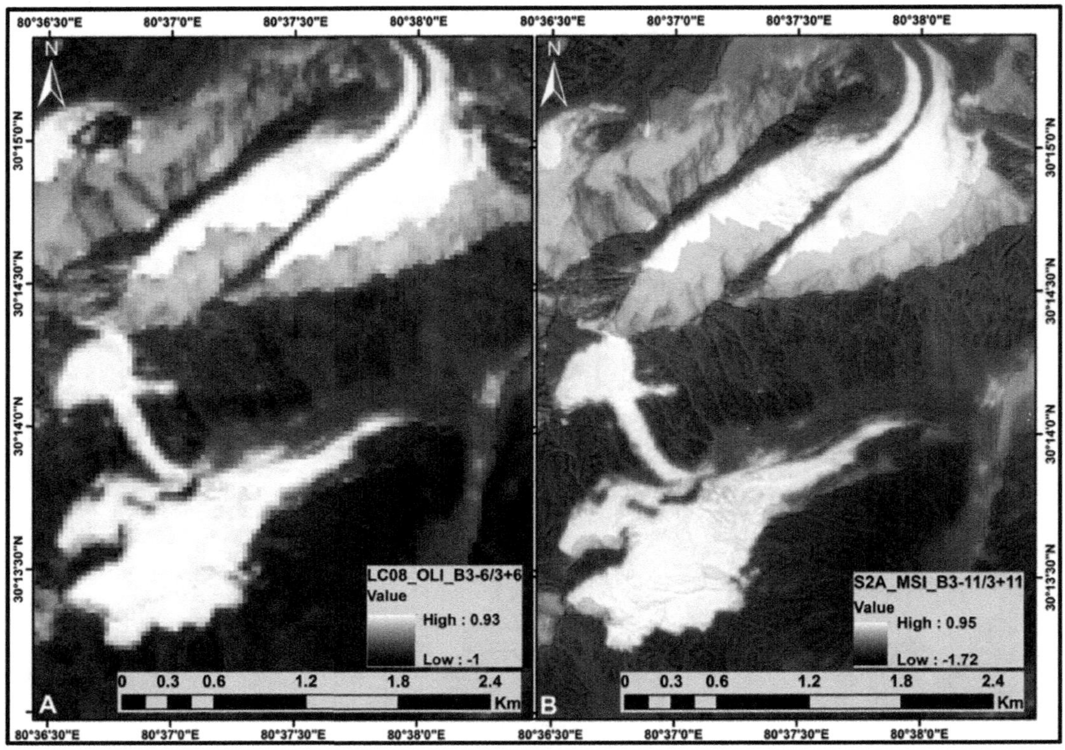

Fig. 4.15 NDSI images: **a** Landsat OLI (band 3-band 6), **b** Sentinel-2A MSI (band 3-band 11) (*Source* WRD-IIR)

ice and surrounding rocky surface. NDSI is highly sensitive for seasonal snow whereas, band ratio is capable and more sensitive for snow cover as well as clean glacier ice (Figs. 4.13, 4.14, and 4.15). (**B**) Band ratio (VIS/NIR) and NDSI methods misclassified proglacial areas with lakes whereas, NIR/SWIR is applicable only for clean ice and appropriate for clean glacier ice mapping than Red/SWIR and NDSI (Singh et al. 2020). Also, in shadow areas, band ratio NIR/SWIR worked better than NDSI (Figs. 4.13 and 4.16).

4.3.5 Spectral Transformation

- *Intensity-hue-separation transformation (IHS)*: This is a transformation of RGB images to another different color space, which might be useful. In this category, the IHS color space is the often and most applied color space for the spectral transformation of images (Kääb et al. 2014; Paul et al. 2004).
- *Principal Component Transformation (PCT)*: In this technique, the image with Multispectral bands can highly be correlated to a similar material of an entire wavelength range, and it occurred due to the topographic effects. PCT targeted the transforming of the original scene linearity to overcome the inter-band correlation.
- *Decorrelation Stretching*: It can be applied to the PCT or IHS transformations to minimize the unessential information of a multispectral image (Gillespie et al. 1986). By using this method, it is better for the visual analysis of the multispectral satellite imagery and incorporate some manual digitization and supported for further classification.

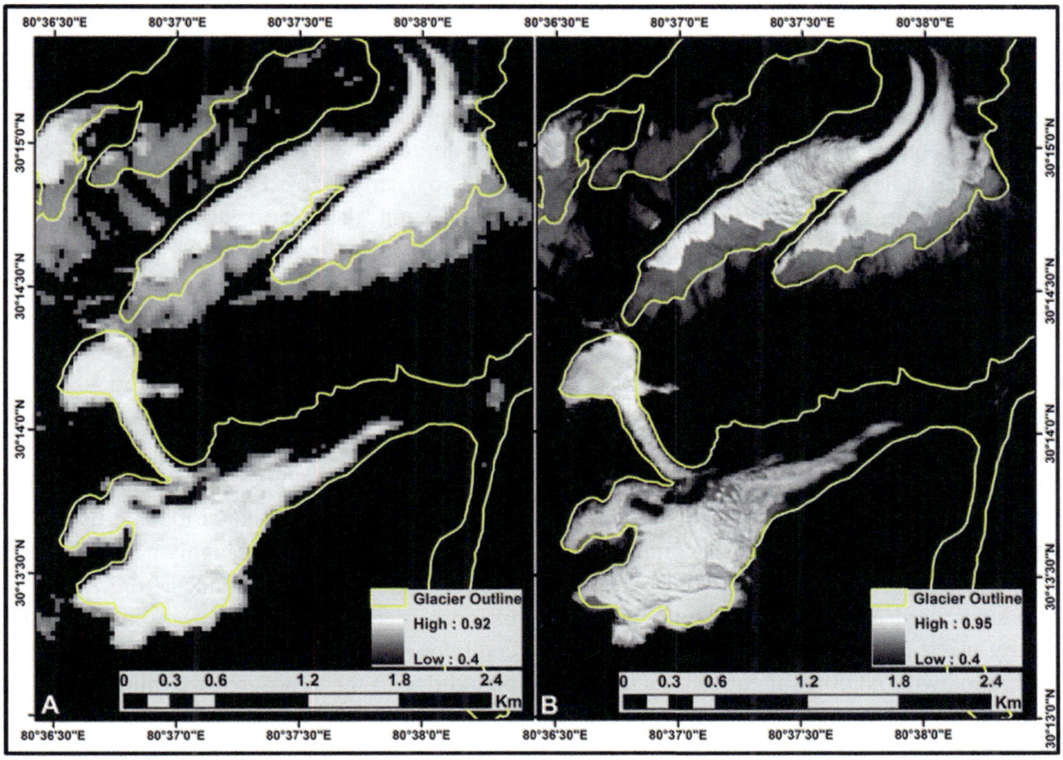

Fig. 4.16 Snow areas extracted from NDSI image using 0.4 threshold value **a** Landsat OLI, **b** Sentinel-2A MSI (*Source* WRD-IIRS)

4.3.6 Unsupervised Classification

Unsupervised classification is the most robust method for relatively homogeneous terrain with some examples (clean-ice glaciers). However, this method suffers from demerits in variable terrain with several classes such as clean ice, debris-covered ice, dirty ice, and shadow (Paul et al. 2002, 2004). This method is user-dependent to assign the classes for different categories that led to the classification problem for the low difference in terrain conditions, especially in ablation areas where ice has a low difference between the debris-mixed and dirty-ice conditions.

4.3.7 Supervised Classification

The supervised classification works very precisely in high mountain terrain by spectral separation. Under the category of unsupervised classification, maximum-likelihood and spectral angle mapper methods can be applied for the spectral differentiation of the terrain. It is still a challenging task to map debris-covered ice using all spectral classifications. However, the best outputs with supervised classification can be obtained by applying visual, SWIR, and TIR bands. Generally, the spectral signatures trained from one image could not be used to other imagery because this minimizes the automation capability or needs radiometric adjustments. Some studies on the supervised classification of glaciers have been carried out by different scholars (Gratton et al. 1990; Paul 2000; Paul et al. 2004) applied spectral unmixing to classify snow and ice.

4.3.8 Artificial Neural Networks (ANN)

The application of the ANN classification is not bonded to the multispectral temporal data, as to the topographic parametric information that can also be applied. The ANN classification was tested in a single domain (Brown et al. 1998, Bishop et al. 1999) for glacier and permafrost terrain. The major applicability of the ANN in cryospheric studies is based on the combination of the multidimensional data (Paul et al. 2004).

4.3.9 Combinations

Several classification procedures often must be a combination of multispectral data by fusion rationing approaches. As mentioned above, like supervised and unsupervised classification outputs, results of other applied methods may be combined to obtain appropriate outcomes. Sidjak and Wheate (1999) obtained good outcomes from supervised maximum likelihood classification by using 2–4 bands of TM and TM4/TM5 band ratios with NDSI as input for glacier mapping. After applying all automated approaches for glacier mapping, manual correction is still needed for the accurate glacier boundary. The joint classifications like band ratios applied for the glacier study provide misclassifications in the outcomes of the edge in vegetation, shadow, and debris cover, which can be removed by using various methods such as NDSI, NDVI, and NDWI.

4.3.10 Debris Cover Mapping

As described above, multispectral imagery can be applied to map clean-ice glaciers automatically whereas, in Himalayan glaciers due to the large volume and mainly debris-laden ice avalanche, rocks fall over the glacier surface from steep valley walls where a majority of glaciers are debris glaciers (Shroder et al. 2000). Based on a thermal band to map the debris-covered glacier, a large number of studies have been carried out in the Indian region of the Himalayas aided by DEM. Some of these works are made by Ranzi et al. (2004) by using ASTER and Landsat images; surface temperature for debris cover by Bhambri and Bolch (2009); slope, plan, and profile curvature to delineate the debris-covered carried out by Bishop et al. (2001). Paul et al. (2004) presented a semi-automatic algorithm to map debris cover area. In the Swiss Alps, the semi-automatic algorithm to map debris cover area was derived by Paul et al. (2004) using multispectral data-based classification. In the Mount Everest region, Bolch et al. (2007) applied a complex morphometric-based approach based on the ASTER DEM and ASTER thermal bands. All these methods for debris cover glacier mapping are region specific and may not be suitable for another area due to the terrain conditions.

Kulkarni et al. (2005, 2007) delineate debris cover glaciers by the manual method using IRS satellite data. Such studies highly need cloud-free imagery and should be obtained from August to October, depending on the seasonal snowfall conditions in the basin at the end of the ablation period due to the exposure of terminal moraines and a clear view of snout positions. But applying all automated methods for debris-covered glacier mapping, the remote sensing methods need to check the ground conditions by field visits because sometimes the terminal moraines have a thin layer of vegetation that affects the automated outcomes.

The global glacier inventory such as the RGI has used glacier mapping methods as discussed in the above sections. The latest version 6 of RGI has given consistent maps of glaciers. The RGI v.5 and 6-based glacier map of High Mountain Asia is shown in Fig. 4.17.

4.4 Glacier Mapping and Monitoring in India

In India, glacier mapping is done initially by SoI and later updated by GSI using limited aerial photographs and field surveys. With the advancements in satellite remote sensing, glacier inventory was updated by many research institutes in India and abroad (Thakur et al. 2017). Kulkarni et al. (2007) used Indian earth observation data along with SoI toposheets to map the major glaciers of Himachal Pradesh, and found

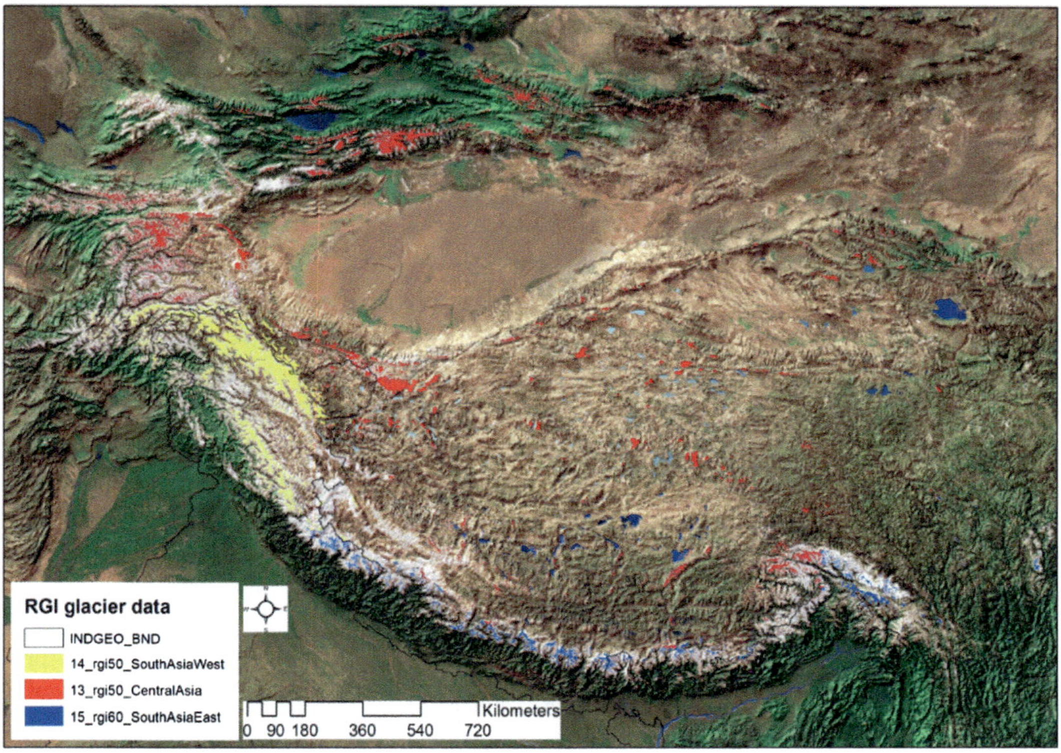

Fig. 4.17 Example of glacier extents by RGI for the Himalayas and Central Asia, overlaid over ESRI Earth mosaic (*Source* WRD-IIRS, glacier boundaries from RGI inventory and global base map from ESRI)

the total glacier area loss of 21%. The inventory of the entire Himalayas was created using the NDSI method (Sharma et al. 2013; Sood et al. 2020b). The historical Corona aerial images with current RS data for mapping and quantifying the glacier changes in the Garhwal Himalaya have been used by many researchers (Bhambri et al. 2011; Bhardwaj et al. 2015; Kumar et al. 2020; Sood et al. 2020a; Singh et al. 2020). The derived maps are useful for the estimation of the equilibrium line altitude (ELA) of a glacier. The use of SAR data for mapping and monitoring of glaciers is proving to be effective (Winsvold et al. 2018; Thakur et al. 2018), mainly since the availability of low cost and free SAR data from satellites such as ENVISAT ASAR, RISAT-1, and Sentinel-1 SAR missions (Thakur et al. 2017).

4.5 Glacial Lakes

The sufficient amount of water mass that originated due to glacier activities especially the retreat located over, besides, or/and in front of the glacier is regarded as glacial lakes (Campbell 2005). Lakes in the vicinity of snout could be dangerous due to their high probability of breaching and the high volume of the stored water (Campbell 2005). Breaching of such glacier lakes results in instantaneous discharge and causes huge flash floods in the downstream area (Huggel et al. 2002; ICIMOD 2007; ICIMOD 2010; Jain et al. 2012). To monitor these hazards, it is critical to creating an inventory of all such glacier lakes (Thakur et al. 2016). Most of such glacier inventories are created using

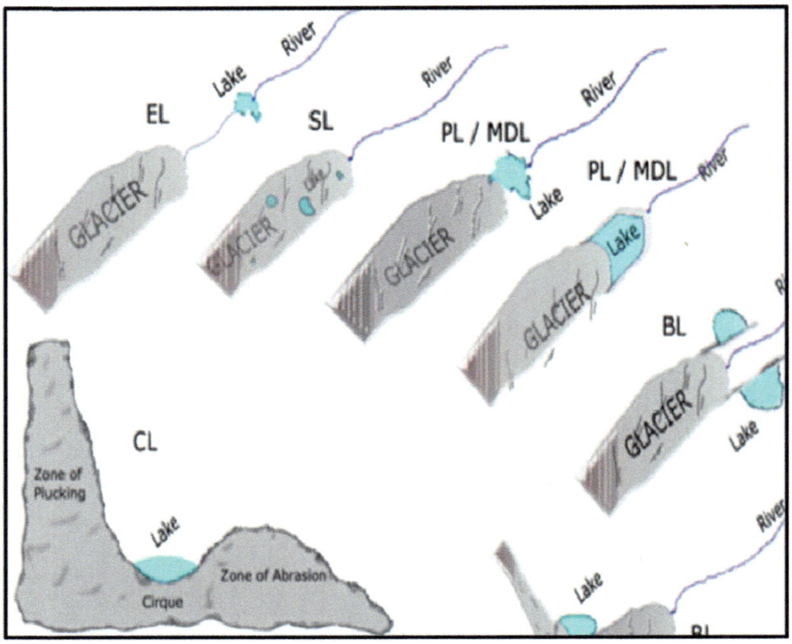

Fig. 4.18 Various types of glacier lakes; SL—Supraglacial Lake; PL—Proglacial Lake; EL—Erosion Lake; CL—Cirque Lake; MDL—Moraine-Dammed Lake; BL—Blocked Lake (*Source* Raj and Kumar 2016)

remote sensing data and through their field investigations subsequently (Thakur et al. 2016; Aggarwal et al. 2017). Apart from creating an inventory, their monitoring is essentially required to quantify the morphological and aerial extent changes. Based on their location, the lakes are classified as erosion, valley trough, cirque, blocked, moraine-dammed, and supraglacial lakes (Campbell 2005; Raj and Kumar 2016) (Fig. 4.18).

4.5.1 Erosion Lakes

Glacial erosion lakes are generated after the retreat of a glacier with time after depression. The cirque and trough Valley-type lakes are considered to mostly be stable (Campbell 2005; Raj and Kumar 2016). The erosion lakes are located far away from the present glaciated regime (Fig. 4.18).

4.5.2 Supraglacial Lakes

The Supraglacial lakes (SL) are generally developed in the area 50 to 100 meters in dimension. SL can be developed at any location on the glacier, but their extent is less than half of the diameter of the Valley glacier (Raj and Kumar 2016). They have the following characteristics such as shifting, merging, and draining of the lakes with a high level of potential energy which may be critical from the GLOF point of view (Fig. 4.18).

4.5.3 Moraine Dammed Lakes

During the recession period of the glacier, ice generally melts at the frontal part of the glacier, resulting in the formation of many supraglacial ponds at the glacier tongue. These sometimes enlarge on interconnecting with each other and tend to deepen further (Campbell 2005) leading to the formation of a moraine-dammed lake (Aggarwal et al. 2017). The lake gets filled with the water from its surrounding area and gives rise to two types of moraines such as ice-cored moraine and an ice-free moraine. Due to continuous melt, the lake becomes more deep and wide (Aggarwal et al. 2017). However, when the ice stored in moraines and under the lake remain with the bedrock and the moraine (Fig. 4.18).

4.5.4 Blocking Lakes

Blocking lakes may be formed in two ways: either the main glacier blocks the branch valley, or the glacier branch blocks the main. Further, the snow avalanche, collapse, and debris flow may also result in blockage, and the formation of the blocking lakes may be due to snow avalanche and debris flow blockage (Campbell 2005) (Fig. 4.18).

4.5.5 Ice-Dammed or Ice-Blocked Lakes

An advancing glacier, when intercepted by a tributary/tributaries pouring into the main glacier valley, leads to the generation of an ice-dammed lake at the side(s) of a glacier and generally is small in size (Aggarwal et al. 2017). It is to be noted that the glacier lakes are generally generated by a fluctuation in the level of the glacier. The moraine-dammed glacier lakes are built with time and disappear as well (Aggarwal et al. 2017). They may disappear either due to their breach or debris fill the lakes (Campbell 2005). Further, the advancement of the mother glacier due to its physiographic or climatic factors may fill the glacier lake with ice again. Such glacier lakes are ephemeral in nature and not much stable. The breach of such glaciers also imposes a threat to the basin downstream (Fig. 4.19a–c).

4.5.6 Mapping and Monitoring of Glacier Lakes

The regular study of glacier lakes using field survey methods is very difficult due to their geographical and topographical locations in high mountains. The most common method to map the glacier lakes from satellite image is the NDWI-based method (McFeeters 1996), which is based on the sensitivity of water to Green (reflects) and Near InfraRed bands (absorbs).

$$NDWI = \frac{GREEN - NIR}{GREEN + NIR}$$

where all positive NDWI corresponds to surface water and negative values as non-water. As this threshold value was not enabled to discriminate between built-up surfaces and water pixels, various modifications have been made, such as Modified NDWI (Xu 2006). The glacier lake inventory of all major glacier lakes of the Himalayas is done regularly by NRSC, Hyderabad (NRSC 2015), and Central Water Commission, CWC, New Delhi, using IRS datasets. Apart from the regular monitoring of glacier lakes be central government agencies, various academic institutes, and research centers also work on mapping of such glacier lakes using optical (Fig. 4.20) and other remote sensing datasets (Campbell 2005; Bolch et al. 2008; Ashraf et al. 2012; Raj et al. 2013; Aggarwal et al. 2017). As optical data limits its application in cloud cover duration, cloud penetration capable radar imageries can be used to increase all-weather flood monitoring efficiency. Therefore, the SAR data can be used for glaciers, glacier lakes, and flood mapping in all types of weather and seasons, irrespective of cloud or snow cover (Bhatt et al. 2016; Thakur et al. 2017; Wangchuk and Bolch 2020).

4.6 Conclusions

Snow cover over a large area is an important reservoir of freshwater. Its assessment is very critical from a hydrological and climate change point of view. It was realized that geospatial technology is an important tool in mapping snow cover easily as compared to traditional snow cover area mapping. Moreover, the availability of reasonably high spatial resolution, and optical and microwave RS data in the public domain has provided the impetus to study snow and its geophysical parameters, which was earlier difficult to study. Along with snow cover, the

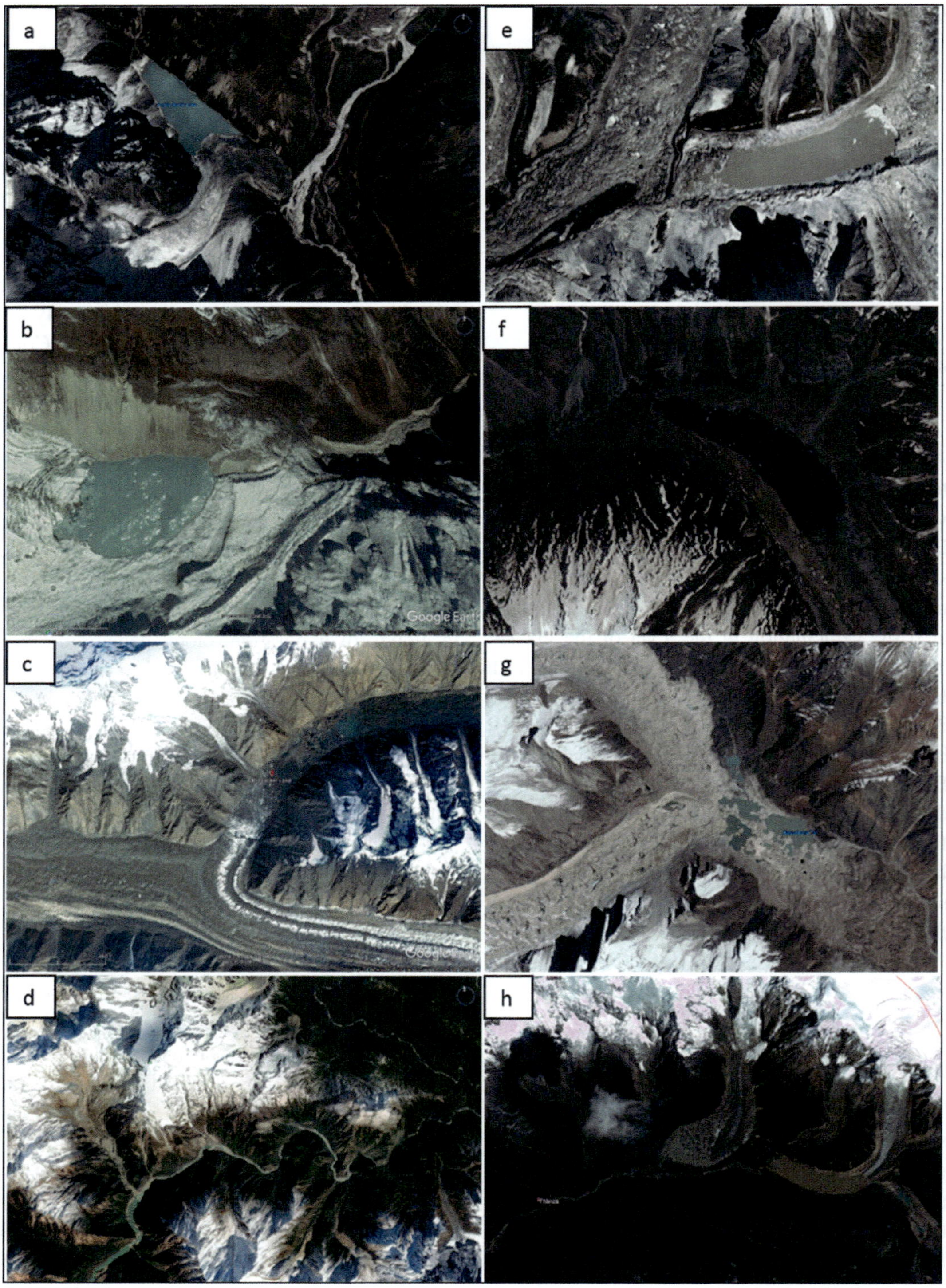

Fig. 4.19 An example of glacier lakes as seen in RS images; **a**, **b**, and **c** Glacier Ice-blocked lakes of East Nepal and Kyrgyzstan; **d** Landslide blocked lake of Tibet, China; **e** Moraine-blocked Pro Glacier and Cirque lake of East Nepal; **f** Moraine-blocked glacier of Lahul, H. P.; **g** Supraglacial lake of Dhauliganga, UK; and **h** Cascade series of moraine-dammed lakes of Bhutan (*Source* Digital Globe, Google Earth)

Fig. 4.20 Example of glacier lake types in Sikkim and map of all glacier lakes of Sikkim (*Source* Aggarwal et al. 2017)

geophysical parameters such as wet/dry snow, snow density, snow liquid content, snow grain size, and change in snow water equivalent now can easily be estimated. Further, these estimations may improve the simulation of the hydrological response of the snow-fed river basins. Remote sensing-based spatio-temporal glacier maps provide comprehensive data to monitor the glacier dynamics. Challenges for automated glacier mapping are related to the clean-ice and debris-covered glacier with a comparison of varying datasets and methodologies. There are still some improvements required along with possible improvements in resolution to overcome the challenges in automated glacier mapping from the research community. SoI and GSI topographic maps are not so accurate for precise results of glaciated terrain due to the lack of sophisticated instruments and skilled approaches for surveying in some cases. Nowadays, huge data is publically available in the form of reports and documents for sharing and exploiting existing tools and techniques that may overcome the present challenges of the study of the glacier before the research community. Classification approaches are primarily based on the multispectral satellite data and are well developed and established by the different research outcomes. The band ratios are simple, robust, and fast methods to map glacier outlines and their different zones. However, these ratios are needed for manual corrections and use appropriated threshold to the automatic optimization of manual editing. Threshold sensitivity for the clean-ice mapping is based on the scene condition, i.e., cloud cover, seasonal snow, shadow, illumination angle, and ground orientation of the glacier. The manual method for glacier mapping may be reduced but cannot be underestimated and never becomes redundant. In summing up, despite the findings from the multispectral imagery may be a very effective tool and approaches for the glacier mapping by remote sensing methods.

Along with mapping and monitoring snow and glaciers, it is also highly imperative to make a good study on glacier lakes due to changing weather and seasonal temporal variation in the Indian region. The critical glacier lakes may result in glacier lake outburst flood. Geospatial technology has been extensively used for mapping glacier lakes. The availability of temporal (at regular intervals) remote sensing data helps in analyzing the glacier lakes whether they are stable, sub-critical, or critical. It helps a disaster mitigation planner to plan their activities in case of a isaster due to GLOF. Therefore, it may be concluded that the role of geospatial technology is indispensable and most critical in cryospheric studies.

References

Aggarwal S, Rai SC, Thakur PK, Emmer A (2017) Inventory and recently increasing GLOF susceptibility of glacial lakes in Sikkim, Eastern Himalaya. Geomorphology 295:39–54

Amlien J (2008) Remote sensing of snow with passive microwave radiometers. a review of current algorithms, Report No. 1019. Norsk Regnesentral, Oslo

Andreassen LM, Paul F, Kääb A, Hausberg JE (2008) Landsat-derived glacier inventory for Jotunheimen, Norway, and deduced glacier changes since the 1930s. Cryosphere 2(2):131–145. https://doi.org/10.5194/tc-2-131-2008

Ashraf A, Naz R, Roohi R (2012) Glacial lake outburst flood hazards in Hindukush, Karakoram and Himalayan Ranges of Pakistan: implications and risk analysis. Geomatics Natural Hazards Risk 3(2):113–132

Avery TA, Berlin GL (1992) Fundamentals of Remote Sensing and Airphoto Interpretation, 5th edn. Macmillan publication, New York (NY)

Bader H (1961) The Greenland ice sheet. U.S. Army Cold Regions Research and Engineering Laboratory, Corps of Engineers, USA, Hanover, N.H.

Bayr KJ, Hall DK, Kovalick WM (1994) Observations on glaciers in the eastern austrian alps using satellite data. Int J Remote Sens 15(9):1733–1742

Bhambri R, Bolch T (2009) Glacier mapping: A review with special reference to the Indian Himalayas. Prog Phys Geogr 33(5):672–704

Bhambri R, Bolch T, Chaujar RK, Kulshreshta SC (2011) Glacier changes in the Garhwal Himalaya, India 1968–2006 based on remote sensing. J Glaciol 57 (203):543–556

Bhardwaj A, Joshi PK, Sam L, Singh MK, Singh S, Kumar R (2015) Applicability of Landsat 8 data for characterizing glacier facies and supraglacial debris. Int J Appl Earth Obs Geoinf 38:51–64

Bhatt CM, Rao GS, Farooq M, Manjusree P, Shukla A, Sharma SVSP, Kulkarni SS, Begum A, Bhanumurthy V, Diwakar PG, Dadhwal VK (2016) Satellite-based assessment of the catastrophic Jhelum floods of September 2014 Jammu and Kashmir India. Geomatics Natural Hazards and Risk 8(2):309–327

Bishop MP, Bonk R, Kamp U, Shroder JF (2001) Terrain analysis and data modeling for alpine glacier mapping. Polar Geography 25(3):182–201. https://doi.org/10.1080/10889370109377712

Bishop MP, Shroder JF, Hickman BL (1999) Spot panchromatic imagery and neural networks for information extraction in a complex mountain environment. Geocarto Int 14(2):19–28

Bishop MP, Shroder JF, Hickman BL, Copland L (1998) Scale-dependent analysis of satellite imagery for characterization of glacier surfaces in the Karakoram Himalaya. Geomorphology 21:217–232

Bolch T, Buchroithner MF, Kunert A, Kamp U (2007) Automated Delineation of Debris-Covered Glaciers Based on ASTER Data. Geoinformation in Europe (Proc. of 27th EARSel Symposium, 04–07 June 2007), Bozen, Italy., (June), 403–410. https://doi.org/10.1016/j.fluid.2003.02.001

Bolch T, Buchroithner MF, Peters J, Baessler M, Bajracharya S (2008) Identification of glacier motion and potentially dangerous glacial lakes in the Mt. Everest region, Nepal using spaceborne imagery. Nat Hazards Earth Syst Sci 8:1329–1340

Bolch T, Kamp U (2006) Glacier mapping in high mountains using DEMs, Landsat and ASTER data. Grazer Schriften Der Geographie Und Raumforschung, (1): 37–48. Retrieved from http://web.unbc.ca/~bolch/publications/BolcKamp06_GeoRaum.pdf

Brown DG, Lusch DP, Duda KA (1998) Supervised classification of types of glaciated landscapes using digital elevation data. Geomorphology 21(3–4):233–250. https://doi.org/10.1016/S0169-555X(97)00063-9

Campbell JG (2005) Inventory of glaciers, glacial lakes and the identification of potential glacial lake outburst floods (GLOFs) affected by global warming in the mountains of India, Pakistan and China/Tibet Autonomous Region, Technical Report of APN Project 2004-CMY- Campbell

Chang ATC, Foster L, Hall DK (1987) Nimbus-7 SMMR derived global snow cover parameters. Annals Glaciol 19:39–44

Chang ATC, Foster JL, Hall DK (1992) The use of microwave radiometer data for characterizing snow storage in western China. Annals Glaciol 16:215–219

Chen C, Nijssen B, Guo J, Tsang L, Wood AW, Hwang J, Lettenmaier DP (2001) Passive microwave remote sensing of snow constrained by hydrological simulations. IEEE Trans Geosci Remote Sens 39:1744–1756

Choudhury BJ, Chang ATC (1981) The albedo of snow for partially cloudy skies. Boundary-Layer Meteorol 20:371–389

Cogley JG, Hock R, Rasmussen LA, Arendt AA et al (2011) Glossary of glacier mass balance and related terms, IHP-VII Technical Documents in Hydrology No. 86, IACS Contribution No. 2, UNESCO-IHP, Paris

Cull S, Arvidson RE, Mellon M, Wiseman S, Clark R, Titus T, Morris RV and McGuire P (2010) Seasonal H2O and CO2 ice cycles at the Mars Phoenix landing site: 1. Prelanding CRISM and HiRISE observations. Journal of Geophysical Research, 115, E00D16. https://doi.org/10.1029/2009je003340

Dhir RD (1951) Feasibility of snow survey in the Himalayas. International Association of Hydrological Sciences IAHS Publication 32:305–314

Dozier JF (1989) Spectral signature of alpine snow cover from the Landsat Thematic Mapper. Remote Sens Environ 28:9–22

Dozier J, Schneider SR, McGinnis DF (1981) Effect of grain size and snowpack water equivalence on visible and near-infrared satellite observations of snow. Water Resour Res 17(4):1213–1221

Foster JL, Hall DK, Chang ATC (1984) An overview of passive microwave snow research and results. Rev Geophys Space Phys 22:195–208

Gillespie AR, Kahle AB, Walker RE (1986) Color enhancement of highly correlated images. I. Decorrelation and HSI contrast stretches. Remote Sens Environ 20(3): 209–235. https://doi.org/10.1016/0034-4257(86)90044-1

Gratton DJ, Howarth PJ, Marceau DJ (1990) Combining DEM Parameters with Landsat MSS and TM Imagery in a GIS for Mountain Glacier Characterization. IEEE Trans Geosci Remote Sens 28(4):766–769. https://doi.org/10.1109/TGRS.1990.573023

Grody NC (1991) Classification of snow cover and precipitation using the Special Sensor Microwave/Imager (SSM/I). J Geophys Res 96 (D4):7423–7435

Gupta RP, Haritashya UK, Singh P (2005) Mapping dry/wet snow cover in the Indian Himalayas using IRS multispectral imagery. Remote Sens Environ 97 (4):458–469

Haeberli W, Cihlar J, Barry RG (2000) Glacier monitoring within the global climate observing system. Annals Glaciol 31:241–246

Hall DK, Chang ATC, Siddalingaiah H (1988) Reflectances of glaciers as calculated using Landsat-5 Thematic Mapper data. Remote Sens Environ 25(3):311–321. https://doi.org/10.1016/0034-4257(88)90107-1

Hall DK, Foster JL, Verbyla DL, Klein AG, Benson CS (1998) Assessment of snow-cover mapping accuracy in a variety of vegetation-cover densities in central Alaska. Remote Sens Environ 66(2):129–137

Hall DK, Kelly R EJ, Foster JL and Chang ATC (2005) Estimation of Snow Extent and Snow Properties. Book Chapter In, Encyclopedia of Hydrological Sciences. Edited by M G Anderson, 811–829

Hall DK, Martinec J (1985) Remote sensing of ice and snow. Chapman and Hall, New York

Hall DK, Riggs GA, Salomonson VV (1995) Mapping Global Snow Cover using Moderate Resolution Imaging Spectroradiometer (MODIS) Data. Glaciological Data 1(95):33–37

Hall DK, Williams RS, Bayr KJ (1992) Glacier recession in Iceland and Austria. Eos, Trans Amer Geophy Union 73(12):129–141. https://doi.org/10.1029/91EO00104

Huggel C, Kääb A, Haeberli W, Teysseire P, Paul F (2002) Remote sensing based assessment of hazards from Glacier Lake outbursts: a case study in the Swiss Alps. Canadian Geotech J 39(2):316–330

ICIMOD (2007) Impact of climate change on Himalayan Glaciers and Glacial Lakes: Case studies on GLOF and Associated Hazards in Nepal and Bhutan

ICIMOD (2010) Formation of Glacial Lakes in the Hindu Kush- Himalayas and GLOF Risk Assessment

Jain SK, Lohani AK, Singh RD, Chaudhary A, Thakural N (2012) Glacial lakes and glacial lake outburst flood in a Himalayan basin using remote sensing and GIS. Nat Hazards 62:887–899. https://doi.org/10.1007/s11069-012-0120-x

Jain SK, Rathore DS, Chaudhary A, Ramasastri KS (2001) Snow cover estimation in Himalayan basins using remote sensing. National Institute of Hydrology (NIH), Roorkee, technical report, CS/AR-10/2000-2001, pp. 51

Kääb A, Bolch T, Casey K., Heid, T., Kargel, J. S., Leonard, G. J., Raup, B. H. (2014). Glacier Mapping and Monitoring Using Multispectral Data. Global Land Ice Measur Space, 75–112

Kaushik S, Joshi PK Singh T (2019) Development of glacier mapping in Indian Himalaya: a review of approaches. Int J Remote Sens40(17):6607–6634 https://doi.org/10.1080/01431161.2019.1582114

König M, Winther J, Isaksson E (2001) Measuring snow and glacier ice properties from satellite. Rev Geophys 39:1–27

Kulkarni AV (1991) Glacier inventory in Himachal Pradesh using satellite images. J Indian Soc Remote Sens 19:195–203

Kulkarni AV, Bahuguna IM, Rathore BP, Singh SK, Randhawa SS, Sood RK, Dhar S (2007) Glacial retreat in Himalaya using Indian remote sensing satellite data. Curr Sci 92(1):69–74

Kulkarni AV, Rathore BP (2003) Snow cover monitoring in Baspa basin using IRS WiFS data. Mausam 54:335–340

Kulkarni AV, Rathore BP, Mahajan S, Mathur P (2005) Alarming retreat of Parbati glacier, Beas basin. Himachal Pradesh Curr Sci 88(11):1844–1850

Kumar D, Singh AK, Taloor AK, Singh DS (2020) Recessional pattern of Thelu and Swetvarn glaciers between 1968 and 2019. Quat. Int, Bhagirathi basin, Garhwal Himalaya, India. https://doi.org/10.1016/j.quaint.2020.05.017

Leonard KC, Fountain AG (2003) Map-based methods for estimating glacier equilibrium-line altitudes. J Glaciol 49(166):329–336

Lievens H, Demuzere M, Marshall H et al (2019) Snow depth variability in the Northern Hemisphere mountains observed from space. Nat Commun 10:4629. https://doi.org/10.1038/s41467-019-12566-y

Lillesand T, Kiefer RW, Jonathan (2015) Remote sensing and image interpretation, 7th Edition. Chipman ISBN: 978-1-118-34328-9. Wiley, Hoboken, NJ, Xii, 720 p

Martinec J, Rango A (1981) Areal distribution of snow water equivalent evaluated by snow cover monitoring. Water Resour Res 17:1480–1488

McFeeters SK (1996) The use of the normalized difference water index (NDWI) in the delineation of open water features. Int J Remote Sens 17:1425–1432

Nagler T, Rott H (2000) Retrieval of wet snow by means of multitemporal SAR data. IEEE Trans Geosci Remote Sens 38(2):754–765

Nikam BR, Garg V, Gupta PK, Thakur PK, Kumar AS, Chouksey A, Aggarwal SP, Dhote P, Purohit S (2017) Satellite-based mapping and monitoring of heavy snowfall in North Western Himalaya and its hydrologic consequences. Curr Sci 113(12):2328–2334

NRSC (December 2015) Report on "Monitoring of Glacial Lakes/Water Bodies in the Himalayan Region of Indian River Basins during 2015", Technical Report Published by National Remote Sensing Centre, Hyderabad

Oza SR, Bothale RV, Rajak RD, Jayaprasad P, Maity S, Thakur PK, Tripathi N, Chouksey A, and Bahuguna IM (2019) Assessment of cryospheric parameters over the Himalaya and Antarctic regions using SCATSAT-1 enhanced resolution data. Curr Sci Scatsat-1 special section 117(6):1002–1013

Paul F (2000) Evaluation of different methods for glacier mapping using landsat tm. EARSeL E Proc 1(1):239–245. https://doi.org/10.1080/10106040008542173

Paul F (2002) Combined technologies allow rapid analysis of Glacier changes. Eos 83(23):3–6. https://doi.org/10.1029/2002EO000177

Paul F, Huggel C, Kääb A (2004) Combining satellite multispectral image data and a digital elevation model for mapping debris-covered glaciers. Remote Sens Environ 89(4):510–518. https://doi.org/10.1016/j.rse.2003.11.007

Paul F, Kääb A (2005) Perspectives on the production of a glacier inventory from multispectral satellite data in Arctic Canada: Cumberland Peninsula, Baffin Island. AnnalsGlaciol 42:59–66

Paul F, Kääb A, Maisch M, Kellenberger T, Haeberli W (2002) The new remote sensing derived Swiss glacier inventory: I. Methods. Annals Glaciol 34:362–366

Paul F, Winsvold SH, Kääb A, Nagler T, Schwaizer G (2016) Glacier remote sensing using Sentinel-2. part II: Mapping glacier extents and surface facies, and comparison to Landsat 8. Remote Sens 8(7):575 https://doi.org/10.3390/rs8070575

Pellikka P, Rees WG (2009) Remote sensing of glaciers. CRC Press, 330. https://doi.org/10.1201/b10155

Prasad VH, Roy PS (2005) Estimation of snowmelt runoff in Beas Basin, India. Geocarto Int 20(2):41–47

Rabatel A, Rabatel A, Dedieu J, Thibert E, Vincent C, Letre A (2008) Twenty-five years of equilibrium-line altitude and mass balance reconstruction on the Glacier Blanc, French Alps (1981–2005), using remote- sensing method and meteorological data. J Glaciol 54(185):307–314. https://doi.org/10.3189/002214308784886063

Racoviteanu AE, Paul F, Raup B, Jodha S, Khalsa S, Armstrong R (2009) Challenges and recommendations in mapping of glacier parameters from space: results of the 2008 Global Land Ice Measurements from Space (GLIMS) workshop. Boulder, Colorado, USA 50(53):53–69

Racoviteanu AE, Williams MW, Barry RG (2008) Optical remote sensing of glacier characteristics: a review with focus on the Himalaya. Sensors 8(5):3355–3383. https://doi.org/10.3390/s8053355

Raina VK, Srivastava D (2008) Glacier atlas of India. Geological Society of India, Bangalore, p 316

Raj KBG, Kumar KV (2016) Inventory of Glacial Lakes and its evolution in Uttarakhand Himalaya using time series satellite data. J Indian Soc Remote Sens 44:959–976

Raj BKG, Kumar VK, Remya SN (2013) Remote sensing-based inventory of glacial lakes in Sikkim Himalaya: semi-automated approach using satellite data. Geomatics Natl Hazards Risk 4(3):241–253

Ramamoorthi AS, Subba Rao P (1981) Application of satellite technology for forecasting snow melt runoff of perennial rivers of India. Proc. of Second Asian Conference on Remote Sensing, Beijing, China

Rango A, Salomonson VV (1975) Employment of satellite snow cover observations for improving seasonal runoff estimates. In Proceedings of Workshop on Operational Applications of Satellite snow Cover Observations, Washington DC, 1975, NASA-SP-391: 157–174

Ranzi R, Grossi G, Iacovelli L, Taschner S (2004) Use of multispectral ASTER images for mapping debris-covered glaciers within the GLIMS project. Int Geosci Remote Sens Symposium (IGARSS) 2:1144–1147. https://doi.org/10.1109/igarss.2004.1368616

RGI Consortium (2017) R. G. I. (RGI)—A. D. of G. G. O. V. 6. 0. (2017) Randolph Glacier Inventory—A Dataset of Global Glacier Outlines: Version 6.0. GLIMS Technical Report RGI, 1(1):1188–1197. https://doi.org/10.1111/j.1469-7610.2010.02280.x

Rott H (1994) Thematic studies in alpine areas by means of polarimetric SAR and optical imagery. Adv Space Res 14(3):217–226. https://doi.org/10.1016/0273-1177(94)90218-6

Rott H, Markl G (1989) Improved snow and glacier monitoring by the Landsat Thematic Mapper, in Proceedings of a Work- shop on Earthnet Pilot Project on Landsat TM Applications, Frascati, Italy, Dec. 1987, Rep. ESA SP-1102, pp 3–12, Eur. Space Agency, Paris, 1989

Saraf AK, Foster JL, Singh P, Tarafdar S (1999) Passive microwave data for snow-depth and snow-extent estimations in the Himalayan Mountains. Int J Remote Sens 20(1):83–95

Sharma AK, Singh SK, Kulkarni AV, Ajai (2013) Glacier inventory in Indus, Ganga and Brahmaputra basins of the Himalaya. National Academy Science Letters, ISSN 0250-541X, 36(5): 497–505

Shroder JF, Bishop MP, Copland L, Sloan VF (2000) Debris-Covered Glaciers and Rock Glaciers in the Nanga Parbat Himalaya, Pakistan. Geografiska Annaler. Series A, Physical Geography 82(1): 17–31. Retrieved from http://www.jstor.org/stable/521438

Sidjak RW, Wheate RD (1999) Glacier mapping of the Illecillewaet icefield, British Columbia, Canada, using Landsat TM and digital elevation data. Int J Remote

Sens 20(2):273–284. https://doi.org/10.1080/014311699213442
Singh P, Kumar N (1997) Impact assessment of climate change on the hydrological response of a snow and glacier melt runoff dominated Himalayan river. J Hydrol 193:316–350
Singh KK, Mishra VD (2006) Snow cover study of northwest Himalaya using passive microwave remote sensing data. Proc. SPIE 6410, Microwave Remote Sensing of the Atmosphere and Environment V, 641014, 11 p, https://doi.org/10.1117/12.693942
Singh KK, Mishra VD, Garg RK (2007) Microwave response of seasonal snow cover measured by using a ground-based radiometer at 6.93 and 18.7 GHz frequencies and at dual polarization. J Indian Soc Remote Sens 35:243–251
Singh S, Sood V, Taloor AK, Prashar S, Kaur R (2020b) Qualitative and quantitative analysis of topographically derived CVA algorithms using MODIS and Landsat-8 data over Western Himalayas. Quat Int, India. https://doi.org/10.1016/j.quaint.2020.04.048
Singh DK, Thakur PK, Naithani BP, Kaushik S (2020) Quantifying the sensitivity of band ratio methods for clean glacier ice mapping. Spatial Inform Res. https://doi.org/10.1007/s41324-020-00352-8
Sood V, Gusain HS, Gupta S, Taloor AK, Singh S (2020a) Detection of snow/ice cover changes using subpixel-based change detection approach over Chhota-Shigri glacier. Quat Int, Western Himalaya, India. https://doi.org/10.1016/j.quaint.2020.05.016
Sood V, Singh S, Taloor AK, Prashar S, Kaur R (2020b) Monitoring and mapping of snow cover variability using topographically derived NDSI model over north Indian Himalayas during the period 2008-19. Applied Computing and Geosci. https://doi.org/10.1016/j.acags.2020.100040
Takala M, Luojus K, Pulliainen J, Derksen C, Lemmetyinen J, Karna JP, Koskinen J, Bojkov B (2011) Estimating northern hemisphere snow water equivalent for climate research through assimilation of space-borne radiometer data and ground-based measurements. Remote Sens Environ 115(20):3517–3529
Taloor AK, Kotlia BS, Jasrotia AS, Kumar A, Alam A, Ali S, Kouser B, Garg PK, Kumar R, Singh AK, Singh B (2019) Tectono-climatic influence on landscape changes in the glaciated Durung Drung basin, Zanskar Himalaya, India: a geospatial approach. Quat Int 507:262–273
Thakur PK, Aggarwal SP, Garg PK, Garg RD, Snehmani PA, Kumar S (2012) Snow physical parameter estimation using space based SAR. Geocarto Int 27(3):263–288. https://doi.org/10.1080/10106049.2012.672477
Thakur PK, Aggarwal S, Aggarwal SP, and Jain SK (2016). One dimensional hydrodynamic modeling of GLOF and impact on hydropower projects in Dhauliganga River using remote sensing and GIS applications. Springer's Natural Hazards Journal. pp. 1–19. https://doi.org/10.1007/s11069-016-2363-4
Thakur PK, Aggarwal SP, Arun G, Sood S, Kumar AS, Mani S, Dobhal DP (2017a) Estimation of snow cover area, snow physical properties and glacier classification in parts of Western Himalaya using C-band SAR data. J Indian Soc Remote Sens 45(3):525–539
Thakur PK, Garg V, Nikam BR, Chouksey A, Aggarwal SP Dhote PR (2017) Cryospheric studies in indian himalayan and polar region: current status, advances and future prospects of remote sensing. Proceedings of the National Academy of Sciences, India Section A: Physical Sciences 87(4):593–616
Thakur PK, Garg V, Nikam BR, Singh S, Jasmine, Chouksey A, Dhote PR, Aggarwal SP, Chauhan P and Kumar AS (2018). Snow Cover and Glacier Dynamics Study Using C-and L-Band SAR Datasets in Parts of North West Himalaya. Int. Arch. Photogramm. Remote Sens. Spatial Inf. Sci., XLII-5, 375–382. https://doi.org/10.5194/isprs-archives-XLII-5-375-2018
Thakur PK and Prasad VH (2006) Study of Frosting Patterns, Polar Ice of Mars, Terrestrial analog studies of Gangotri Glacier of India using Remote Sensing and Spectral Nature of C02/H2O ice. In: The International "Fourth Mars Polar Science Conference" at SLF Davos, Switzerland, 2–6 October 2006, 2 p
Warren SG (1982) Optical properties of snow. Rev Geophys 20(1):67–89
Wangchuk S, Bolch T (2020) Mapping of glacial lakes using Sentinel-1 and Sentinel-2 data and a random-forest classifier: Strengths and challenges. Science of Remote Sens 2:1–13. https://doi.org/10.1016/j.srs.2020.100008
Williams RS, Hall DK, Sigurdsson O, Chien JYL (1997) Comparison of satellite-derived with ground-based measurements of the fluctuations of the margins of Vatnajökull, Iceland, 1973–1992. Ann Glaciol 24:72–80
Winsvold SH, Kääb A, Nuth C, Andreassen LM, Ward JJ, van P, Schellenberger T (2018) Using SAR satellite data time series for regional glacier mapping. The Cryosphere 12:867–890
Xiong C, Shi J, Lemmetyinen J (2014) Refinement of the X and Ku band dual-polarization scatterometer snow water equivalent retrieval algorithm. In Proceedings of the IEEE International Geoscience and Remote Sensing Symposium (IGARSS), Quebec City, QC, Canada, 13–18 July 2014; pp 2419–2422
Xu H (2006) Modification of normalised difference water index (NDWI) to enhance open water features in remotely sensed imagery. Int J Remote Sens 27:3025–3033
Yueh S, Cline D, Elder K (2008) POLSCAT Ku-band Radar Remote Sensing of Terrestrial Snow Cover. IGARSS 2008–2008 IEEE International Geoscience and Remote Sensing Symposium. https://doi.org/10.1109/igarss.2008.4779276

Dr. Praveen K. Thakur is currently working as Scientist/Engineer 'SF' at Water Resources Department (WRD) of Indian Institute of Remote Sensing (IIRS), Indian Space Research Organization (ISRO)—Dehradun. He completed his B. Tech. In Civil Engineering from National Institute of Technology (NIT), Hamirpur (June 2001 with honours), Master of Technology (M. Tech.) in Water Resources Engineering from Indian Institute of Technology (IIT), Delhi (December 2002) and Ph. D. In Geomatics Engineering (September 2012) from IIT Roorkee. He joined IIRS Dehradun in 2004 and has more than 17 years of professional experience in the field of water resources and geospatial applications. He has published 63 research papers in peer-reviewed international/national journals, 10 book chapters, 1 book and more than 55 papers in the conferences/symposiums. His current research interests are geospatial technology application in water resources; microwave remote sensing for snow, ice and floods studies; hydrological, hydrodynamic and groundwater modelling; and data assimilation and planetary remote sensing for ice and water cycle studies. He has contributed as principle investigator in eight research projects as Co-PI in four projects sponsored by ISRO, and four announcement of opportunity projects by ESA, DLR and JAXA, respectively. He is science team member of NASA-ISRO SAR (NISAR) mission and Surface Water and Ocean Topography (SWOT) mission. He has guided more than 33 M. Sc./M. Tech./post-graduate diploma thesis, 3 Ph.D. students and currently supervising 2 Ph.D. students. He is life member of six professional societies related to remote sensing (ISRS), water resources (ASCE and IWRS) and hydrometeorology (IAHS, IMS and IAH). He is a recognized reviewer for many reputed national and international journals in the field of remote sensing and hydrology. He is the recipient of ISRO young scientist merit award for the year 2014

Dr. Vaibhav Garg is presently working as Scientist at Water Resources Department, Indian Institute of Remote Sensing (IIRS), ISRO, Dehradun, India. His field of interests are large-scale hydrological modelling and application of geospatial technology in water resources problems. He did his doctoral research at the Department of Civil Engineering, IIT, Bombay, India. He had completed his Master's in water resources engineering from the Department of Civil Engineering, Malaviya National Institute of Technology Jaipur, India. He has also worked with the National Institute of Hydrology, Roorkee; IIRS, Dehradun and IIT Bombay, Mumbai, with research fellow position. He has more than 10 years of professional experience in the field of water resources. Till date, he has published 40 peer-reviewed journal publications and 06 book chapters dealing with water resources problems. He is a life member of International Association of Hydrological Sciences; Indian Society of Hydraulics; Indian Society of Remote Sensing and Indian Meteorological Society and International Association for Water, Environment, Energy and Society

Dr. Bhaskar R. Nikam is presently working as Scientist/Engineer 'SF' in Water Resources Department of Indian Institute of Remote Sensing, Indian Space Research Organisation, Dehradun, India. He has done his Ph.D. in the field of Water Resources Management from Indian Institute of Technology, Roorkee. His research interests are in the field retrieval of hydrological parameters using remote sensing, application of geospatial technology in irrigation water management, watershed process, hydrological modelling, climate change studies, etc. He has participated in many national and organization-level research and operational projects related to the applications of remote sensing to real-world problems of water sector. Apart from research activities, he is also involved in education and capacity-building activities of Indian Institute of Remote Sensing (IIRS) and Centre for Space Science and Technology Education in Asia and the Pacific (affiliated to United Nations). He was visiting faculty at Asian Institute of Technology, Bangkok, Thailand. He has around 110 publications to his credit including 37 publications in peer-reviewed journals' four book chapters

Dr. S. P. Aggarwal holds B. Tech. in agricultural engineering from Allahabad University; PG and Ph.D. from IARI, New Delhi and post-doctoral research from IHE and ITC, Netherlands. Currently, he is Head of Water Resources Department, Indian Institute of Remote Sensing, ISRO, Dehradun. He has more than 20 years' experience in remote sensing and GIS applications in water resources management. He is also programme coordinator of Centre for Space Science and Technology Application in Asia and the Pacific, UN-affiliated Centre, Dehradun. He is Associate Editor of Journal of the Indian Society of Remote Sensing. He is Secretary of the International Society for Photogrammetry and Remote Sensing, Technical Commission V. He has been conferred with Eminent Engineers Award in 2014 by the Institution of Engineers of India, Uttarakhand State Centre and President Appreciation Medal by the Indian Society of Remote Sensing for 2016. He has published 100 research papers in peer-reviewed national/international journals and symposia

Dr. Suruchi Aggarwal is currently working as a Geodata Engineer and Advisor for the Hole Municipality in Norway. She obtained her Ph.D. from the University of Delhi in 2018 and her work was a collaboration between University of Delhi and the Indian Institute of Remote Sensing (IIRS), Indian Space Research Organisation (ISRO) in Dehradun. During her Ph.D., she worked on Hydrodynamic Modelling of GLOF Hazard and Impact on Hydropower Projects in Tista Basin, Sikkim, Himalaya. After her Master's in Geography from Kurukshetra University in 2010, she obtained a post-graduate Diploma in Natural Hazards and Disaster Risk Management (PGD-NHDRM) with specialization in Hydrometeorology by the IIRS-ISRO, Dehradun, India, and Faculty of Geo-Information Science and Earth Observation (ITC), University of Twente, The Netherlands, in 2011. In 2013, she got selected for Indo-Swiss Capacity Building Programme on Himalayan Glaciology hosted by School of Environmental Sciences, Jawaharlal Nehru University (JNU), New Delhi, India (1–27 April 2013), organized by the Embassy of Switzerland. She has working experience as an University Lecturer in Swami Shraddha Nand College (SSNC), University of Delhi, in 2011-12. After moving Norway, she has been working as a GIS expert for the Section for Geoinformation/Department of Hydrology at the Norwegian Water Resources and Energy Directorate (NVE), where she worked with analysing and mapping of flood zones and glacial lakes for Norway. She has four publications to her credit in journals and conferences

Mr. Dhanendra Singh is a Ph.D. scholar in the Department of Geography, HNB Garhwal University (Central), Srinagar-Garhwal, Uttarakhand. He received MA (Geography) from CSJMU Kanpur; M. Phil (Earth Sci.) from MGCGV Chitrakoot, Madhya Pradesh and a PG diploma in remote sensing and GIS from the University of Allahabad, Prayagraj. He cleared UGC-NET in Geography in 2016. His research interests include snow and glacier studies, remote sensing, climate science and geomorphology. Currently, he is doing Ph.D. in collaboration with the Indian Institute of Remote Sensing (IIRS-ISRO) Dehradun. Before joining Ph. D., he was associated with the SAC-ISRO project on Snow and Glacier Studies Phase-II at RSAC-UP for monitoring snow cover dynamics in the Indus Basin of Jammu and Kashmir. He has participated in different training programmes in glacier studies, climate change and remote sensing where he learned many aspects of snow and glacier monitoring techniques in the field as well as using remote sensing and GIS. He has presented two papers in international workshops/seminars and two research papers in peer-reviewed journals. He is also engaged in monitoring glacier dynamics in Svalbard, North Pole. He is also representing India at the Association of Polar Early Career Scientists (APECS) organization as a council member

5 Remote Sensing Based Assessment of Glacier Resources in Parts of Ladakh Mountain Range, a Trans-Himalayan Region

Riyaz Ahmad Mir

Abstract

The present study describes the status of glacier resources of Ganglas, Phyang, Khalsar, Rong, and No. -5 catchments of the Ladakh Mountain Range (LMR) of upper Indus Basin, a Trans-Himalayan region. In this region, the Leh town and its surrounding areas rely primarily for water supplies on the streams, springs, and groundwater fed by the glacier meltwater covering high altitude areas of LMR. However, during recent years, the demand for water supplies in this area has increased rapidly due to the rapid population growth and urbanization, growing economic development, and higher influx of tourists. Therefore, for the assessment of these glacier resources, the Survey of India (SoI) maps, Landsat data (TM, ETM+, OLI/TIRS), Google earth images, and ASTER DEM have been used. An inventory of 90 glaciers covering an area of 21.1 km^2 comprising 2.6% of the total area of the study basins has been generated for 2017 (OLI/TIRS). The glaciers are small in size (mean size − 0.24 km^2), high altitude (mean elevation − 5570 m amsl), northerly facing (NW-NE), and moderately steep (10°–40°). Using SoI as a base map, the glaciers indicated a loss of 21.2% (5.1 km^2) at a rate of 93 m^2/year from 1962 to 2017. However, using the TM-2000 scene as a base map, the glaciers indicated a loss of 12.5% (2.7 km^2) at a rate of 160 m^2/year from 2000 to 2017, a recent time period. The small glaciers (size <0.12 km^2) indicated a loss of 17.9% whereas the glaciers lying in the elevation zone of 5800–6000 m (amsl) and above indicated a loss of 44.1%. The glaciers with a steep slope (50°–60°) and southerly aspect indicated a loss of 18.9% and 20.7%, respectively. Overall, the small and high altitude glaciers with a southerly aspect and steep slope indicated a higher area loss. This glacier loss may have a strong influence on the downstream water resources and supplies of the area. Nevertheless, these observations may help in planning and developing better strategies for the management of various sources of water supplies in this area.

Keywords

Galcier · Ladakh mountain range · Landsat · Remote sensing · Trans-Himalaya

R. A. Mir (✉)
Department of Earth Sciences, Indian Institute of Technology Roorkee, Roorkee 247667, India
e-mail: riyazgsi@gmail.com

Geological Survey of India, UT: J&K, Ladakh 190008, Srinagar, India

A. K. Taloor et al. (eds.), *Water, Cryosphere, and Climate Change in the Himalayas*, Geography of the Physical Environment, https://doi.org/10.1007/978-3-030-67932-3_5

5.1 Introduction

The Hindu Kush-Karakoram-Himalayan (HKKH) region contains the largest glacial system (~40,800 km^2) outside the Polar Regions (Bolch et al. 2012). This glacier system of this region influences the overall hydrological system of the Indus River in low land areas. It provides water resources to recharge the river-fed aquifers, hydro-power generation, horticultural and agricultural activities, ecosystems, and ultimately contribute to the sea-level increase (Dyurgerov and Meier 2005; Mir et al. 2017). Nonetheless, the glacial systems being physically complex and dynamic are also very sensitive to climatic variations and changes (Benn and Evans 2010). Therefore, the dynamics of glacial systems is considered as a key indicator for assessing climatic variations particularly in the areas where climatic observation facilities are very rare and unavalible (Barry 2006; Bolch et al. 2012; Mir et al. 2014).

Generally, the recent and rapid shrinkage of glacial resources of the HKKH region during the last five decades has been attributed to the growing temperatures and declining precipitation (Yao et al. 2012; Yang et al. 2014; Jain and Mir 2017; Mir et al. 2017; Taloor et al. 2019). However, due to the longest extension and wide area distribution of HKKH region, the glacier recession and shrinkage pattern is widely heterogeneous. For instance, majority of the glaciers in the monsoon-affected regions of the eastern and central Himalayan region between Himachal Pradesh toward the west and Bhutan toward the east are receding at different rates in line with the global glacier recession (Bolch et al. 2012; Bhambri and Bolch 2009; Mir and Majeed 2018; Bisht et al. 2020; Kumar et al. 2020). But, the glacier recession and areal changes are less pronounced in the western Himalayan region (Mir et al. 2018; Chand and Sharma 2015; Schmidt and Nüsser 2009; Pandey et al. 2011; Sood et al. 2020; Singh et al.2020). Contrary to it, the Karakoram glaciers falling in the cold-arid climatic regime have been in a balance/or stable state since the 1970s or have even grown/or advanced since the 1990s (Kaab et al. 2015). This anomalous behavior and phenomenon of the Karakoram glaciers has been termed as 'Karakoram Anomaly' (Hewitt 2005, 2014; Rankl and Kienholz 2014; Frey et al. 2014; Bolch et al. 2017). Similar behavior of glacial systems has also been reported in the Pamir and West Kunlun Shan in the extreme western Himalayan region (Kaab et al. 2015). But, the glacial systems of Ladakh Mountain Range (LMR) are located at the interface between receding glaciers of the Greater Himalayan Range (GHR) and advancing glaciers of Karakoram Mountain Range (KMR). In this area, more than 1800 glaciers covering an area of 997 km^2 have been inventoried in the central Ladakh region. In the recent past, a few studies have documented the recent glacier changes also (e.g., Schmidt and Nüsser 2009, 2012).

Historically, the Ladakh region relies for water supplies on glacier fed water resources such as the streams, springs, and to a large extent on local groundwater. However, during the past few years, some of these traditional resources of water supply particularly the streams have become unviable which is linked to the recent glacier shrinkage in the upper catchments of the area (Norphel 2009). As per the views of the local inhabitants, the glaciers in this region are also receding drastically (Vince 2009) with a significant impact on the production and availability of meltwater (Smiraglia et al. 2007). Further, it is reported that in near future with low and decreasing summer precipitation, snow and ice melt may be the only water resource available (Thayyen and Gergan 2010). In this region, the glaciers are typically restricted to high altitudes because of the semi-arid conditions. The sizes of these glaciers are also relatively small in this region. Despite these facts, the water stored in these small size glaciers determines the potential for irrigation of crop cultivation and horticulture. The agricultural and horticultural activities form the basis for regional food security and socio-economic development (Labbal 2000; Dame and Nüsser 2011; Nusser et al. 2012; Nusser and Baghel 2016). However, the recent climatic shifts in the region may also influence widely the glacier water storage and glacier melt runoff production (Barnett and Adam 2005; Parveen et al. 2015; Singh et al. 2017; Sarkar et al. 2020) in this area. Simultaneously, the demand for water supplies has been increasing in the region as a result of rapid population growth and urbanization, economic development, and higher tourist influx growth (Deen 2009). During recent years, the tourism in Ladakh region has thrived and recorded an increase in the number of visitors including both the domestic and foreigners. According to a report of Department of Tourism, a total of 227312 tourists have visited Leh district from January till September 15 in 2017. Therefore, in addition to the ongoing climate warming, the glacial resources are likely to be affected more due to the higher thriving of tourism vis a vis demand in this area.

Thus, keeping in view the above discussion, the present study was carried out in parts of LMR focusing particularly around Leh town and its surrounding area, i.e., Phyang, Khardung La, Khalsar, etc. The selected areas are the hotbed of the tourism movement, business, and other anthropogenic activities. The assessment of glacier resources under the influence of climate change and tourism development in this part of the region is of utmost significance and importance. The main objectives of the present study are, therefore, to inventory the glaciers and to monitor glacier changes and its controlling factors during recent decades using satellite data sources.

5.2 Study Area

The study area is a part of the Ladakh Mountain Range (LMR) in the Trans-Himalayan region, sandwiched between the Karakoram Mountain Range (KMR) and Himalayan Mountain Range (HMR). The LMR is about 370 km long trending W-NW between the Indus and Nubra Rivers with the highest peaks generally between 5000–6000 m amsl (Khan et al. 2014). The study area spreads between 37° 07′ to 34° 30′ N latitudes and 77° 29′ to 77° 48′ E longitudes (Fig. 5.1). The altitude of the area varies from 3135–6219 m amsl. The present selected study area consists of five sub-basins that are the Ganglas-1, Phyang-2, Khalsar-3, Rong-4, and No.-5 draining the parts of the LMR. The sub-basins cover an area of 162.6 km^2, 125.1 km^2, 162.6 km^2, 285.7 km^{2,} and 89.8 km^{2,} respectively. For the sub-basin No. 5, no name exists in the literature as well as no the toposheet and there is also not any village in the surroundings of it. Therefore, the id as No. 5 has been assigned to this catchment in this study. The glacier meltwater of sub-basins 1 and 2 drains southwestern slopes of LMR before discharging into the Indus River directly whereas the sub-basins 3, 4, and 5 drain its northeastern slopes and discharge into the Shyok River—a major tributary of the Indus River System joining with the Indus River downstream in Pakistan. The total area of all five selected catchments is about 825.8 km^2 with 2.6% of glacier cover. Out of the selected sub-basins, the main town of Leh is located within the Ganglas catchment. The glacial and snowmelt fed streams are a lifeline for the population constrained on the landforms of glacio-fluvial moraines within the broad glacial valleys, on the huge scree deposits along the foothills, and on the active mega alluvial fans within the current river valleys. The glacier fed streams flow during the period from May to October and later during the winter period, the streams freeze due to the lowermost temperatures in the region.

5.2.1 Climate of the Area

The LMR lies in the semi-arid climatic regime produced as a result of rain shadow effect of the HMR and KMR that are affected by two distinct climatic regimes. For example, in the HMR, about 80% of precipitation is produced by the summer monsoon whereas the westerly-dominated KMR receives two-thirds of its precipitation through winter westerlies (Bookhagen and Burbank 2010). It is reported that the mean annual precipitation of the upper Indus Basin generally declines from west to east and south to north (Dahri et al. 2016). Roughly 1/3 of the total

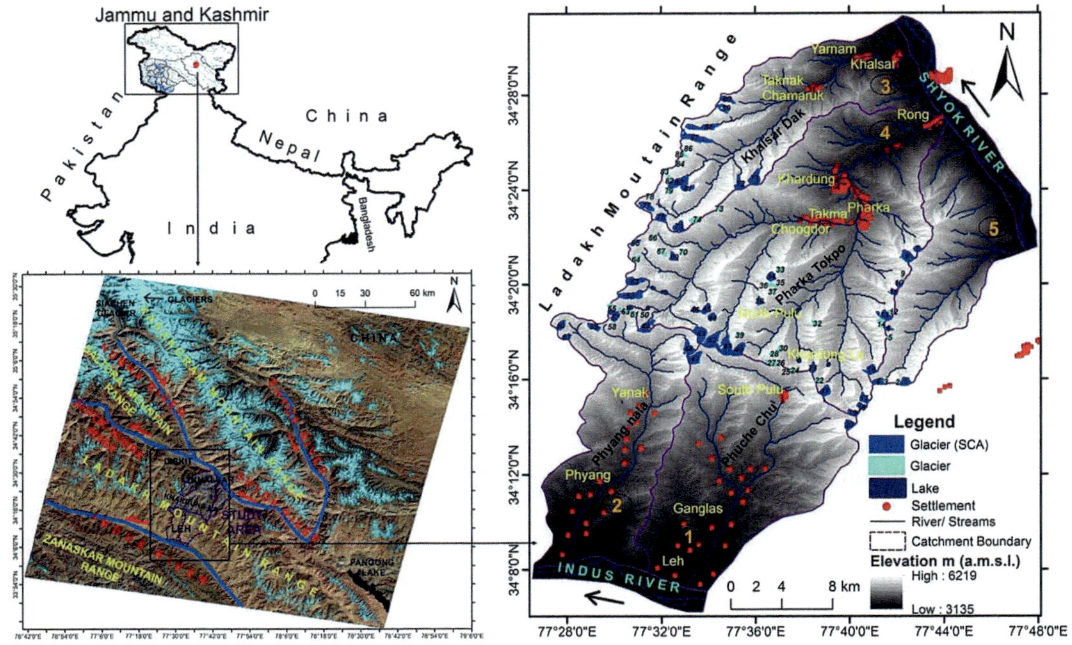

Fig. 5.1 Location map of the study area. The study area consists of five sub-basins of the Trans-Himalayan Ladakh Mountain Range designated as 1, 2, 3, 4, and 5. The sub-basins 1 and 2 drain the southwestern slopes of the LMR and discharge into the Indus River while the sub-basins 3, 4, and 5 drain the northeastern slope of the LMR and discharge into the Shyok River. The images in the background of the location map are Landsat ETM + (2017) and ASTER DEM (*Source* ASTER DEM)

annual precipitation is produced by westerly disturbances between December and February (Schmidt and Nüsser 2012). The dominant snowfall occurs during the winter season that contributes snow and ice to the glacial systems of this area. However, the precipitation in the form of snowfall occurs throughout the year at higher altitudes. In this area, the local weather stations are extremely limited in availability as well as quality (Crook and Osmaston 1994) and are entirely located below the elevation of glacierised areas (Kapnick et al. 2014). In the region, the long term mean annual precipitation recorded at valley bottom at Leh (3500 m amsl) is only 115 mm. Mean monthly temperatures ranges from −7.2 °C in January to 17.9 °C in July. In the Ladakh region, the glaciers are of usually continental type because of the low mean annual air temperature and low annual precipitation (Owen et al. 1998).

5.3 Data Used

In this study, the glaciers of five sub-catchments of the Indus River system draining the parts of LMR have been selected for assessment from 1962 to 2017. For this purpose, the study generated a database of glaciers for different time periods using the Survey of India (SoI) toposheets, low to moderate resolution Landsat data series, high-resolution Google Earth images provided in virtual Google Earth, and ASTER DEM. The details of the data sources are given in (Table 5.1). The glacier outline database was generated for 1962, 2000, 2011, and 2017 using SoI maps of 1962, multispectral Landsat Thematic Mapper (TM) of 1998, 2000, and 2011, Enhanced Thematic Mapper Plus (ETM +) of 2017, Operational Land Imager (OLI/TIRS) of 2017, and Google Earth images of 2018. The

Landsat data series are freely available on the earth explorer web server of usgs.gov.in. It is important to mention that for the extraction of the data, the images used were selected on the criteria of being with minimal snow and cloud covered and of late ablation season (August in Ladakh). It is due to this reason that three available scenes of low-resolution Multispectral Scanner (MSS) for 1976, 1977, and 1978 were avoided. For 1998 and 2000 years, two scenes of TM were available, but to maintain the consistency with the least ambiguity in the data, only one scene of 2000 was selected for the processing. But, the available scene of 1998 being close to 2000 was used as an additional source of information for proper and correct mapping/digitizing and maintaining the glacier outlines. Similarly, for assisting the mapping of glaciers from 2017 (OLI/TIRS) scene, another scene of ETM of 2017 and higher resolution Google earth images were used to assist the glacier demarcation.

5.4 Methods

In this study, the SoI maps were first scanned and georeferenced. Then, using the GIS platform, the SoI maps were primarily used for the registration of the 2000 Landsat TM image having a relatively higher resolution of 15 m. For coregistration of the SoI and TM image, about >100 ground-control points (GCP) were used. The GCPs were evenly distributed within the whole study area. Subsequently, all other data images were coregistered with 2000 Landsat TM image. The root mean square error (RMSE) of coregistration of TM and ETM + scenes was found to be 15 m (0.5). For further processing, the images were also geometrically rectified to the same and common projection system of WGS 84 UTM Zone 43.

In this area, the glaciers are debris free and very clean in nature, therefore, the initial mapping and demarcation of glaciers was carried out automatically using band ratio techniques. A band ratio technique of near-infrared and shortwave infrared channels (i.e., TM4/TM5) of the images was used in this study (e.g., Paul et al. 2009; Mir et al. 2017). It was followed by the application of a 3 × 3 median filter to eliminate any isolated pixels (Paul et al. 2002). Then, binary images were generated using a given threshold of 2.2 for all Landsat images followed by (binary images) conversion into a vector format. Any misclassified water bodies, shadows, isolated rocks, and any other erroneous glacier

Table 5.1 Remote sensing data and its source used in this study

Dataset	Scale/resolution (m)	Survey/acquisition Date	ID	Path/Row	Source
SoI maps	1:50,000	1962	52F/11&F/12		SoI
MSS	60	21/11/1976	LM21580361976326AAA03	158/036	USGS
MSS	60	18/06/1977	LM21580361977230AAA03	158/036	USGS
MSS	60	18/07/1978	LM31590361978199AAA02	159/036	USGS
TM	30	16/09/1998	LT51470361998259BIK02	147/036	USGS
TM	15	31/10/2000	ID_12 p145r035_7f19991031	145/035	USGS
TM	30	03/08/2011	LT51470362011215KHC00	147/036	USGS
ETM+	30	22/09/2017	LE71470362017239SGI00	147/036	USGS
OLI/TIRS	30	06/10/2017	LC81470362017279LGN00	147/036	USGS
Google Earth	>5	2018			
ASTER GDEM v2.0	30	2017	–	–	USGS

Source USGS

areas associated with vector data were then easily eliminated by post-processing (i.e., manual exclusion/inclusion of wrongly classified/missing areas). Additionally, the ASTER GDEM was also been used for the correction of glacier outlines through better visualization of the glacier extents and mapping of the ice divides. Wherever possible, visual inspection of the glaciers, ice divides, and associated glacial features on high-resolution Google Earth (≤ 5 m) was used as an additional source of information for correct mapping. Any mismatch observed was eliminated manually (Mir et al. 2018). From SoI maps, the glaciers were delineated manually by onscreen digitization. It is important to note that the available Landsat data allowed monitoring of glacier from the 2000 but the SoI maps extend the study period to the 1962s.

For further processing, glacier polygons smaller than 0.01 km^2 were excluded from the analysis. It is important to note that the errors might have been introduced due to the seasonal snow cover (if any) and the expected larger uncertainty of delineation of small glaciers. Therefore, to reduce those errors, the comparison of glacier areas was restricted to those larger than 0.10 km^2. Glacier characteristics were calculated for the 2017 scene using the ASTER GDEM v2.0. The glacier characteristics include the estimation of mean elevation, aspect, and slope. Additionally, the glacial lakes associated with glaciers were also delineated. The glacial lakes were not processed further due to their smaller size. In order to evaluate the influence of topography on glacier distribution, bar charts have been plotted for the analysis.

5.4.1 Uncertainty in the Study

The quantification of glacier mapping errors is crucial to ascertain the accuracy and significance of results (Mir et al. 2017, 2018; Bisht et al. 2019, 2020). In this study, the errors of coregistration and glacier boundary delineations were considered that might have resulted in different levels of accuracy. The measurement of uncertainty of area extent (U_{area}) was determined by the buffer method for each glacier (Granshaw and Fountain 2006). A buffer distance of 10 m that is equal to the width of the digitizing error (RMSR) was created. The uncertainty was calculated as an average of the ratio of original glacier areas to the areas with a buffer increment. Overall, in the present scenario, an average uncertainty of $\pm$ 0.020 km^2 (2.0%) was determined for each time period under study. Additionally, the high-resolution GE images were also used to check the accuracy by repeating the digitization of a few selected large-sized glaciers. This method resulted in an accuracy of approximately $\pm$ 2 m ($\leq \pm 1\%$).

5.5 Glacier Characteristics

A total of 90 glaciers having an area larger than 0.008 km^2 were mapped in the study area as per the data of 2017 (OLI/TIRS). The glaciers covered a total area of 21.1 km^2 which is about 2.6% of the total area of the study area. The glaciers are generally small size with a mean size of 0.24 km^2. The size of the glaciers varies from a maximum area of 0.91 km^2 (G. No. 41) and a minimum area of 0.008 km^2 (G. No. 9). Glaciers having smaller size are higher in number than large-sized glaciers thereby indicating that the number of glaciers declines with the increase in size. Previously, only 4% of the glaciers having a size larger than 2 km^2 have been reported in the region of whole LMR (Smidtch and Nüsser 2017). Glacier elevations are largely distributed between 5200–5900 m amsl, with a mean glacier elevation of 5570 m amsl. The termini of the glaciers are located around an altitude varying from 5205–5772 m amsl whereas the mean altitude varies from 5378–5873 m amsl. In this area, the glaciers are characterized by high altitudinal termini and relatively small elevation range of 595 m amsl. For the glaciers of the whole LMR, an elevation range between 2000–6000 m amsl has been reported previously (Chudley et al. 2017). In this area, about 63.2% of glacier area lies in the elevation zone of 5400–5600 m amsl thereby reflecting that the glaciers tend to have higher elevations. The average maximum

Table 5.2 General characteristics of the glaciers present in the study area derived from Landsat OLI/TIRS image (2017)

Glaciers attributes	
Total number	90
Total area (km^2)	21.1
Average elevation minimum (m amsl)	5416
Average elevation maximum (m amsl)	5706
Average elevations mean (m amsl)	5570
Average mid-altitude (m amsl)	595
Maximum altitude (m amsl)	5873
Minimum altitude (m amsl)	5205
Altitude range (m amsl)	327
Mean size (km^2)	0.24
Mean slope (°)	25
Mean aspect (°)	SE
Nature	Clean

Source USGS

elevation of 5706 m amsl and average minimum elevation of 5416 m were identified for the glaciers. Moreover, the larger glaciers tend to have a lower minimum elevation, higher maximum elevation, and shallower gradient than smaller glaciers. The mean Equilibrium Line Altitude (ELA) is located around a mean altitude of 5500 m amsl in this area.

In the study area, glaciers tend to have mostly northerly aspect. About 73.7% of glaciers are having a mean aspect greater than 315° or less than 45° which indicated that the glaciers are spread between the northwest through north to northeast aspects (i.e., NW-NE), whereas a small percentage (10%) of glacier cover is present on southern slopes. These observations are similar to Schimdt and Nüsser (2017) who have reported that about 73% of the glaciated area of the LMR are located on NW-NE facing slopes, whereas south-facing slopes have a small ice cover of about 10%. The slope of the glaciers varies from gentle slope to steep slope. About 73.7% of glacier area is spread within the slope range of 10°–40°. Furthermore, the glaciers in this area are generally clean type with an insignificant debris cover. Chudley et al. (2017) have reported a negligible amount of 3.49% debris cover on the glaciers of the whole LMR. All the debris-free glaciers in this area illustrate the importance of snowfall as the main source of ice to the glaciers which is in contradiction to a significant impact of snow redistribution by avalanches in the areas of the greater Himalayan range. The detailed characteristics of the glacier are given in (Table 5.2).

5.6 Glacier Area Changes from 1962/2000–2017

In this study, only 47 glaciers having an area greater than 0.12 km^2 have been analyzed to understand the glacier changes and dynamics. From the change detection studies, it has been observed that during 1962, the total area of glaciers was 24.1 km^2. However, during the recent year of 2000, the glacier area has decreased to 21.7 km^2. Similarly, the glacier area has further reduced to 19.9 km^2 and 19 km^2 during 2011 and 2017, respectively. In general, it is observed that the glaciers have lost the area very periodically from 1962–2017 (Fig. 5.2a). In nutshell, a reduction in glacier area of about 21.2% (5.1 km^2) at a rate of 93 m^2/year is observed from 1962 to 2017. During different time periods from 1962 to 2000, the glaciers have lost an area of 9.9% (2.4 km^2) at a rate of 65 m^2/year. Similarly, during a recent time interval from

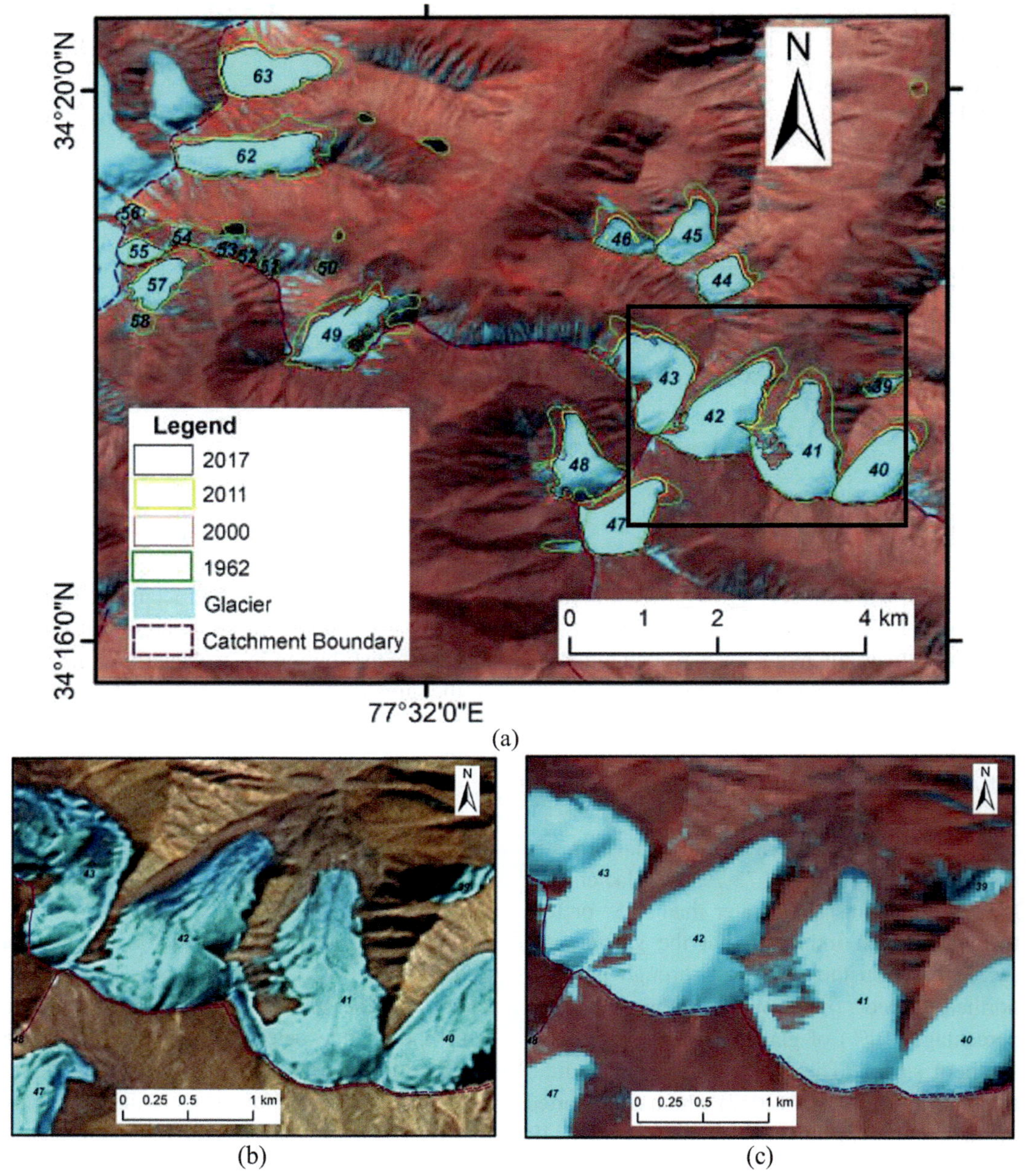

Fig. 5.2 **a** Satellite images showing glacier area loss during different time periods from 1962 to 2017. The subset figures **b** is an image of TM-2000 and **c** is an image of OLI-2017 showing a closer view of the changes of glaciers during these two time periods (2000–2017) (*Source* USGS)

2000 to 2011, a total area loss of 8.3% (1.8 km^2) at a rate of 164 m^2/year has been observed followed by a loss of 4.6% (0.9 km^2) at a rate of 154 m^2/year during 2011–2017. Keeping in view the higher uncertainty associated with SoI maps and to have a better insight into the glacier changes, a recent time period from 2000 to 2017 comprising 17 years has also been considered for assessment of glacier changes. The results indicated a loss of 12.5% (2.7 km^2) at a rate of 160 m^2/year. The loss of 12.5% (2000–2017) is much lower than a loss of 21.2% (1962–2017).

Table 5.3 Changes in glacier area during different time periods from 1962 to 2017 for the selected glaciers

Years	Total area km^2	Glacier area changes				
		Time intervals	Area (km^2)	Rate (km^2/a)	Area (%)	Rate in % (km^2/a)
1962	24.1	1962–2000	2.4	0.065	9.9	0.27
2000	21.7	2000–2011	1.8	0.164	8.3	0.75
2011	19.9	2011–2017	0.9	0.154	4.6	0.77
2017	19.0	2000–2017	2.7	0.160	12.5	0.74
–	–	1962–2017	5.1	0.093	21.2	0.39

Source USGS and Survey of India

However, the rate of glacier loss of 93 m^2/year during 1962–2017 is observed to be much lower than a rate of 160 m^2/year observed during 2000–2017 (Fig. 5.2b, c). Glacier wise, the G No. 8 indicated the lowest area loss of 0.95%, whereas G No. 89 indicated a higher loss of 95.15% considering the time period from 1962 to 2017. However, considering a time period from 2000 to 2017, the G No. 89 indicated the lowest loss of 2.5%, whereas the G No. 19 indicated a higher loss of 31.6%. The glaciers showing higher area loss have a smaller size. Overall, from these observations, it is clearly inferred that the glacier area loss has been continuous from the 1960s with an increasing tendency toward higher loss during recent years. For example, a higher area loss of 12.5% at a rate of 160 m^2/year during recent years is much higher than the area loss of 9.9% at a rate of 65 m^2/year during 1962–2000. However, the small discrepancies in between over the studied time periods are attributed to the heterogeneous selection of time intervals in the study (Table 5.3).

5.7 Relationship Between Glacier Changes with Glacier Area and Topographic Features

Like other Himalayan glaciers, the influence of glacier area and other topographic features (Mir et al. 2017; Mir 2018) can also be very crucial in modulating the observed glacier changes in the present area. The study indicated that on an average, the small glaciers (< 0.12 km^2) lost relatively higher area than the large glaciers observed from 2000 to 2017. However, the statistical relation between glacier area changes (percentage wise) with the glacier area revealed a weak but positive correlation (Fig. 5.3a). The weak correlation is attributed to the fact that in this region, generally, 95% of the glaciers are smaller in size that may probably be buffering the influence of glacier size on its changes. The relationship also showed a lower percentage of area loss for bigger glaciers and a wider distribution of glacier changes among smaller glaciers, which is attributed to the larger number of small glaciers as well as diverse aspect and altitude ranges of these glaciers in the area. Further, the study also indicated that large glaciers are also losing significant areas. The statistical analysis between the glacier size and area loss in absolute units (Fig. 5.3b) of individual glaciers showed a good correlation (R^2 = 0.57).

In addition to size, the glacier area loss is also dependent on the elevation/altitude of the glaciers. As per the hypsometric curve, the maximum glacier area (55.6%) is distributed in the elevation zone of 5400–5600 m amsl, followed by 24.9% in the elevation zone of 5600–5800 m amsl, and 15.7% in the elevation below the 5400 m amsl. A small percentage of glacier area about 3.3% and 0.53% is distributed in the higher elevation zones of 5800–6000 m amsl and above the elevation of 6000 m amsl. Moreover, a similar elevation area distribution pattern of the glacier area is observed for all the studied years/periods. However, it is a very important observation that the glacier area changes have been contrary to the glacier area-altitude

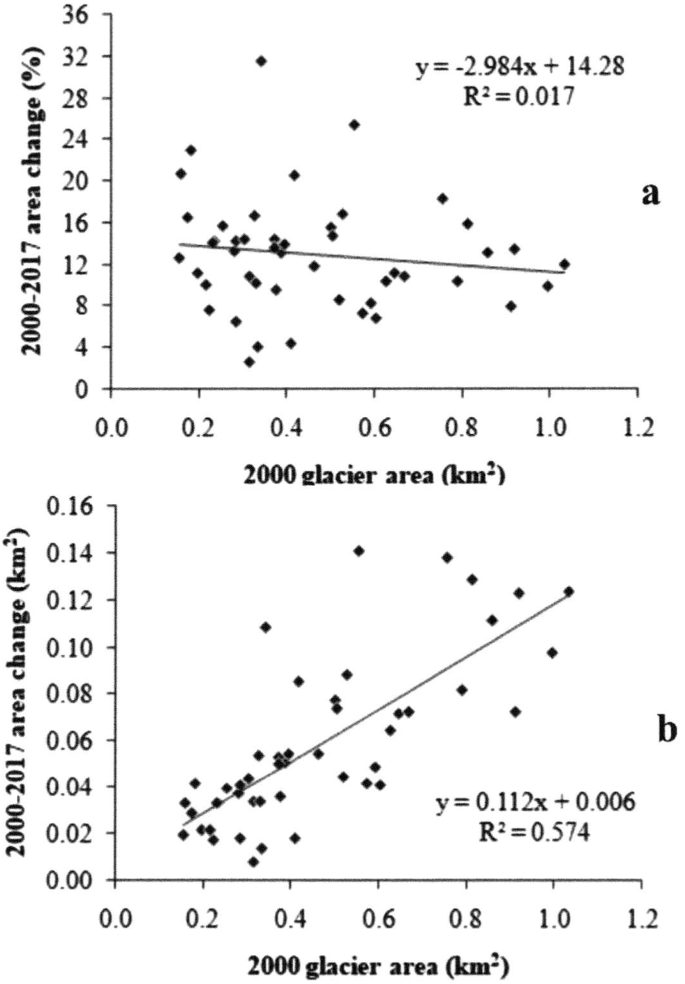

Fig. 5.3 Relationship between **a** glacier size and glacier area loss (%) and **b** glacier size and glacier area loss (km^2) of the study area (*Source* USGS)

distribution. That is, the higher elevation ranges have lost high glacier area during the studied time intervals such as from 2000–2011, 2000–2017, and 2000–2017. The glaciers above an elevation of 5800 m amsl have indicated higher area loss. Similarly, the glaciers lying below an elevation of 5400 m amsl also indicated higher area loss. The glaciers lying in the elevation range of 5400 to 5800 m amsl revealed a moderate glacier area loss. Overall from 2000 to 2017, the glaciers lying below an elevation of 5400 m amsl showed about a loss of 25.6%, whereas the glaciers lying in the elevation zone of 5800–6000 m amsl and above indicated an area loss of 17.9% and 44.1%. The glaciers in the elevation zones of 5400–5600 m amsl and 5600–5800 m amsl have indicated a loss of 9.6% and 7.6%, respectively (Fig. 5.4a).

As per the slope distribution, the maximum area of glaciers that is 29.2% and 28.4% is present within the slope range of 10°–20° and 20°–30°, respectively. The slope range of 30°–40° contains 18.4% of glacier area, whereas slope range of less than 10° contains 12.8% of glacier area. The higher slope range of 40°–50° contains 8.4% of glacier area, whereas the slope ranges above 50° contain about 2.7% of the glacier area. Overall about 88.8% of glacier area is distributed in the slope zone below 40° which is moderately steep. A similar pattern is observed during all the studied years and in general, with an increase in slope, the area of the glaciers decreases.

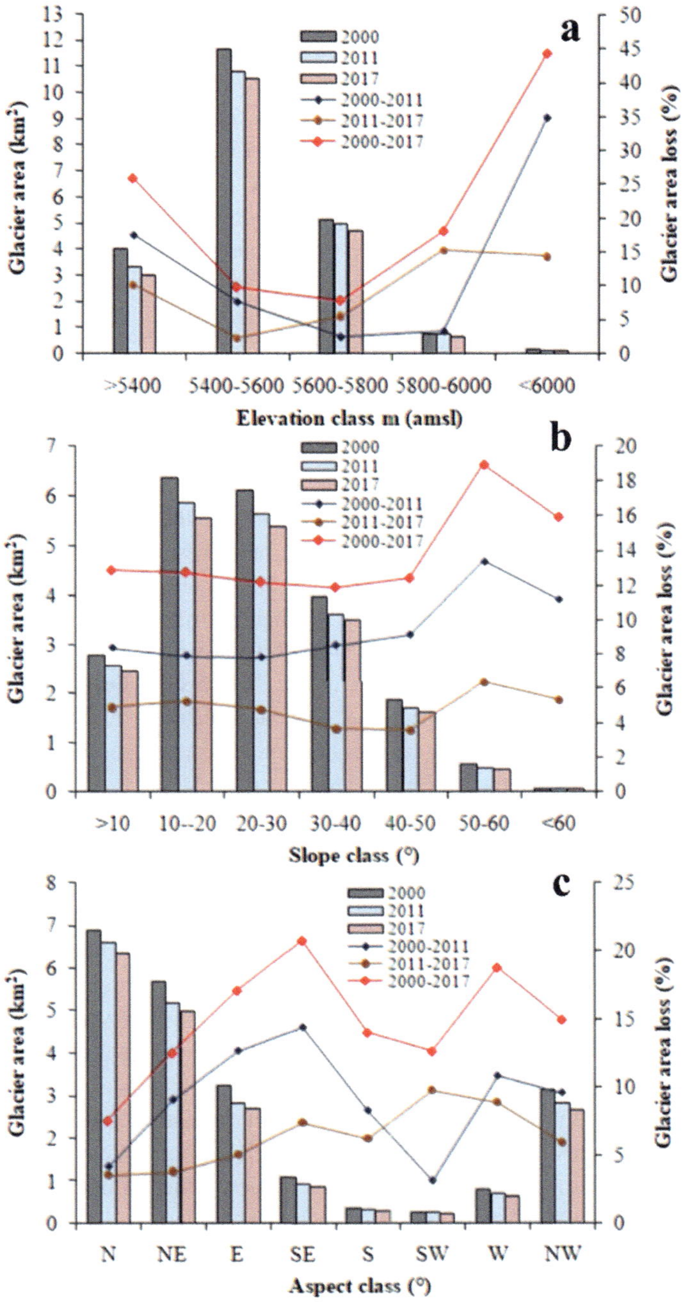

Fig. 5.4 Relationship between the topographic factors and percent glacier area loss **a** elevation vs glacier area loss **b** slope vs glacier area loss **c** aspect vs glacier area loss for the study area (*Source* USGS)

However, the glacier area change exhibited a positive relationship with the slope that is with an increase in slope, the glacier area change also increases. For example, in the slope zone of 30°–40°, 11.9% of area change is observed whereas below the 30° slope, an area change of about 12.5% is observed. However, the glacier area change has increased consistently from 12.4% in the slope zone of 30°–40° to 18.9% in the slope range of 50°–60° and 15.9% above an elevation of 60°. A similar pattern of area change with slope is observed during a time interval of 2000–2011, which is however moderately distorted during a recent time interval of 2011–2017, thereby indicating a dominant control of slope on glacier changes from 2000 to 2011 (Fig. 5.4b).

As per the aspect, the maximum glacier area (34.9%) is distributed in the north aspect, followed by 26.4% in the northeast aspect, 14.1% in the northwest aspect, and 14.2% in the eastern aspect. The rest of the slope aspect covered less than 2% of the glacier area. A similar pattern in aspect wise distribution of glacier area is observed during all the studied time periods. There has been a significant effect of aspect on glacier area changes as well. For example, the glaciers with northerly aspect have lost less area as compared to the other aspects. During the time period from 2000 to 2017, the glacier loss varied from 7.5% in the north aspect to 14.9% in the northwest aspect. The east, southeast, and western aspects revealed a higher area loss of 17.1%, 20.7%, and 18.7%, respectively. During the time periods of 2000–2011 and 2011–2017, an almost similar pattern in glacier area loss has been observed with few exceptions such as, during 2000–2017, the north aspect revealed lower glacier area changes (7.5%), but during 2000–2011, the north and southwest aspect indicated lower changes of 4.1% and 3.1%, respectively. But, during 2011–2017, the north and northeast revealed lower changes of 3.5% and 3.7%, respectively (Fig. 5.4c).

5.8 Discussion

The present study has been carried out in the upper Indus Basin covering parts of LMR falling in the semi-arid Trans-Himalayan region. The study area consists of five sub-basins designated as Ganglas, Phyang, Khalsar, Rong, and No.-5. In this region, there are a large number of small to medium size drainage basins on northern as well as southern slopes of the LMR flowing either directly into the Indus River system or its major northern tributary-Shyok River. In this study, an inventory of 90 glaciers having an area larger than 0.008 km^2 has been generated for the year 2017 using remote sensing data and geographic information system techniques. The inventoried glaciers are spread on a small area of about 21.1 km^2 (2.6%) of the selected total catchment area of the study area. But, it is very interesting to mention that about 63.2% of glacier area is present in a higher elevation range of 5400–5600 m amsl. As observed in this study, the mean ELA of the glaciers is also located around higher elevations of 5500 m amsl.

Furthermore, the glacier changes have been studied from 1962 to 2017 using SoI toposheets and Satellite data. About 47 glaciers having an area greater than 0.12 km^2 have been selected for change detection studies. The study indicated a loss of glacier area of about 21.2% (5.1 km^2) at a rate of 93 m^2/year from 1962 to 2017. While considering the recent time period of 2000–2017, the study indicated a lower loss of 12.5% (2.7 km^2) at a rate of 160 m^2/year. The higher area loss during 1962–2017 may be attributed to the use of SoI maps for which a number of studies have reported uncertainty problems. Furthermore, the small glaciers (<0.12 km^2) lost relatively higher area than the large glaciers during 2000–2017 thereby indicating their marked sensitivity to climate change. In this region also, a few studies have reported a rising pattern of the temperature and declining pattern of precipitation (Chevuturi et al. 2016). But, the precipitation decline in this region may be considered a significant factor controlling the glacier area loss in addition to rising temperature. It is because, the solid precipitation, i.e., snowfall, is considered as the main input and contributor of ice to the glaciers. Furthermore, it is also observed that the glacier changes in this region have been influenced by the glacier size and slope discontinuously during different time periods. However, the influence of the elevation on the glacier changes has been very strong as is understood through the analysis of area-altitude distribution. The glacier area-altitude distribution has revealed that the higher elevation ranges of 5800–6000 and above lost a higher glacier area of 17.9% and 44.1% than the other lower elevation ranges. Similarly, the southern aspects that receive higher solar radiation also indicated higher glacier area loss of 20.7% (SE).

Overall, these observations clearly indicated that the glacier area loss has been continuous from the 1960s with an increasing tendency toward higher loss during recent years. Using the SoI maps, the study indicated a loss of 21.2% (5.1 km^2) at a rate of 93 m^2/year from 1962 to 2017, which is almost similar to a number of glacier changes observed in the HKH (Kulkarni 2011; Mir et al. 2017, 2018). For instance, Mir et al. (2017) using the Landsat satellite data reported 18.1% of glacier area loss from 1976 to 2011 in the Baspa basin in western Himalayan region. Kulkarni (2012) assessed the glaciers of Goriganga, Bhagirathi, Bhut, Warwan, Zanskar, Chandra, and Miyar basins of western Himalaya using SoI topographic maps, and Indian remote sensing satellite (IRS) images and reported a deglaciation of 19%, 14%, 10%, 21%, 9%, 20%, and 8% from 1962 to 2001, respectively. In addition, based on the SoI toposheets and Landsat satellite images, Mir et al. (2014) reported a deglaciation of 26.1% during 1966–2011 for the 34 glaciers in Tirungkhad Sub-basin of Satluj River. Mir et al. (2018) have reported a shrinkage of 2.6 ± 0.56 km^2 (19.2 ± 4.1%) for Dalung glacier and 3.4 ± 0.65 km^2 (12.7 ± 2.4%) for the Padam glacier located in Zanskar Himalaya during a time period of 1962–2015 using satellite and SoI maps. Mir (2018) reported a loss of 19.09 ± 3.3% at a rate of 35 m^2/year from 1962 to 2016 for the Kolahoi glacier of Kashmir Himalaya. Majeed et al. (2020) in a recent study have reported a higher glacier area loss of 29.7% at a rate of 16.5 m/year (Sonapani Glacier) in Himachal, western Himalayan region over a century from 1906–2016. However, in this study, the glacier loss of 12.5% (2.7 km^2) at a rate of 160 m^2/year estimated during 2000–2017 on the satellite data sources is significantly lower than the loss estimated on the basis of SoI maps. The satellite-based loss of 12.5% is almost similar to a loss of 14% (3 m/year) reported from 1969 to 2010 in the adjacent area of Kang Yatze Massif of the region (Schmidt and Nüsser 2012). The satellite-based glacier changes are also similar to the previous glacier area changes of 12.8% and 13% reported for the glaciers of this region previously (Chudley et al. 2017; Schmidt and Nüsser 2017).

5.9 Conclusions

In this study, an inventory of 90 glaciers covering an area of 21.1 km^2 which is about (2.6%) of the total selected catchment area has been generated using remote sensing satellite data of 2017. The glaciers are generally small with a mean size of 0.24 km^2, high altitude with a mean elevation of 5570 m amsl, northerly facing, and moderately steep with 73% of glaciers spread between 10°–40° slope range. In addition, the mean Equilibrium Line Altitude (ELA) is located at and around a higher elevation of 5500 m amsl. Due to the small size of glaciers, only 47 glaciers (size > 0.12 km^2) have been selected for change detection studies. The results indicated a glacier area loss of about 21.2% (5.1 km^2) at a rate of 93 m^2/year from 1962 to 2017 based on SoI base maps. However, considering a recent time period of 2000–2017 based on satellite base maps, the study indicated a loss of 12.5% (2.7 km^2) at a rate of 160 m^2/year. The results revealed that the small glaciers (< 0.12 km^2) lost relatively higher area from 2000 to 2017. The area-altitude distribution revealed that the elevation zone of 5800–6000 and above have lost a higher glacier area of 17.9% and 44.1%, respectively. Similarly, the steep slope and southerly aspect of glaciers also controlled the higher glacier area losses. For example, the slope zone of 50°–60° lost an area of 18.9% and the southeast aspect lost an area of 20.7%. Overall, the higher glacier area changes during recent years and the higher sensitivity of small glaciers to the changes in this area are of serious concern. The observed glacier changes in this area may have a strong influence on the downstream water supplies and resources and other allied sectors. Since the demand for the water supply is drastically increasing in this area in response to the rapid population growth, rapid urbanization, economic development, and especially increased influx of tourists, therefore, the presented glacier change observations in this study can be suitably used to develop and adopt the strategies for the better management of these limited glacier ice reserves vis-a-vis water resources in this area.

Acknowledgements The author would like to thank NASA for making the Landsat and ASTER GDEM datasets freely available under the USGS and GLCF web servers. Special thanks are due to the Google Earth Community for providing the high-resolution images through the virtual Google Earth application and Survey of India for providing SoI topographic maps. The work was carried out during Ph. D. program.

References

Barnett TP, Adam JC, Lettenmaier DP (2005) Potential impacts of a warming climate on water availability in snow-dominated regions. Nature 438:303–309

Barry RG (2006) The status of research on glaciers and global glacier recession: a review. ProgPhysGeogr 30:285–306

Benn DI, Evans DJA (2010) Glaciers and Glaciation. 2nd ed. Hodder Arnold Publication

Bhambri R, Bolch T (2009) Glacier mapping: a review with special reference to the Indian Himalayas. ProgPhysGeogr 35:672–704

Bisht H, Kotlia BS, Kumar K, Joshi LM, Sah SK, Kukreti M (2020). Estimation of the recession rate of Gangotri glacier, GarhwalHimlaya (India) through kinematic GPS survey and satellite data. Environ Earth Sci 13(6):1–12. https://doi.org/10.1007/s12665-020-090780

Bisht H, Rani M, Kumar K, Sah S, Arya PC (2019) Retreating rate of Chaturangi glacier, Garhwal Himalaya, India derived from kinematic GPS survey and satellite data. CurrSci 116:304–311. https://doi.org/10.18520/cs/v116/i2/304-311

Bolch T, Kulkarni A, Kääb A, Huggel C, Paul F, Cogley JG, Frey H, Kargel JS, Fujita K, Scheel M, Bajracharya S, Stoffel M (2012) The state and fate of himalayan glaciers. Sci 336(6079):310–314

Bolch T, Pieczonka T, Mukherji K, Shea J (2017) Brief communication: glaciers in the Hunza Catchment (Karakoram) have been nearly in balance since the 1970s. Cryosph 11:531–539

Bookhagen B, Burbank DW (2010) Toward a complete himalayan hydrological budget: spatiotemporal distribution of snowmelt and rainfall and their impact on river discharge. J Geophys Res: Earth Surface (2003–2012):115 (F3)

Chand P, Sharma MC (2015) Glacier changes in the Ravi Basin, north-western Himalaya (India) during the last four decades (1971–2010/13). Glob Planet Chang 135:133–147

Chevuturi A, Dimri AP, Thayyen RJ (2016) Climate change over Leh (Ladakh), India. Theor App Clim, 1–15

Chudley TR, Miles ES, Willis IC (2017) Glacier characteristics and retreat between 1991 and 2014 in the Ladakh Range. Jammu Kashmir Remot Sen Lett 8(6):518–527

Crook J, Osmaston H (1994) Himalayan Buddhist villages: environment, resources, society and religious Life in Zangskar, Ladakh. MotilalBanarsidass Publications, Delhi

Dahri ZH, Ludwig F, Moors E, Ahmad B, Khan A, Kabat P (2016) An appraisal of precipitation distribution in the high-altitude catchments of the Indus Basin. Sci Total Environ 548(549):289–306

Dame J, Nüsser M (2011) Food security in high mountain regions: agricultural production and the impact of food subsidies in Ladakh, northern India. Food Sec 3:179–194

Deen M (2009) Role of LEHO in Managing Water Resources of LehAmruta Sudhalkar, 1st August

Dyurgerov MB, Meier MF (2005) Glaciers and the changing earth system: a 2004 snapshot, vol 58. Institute of Arctic and Alpine Research, University of Colorado, Boulder

Frey H, Machguth H, Huss M, Huggel C, Bajracharya S, Bolch T, Kulkarni AV, Linsbauer A, Salzmann N, Stoffel M (2014) Estimating the volume of glaciers in the Himalayan-Karakoram Region using different methods. Cryosph 8:2313–2333

Granshaw FD, Fountain AG (2006) Glacier change (1958–1998) in the North Cascades National Park Complex, Washington, USA. J Glaciol 52:251–256

Hewitt K (2005) The Karakoram Anomaly? Glacier expansion and the elevation effect. Karakoram Himalaya. Mt Res Dev 25(4):332–340

Hewitt K (2014) Karakoram glaciers and climate change. Glaciers of the Karakoram Himalaya; Springer: Dordrecht. The Netherlands; Heidelberg, Germany, pp 291–326

Jain SK, Mir RA (2017) Glacier and glacial lake classification for change detection studies using satellite data: a case study from Baspa basin, western Himalaya. GeocartInt 1–24

Kääb A, Treichler D, Nuth C, Berthier E (2015) Brief Communication: Contending Estimates of 2003–2008 Glacier Mass Balance over the Pamir–Karakoram–Himalaya. The Cryosph 9(2):557–564

Kapnick SB, Delworth TL, Ashfaq M, Malyshev S, Milly PCD (2014) Snowfall less sensitive to warming in karakoram than in himalayas due to a unique seasonal cycle. Nature Geosci 7(11):834–840

Khan A, Richards KS, Parker GT, McRobie A, Mukhopadhyay B (2014) How large is the upper Indus Basin? The pitfalls of auto-delineation using DEMs. J Hydrol 509:442–453

Kulkarni AV (2012) Monitoring Himalayan cryosphere using remote sensing techniques. J Indian InsSci 90 (4):457–469

Kulkarni AV, Rathore BP, Singh SK, Bahuguna IM (2011) Understanding changes in Himalayan cryosphere using remote sensing technique. Int J Remot Sen 32(3):601–615

Kumar D, Singh AK, Taloor AK, Singh DS (2020) Recessional pattern of Thelu and Swetvarn glaciers between 1968 and 2019. Quat Int, Bhagirathi basin, Garhwal Himalaya, India. https://doi.org/10.1016/j.quaint.2020.05.017

Labbal V (2000) Traditional oases of Ladakh: A case study of equity in water management. In Sharing water. Irrigation and Water Management in the Hindukush-Karakorum-Himalaya; Kreutzmann H (ed) Oxford University Press: Karachi, Pakistan, 161–183

Majeed Z, Mukhtar AM, Mir RA, Kumar P, Krihna K (2020) Sonapani Glacier Recession over a Century from 1906-2016, Chandra Basin, Himachal Himalaya. J GeolSoc India 95:36–44

Mir RA (2018) Recent changes of two parts of Kolahoi Glacier and its controlling factors in Kashmir basin, western Himalaya. Remot Sen ApplSoc Env 11:265–281

Mir RA, Jain SK, Jain SK, Thayyen RJ, Saraf AK (2017) Assessment of recent glacier changes and its controlling factors from 1976 to 2011 in Baspa basin, western Himalaya. ArctAntar Alp Res 49(4):621–647

Mir RA, Jain SK, Lohani AK, Saraf AK (2018) Glacier recession and glacial lake outburst flood studies in Zanskar basin, western Himalaya. J Hydrol 564 (2018):376–396

Mir RA, Jain SK, Saraf AK, Goswami A (2014) Glacier changes using satellite data and effect of climate in Tirungkhad basin located in Western Himalaya. GeocartInt 29:293–313

Mir RA, Majeed Z (2018) Frontal recession of Parkachik Glacier between 1971-2015, Zanskar Himalaya using remote sensing and field data. GeocartInt 33(2):163–177

Norphel C (2009) History of Water Management in Urban LehAmrutaSudhalkar. 1st August

Nüsser M, Baghel R (2016) Local knowledge and global concerns: Artificial glaciers as a focus of environmental knowledge and development interventions. In Ethnic and Cultural Dimensions of Knowledge; Meusburger P, Freytag T, Suarsana L, (eds), Springer: Dordrecht, The Netherlands; Heidelberg, Germany, 191–209

Nüsser M, Schmidt S, Dame J (2012) Irrigation and development in the upper Indus Basin: Characteristics and recent changes of a socio-hydrological system in Central Ladakh, India. Mt Res Dev 32:51–61

Owen LA, Derbyshire E, Fort M (1998) The quaternary glacial history of the Himalaya. QuatProc 6:91–120

Pandey AC, Gosh S, Nathawat MS (2011) Evaluating patterns of temporal glacier changes in Greater Himalayan Range, Jammu & Kashmir, India. GeocartoInt 26:321–338

Parveen S, Winiger M, Schmidt S, Nüsser M (2015) Irrigation systems of upper Hunza (Karakoram) between persistence and changes. Erdkunde 69:69–85

Paul F, Andreassen LM (2009) A new glacier inventory for the Svartisen region, Norway, from Landsat ETM + data: challenges and change assessment. J Glaciol 55(192):607–618

Paul F, Kääb A, Maisch M, Kellenberger T, Haeberli W (2002) The new remote-sensing-derived swiss glacier inventory: I. Methods Ann Glaciol 34(1):355–361

Rankl M, Kienholz C, Braun M (2014) Glacier changes in the Karakoram Region mapped by multimission satellite imagery. Cryosph 8:977–989

Sarkar T, Kannaujiya S, Taloor AK, Ray PKC, Chauhan P (2020) Integrated study of GRACE data derived interannual groundwater storage variability over water stressed Indian regions. Groundwater Sust Develop 10: P100376. https://doi.org/10.1016/j.gsd.2020.100376

Schmidt S, Nüsse M (2009) Fluctuations of Raikot Glacier during the past 70 years: a case study from the Nanga Parbat Massif, northern Pakistan. J Glaciol 55:949–959

Schmidt S, Nüsser M (2012) Changes of high altitude glaciers from 1969 to 2010 in the Trans-Himalayan Kang Yatze Massif, Ladakh, northwest India. ArctAntarct Alp Res 44:107–121

Singh AK, Jasrotia AS, Taloor AK, Kotlia BS, Kumar V, Roy S, Ray PKC, Singh KK, Singh AK, Sharma AK (2017) Estimation of quantitative measures of total water storage variation from GRACE and GLDAS-NOAH satellites using geospatial technology. Quat Int 444:191–200. https://doi.org/10.1016/j.quaint.2017.04.014

Singh S, Sood V, Taloor AK, Prashar S, Kaur R (2020) Qualitative and quantitative analysis of topographically derived CVA algorithms using MODIS and Landsat-8 data over Western Himalayas. Quat Int, India. https://doi.org/10.1016/j.quaint.2020.04.048

Sood V, Gusain HS, Gupta S, Taloor AK, Singh S (2020) Detection of snow/ice cover changes using subpixel-based change detection approach over Chhota-Shigri glacier, Western Himalaya, India. QuatInt. https://doi.org/10.1016/j.quaint.2020.05.016

Smiraglia C, Mayer C, Mihalcea C, Diolaiuti G, Belo M, Vassena G (2007) Ongoing variations of Himalayan and Karakoram glaciers as witnesses of global changes: recent studies on selected glaciers. In: Baudo R, Tartari G, Vuillermoz E (eds) Mountains Witnesses of Global Changes Research in the Himalaya and Karakoram: Share-Asia Project. Elsevier, Amsterdam, pp 235–248

Taloor AK, Kotlia BS, Jasrotia AS, Kumar A, Alam A, Ali S, Kouser B, Garg PK, Kumar R, Singh AK, Singh B (2019) Tectono-climatic influence on landscape changes in the glaciated Durung Drung basin, Zanskar Himalaya, India: a geospatial approach. QuatInt. 507:262–273

Thayyen RJ, Gergan JT (2010) Role of glaciers in watershed hydrology: a preliminary study of a Himalayan Catchment. Cryosph 4:115–128
Vince G (2009) Glacier Man. Science-News Focus 326:629–661
Yang K, Wu H, Qin J, Lin C, Tang W, Chen Y (2014) Recent climate changes over the Tibetan Plateau and their impacts on energy and water cycle: a review. Glob Planet Chang 112:79–91
Yao T, Thompson L, Yang W, Yu W, Gao Y, Guo X, Yang X, Duan K, Zhao H, Xu B, Pu J (2012) Different glacier status with atmospheric circulations in Tibetan Plateau and surroundings. Nature Climat Chang 2 (9):663–667

Dr. Riyaz Ahmad Mir is working as a Sr. Geoscientist in Geological Survey of India for the last 8 years. He has been honored with high ranks in joint CSIR-UGC (NET+JRF) examination and UPSC Geologist/Geoscientists Examination. After receiving his Master's degree in Geology from Bundelkhand University, Jhansi, he completed his M.Phil. in Hydrogeochemistry from University of Kashmir, Srinagar and his Ph.D. in Geoinformatics from Indian Institute of Technology (IITR), Roorkee. He has also worked with the National Institute of Hydrology (NIH), Roorkee for his Doctorate degree. He has been awarded with best Paper Presentation (SiD-2018) at Cluster University, Srinagar. He has a good working experience in the areas of hydro-and-sediment geochemistry, remote sensing and GIS and its applications in glacier, Glacier Lake Outburst Flood (GLOF), landslide studies and its modelling, climate change studies and its effect of water resources. He has also a vast field expertise in Geochemical Mapping (GCM), Specialized Geological Thematic Mapping (STM) and landslide zonation and susceptibility mapping, etc. He has published about 20 research articles in refereed international/national journals, about 10 extended abstracts in national and international conferences/symposiums and 1 book chapter (Springer publication). He has also completed about 7 Geological Reports of different research projects at GSI. He has published two articles in Hindi Magazine 'Tawi' published by GSI, J&K and is also writing to daily locals regarding geoscientific issues. He has delivered several guest lectures at Degree College and Intermediate level pertaining to the role of Earth Sciences in society

Estimation of Geodetic Mass Balance for Bada Shigri Glacier and Samudra Tapu Glacier in Chandra Basin, India

6

M. Geetha Priya, Ishmohan Bahuguna, D. Krishnaveni, and Suresh Devaraj

Abstract

Bada Shigri and Samudra Tapu are the two largest glaciers of Chandra basin, a sub-basin of Western Himalayas located in Lahaul and Spiti district of Himachal Pradesh, India. The length of the Bada Shigri and Samudra Tapu glaciers are around 30 km and 17 km with snout elevation at about 4000 m and 4200 m above mean sea level (amsl), respectively. The mass balance of these two representative glaciers of the Chandra basin was calculated using the geodetic method for the years 2000–2011. SRTM and ASTER DEMs are used for calculating the change in elevation based on DEM subtraction after pre-processing of data and appropriate bias corrections. The Randolph Glacier Inventory (RGI-5.0) and LISS-III data of IRS-P6 are used for delineating the glacier boundaries manually. An average elevation changes of approximately 10.1 m and 12.3 m has been observed for the 11 years time period, resulting in a cumulative mass balance of −9.09 and −11.07 mass water equivalent (mwe) for Bada Shigri glacier and Samudra Tapu glacier, respectively. An average volume of ice loss observed for the glaciated region of 232 km^2 (area for both glaciers) is approximately 1.15 km^3 during the study period.

Keywords

Bada shigri · DEM · Glaciers · Mass balance · Remote sensing · Samudra tapu

M. Geetha Priya (✉)
School of Electronics Engineering, Vellore Institute of Technology, Chennai 600127, India
e-mail: geetha.sri82@gmail.com

I. Bahuguna
Earth, Ocean Atmosphere Planetary Sciences and Applications Area (EPSA), Space Applications Centre (SAC), ISRO, Ahmedabad 380015, India
e-mail: imbisro@gmail.com

D. Krishnaveni · S. Devaraj
CIIRC- Centre for Incubation, Innovation, Research and Consultancy, Jyothy Institute of Technology, Bangalore 560082, India
e-mail: mailkveni@gmail.com

S. Devaraj
e-mail: devaraj.suresh1991@gmail.com

6.1 Introduction

Asian glaciers are an important regional buffer against drought. Some parts of the Himalayan region receive snow precipitation annually specially in winters, and throughout the year at high altitudes (Bhat et al. 2019; Sabin et al. 2020, Kumar et al. 2020; Sood et al. 2020). The winter accumulated snow-fed glaciers that melt during summer to nourish the rivers originating from Himalayan basins catering to the needs of people living at higher altitudes as well as in the

A. K. Taloor et al. (eds.), *Water, Cryosphere, and Climate Change in the Himalayas*,
Geography of the Physical Environment, https://doi.org/10.1007/978-3-030-67932-3_6

downstream regions (Shrestha and Aryal 2011; Bisht et al. 2020a). Hence, assessment of glacier mass balance is crucial to determine the health of glaciers due to climate change implications (Singh et al. 2018). Mass balance refers to a change in the mass of a glacier over a specified period of time (Che et al. 2019). Mass balance is the gain of ice during winter accumulation and loss of ice during summer ablation from the glacier system (Maurer et al. 2019). Glaciers record negative mass balance when they tend to lose more mass than they receive and positive mass balance when they tend to gain more mass than they lose (Jia et al. 2020). The average mass balance of the glaciers observed for a long-term observation series around the world continues to be negative (WGMS 2019). Ground truth data with in situ measurements for glacier mass balance studies are restricted due to the harsh and undulating rugged terrain environment of the Himalayan region (Rashid and Abdullah 2016; Taloor et al. 2019; Bisht et al. 2020b; Singh et al. 2020).

Geodetic methods are an indirect approach used for the determination of the mass balance of a glacier (Suresh and Yarrakula 2018). Maps of a glacier made at two different times can be compared and the difference in surface elevation observed can be used to determine the changes in mass balance with time (Farias-Barahona et al. 2020). The geodetic mass balance is calculated from the volume change derived from topographic data (Basantes et al. 2018). For the estimation of the mass balance, two digital elevation models (DEMs) acquired at different date's t1 and t2, usually at the end of the ablation period, are used. Time period, $\Delta t = t2 - t1$, for estimation of mass balance can vary from a time limit of one year to many decades. The volume change ΔV in the period Δt is then calculated for the entire glacier either from the contour lines of elevation (Lang and Patzelt 1971) or with a raster method (Unk et al. 1997). The multiplication of the volume change ΔV with the mean density ρ results is the mass balance within the considered period (Priya and Krishnaveni 2019). Classical geodetic work of the highest (mm) precision was demonstrated in the 1970s for purposes of measuring horizontal crustal strain over regional scales (Savage 1983). In addition, the precise recession rate (mm level accuracy) of the Indian Himalayan glaciers through the geodetic method (Kinematic GPS survey) was also carried out by a few researchers (Kumar et al. 2020; Bisht et al. 2019, 2020b). However, mass balance estimation over a long period (decadal scale) is more significant and prone to less errors when compared to annual mass balance estimation using the geodetic method (Fischer 2010).

Adopting the above-discussed methodology, investigation of glacier mass changes over the Himalayan region has been taken up as a topic of research. Mass change estimation using the geodetic method over the Himalayan region was carried out using Shuttle Radar Topographic Mission (SRTM) DEM and Advanced Spaceborne Thermal Emission and Reflection Radiometer (ASTER) DEM for a period of 11 years (2000–2011).

6.2 Materials and Methods

6.2.1 Study Area

The two glaciers namely Bada Shigri and Samudra Tapu of Chandra basin in Lahaul and Spiti district, Himachal Pradesh are the study areas proposed for the estimation of mass balance. The geographical location of the study area is shown in Fig. 6.1. Bada Shigri (137 km^2) is the largest glacier and Samudra Tapu (95 km^2) is the second largest glacier in the upper Chandra basin (Dobhal and Kumar 1996). The existing snout position of Bada Shigri and Samudra Tapu glaciers are approximately at 4000 m elevation 32°06′N and 77°44′E at a distance of about 4 km from the Chandra river bed and 4200 m elevation 32°30′ N and 77°32′ E about 10 km southwest of Chandra Tal Lake (Kulkarni et al. 2006), respectively.

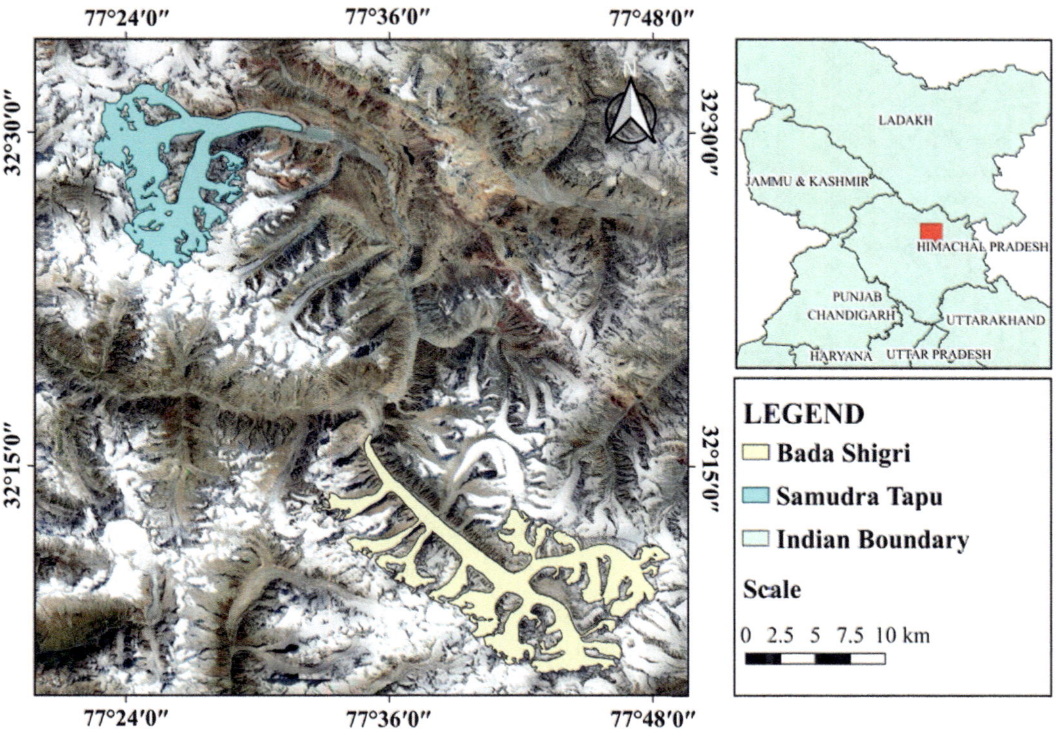

Fig. 6.1 Geographical location of the study area (*Source* Google Earth and RGI-5)

6.2.2 Data Used

Glacier boundary shapefiles obtained from Randolph Glacier Inventory 2005 (RGI-5) are modified for glacier area, ablation and accumulation region delineation based on transient snow line estimated from LISS-III sensor of Resourcesat-1 (IRS-P6) launched during 2003. The DEMs generated from SRTM (acquired during February 2000) and ASTER (acquired during October 2011) with a spatial resolution of 30 m were considered for geodetic mass balance estimation. The SRTM and ASTER DEMs for the proposed study area were downloaded from the United States Geological Survey earth explorer (http://earthexplorer.usgs.gov). LISS-III images of Resourcesat-1 (IRS—P6) acquired on December 2, 2011 was utilized for the extraction of the snowline. Due to the unavailability of cloud-free images covering the study area during the ablation period (June-September) of the year 2011, Meteorological data from Indian Meteorological Department (IMD) observatory located at Lahaul and Spiti district, Himachal Pradesh had been procured for the above mentioned period, to ensure that significant snowfall has not occurred during the above mentioned period affecting the snowline estimation. The LISS-III images have a spatial resolution of 23.5 m covering a swath of 140 km with a revisit period of 24 days. The proposed individual glaciers were covered by four tiles of LISS-III images which were mosaicked for further processing of snowline estimation. Figure 6.2 represents the LISS-III (IRS-P6) FCC image for Bada Shigri and Samudra Tapu glaciers. The datasets used in the present study are listed in Table 6.1.

6.2.3 Methodology

Data from LISS-III sensor on-board Resourcesat-1 satellite is the main source of data to delineate the snowline at the end of the ablation season. The glacier boundary from RGI-05 was modified for the glacier area and delineation of the

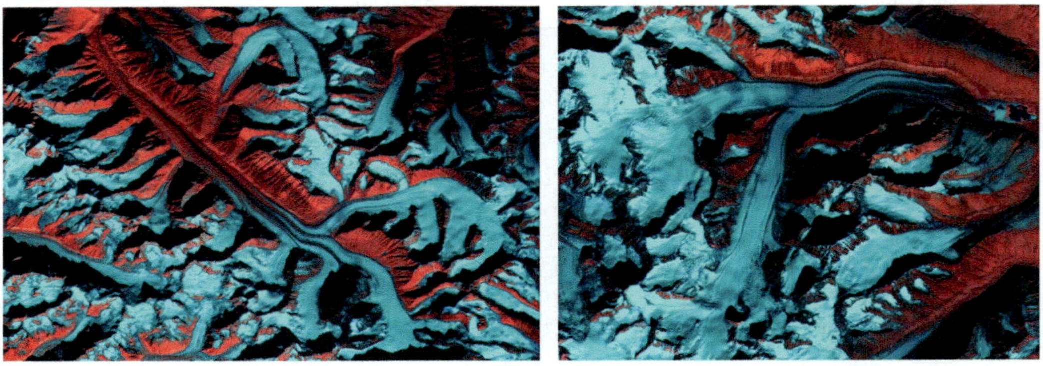

(a) Bada Shigri (b) Samudratapu

Fig. 6.2 Mosaicked images of LISS-III in FCC for the study area (*Source* LISS-III, IRS-P6)

Table 6.1 Satellite data used for monitoring of glaciers of Chandra basin

Sensor type	Scene ID	Acquisition time	Spatial resolution (m)
LISS-III	L3-NI43X11-095-048-02dec11	02/12/2011	23.5
LISS-III	L3-NI43X12-095-048-02dec11		
LISS-III	L3-NI43X15-095-048-02dec11		
LISS-III	L3-NI43X16-095-048-02dec11		
LISS-III	L3-NI43X10-095-048-02dec11		
LISS-III	L3-NI43X11-095-048-02dec11		
LISS-III	L3-NI43X06-095-048-02dec11		
LISS-III	L3-NI43X07-095-048-02dec11		
SRTM DEM	–	February 2000	30
ASTER DEM	–	October 2011	

(*Source* https://www.nrsc.gov.in/)

snowline. Figure 6.3 shows the workflow adopted for the mass balance estimation using the geodetic method. SRTM DEM has been derived from SAR interferometry and ASTER DEM has been derived from stereophotogrammetry (two different sources namely microwave SAR and multispectral imager). The SRTM and ASTER DEMs are of EGM-96 geoidal height model and are reprojected to UTM zone 44 N based on the study area for the present study (Vaishnavi et al. 2020). Co-registration of SRTM and ASTER DEMs has been carried out for minimizing Δx and Δy positional and Δz elevation offsets for stable (non-glaciated) regions (Nuth and Kaab 2011; Gardelle et al. 2012).

After successful co-registration of both the DEMs along and across the track, computation of elevation differences between the two DEMs for 11 years of period has been carried out. Elevation changes above +100 m and below −100 m represent outliers due to data gaps and DEM edges (Berthier et al. 2010; Bolch et al. 2011). Hence, the outliers were removed by the masking process.

Glacier boundary modified from RGI-05 based on LISS-III images is considered and the glacier area is divided into two zones namely, ablation and accumulation zones. In addition to the removal of outliers, the pixel values representing elevation difference more than three

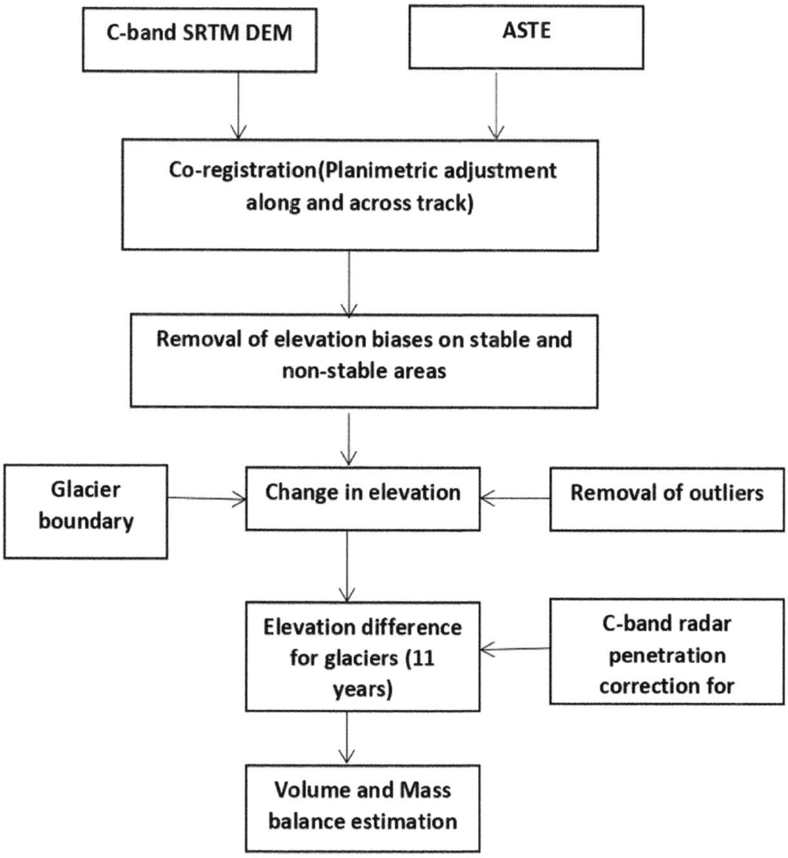

Fig. 6.3 Methodology adopted in the present study (*Source* Authors contribution)

standard deviations from the mean elevation difference from each zone were discarded (Gardelle et al. 2012). Microwave SAR sensors with smaller frequencies have higher wavelength resulting in penetrating through target objects before backscattering (Suresh and Yarrakula 2019). SRTM DEM has been derived from microwave SAR C-band data using the SAR Interferometry technique (Suresh and Yarrakula 2018). An average radar penetration correction of 2.4 ± 0.4 (m) over the accumulation zone of glaciers in the Lahaul and Spiti region (Gardelle et al. 2012; Kääb et al. 2012) has been applied for compensating as SRTM DEM C-band data was acquired in the winter season. After the process of co-registration, differencing the DEMs, outlier's removal and penetration correction, the estimation of volume and mass balance was carried out for the glaciers.

6.3 Results and Discussion

Changes in the elevation for the two glaciers namely Bada Shigri and Samudra Tapu were computed for a period of 11 years (2000–2011) using SRTM and ASTER DEMs. The average elevation of snowline was estimated from LISS-III images for the two glaciers during the end of ablation time in the year 2011. For Bada Shigri glacier and Samudra Tapu glacier, the average elevation of the snowline was found to be at 4800 m and 5248 m, respectively. The estimated snowline elevation was used to delineate ablation and accumulation zones for each glacier separately. Radar penetration correction of 2.4 ± 0.4 m was applied for accumulation zones to compensate for SRTM C-band. An average elevation change of each zone is later considered

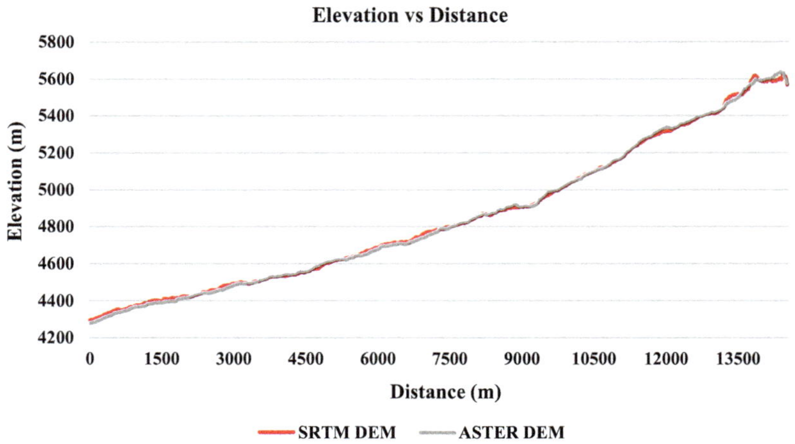

Fig. 6.4 Elevation profile along the Glacier (*Source* Authors contribution)

as the total elevation change for the entire glacier. Figure 6.4 displays a sample elevation profile obtained along the glacier. Figures 6.5 and 6.6 represent the elevation changes for the Bada Shigri and Samudra Tapu glaciers, respectively.

Over the duration of 11 years (2000–2011), the glaciers show a significant decrease in surface elevation and thinning. An average thinning of 11.2 m for two glaciers has been observed for a glaciated region of 232 km^2. The average elevation changes for each zone are then considered as the total elevation change for the entire glacier. The change in elevation was multiplied with the glacier area for computing volume change. The cumulative volume of ice loss is approximately 1.131 km^3 and 1.169 km^3 for Bada Shigri and Samutra Tapu glaciers, respectively, over a period of 11 years (2000–2011). This indicates that an average volume of ice loss for the above said period is 1.15 km^3 resulting in an average annual ice loss of 0.1 km3. For calculation of mass balance, an ice density of 900 kg m^{-3} was assumed (Slobbe et al. 2009).

The change in volume was multiplied with ice density to estimate the geodetic mass balance of the two glaciers. The estimated cumulative mass balance for the time frame 2000–2011 was approximately −9.09 mwe and −11.07 mwe for Bada Shigri and Samudra Tapu glaciers, respectively, as shown in Table 6.2. The estimated geodetic mass balance is −0.83 mwe/year and −1mwe/year for the 11 years period of investigation for Bada Shigri and Samudra Tapu glaciers, respectively, with a standard error of ± 0.33. Comparison of mass balance obtained from other existing methods of geodetic (Tandem X DEM), glaciology and Accumulation Area Ratio (AAR) and Temperature Index (TI) are shown in Table 6.3. Values indicate the appropriateness of the present study and methodology with a high positive correlation (R^2 = 0.91) (Fig. 6.7). More thinning and mass loss has been observed in the Samudra Tapu glacier in comparison with the Bada Shigri glacier. This could have resulted due to the orientation and topographical conditions of the Samudra Tapu glacier. Also, the controlling environmental factors vary from glacier to glacier and basin to basin. The results obtained in this research work lead to overestimation or underestimation of volume and mass change due to uncertainties in assuming the following: Ice density, snow density, application of ice density for accumulation area for mass calculation, Radar wave penetration correction, Outlier values, Data gaps, Voids in SRTM DEM and Glacier area.

6.4 Conclusion

The present study is an attempt to estimate the mass balance of the two benchmark and largest glaciers of the Chandra basin. The geodetic method of mass balance estimation is one of the robust techniques compared to other existing

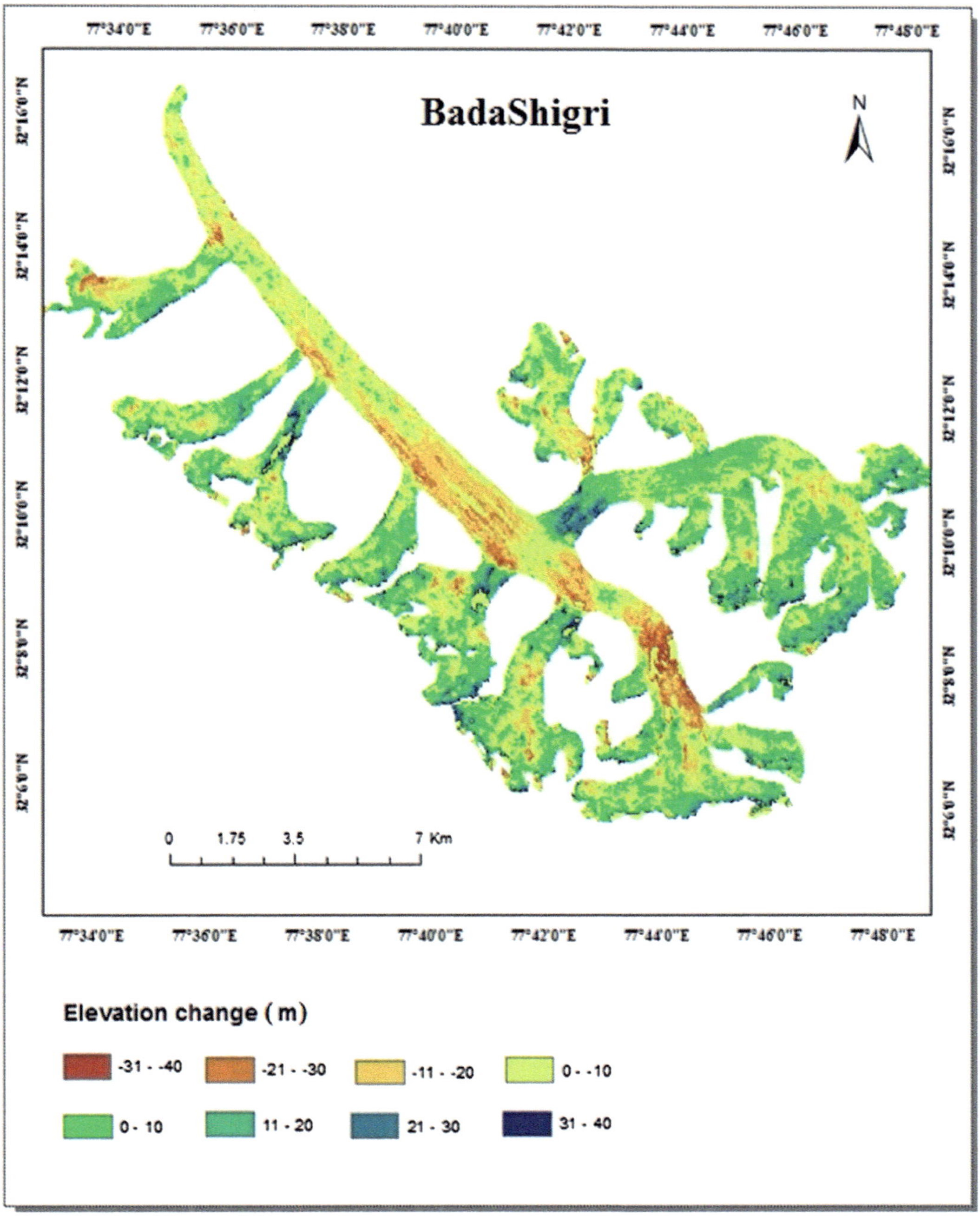

Fig. 6.5 Change in the elevation for Bada Shigri glacier (*Source* Authors contribution)

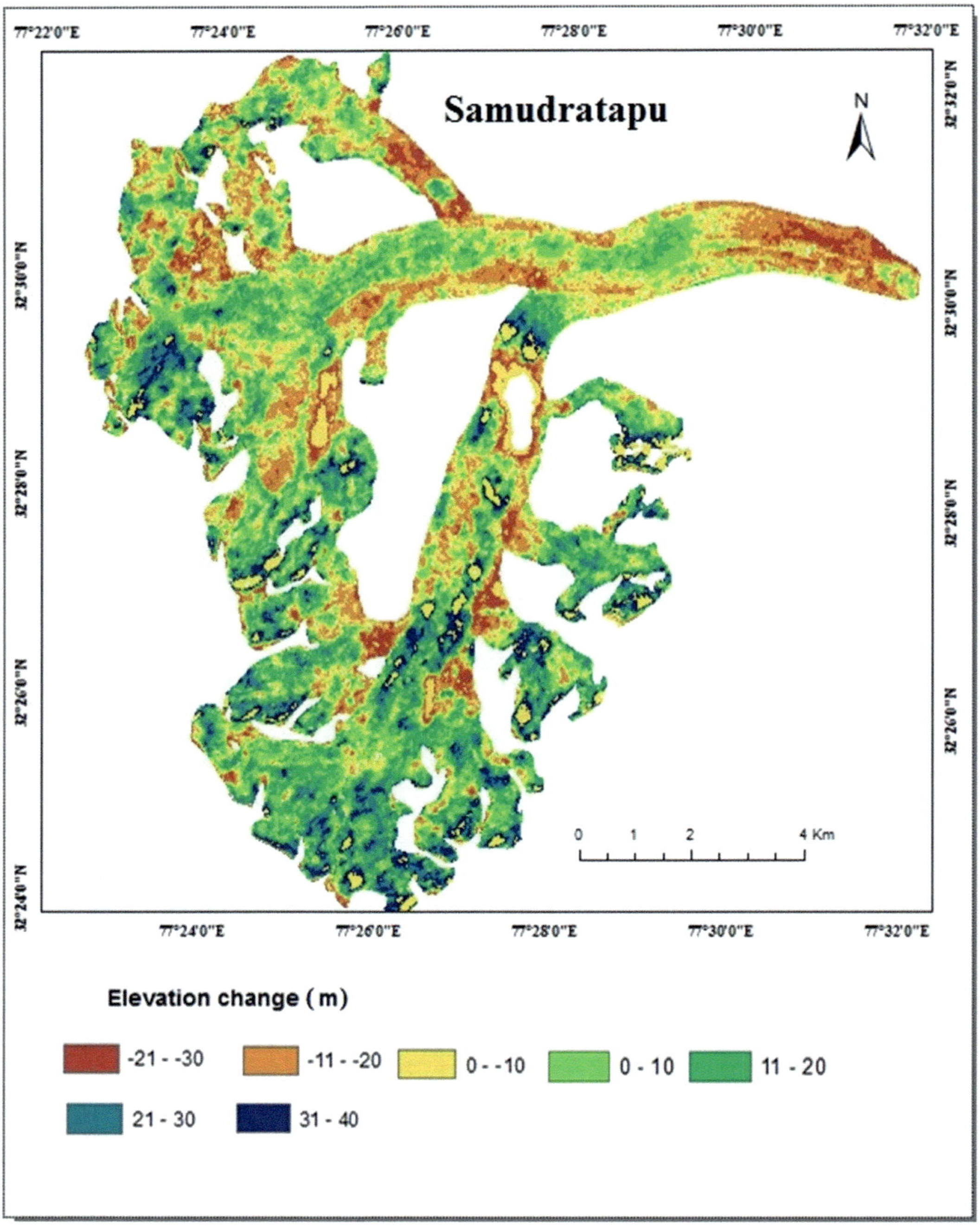

Fig. 6.6 Change in the elevation for Samudra Tapu glacier (*Source* Authors contribution)

Table 6.2 The basic information of the Bada Shigri and Samudra Tapu glacier for the period of 11 years (2000–2011) (*Source* Authors contribution)

Glacier name	Bada Shigri	Samudra Tapu
Slope in degrees	12.3	12.3
Aspect in degrees	355	48
Central longitude	77.5946	77.4146
Central latitude	32.1659	32.4875
Area (sq.km) during study period	137	95
Average elevation difference (m) for 11 years	−10.1	−12.3
Cumulative volume of ice loss (cu.km)	1.131	1.169
Cumulative mass balance (mwe)	−9.09	−11.07
Annual mass balance (mwe)	−0.83	−1.00
Average snowline elevation (amsl)	4800	5248

Table 6.3 Comparison of MB from different studies (*Source* Authors contribution)

Method	Geodetic		TI/AAR method	Glaciology	
Glacier name	MB rate from the present study (2000–2011) (mwe y −1)	MB rate from other studies (2000–2013) (mwe y−1) (Vijay and Braun 2016; Ramsankaran et al. 2019)	MB rate from the other studies (1984–2012) (mwe y−1) - (Tawde et al. 2016)	MB rate from the other studies (2016–2017) (WGMS 2019)	MB rate from the other studies (2017–2018) (WGMS 2019)
Bada Shigri	−0.83 ± 0.33	−0.66 ± 0.32	−0.7 ± 0.46	−0.56	−0.82
Samudra Tapu	−1.00 ± 0.63	−0.69 ± 0.33	−0.8 ± 0.46	−1.12	−1.56

methodologies. Geodetic mass balance for two glaciers has been estimated using SRTM and ASTER DEMs for a time frame of 2000–2011. The estimated geodetic mass balance is −0 mwe/year and −1 m w e/year for the 11 years period of investigation for Bada Shigri and Samudra Tapu glaciers, respectively. An average thinning of 11.2 m for two glaciers has been observed for a glaciated region of 232 km^2. An average volume of ice loss for the above said period is 1.15 km^3 resulting in an average annual ice loss of 0.1 km^3. An average mass loss of 0.915 mwe/year has been observed for the entire study period. The study also shows the effectiveness of remote sensing based methods for glaciological studies in the Himalayan region, because of tough terrain and logistic difficulties occurred in the field-based observation methods.

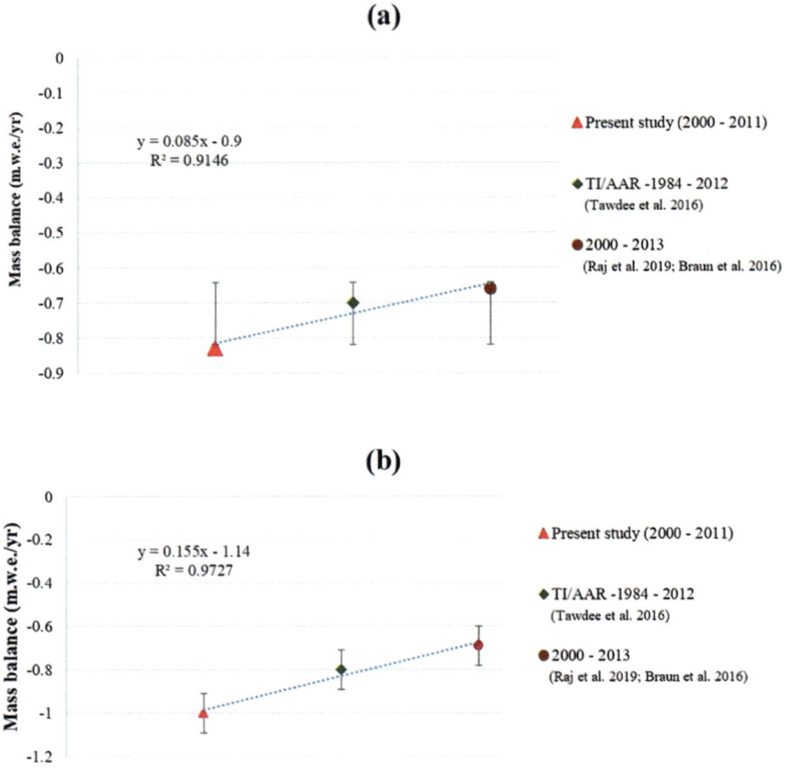

Fig. 6.7 Regression analysis of Mass Balance **a** Bada Shigri **b** Samudra Tapu (*Source* Authors contribution)

Acknowledgements This research work is being supported by the Space Applications Centre- ISRO, Ahmadabad under the project "Integrated Studies of Himalayan Cryosphere (ISHC) using space based Inputs". The authors gratefully acknowledge the support and cooperation given by Dr.A.S. Rajawat, Group Director, GHCAG/EPSA, SpaceApplications Centre, ISRO, Ahmedabad and Dr. Krishna Venkatesh, Director, CIIRC, Jyothy Institute of Technology, Bangalore.

References

Basantes R, Rabatel A, Vincent C, Sirguey P (2018) An optimized method to calculate the geodetic mass balance of mountain glaciers. J Glaciol. https://doi.org/10.1017/jog.2018.79

Berthier E, Schiefer E, Clarke G, et al (2010) Contribution of Alaskan glaciers to sea-level rise derived from satellite imagery. Nat Geosci 392–95. https://doi.org/10.1038/ngeo737

Bhat MS, Alam A, Ahmad B, Kotlia BS, Farooq H, Taloor AK, Ahmad S (2019) Flood frequency analysis of river Jhelum in Kashmir basin. Quat Int 507:288–294

Bisht H, Kotlia BS, Kumar K, Arya PC, Sah SK, Kukreti M et al (2020a) Estimation of suspended sediment concentration and meltwater discharge draining from the Chaturangi glacier. Garhwal Himalaya. Arab J Geosci 13(6):1–12. https://doi.org/10.1007/s12517-020-5204-w4

Bisht H, Kotlia BS, Kumar K, Joshi LM, Sah SK, Kukreti M (2020b). Estimation of the recession rate of Gangotri glacier, Garhwal Himlaya (India) through kinematic GPS survey and satellite data. Environ Earth Sci 13(6):1–12. https://doi.org/10.1007/s12665-020-090780

Bisht H, Rani M, Kumar K, Sah S, Arya PC (2019) Retreating rate of Chaturangi glacier, Garhwal Himalaya, India derived from kinematic GPS survey and satellite data. CurrSci 116: 304–311. https://doi.org/10.18520/cs/v116/i2/304-311

Bolch T, Pieczonka T, Benn DI (2011) Multi-decadal mass loss of glaciers in the Everest area (Nepal Himalaya) derived from stereo imagery. Cryosph 5 (2):349–358. https://doi.org/10.5194/tc-5-349-2011

Che Y, Zhang M, Li Z, et al (2019) Energy balance model of mass balance and its sensitivity to meteorological variability on Urumqi River Glacier No. 1 in the Chinese Tien Shan. Sci Rep 9(1):13958. https://doi.org/10.1038/s41598-019-50398-4

Dobhal D, Kumar S (1996) Inventory of glacier basins in Himachal Himalaya. J Geol Soc India 48671–681

Farias-Barahona D, Sommer C, Sauter T et al (2020) Detailed quantification of glacier elevation and mass changes in South Georgia. Environ Res Lett 15 (3):34036. https://doi.org/10.1088/1748-9326/ab6b32

Fischer A (2010) Comparison of direct and geodetic mass balances on a multi-annual time scale. Cryosph Discuss 5. https://doi.org/10.5194/tc-5-107-2011

Gardelle J, Berthier E, Arnaud Y (2012) Slight mass gain of Karakoram glaciers in the early twenty-first century. Nat Geosci 5(5):322–325. https://doi.org/10.1038/ngeo1450

Jia Y, Li Z, Jin S, et al (2020) Runoff changes from Urumqi Glacier no. 1 over the past 60 years, Eastern Tianshan, Central Asia. Water (Switzerland) 12(5):1–15. https://doi.org/10.3390/W12051286

Kääb A, Berthier E, Nuth C et al (2012) Contrasting patterns of early twenty-first-century glacier mass change in the Himalayas. Nature 488(7412):495–498. https://doi.org/10.1038/nature11324

Kulkarni AV, Dhar S, Rathore BP et al (2006) Recession of samudra tapu glacier, chandra river basin, Himachal Pradesh. J Indian Soc Remote Sens 34(1):39–46. https://doi.org/10.1007/BF02990745

Kumar D, Singh AK, Taloor AK, Singh DS (2020) Recessional pattern of Thelu andSwetvarn glaciers between 1968 and 2019. Quat Int, Bhagirathi basin, Garhwal Himalaya, India. https://doi.org/10.1016/j.quaint.2020.05.017

Lang H, Patzelt G (1971) Volume changes of the back ice (Otztal Alps) compared to the mass change in the period 1953-64. J Glacial Sci Glacial Geol 7(1–2):229–238

Maurer JM, Schaefer JM, Rupper S, Corley A (2019) Acceleration of ice loss across the Himalayas over the past 40 years. Sci Adv 5(6). https://doi.org/10.1126/sciadv.aav7266

Nuth C, Kääb A (2011) Co-registration and bias corrections of satellite elevation data sets for quantifying glacier thickness change. Cryosph 5(1):271–290. https://doi.org/10.5194/tc-5-271-2011

Priya MG, Krishnaveni D (2019) An Approach to measure snow depth of winter accumulation at basin scale using satellite data. Int J Comput Inf Eng 13 (2):70–74

Ramsankaran R, Pandit A, Parla A (2019) Decadal estimates of surface mass balance for glaciers in chandra basin, western himalayas, India—A Geodetic Approach, pp 109–125

Rashid I, Abdullah T (2016) Investigation of temporal change in glacial extent of Chitral watershed using Landsat data: a critique. Environ Monit Assess 188. https://doi.org/10.1007/s10661-016-5565-z

Sabin TP, Krishnan R, Vellore R et al (2020) Climate Change Over the Himalayas BT—Assessment of Climate Change over the Indian Region: A Report of the Ministry of Earth Sciences (MoES), Government of India. In: Sanjay J, Gnanaseelan C et al (eds) Krishnan R. Springer Singapore, Singapore, pp 207–222

Savage JC (1983) Strain accumulation in Western United States. Annu Rev Earth Planet Sci 11(1):11–41. https://doi.org/10.1146/annurev.ea.11.050183.000303

Shrestha AB, Aryal R (2011) Climate change in Nepal and its impact on Himalayan glaciers. Reg Environ Chang 11(1):65–77. https://doi.org/10.1007/s10113-010-0174-9

Singh S, Kumar R, Dimri AP (2018) Mass Balance Status of Indian Himalayan Glaciers: A Brief Review. Front Environ Sci 630. https://doi.org/10.3389/fenvs.2018.00030

Singh S, Sood V, Taloor AK, Prashar S, Kaur R. Qualitative and quantitative analysis of topographically derived CVA algorithms using MODIS and Landsat-8 data over Western Himalayas, India. Quat Int. https://doi.org/10.1016/j.quaint.2020.04.048

Slobbe DC, Ditmar P, Lindenbergh RC (2009) Estimating the rates of mass change, ice volume change and snow volume change in Greenland from ICESat and GRACE data. Geophys J Int 176(1):95–106. https://doi.org/10.1111/j.1365-246X.2008.03978.x

Sood V, Gusain HS, Gupta S, Taloor AK, Singh S (2020) Detection of snow/ice cover changes using subpixel-based change detection approach over Chhota-Shigri glacier, Western Himalaya, India. Quat Int. https://doi.org/10.1016/j.quaint.2020.05.016

Suresh D, Yarrakula K (2018) Subsidence monitoring techniques in coal mining: Indian scenario. Indian J Geo-Marine Sci 47(10):1918–1933

Suresh D, Yarrakula K (2019) InSAR based deformation mapping of earthquake using Sentinel 1A imagery. Geocarto Int 1–10. https://doi.org/10.1080/10106049.2018.1544289

Taloor AK, Kotlia BS, Jasrotia AS, Kumar A, Alam A, Ali S, Kouser B, Garg PK, Kumar R, Singh AK, Singh B (2019) Tectono-climatic influence on landscape changes in the glaciated Durung Drung basin, Zanskar Himalaya, India: a geospatial approach. Quat Int 507:262–273

Tawde S, Kulkarni A, Govindasamy B (2016) Estimation of glacier mass balance on a basin scale: an approach based on satellite-derived snowlines and a temperature index model. Curr Sci 1111977. https://doi.org/10.18520/cs/v111/i12/1977-1989

Unk M, Morelli R, Stahel W (1997) Mass Balance of Griesgletscher 1961–1994: Different Methods of Determination. J Glacial Sci Glacial Geol 41–55

Vaishnavi B, Yarrakula K, Karthikeyan J (2020) Flood inundation mapping of lower godavari river basin using remote sensing and GIS. Indian J Ecol 47(1):30–35

Vijay S, Braun M (2016) Elevation Change Rates of Glaciers in the Lahaul-Spiti (Western Himalaya, India) during 2000–2012 and 2012–2013. Remote Sens 8 (12). https://doi.org/10.3390/rs8121038

WGMS (2019) World Glacier Monitoring Service. https://wgms.ch/latest-glacier-mass-balance-data/. Accessed 12 Sep 2020

Er. M. Geetha Priya, alumni of Anna University obtained her Bachelors (B.E.), Master's (M.E.) and Doctoral degree from Anna University, Chennai in the faculty of Engineering. She is currently working as Professor at School of Electronics Engineering, Vellore Institute of Technology, Chennai. Her areas of research include low power design, reversible logic, energy systems, RADAR, UAV, optical and microwave remote sensing. She has more than 15 years of experience in different domains of academics and research with numerous research publications in peer-reviewed high-impact journals to her credit besides two products and three patent filings. She was a member of the Indian Scientific Expedition to the Arctic (North Pole)—Summer-2019 logistically funded by NCPOR, MOES, GoI and became the first Indian team to fly UAVs (Drones) in the North Pole to undertake remote sensing activity. She has also extensively undertaken field visits in the Himalayas and Polar Regions for ground truth verification. Besides she is also a trained mountaineer having been trained at the Himalayan Mountaineering Institute. She has completed several funded research projects successfully. Currently, she is actively carrying out research in the area of remote sensing which includes Cryosphere, earth science and disaster management studies

Dr. Ishmohan Bahuguna, Scientist/Engineer G at Space Applications Centre, Ahemedabad, India, has contributed significantly in application of Remote Sensing in Geohydrological, Geosciences and Cryospheric applications specially in monitoring of Himalayan snow and glaciers for the last 33 years. He has executed many national-level projects along with about 15 collaborating agencies and many expeditions to Himalayan glaciers for ground validation and data collection. His work on DEM generation from stereo data for finding changes in glacier dimensions in early years of this century was pioneering work in glaciology. He has trained number of professionals working in Cryosphere science

Dr. D. Krishnaveni obtained her Bachelor of Technology in Electronics and Communication from Nagarjuna University, Andhra Pradesh; Master of Technology from B.M.S. C.E., Bangalore, Visvesvaraya Technological University (V.T.U), Karnataka, India and Ph.D. degree from V.T.U. Presently, she is working as Assistant Professor in the Department of Electronics and Communication Engineering, Jyothy Institute of Technology, Bangalore, Karnataka, India. Her areas of research are reversible logic design, quantum computing Digital Design and satellite image processing. She is actively involved in research projects related to Polar Cryosphere

Dr. Suresh Devaraj, Remote Sensing expert, has received his Bachelor's degree in Civil Engineering from Vel Tech, Chennai, and Master's in Remote Sensing and GIS from the Regional Centre of Anna University, Tirunelveli. He obtained his Ph.D. in Landslides using InSAR techniques from Vellore Institute of Technology, Vellore, India. He worked as a Geomatics Analyst in kCube Consultancy Services Pvt. Ltd., Bangalore where he worked on several projects for NLC, MSSRF, TN slum clearance board, etc. He has more than 5 years of industrial and research experience and has worked on multispectral, microwave, LIDAR and UAV datasets. He is presently working on InSAR-based research activities in the Himalayan and Arctic Regions

Analysis of Snow Dynamics in Beas River Basin, Western Himalaya Using Combined Terra–Aqua MODIS Improved Snow Product and in Situ Data During Twenty-First Century

7

Dhiraj Kumar Singh, Hemendra Singh Gusain, Sanjay Kumar Dewali, Reet Kamal Tiwari, and Ajay Kumar Taloor

Abstract

Study of snow dynamics is an essential parameter for scientific studies such as climate change, cryospheric hazard mapping, energy budget assessment, management of water assets, etc. In this paper, an analysis of snow dynamics in the Beas river basin, Western Himalaya, India has been carried out using Moderate Resolution Imaging Spectroradiometer (MODIS) satellite images and in situ data during more than a decade winter period (November–April) from 2003–2017. MODIS sensor images and 8-days composite snow products have large uncertainty in mountain regions because of cloud cover and sensor limitation. Therefore, in this paper, a combined Terra–Aqua MODIS satellite-derived improved snow product version 6 (MOYDGL06*) has been used for the estimation of snow cover area (SCA) during the years 2003–2017. SCA in the study area varied from ~19% (November, 2016) to ~98% (February, 2015) during the era. It was found that SCA and total precipitation are decreasing at the rate of 3.2 km^2 and 64.7 cm, while the mean temperature is increasing at the rate of 0.16 °C, respectively, for the period 2003–2010. However, a similar trend was found during 2010–2017, SCA and total precipitation are decreasing at the rate of 25.39 km^2 and 44.9 cm, while the mean temperature is increasing at the rate of 0.35 °C, respectively. The satellite-extracted SCA trend was in correlation with in situ observed climate parameters. Moreover, SCA variability has been explored for different winter season months. The paper highlights the decreasing SCA and total perception trend, while increasing mean temperature trend during the twenty-first century and indicates that climate change is probably one of the major factors.

Keyword

Beas river basin · Climate change · MODIS · Snow product · Snow dynamics

D. K. Singh (✉) · H. S. Gusain · S. K. Dewali
Snow and Avalanche Study Establishment- RDC (DRDO), Chandigarh 160037, India
e-mail: erdhiraj.singh@gmail.com

H. S. Gusain
e-mail: gusain_hs@yahoo.co.in

S. K. Dewali
e-mail: deosanjay04@gmail.com

R. K. Tiwari
Indian Institute of Technology (IIT) Ropar, Rupnagar 140001, India
e-mail: reetkamal@iitrpr.ac.in

A. K. Taloor
Department of Remote Sensing & GIS, University of Jammu, Jammu 180006, India
e-mail: ajaytaloor@gmail.com

A. K. Taloor et al. (eds.), *Water, Cryosphere, and Climate Change in the Himalayas*,
Geography of the Physical Environment, https://doi.org/10.1007/978-3-030-67932-3_7

7.1 Introduction

During the last two decades, climate change and global warming over different areas of the cryospheric communities have upraised the concern about the snow cover dynamics and glaciers reduction (Negi et al. 2018). Spatial and temporal snow dynamics over a period is an indicator of fluctuations in the environment and acts as a pointer for climate transformation. Various researchers (Yaning et al. 2008; Brown et al. 2011; Saavedra et al. 2016; Azmat et al. 2016; Shukla et al. 2017; Negi et al. 2018; Shafiq et al. 2018; Singh et al. 2018; Bisht et al. 2019; Taloor et al. 2019; Kumar et al. 2020; Taloor et al. 2020) across the cryospheric world have mentioned the linkage between the snow cover area (SCA) and climatic parameters variation as a pointer of environment change. The seasonal SCA is also crucial for understanding the solar energy balance, water cycles supervision, Earth's hydrology, etc. as these processes control the deviation in regional and global climate (Frei et al. 1999; Salomonson and Appel 2004; Arndt et al. 2010; Gurung et al. 2011; Sharma et al. 2012, 2014; Negi et al. 2018; Singh et al. 2020; Sood et al. 2020).

Himalayas are the largest reservoir of snow, ice, and glaciers after the polar region. Himalayan SCA directly affects the Indian state's economy and development across society (Kulkarni et al. 2007). It contributes ~22% of hydropower and ~17% of agriculture in the term of gross domestic product (GDP) (Sharma et al. 2014). However, due to the high altitude, roughness, and inaccessibility of Himalayan regions, it is very tough to gather information on SCA using traditional measurement techniques (Bisht et al. 2018). Therefore, satellite-based imaging techniques provide key to monitor regular SCA variations and to measure snow cover parameters.

Over a half era, passive optical satellite images have been widely used for continuous monitoring of snow due wide coverage (Hüsler et al. 2014). However, it has limitations to collect the snow cover information on a regular basis on cloudy days (Singh et al. 2019). To overcome this, 8-day Moderate Resolution Imaging Spectroradiometer (MODIS) composite snow cover product was generated to reduce the cloud cover over snow-covered region (Hall et al. 2002). The 8-day product significantly decreases the amount of cloud cover; however, during monsoon and winter seasons, some clouds exist (Liang et al. 2008). The existence of clouds over snow and obscuration of old snow and glacier ice due to their low albedo cause the underestimation of the snow and ice cover extent and must be discarded (Wang et al. 2008). Low spatial resolution and large solar zenith angle in the term of contrast primarily cause overestimation of snow and ice (Li et al. 2016; Huang et al. 2017; Hou et al. 2019). Further, limitation of MODIS senor images reduces the accuracy of snow cover and consistently poor in the evergreen forests and the early melt season (Hall and Riggs 2007). Muhammad and Thapa (2020) have estimated the uncertainty in Terra and Aqua MODIS 8-d composite C6 products for high-mountains of Asia from 2002 to 2018. They observed MODIS products comprising 46% overestimation and 3.66% underestimation, mainly caused by sensor limitations and cloud cover, respectively.

During the past decade, remote sensing images and products have been used to study snow cover variability of Western Himalayan cryosphere (Jain et al. 2009; Kulkarni et al. 2010; Gurung et al. 2011; Sharma et al. 2012, 2014; Kour et al. 2015; Barman et al. 2015; Birajdar et al. 2016; Gurung et al. 2017; Negi et al. 2018; Shafiq et al. 2018; Singh et al. 2018). Several scientific communities and researchers (Vikhamar and Solberg 2003; Cherry et al. 2005; Lemke et al. 2007; Jain et al. 2008; Muntán et al. 2009; Zhao et al. 2009; Sharma et al. 2014; Kour et al. 2015) have conveyed negative trend in Himalayan regions seasonal SCA, mainly in the spring season. Sharma et al. (2014) have described a negative SCA trend in Indian Himalayan basins, i.e., Bhaga, Chenab, Baspa, and Beas river basin. A similar SCA trend has also been published by several investigators (Kulkarni et al. 2007, 2010; Sharma et al. 2012; Kour et al.

2015) for different river basins using MODIS images, 8-day product and 10-day product. A positive SCA trend in Indian Himalaya and the Karakoram has been published by Tahir et al. (2015) using MODIS 8-day snow cover product. Shafiq et al. (2018) have calculated snow cover variability in Kashmir Himalayas, India using 8-day MODIS snow cover product from 2000 to 2016. They observed a positive trend in SCA and a negative trend in temperature during the data period. Singh et al. (2018) have estimated snow cover variability in Indian Himalayan climate zones using MODIS senor images generated 10-day snow cover product during 2001–16. They observed a shift in SCA trends after 2010 in Indian Himalaya and climatic zones and a slowdown in snow/ice cover reduction during the recent years. Due to large topographical variability and difference in studies period, Himalayan regions are showing different SCA trends as observed from the literature. However, in the Indian Himalayan parts, despite its importance, very few studies have been reported. The above-mentioned studies have limitations in the term of accuracy due to cloud cover and sensor limitations.

Muhammad and Thapa (2020) have reduced the uncertainty in snow cover product by combining Terra–Aqua MODIS snow cover products (MOYDGL06*) using a seasonal, temporal, and spatial filter. Authors compared this product with Landsat-8 data and observed improvement in accuracy by 10%. Keeping in view of the above studies, the aim of the present study is to estimate snow cover variability using combined Terra–Aqua MODIS (MOYDGL06*) improved snow product. The objective of this paper is to investigate the long-term monthly, and annual variability in snow dynamics in the terms of SCA and climatic variables (i.e., maximum and minimum temperature and total precipitation) for Beas River basin, India during 2003–2017. The non-parametric Mann–Kendall and Sen's slope and t-test have been used to estimate and quantify the trend.

7.2 Study Area

The present study is focused on the Beas River basin, India which is a part of the Indus River system that lies in the Lower Indian Himalayan zone and is shown in Fig. 7.1. It originates from a small glacier named as 'Beas Kund' at an elevation of 3900 m above the mean sea level (amsl) in the eastern slope of Rohtang pass, Himachal Pradesh, India. The study area is very popular across the world for its charming beauty, recreation activities, and ski slopes. Millions of holidaymakers visit this area for tracking, skiing, mountaineering, etc., every year. It provides water for irrigation, drinking, power generation, etc., to Himachal Pradesh and Punjab. The mean elevation and slope of the study area are 3517 mamsl and 30°, respectively. During the winter period, it receives heavy snowfall at above 2000 m amsl. Snow and Avalanche Study Establishment (SASE), Chandigarh, one of the laboratories of Defence Research and Development Organisation, India has installed an observation station named "Dhundi" in the study area for the collection of snow-met parameters at an elevation of 3050 amsl (Fig. 7.1). According to Buhler et al. (2013), the mean snowfall is 11.5 m and the mean monthly temperature varied over the past two decades from 3.1 °C to −1.2 °C during the winter season from November to April. The region links the HP with Leh (Union territory of Ladakh) of India through Atal Tunnel (formerly known as Rohtang Tunnel) and receives heavy snowfall and avalanches during the winter spell of every year.

7.3 Materials and Methods

Mapping techniques in snow-covered regions have been started regularly with the launch of Terra and Aqua MODIS satellite by the National Administrative Space Application (NASA) in 1999 and 2002, respectively. It has 36 spectral bands within the wavelength of 0.4–14.4 μm and

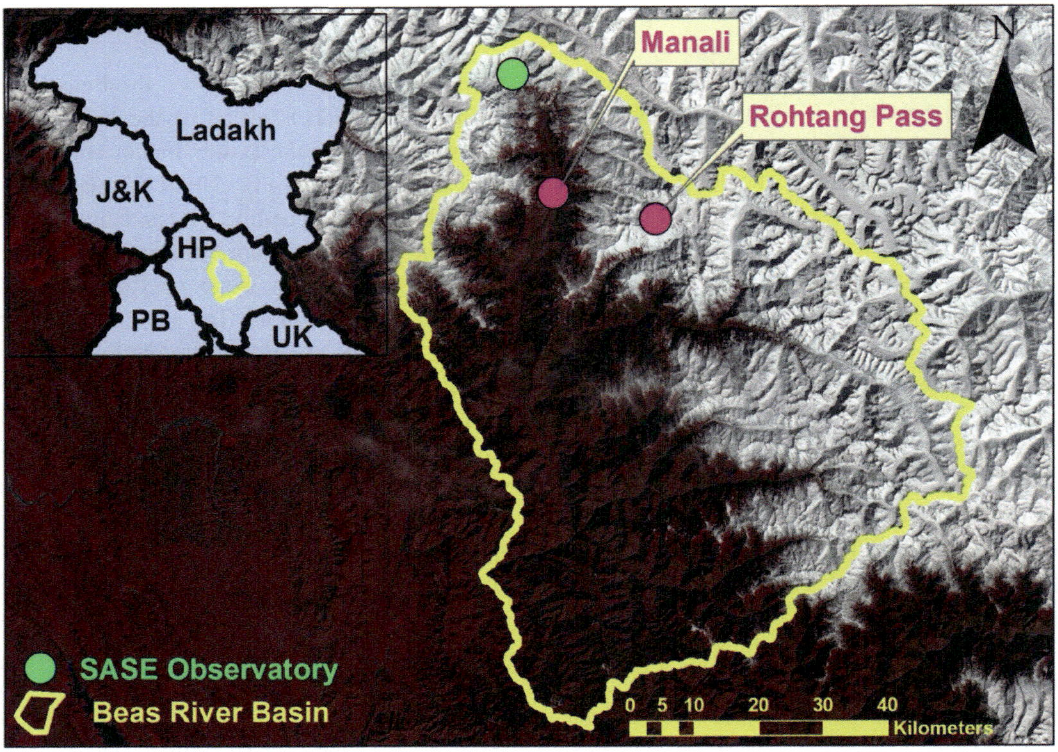

Fig. 7.1 Study area map of Beas River basin, India depicting SASE observatory named "Dhundi" and some tourist locations symbolized with pink color (i.e., Manali and Rohtang Pass) (*Source* https://earthexplorer.usgs.gov)

delivers cloud and land surface data. The MODIS daily, 8-day, and monthly products are available at 500 m to 5 km spatial resolution. 8-day composite snow cover product was generated to remove the presence of cloud cover by the integration of consecutive 8-day of MODIS images (Hall et al. 2002). In 8-day snow product, a significant amount of clouds remain which cause underestimation of SCA and must be replaced by snow pixel (Yu et al. 2016). Moreover, the overestimation was reduced by combining the Terra–Aqua MODIS snow product to calculate snow more accurately.

In this paper, an improved Terra–Aqua MODIS 8-day snow-covered product has been used to estimate the SCA of the present study during the winter period (November to April) of 2003–17 and can be download from https://doi.pangaea.de/10.1594/PANGAEA.901821 website. The product (MOYDGL06*) was developed by combining and merging Terra–Aqua 8-day MODIS (MOD10A2.006* and MYD10A2.006*) snow product version 6 with Randolph Glacier Inventory (RGI6.0). This data is specifically generated for High Mountain Asia (HMA) at 500 m spatial resolution with 8-day temporal resolution. In this product, the cloud elimination process consists of different steps and filters (seasonal filtering, temporal filtering, spatial filter, merging Terra- and Aqua-filtered to generated snow product, and integrating glaciers (RGI 6.0) in the improved snow product) and detailed methodology is described by Muhammad and Thapa (2020). The improved final snow product consist of different flags for pixels changed from non-snow to snow or the new way around. The pixels flag values were categorized into six classes, 0 for non-snow; 200 if both original and product have snow pixel; −200 if pixel changes from snow to non-snow in the product; 210 if the non-snow pixel is converted to snow due to cloud cover; and 240 and

250 for pixel covered with debris and debris-free ice, respectively. Further, the snow pixels were extracted from snow products to generate snow cover maps and analysis. The binary snow cover map consists of snow-covered area and non-snow area.

7.3.1 Meteorological Data

In the present study, daily snow-meteorological data of the winter season (November-April) for the period of 2003–2017 (14 years) collated from SASE observatory named “Dhundi” were examined to estimate the temporal variation of climatic parameters. The geographical location of meteorological observatory station is shown in Fig. 7.1. The climatic parameters considered in this study were temperature (maximum, minimum and mean) and precipitation. The mean temperature is calculated by averaging the maximum and minimum temperature values. Further, the statistical analysis was carried out in climatic data and satellite-derived SCA during the winter season of the data period for analyzing the trend. The monthly distribution of mean temperature and total precipitation during the study period is presented in Fig. 7.2.

7.3.2 Statistical Analysis: Mann–Kendall Test

Statistical significance of SCA and climatic parameters trend was analyzed using Mann–Kendall test (MK test). It is used to examine the trend in time series non-parametric data (Mann 1945; Kendall 1975). In this test, no trend in series data is denoted by the null hypothesis (H_o), while the trend in series data is denoted by alternative hypothesis (H_1) and used to test the null hypothesis. The MK test statistics (Zs) is calculated using Eq. (1),

$$\text{if } |Zs| > Z_{\alpha/2} \tag{1}$$

where α denotes the significance level, if $\alpha = 5\%$ then $Z_{0.025} = 1.96$, at that time, the null hypothesis trend is significant. This test can be applied to the non-normally distribution series of data.

Similarly, Sen’s (1968) has developed a non-parametric technique to calculate the slope of the liner trend using Eq. (2),

$$Q_i = median\left(\frac{X_j - Xi}{t_j - t_j}\right) \tag{2}$$

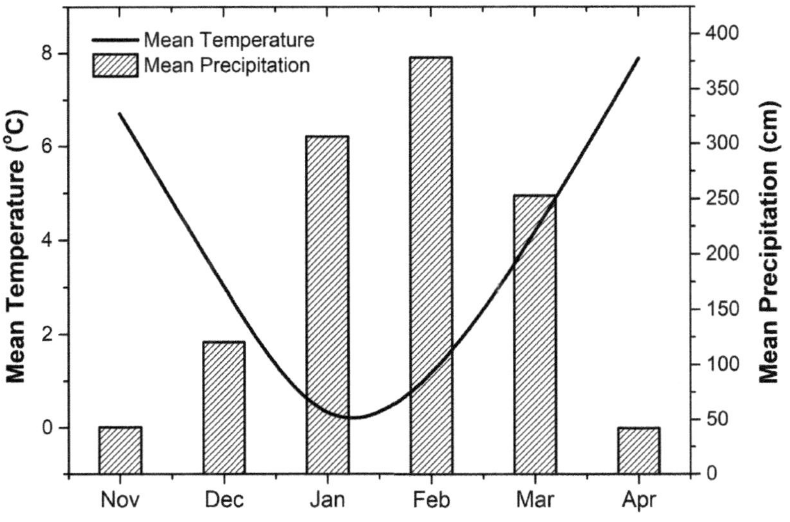

Fig. 7.2 Monthly distribution of mean temperature and precipitation during the winter season of years 2003–2017 (*Source* Snow and Avalanche Study Establishment Chandigarh)

where Q_i denotes the value of Sen's slope for i = 1, 2, 3…N, X_j *and* Xi are the data values at the time t_j and t_j, respectively. In this paper, a positive (negative) value of Sen's slope represents the increasing (decreasing) trend in data at 95% of confidence level.

7.4 Results and Discussion

Figure 7.3 shows the intra and inter temporal annual variation of SCA in the Beas River basin, India during 2003–17. The mean annual SCA during the winter period (November--April) was estimated from 8-day improved snow cover product. The total areal extent of the study area is 5384 km^2 and shown in Fig. 7.1. The SCA changes from month to month due to large variability in snow ablation and widespread snowfall pattern. The highest and lowest SCA has been observed in January/February and November months during the winter season. Snow cover in the study area varied from 20% to 98% and mean SCA from 3,657 km^2 (2003–04) to 4,385 km^2 (2015–16) during the data period. There is a large temporal variability in maximum snow cover extent in each month due to high patchiness in snowfall pattern caused by western disturbance and local topography.

Figure 7.4 shows monthly average SCA maps of the Beas River basin during the winter period of 2003–2017. The maximum and minimum SCA was observed in February (~98%) and April (~78%) months. In the study area, generally October onward, snowfall starts and about ~94% of SCA was observed in month November. January, February, and March months receive the highest snowfall due to western disturbance and results in maximum SCA of about ~98%, ~96%, and ~96%, respectively. However, from April month onward, SCA starts decreasing because of the rise in ambient temperature and ablation.

Figure 7.5a, b shows the temporal variation of mean SCA, mean temperature, and total mean precipitation for the period 2003–2010 and 2010–2017, respectively, in the Beas River basin. A decreasing trend in mean SCA and total mean precipitation was found, while increasing mean temperature was found during the period 2003–2010 and 2010–2017, respectively. The trend was analyzed using t-test at 95% confidence level ($\alpha = 0.05$) and was found statistically insignificant during the data period. It was found that SCA and total precipitation are decreasing at the rate of 3.2 km^2 and 64.7 cm, while the mean temperature is increasing at the rate of 0.16 °C, respectively, for the period 2003–2010.

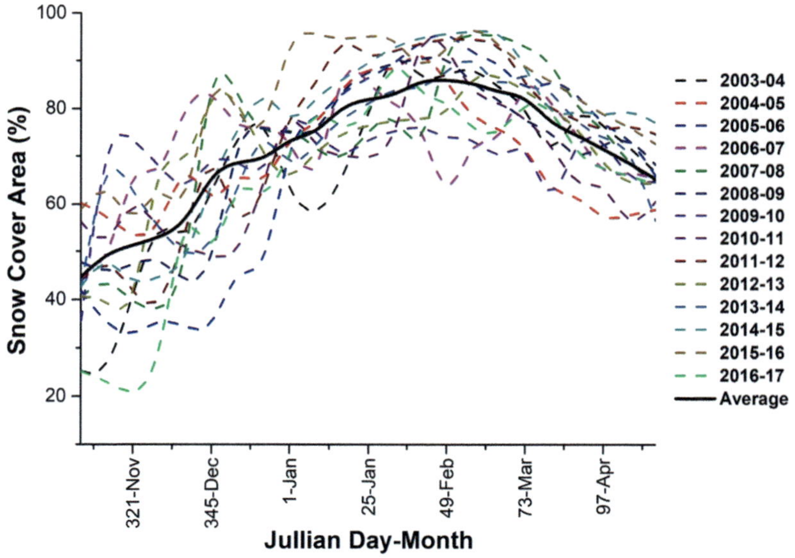

Fig. 7.3 Intra and inter temporal annual variation of SCA in Beas River basin, India during the winter period (November–April) of 2003–2017 (*Source* https://earthexplorer.usgs.gov)

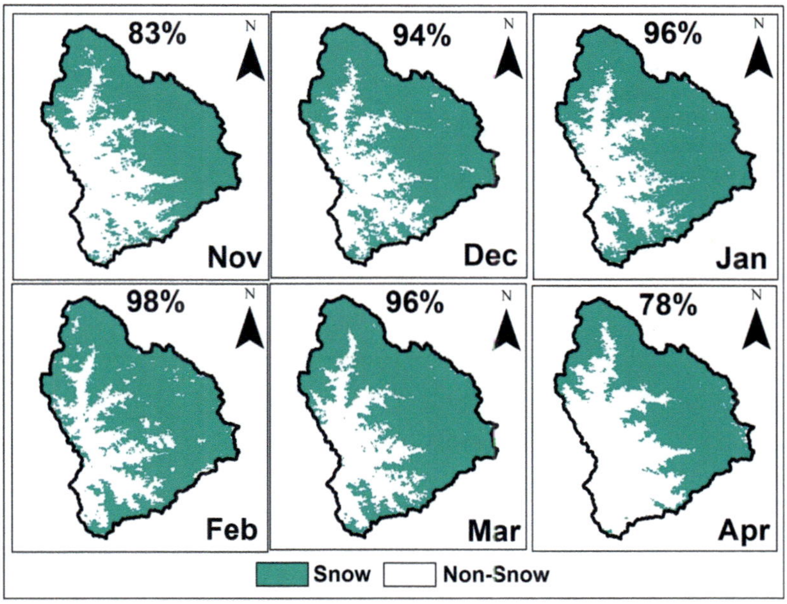

Fig. 7.4 Monthly average SCA maps in Beas River basin during the winter period of 2003–2017 (*Source* https://earthexplorer.usgs.gov)

However, a similar trend was found during 2010–2017, SCA and total precipitation are decreasing at the rate of 25.39 km^2 and 44.9 cm, while the mean temperature is increasing at the rate of 0.35 °C, respectively. The satellite-extracted SCA trend was in correlation with in situ observed climate parameters. The mean temperature in the study area varied from 2.1 (2011–12) to 4.8 °C (2016–2017). The highest mean total precipitation in the study area was observed in the year 2004–2005 of about ~1484 cm. It may be due to widespread snowfall over Western Himalaya in the year 2004–2005 reported by the Ministry of Home Affairs, Government of India and many researchers such as Kour et al. (2015) and Singh et al. (2010 Fig. 7.6a–d shows a monthly variation of SCA, maximum and minimum temperature, and total precipitation during the winter data period. Highest SCA and total precipitation were observed in January, February, and March months of every year, and from April onward, SCA and total precipitation start decreasing due to an increase in ambient temperature in the region. A sharp increase in SCA from January to February in the study area is due to frequent widespread snowfall. This will increase the total precipitation amount and decrease the maximum and minimum temperature (Fig. 7.6b–d). The highest value of SCA, maximum and minimum temperature, and total precipitation in the Beas River basin were found to be 96% in January month, 13.6 °C in March month, 2.1 °C in March month, and 681 cm in February, respectively. However, the lowest value was observed to be 23% in November month, 0.3 °C in January month, −9.3 °C in January month, and 1 cm in December month for SCA, maximum temperature, minimum temperature, and total precipitation, respectively.

The Mann–Kendal and Sen's slope test has been performed to estimate the trend in SCA, maximum and minimum temperature, and total precipitation for the Beas River basin at 95% confidence level. Table 7.1 summarizes the statistical trend analysis of SCA and climatic parameters for different months of the winter period. An increasing trend in SCA has been observed from November to December month, excepy February month in which SCA is decreasing. Although the trends observed were statistically insignificant in SCA during the winter season, an increasing trend in maximum temperature was observed from November to January month and April month, while a decreasing trend was observed in February to

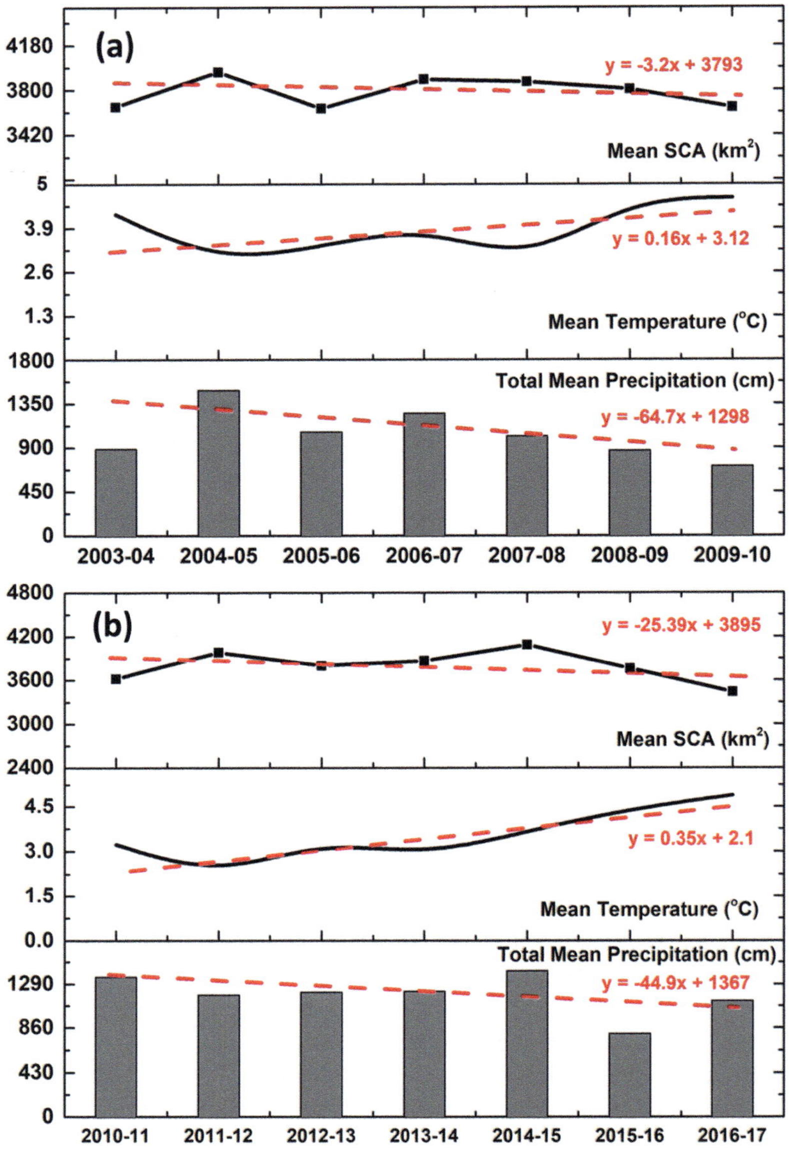

Fig. 7.5 Temporal variation of mean SCA, mean temperature, and total mean precipitation during the period **a** 2003–2010 and **b** 2010–2017 in Beas River basin. Straight dotted line indicates a linear trend in the data (*Source* Snow and Avalanche Study Establishment Chandigarh)

March month. Although the trends observed were statistically insignificant; however, a significant trend was observed in November month at the rate of 0.27 °C. An increasing trend in minimum temperature was found from November to February month, while a decreasing trend from March to April month. Although the trends observed were statistically insignificant in SCA during the winter season, a decreasing trend in total precipitation from November to December has been observed, while an increasing trend in January to April month.

7.5 Conclusions

In the present study, SCA has been estimated using combined Terra–Aqua MODIS satellite-derived snow product version 6 (MOYDGL06*) in Beas river basin, Western Himalaya, India from 2003 to 2017. SCA was estimated at 8-day interval using MODIS snow cover product. The total areal extent of the study area is 5384 km^2. Large intra and inter temporal annual variation was found in SCA over the Beas River basin,

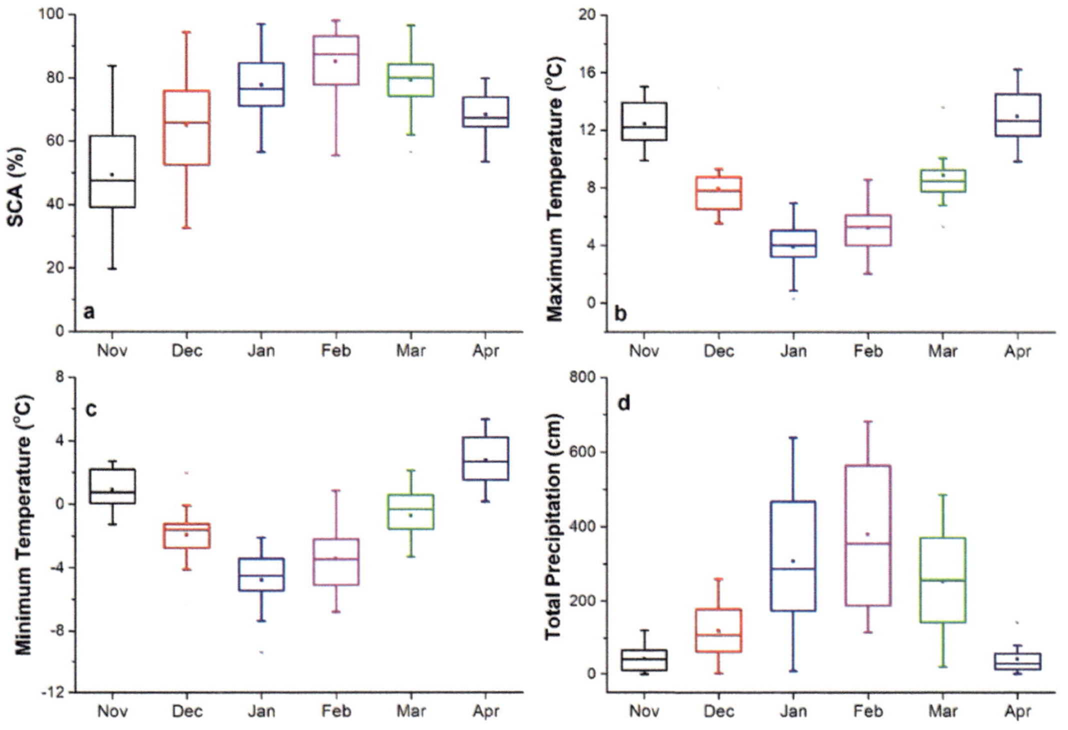

Fig. 7.6 Box-charts denoting winter season monthly variation of **a** SCA, **b** maximum temperature, **c** minimum temperature, and **d** total precipitation during data period (*Source* Snow and Avalanche Study Establishment Chandigarh)

India during the data period. The SCA changes from month to month due to large variability in snow ablation and widespread snowfall pattern. The highest and lowest SCA was found in January/February and November months during the winter period. Snow cover in the study area varied from 20% to 98% and mean SCA from 3,657 km^2 (2003–04) to 4,385 km^2 (2015–16) during the data period. The maximum and minimum SCA was found in February (~98%) and April (~78%) months.

In the study area, generally October onward, snowfall starts and about ~94% of SCA was observed in month November. January, February, and March months receive the highest snowfall due to western disturbance and results in maximum SCA of about ~98%, ~96%, and ~96%, respectively. However, from April month onward, SCA starts decreasing because of the rise in ambient temperature and ablation. It was found that SCA and total precipitation are decreasing at the rate of 3.2 km^2, and 64.7 cm, while the mean temperature is increasing at the rate of 0.16 °C, respectively, for the period 2003–2010. However, a similar trend was found during 2010–2017, SCA and total precipitation are decreasing at the rate of 25.39 km^2 and 44.9 cm, while the mean temperature is increasing at the rate of 0.35 °C, respectively. The satellite-extracted SCA trend was in correlation with in situ observed total precipitation.

Additionally, MK and Sen's slope test was also used for statistical trend analysis of SCA and climatic parameters for different months of the winter period. An increasing trend in SCA has been observed from November to December month, except February month in which SCA is decreasing. Although the trends observed were statistically insignificant in SCA during the winter season, an increasing trend was observed in maximum temperature from November to January month and April month, while a

Table 7.1 Statistical trend analysis of SCA, maximum and minimum temperature, and total precipitation during the winter period of 2003–2017 in Beas River basin, India

Parameters	November	December	January	February	March	April
SCA (%/year)						
Zs	1.13*	0.41*	0.05*	−0.32*	0.40*	0.32*
Qs	0.83	0.29	0.09	−0.29	0.12	0.05
Trend	Increasing	Increasing	Increasing	Decreasing	Increasing	Decreasing
Maximum Temperature (°C/year)						
Zs	**1.86****	1.42*	0.44*	0.05*	1.42*	0.16*
Qs	0.27	0.16	0.07	0.02	0.14	0.04
Trend	Increasing	Increasing	Increasing	Decreasing	Decreasing	Increasing
Minimum Temperature (°C/year)						
Zs	0.99*	0.71*	0.60*	0.88*	1.31*	1.09*
Qs	0.09	0.08	0.1	0.14	0.15	0.13
Trend	Increasing	Increasing	Increasing	Increasing	Decreasing	Decreasing
Total Precipitation (cm/year)						
Zs	0.67*	0.60*	0.1*	0.1*	0.79*	0.16*
Qs	1.7	5.5	3.5	0.92	6.56	1.28
Trend	Decreasing	Decreasing	Increasing	Increasing	Increasing	Increasing

Note Zs = Mann–Kendall test statistic Z; Qs = Sen's slope; all the trends are tested at 95% level; **Significant; *Insignificant
(*Source* Snow and Avalanche Study Establishment Chandigarh)

decreasing trend was observed from February to March month. Although the trends observed were statistically insignificant, however a significant trend was observed in November month at the rate of 0.27 °C. An increasing trend in minimum temperature was found from November to February month, while a decreasing trend from March to April month. Although the trends observed were statistically insignificant in SCA during the winter season, a decreasing trend in total precipitation from November to December has been observed, while an increasing trend from January to April month.

The present paper explored the annual and monthly variability in SCA, temperature, and total precipitation in the Beas River basin and highlights an increasing trend in the study area during recent years. Decreasing SCA and total precipitation, while increasing temperature during recent years, indicates a bad signal climate. Additionally, snow cover variation has also been explored during different winter seasons. The study may be useful in snow avalanche mapping and assessment, hydrology and water resources management, and climate change study.

Acknowledgements The authors are grateful to Director, Snow and Avalanche Study Establishment (SASE) for providing facilities to carry out this work. We are also thankful to all SASE's persons responsible for snow-meteorological data collection and management. The MODIS improved snow cover product (MOYDGL06*) is provided by https://doi.pangaea.de/10.1594/ PANGAEA.901821 and we duly acknowledged this website.

References

Arndt DS, Baringer MO, Johnson MR (2010) State of the climate in 2009. Am MeteorolSoc 91:S1–S222

Azmat M, Liaqat U W, Qamar M U Awan UK (2017) Impacts of changing climate and snow cover on the flow regime of Jhelum River, Western Himalayas. Reg Environ Change 7(3):813–825. https://doi.org/10.1007/s10113-016-1072-6

Barman S, Bhattacharjya RK (2015) Change in snow cover area of Brahmaputra river basin and its sensitivity to temperature. Environ Sys Res 4(1):1–10. https://doi.org/10.1186/s40068-015-0043-0

Birajdar F, Venkataraman G, Samant H (2016) Monitoring Snow Cover Area Using Different Algorithms on Indian Remote Sensing Data. Geostatistical and Geospatial Approaches for the Characterization of Natural Resources in the Environment, pp 749–753. https://doi.org/10.1007/978-3-319-18663-4_115

Bisht H, Arya PC, Kumar K (2018) Hydro-chemical analysis and ionic flux of meltwater runoff from Khangri Glacier, West Kameng, Arunachal Himalaya India. Environ Earth Sci 77:1–16. https://doi.org/10.1007/s12665-018-7779-6

Bisht H, Rani M, Kumar K, Sah S, Arya PC (2019) Retreating rate of Chaturangi glacier, Garhwal Himalaya, India derived from kinematic GPS survey and satellite data. Curr Sci 116:304–311. https://doi.org/10.18520/cs/v116/i2/304-311

Brown RD, Robinson DA (2011) Northern hemisphere spring snow cover variability and change over 1922–2010 including an assessment of uncertainty. The Cryosphere 5(1):219–229

Bühler Y, Kumar S, Veitinger J, Christen M, Stoffel A, Snehmani (2013) Automated identification of potential snow avalanche release areas based on digital elevation models. Nat Haz Earth Sys Sci. 13(5):1321–1335. https://doi.org/10.5194/nhess-13-1321-2013

Cherry J, Cullen H, Visbeck M, Small A, Uvo C (2005) Impacts of the North Atlantic Oscillation on Scandinavian hydropower production and energy markets. Water Resour Manage 19:673–691. https://doi.org/10.1007/s11269-005-3279-z

Frei A, Robinson DA (1999) Northern Hemisphere snow extent: regional variability 1972–1994. Int J Climatol 19(4):1525–1560

Gurung DR, Kulkarni AV, Giriraj A, Aung KS, Shrestha B, Srinivasan J (2011) Changes in seasonal snow cover in Hindu Kush-Himalayan region. Cryosphere Discuss 5:755–777

Gurung D R, Maharjan S B, Shrestha A B, Shrestha M S, Bajracharya S R, Murthy M S R (2017) Climate and topographic controls on snow cover dynamics in the Hindu Kush Himalaya. Int J Climato. 37(10):3873–3882. https://doi.org/10.1002/joc.4961

Hall D K, Riggs G A (2007) Accuracy assessment of the MODIS snow products. Hydrol Proces Int J21 (12):1534–47. https://doi.org/10.1002/hyp.6715

Hall D K, Riggs G A, Salomonson V V, DiGirolamo N E, Bayr K J (2002) MODIS snow-cover products. Remote Sens Environ 83(1–2):181–194. https://doi.org/10.1016/S0034-4257(02)00095-0

Hou J, Huang C, Zhang Y, Guo J, Gu J (2019) Gap-Filling of MODIS fractional snow cover products via non-local spatio-temporal filtering based on machine learning techniques. Remote Sens 11(1):90. https://doi.org/10.3390/rs11010090

Huang X, Deng J, Wang W, Feng Q, Liang T (2017) Impact of climate and elevation on snow cover using integrated remote sensing snow products in Tibetan Plateau. Remote Sens Environ 190:274–288. https://doi.org/10.1016/j.rse.2016.12.028

Hüsler F, Jonas T, Riffler M, Musial J P, Wunderle S (2014) A satellite-based snow cover climatology (1985–2011) for the European Alps derived from AVHRR data. Cryosphere 8(1):73–90. https://doi.org/10.5194/tc-8-73-2014

Jain SK, Goswami A, Saraf AK (2008) Accuracy assessment of MODIS, NOAA, and IRS data in snow cover mapping under Himalayan conditions. Int J Remote Sens 29:5863–5878

Jain SK, Goswami A, Saraf AK (2009) Role of Elevation and Aspect in Snow Distribution in Western Himalaya. Water Resou Manag 237:1–83

Kendall MG (1975) Rank correlation methods. Charles Griffin, London

Kour R, Patel N, Krishna AP (2015) Effects of Terrain Attributes on Snow Cover Dynamics in Parts of Chenab Basin, Western Himalayas. Hydrol Sci J 61 (10):1861–1876. https://doi.org/10.1080/02626667.2015.1052815

Kulkarni AV, Bahuguna IM, Rathore BP, Singh SK, Randhawa SS, Sood RK, Sunil D (2007) Glacial retreat in Himalaya using indian remote sensing satellite data. Curr Sci 92:69–74

Kulkarni AV, Rathore BP, Singh SK, Ajai (2010) Distribution of seasonal snow cover in central and western Himalaya. Annals Glaciol 51(54):123–128

Kumar D, Singh AK, Taloor AK, Singh DS (2020) Recessional pattern of TheluandSwetvarnglaciers between 1968 and 2019, Bhagirathi basin, Garhwal Himalaya, India. Quat Int. https://doi.org/10.1016/j.quaint.2020.05.017

Lemke P, Ren J, Alley RB, Allison I, Carrasco J, Flato G, Fujii Y, Kaser G, Mote P, Thomas RH, Zhang (2007) Observations: changes in snow, ice and frozen ground. Contribution of Working Group I to the Fourth Assessment Report of the Intergovernmental Panel on Climate Change, Cambridge, Cambridge University Press, pp 337–383

Li H, Li X, Xiao P (2016) Impact of sensor zenith angle on MOD10A1 data reliability and modification of snow cover data for the Tarim River Basin. Remote Sens 8(9):750. https://doi.org/10.3390/rs8090750

Liang TG, Huang XD, Wu CX, Liu XY, Li WL, Guo ZG, Ren JZ (2008) An application of MODIS data to snow cover monitoring in a pastoral area: a case study in Northern Xinjiang. China. Remote Sens Environ 112 (4):1514–1526. https://doi.org/10.1016/j.rse.2007.06.001

Mann HB (1945) Nonparametric tests against trend. Econometrica 13:245–259

Muhammad S, Thapa A (2020) An improved Terra-Aqua MODIS snow cover and Randolph Glacier Inventory 6.0 combined product (MOYDGL06*) for high-mountain Asia between 2002 and 2018. Earth Sys Sci Data 12(1):345–356. https://doi.org/10.5194/essd-12-345-2020

Muntán E, García C, Oller P, Martí G, García A, Gutiérrez E (2009) Reconstructing snow avalanches

in the Southeastern Pyrenees. Nat Hazards Earth SystSci 9:1599–1612
Mul S, Ahmed P, Zul I, Joshi PK, Bhat WA (2018) Snow cover area change and its relations with climatic variability in Kashmir Himalayas, India. GeocartoInt 34(6):688–702. https://doi.org/10.1080/10106049.2018.1469675
Negi HS, Shekhar MS, Gusain HS, Ganju A (2018) Winter Climate and Snow Cover Variability Over North-West Himalaya. In: Goel P, Ravindra R, Chattopadhyay S (eds) Science and Geopolitics of The White World. Springer, Cham. https://doi.org/10.1007/978-3-319-57765-4_10
Saavedra FA, Kampf SK, Steven R, Fassnacht SR, Sibold JS (2016) A snow climatology of the Andes Mountains from MODIS snow cover data. Int J Climatol 37 (3):1526–1539. https://doi.org/10.1002/joc.4795
Salomonson VV, Appel I (2004) Estimating fractional snow cover from MODIS using the normalized difference snow index. Remote SensEnviron 89:351–360
Sen PK (1968) Estimates of the regression coefficient based on Kendall's Tau. J American Statis Assoc 63 (324):1379–1389. https://doi.org/10.1080/01621459.1968.10480934
Sharma V, Mishra VD, Joshi PK (2012) Snow cover variation and stream flow simulation in a snow-fed river basin of the Northwest Himalaya. J Mountain Sci 9:853–868
Sharma V, Mishra VD, Joshi PK (2014) Topographic controls on spatio-temporal snow cover distribution in Northwest Himalaya. Int J Remote Sens 35(9):3036–3056
Shukla S, Kansal ML, Jain S K (2017) Snow cover area variability assessment in the upper part of the Satluj River Basin in India. GeocartoInt 32(11):1285–1306. https://doi.org/10.1080/10106049.2016.1206975
Singh DK, Gusain HS, Mishra V, Gupta N (2018) Snow cover variability in North-West Himalaya during last decade. Arabian J Geosci 11(19):1–12. https://doi.org/10.1007/s12517-018-3926-3
Singh KK, Kumar R, Singh D K, Negi H S, Dewali S K, Kedia J (2019) Retrieving snow cover information from AMSR-2 satellite data for North-West Himalaya, India. GeocartoInt 1–17. https://doi.org/10.1080/10106049.2019.1588394
Singh S, Sood V, Taloor AK, Prashar S, Kaur R. Qualitative and quantitative analysis of topographically derived CVA algorithms using MODIS and Landsat-8 data over Western Himalayas, India. Quat Int. https://doi.org/10.1016/j.quaint.2020.04.048
Sood V, Gusain HS, Gupta S, Taloor AK, Singh S (2020) Detection of snow/ice cover changes using subpixel-based change detection approach over Chhota-Shigri glacier, Western Himalaya, India. Quat Int. https://doi.org/10.1016/j.quaint.2020.05.016
Tahir AA, Chevallier P, Arnaud Y, Ashraf M, Bhatti MT (2015) Snow cover trend and hydrological characteristics of the Astore River basin (Western Himalayas) and its comparison to the Hunza basin (Karakoram region). Sci Total Environ 505:748–761
Taloor AK, Kotlia BS, Jasrotia AS, Kumar A, Alam A, Ali S, Kouser B, Garg PK, Kumar R, Singh AK, Singh B (2019) Tectono-climatic influence on landscape changes in the glaciatedDurung Drung basin, Zanskar Himalaya, India: A geospatial approach. Quat Int 507:262–273
Taloor AK, Kumar V, Singh VK, Singh AK, Kale RV, Sharma R, Khajuria V, Raina G, Kouser B, Chowdhary NH (2020) Land use land cover dynamics using remote sensing and GIS techniques in Western Doon Valley, Uttarakhand, India. In: Geoecology of Landscape Dynamics 2020 (pp. 37–51). Springer, Singapore. https://doi.org/10.1007/978-981-15-2097-6_4
Vikhamar D, Solberg R (2003) Snow-cover mapping in forests by constrained linear spectral unmixing of MODIS data. Remote Sens Environ 88:309–323. https://doi.org/10.1016/j.rse.2003.06.004
Wang X, Xie H, Liang T (2008) Evaluation of MODIS snow cover and cloud mask and its application in Northern Xinjiang, China. Remote Sens Environ 112(4): 1497–1513. https://doi.org/10.1016/j.rse.2007.05.016
Yaning C, Changchun XU, Yapeng C, Li Z, Zhonghe P (2008) Response of snow cover to climate change in the Periphery Mountains of Tarim river basin, China, over the past four decades. Annals Glaciol 49:166–172
Yu J, Zhang G, Yao T, Xie H, Zhang H, Ke C, Yao R (2016) Developing daily cloud-free snow composite products from MODIS terra-aqua and IMS for the tibetan plateau. IEEE Transc Geosci Remote Sens 54 (4):2171–2180. https://doi.org/10.1109/TGRS.2015.2496950
Zhao Q, Liu Z, Ye B, Qin Y, Wei Z, Fang S (2009) A snowmelt runoff forecasting model coupling WRF and DHSVM. Hydrol Earth Syst Sci 13:1897–1906

Dr. Dhiraj Kumar Singh received the Ph.D. degree sponsored by Defence Research and Development Organization (DRDO), Chandigarh in Image and Signal Processing (Electronics and Communication Engineering) from Punjab Engineering College (Deemed to be University), Chandigarh, India, in 2020. He received the M.Tech. and B.Tech. in Electronics and Communication Engineering from the Punjab Technical University, India, in 2010 and 2012, respectively. Since 2012, he has more than 8 years of experience in teaching and R & D. Currently, he is working as Research Associate at Snow and Avalanche Study Establishment, DRDO, Chandigarh, India. He has contributed in different projects sponsored by DRDO and Department Science and Technology, India. His research interest includes satellite sensors calibration, remote sensing and GIS, and digital image and signal processing. He is the author and reviewer of many peer-reviewed book chapters and journal articles

Sh. Sanjay Kumar Dewali received M. Tech. in Cold Region Science and Engineering from Govind Ballabh Pant University of Agriculture and Technology, Pantnagar (Uttarakhand). He joined Snow and Avalanche Study establishment (DRDO) as Scientist–B in 2001 and has been actively working in the various DRDO projects in various capacities. His research areas include spatial decision support systems for cryospheric hazard management, development of GIS-enabled system for avalanche area assessment and mapping, development of integrated avalanche warning and navigation system with satellite link, avalanche hazard modelling and multi-scale terrain analysis, application of UAV and LiDAR remote sensing for operational snow information mapping and monitoring, OBIA of multi-source high-resolution images for feature extraction and mapping and remote sensing of snow and avalanches and digital terrain analysis. He has published 25 research papers in various national and international journals and developed many GIS and remote-sensing-based applications for operational use

Dr. Hemendra Singh Gusain received the Ph.D. in Geomatics Engineering (Civil Engineering) from Indian Institute of Technology (IIT) Roorkee, Uttarakhand, India, in 2015. He received the B.S. and M.Sc. degrees in Physics from HNB Garhwal University, Garhwal, India in 1994 and 1996. He also received M. Tech. in Cold Region Science and Engineering from GBPUA & T, Pantnagar, India in 2001. Currently, he is working as a Senior Scientist at Snow and Avalanche Study Establishment, Defence Research and Development Organization (DRDO), Chandigarh, India. He has more than 19 years of experience in R & D. His research interest includes optical and thermal remote sensing, cryosphere–atmosphere interaction, snow avalanche climatology and snow cover properties

Prof. Reet Kamal Tiwari received the Ph.D. degree in Earth Sciences from Indian Institute of Technology (IIT) Roorkee, Uttarakhand, India, in 2015. He received M.Sc. degree in Geology and M. Tech. degree in Remote Sensing and GIS in 2007 and 2009. In 2013, he joined Centre for Glaciology at Wadia Institute of Himalayan Geology, India as Scientist. Since March 2017, he has been Assistant Professor with Civil Engineering Department, Indian Institute of Technology, Ropar, India. He has totally 7 years of experience in teaching and R & D. His research interests include geospatial technology applications in the field of snow, ice and glaciers, climate change, natural resources management, environmental monitoring and planetary sciences

Dr. Ajay Kumar Taloor has obtained his Doctorate in Remote Sensing and GIS applications in hydrogeology from University of Jammu, NAAC accredited A^+ University of India. Thirteen years of research experience in the applications in geospatial technology for natural resources management of land and water resources. He has excelled twice with best paper presentation award in India. Being an expert of remote sensing applications in water science, cryosphere and climate change, tectonic and quaternary geomorphology, he is working on two major research projects on using space-based inputs for glacier mapping and climate change in Himalayas. He has also published many articles in tectonic and quaternary geomorphology in the recent years. He has high scientific temper and strong HR relations in science world, with high professional and managerial skills. He has edited many volumes in the top-rated journals in the Elsevier and Springer publishers, member of Editorial Board of the Quaternary Science Advances, and reviewer of the many top-rated international journals in science world

Moraine Dammed Lakes Inventory in Satluj, Ravi, Chenab and Beas Basins of Himachal Pradesh, India

8

Surjeet Singh Randhawa, Sunil Dhar, Bhanu Prakash Rathore, Rajesh Kumar, Neha Thakur, Pooja Rana, Duni Chand Rana, and Ajay Kumar Taloor

Abstract

Multispectral satellite imageries analyzed that the Himalayan region reflected the fast retreat of most of the glaciers. This further showed that especially the moraine-dammed proglacial lakes are also increasing. Those lakes are a major threat to property and life to downstream locations. Numerous research carried out in the Himalayan region under the changing climate and glacier status has indicated an alarming rise in the number and sizes of the glacial lakes that may be potentially disastrous when converts into the glacial lake outburst flood (GLOF). This study focusses on the glacial lakes mapping at the different time frame in the basins of Chenab, Ravi, Satluj and Beas of Himachal Himalaya using LISS III. The remotely sensed imageries on the GIS platform helped in preparing the moraine dammed glacial lake inventory and assessing the potentiality of becoming the GLOF in these basins. This is of prime importance for preparedness against disasters due to GLOFs. This requires a continuous observation for the pre-disaster preparedness.

Keywords

Climate change · Glaciers · Himalayas · Lake inventory · Moraine dammed lake · Remote sensing

S. S. Randhawa · N. Thakur · P. Rana · D. C. Rana
State Centre on Climate Change under the aegis of H.P. Council for Science Technology & Environment, Shimla 171002, Himachal Pradesh, India
e-mail: ssrandhawa15@gmail.com

N. Thakur
e-mail: neha33thakur44@gmail.com

P. Rana
e-mail: pooja.iirs@gmail.com

D. C. Rana
e-mail: dcrana04@yahoo.co.in

S. Dhar (✉)
Department of Environmental Science, Central University Jammu, Rahya-Suchani Baghla, District Samba, Jammu 181143, Jammu Kashmir, India
e-mail: sunildhar99@yahoo.com

B. P. Rathore
Space Applications Centre, Jodhpur Tekra, Ambawadi Vistar, Ahmedabad 380015, Gujarat, India
e-mail: rathorebp@sac.isro.gov.in

R. Kumar
Department of Environmental Science, School of Earth Sciences, Central University of Rajasthan, N.H. 8, Bandarsindri, Ajmer 305817, Rajasthan, India
e-mail: rajesh.kumar@curaj.ac.in

A. K. Taloor
Department of Remote Sensing and GIS, University of Jammu, Jammu 180006, Jammu Kashmir, India
e-mail: ajaytaloor@gmail.com

A. K. Taloor et al. (eds.), *Water, Cryosphere, and Climate Change in the Himalayas*, Geography of the Physical Environment, https://doi.org/10.1007/978-3-030-67932-3_8

8.1 Introduction

A huge natural reserve of natural resources is found in the mountain ecosystems which governs the economic well-being of the region. The degradation of the mountain ecosystem is responsible for the landslides, desertification GLOF, etc. in the mountain region. The increased rate of disturbances in the mountain ecosystem and related issues of landslides, desertification, GLOF, etc. are the major challenges due to natural and anthropogenic factors (Bhat et al. 2019; Haque et al. 2020).

The Himalayas have a glaciated area of 33000 km^2 comprising the largest freshwater reserves in the form of frozen water other than the Polar Regions (Bahuguna et al. 2003). The increased average temperature of the earth has significantly increased the rate of loss of the glaciated areas since Little Ice Age (Denton and Hughes 1981) and resulted in a loss of glacial cover and its volume. A large number of glaciers in the Himalaya are under retreat in the last few decades (Bhardwaj et al. 2016; Singh et al. 2017, 2020; Kumar et al. 2018, 2020a, b; Bisht et al. 2019; Taloor et al. 2019; Khan et al. 2020; Sood et al. 2020a; Sarkar et al. 2020). Hence, a consistent observation of the glaciers helps address the time series of the retreat of glacier surface area and the mass loss of the glaciers in the Himalaya (Bisht et al. 2019, 2020; Sood et al. 2020b). Besides, the knowledge of meltwater hydrochemistry is very important to address the issues related to the quality of water that is the lifeline for the downstream communities for their livelihood (Bisht et al. 2018).

The Satluj valley witnessed a disastrous flood in the years 2000 under the cloud bursts and induced GLOF, causing severe damage to downstream. Hence, the region requires to have continuous monitoring concerning lakes in the high mountain region with its potentiality of becoming GLOF. An attempt has been made in this work to map the glacierized areas of Chenab, Ravi, Satluj and Beas basins with the help of LISS III for the year 2013 and its physical verification to some of the region. An inventory of moraine dammed glacial lake has been prepared, which can be used as a baseline data for understanding the potential hazards in downstream.

8.2 Study Area

The investigation of the lakes has been carried out in the basin within the Satluj, Chenab, Ravi and Beas of Himachal Pradesh. The rivers Chenab, Ravi and Beas originate from Himachal Himalaya while the river Satluj originates (4500 m) near Rakas and Manasarovar lakes of Tibetan plateau and all these major rivers contribute to the Indus River. For the detailed study, the Satluj basin is divided into three sub-basins (Spiti: sub-basin-1, Lower Satluj: sub-basin-2, Upper Tibet: sub-basin-3). Similarly, the other basins have been divided as Ravi, Beas (Jiwa, Parbati), Chenab (Chandra, Bhaga and Miyar sub-basins) for preparing the lake inventory of the moraine dammed lakes (Fig. 8.1).

8.3 Satellite Data Used

The moraine-dammed glacial lakes inventory of Satluj, Ravi, Beas and Chenab has been carried out by using the satellite data of Indian Remote Sensing (IRS), LISS-3 was used for delineation of all such lakes in the different basins. The details of the satellite imageries used for the analysis have been presented in Table 8.1.

8.4 Methodology

The IRS, LISS-3 imageries were used for the identification of glacial lakes and understanding the potentiality of becoming GLOF in the basins of the study area. The Satluj basin has been covered for its entire region including the

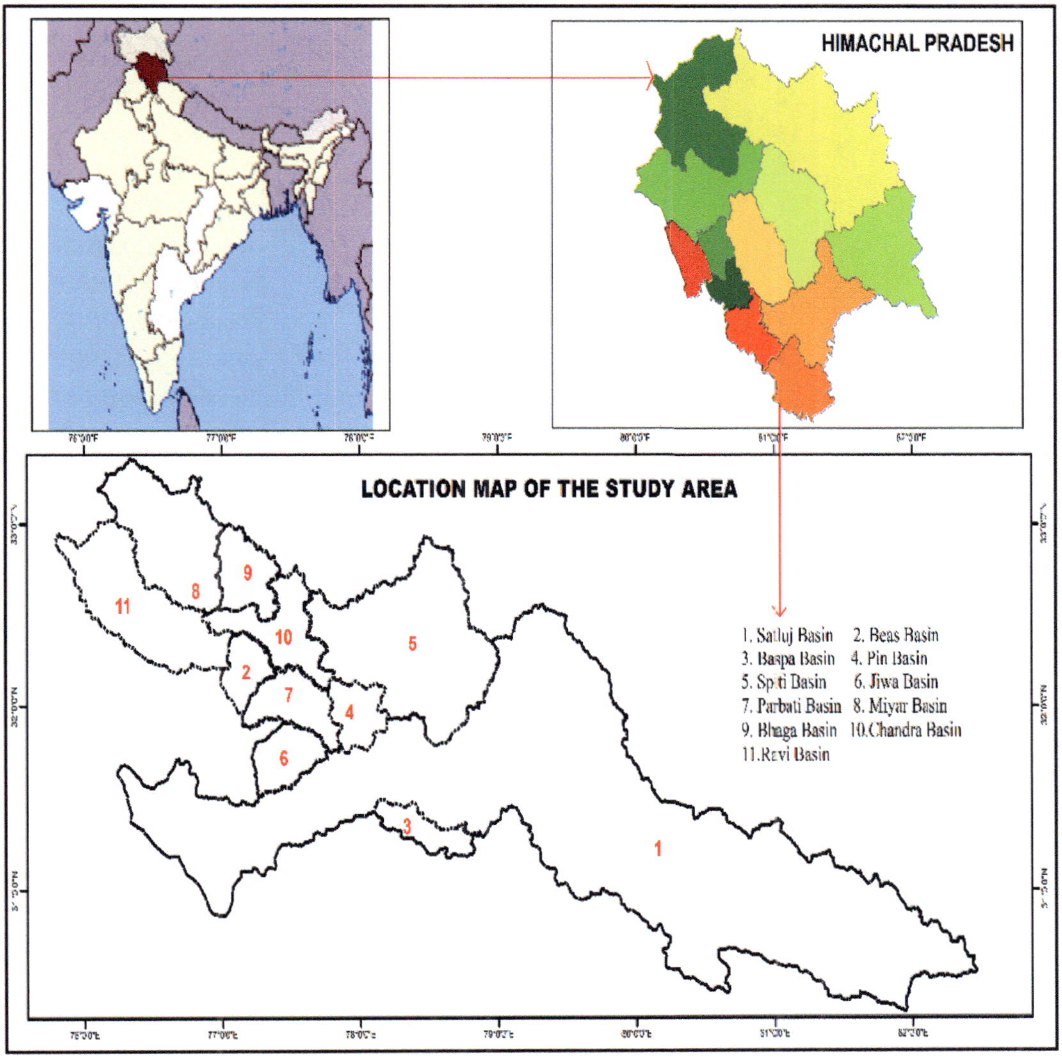

Fig. 8.1 Showing the study area of the different basins (*Source* IRS, LISS-3)

Mansarovar Lake. The better image quality having no cloud cover was used in the interpretation of the lake features. The geometric corrections were performed through a polynomial transformation of third-order. The boundary of basins was overlaid on the imageries. The boundary of lakes was delineated in the respective basins using ERDAS imagine. The deduced polygons were cleaned and polygon layers were built. For the identification, the polygons were assigned different ids for discussion of results. The process flow for the analysis of the data has been presented in Fig. 8.2.

8.5 Results and Discussions

8.5.1 Moraine Dammed Lakes Inventory of Chenab, Ravi, Satluj and Beas Basins in Himachal Pradesh by Using LISS III

Based on the visual interpretation techniques, satellite imageries analysis was carried out to generate updated information on moraine dammed glacier lakes in all the basins in Himachal

Table 8.1 The details of the satellite imageries used for lakes identification (*Source* IRS LISS-3 satellite imageries)

S. No.	Date of Pass	Path-Row	Sensor type
1	03-9-2013	96-48	Resourcesat-2/LISS-3
2	29-7-2013	96-49	Resourcesat2/LISS-3
3	20-9-2013	97-48	Resourcesat-2/LISS-3
4	20-9-2013	97-49	Resourcesat2/LISS-3
5	20-7-2012	98-48	Resourcesat-2/LISS-3
6	20-7-2012	98-49	Resourcesat2/LISS-3
7	01-7-2012	99-49	Resourcesat-2/LISS-3
8	09-10-2013	96-48	Resourcesat2/LISS-3
9	12-7-2013	94-47	Resourcesat-2/LISS-3
10	12-7-2013	96-48	Resourcesat2/LISS-3
11	31-7-2013	94-47	Resourcesat-2/LISS-3
12	29-7-2013	96-48	Resourcesat2/LISS-3
13	12-7-2013	95-48	Resourcesat-2/LISS-3
14	04-7-2013	94-48	Resourcesat2/LISS-3

(*Source* IRS, LISS–3) HWL: High Altitude Wetlands

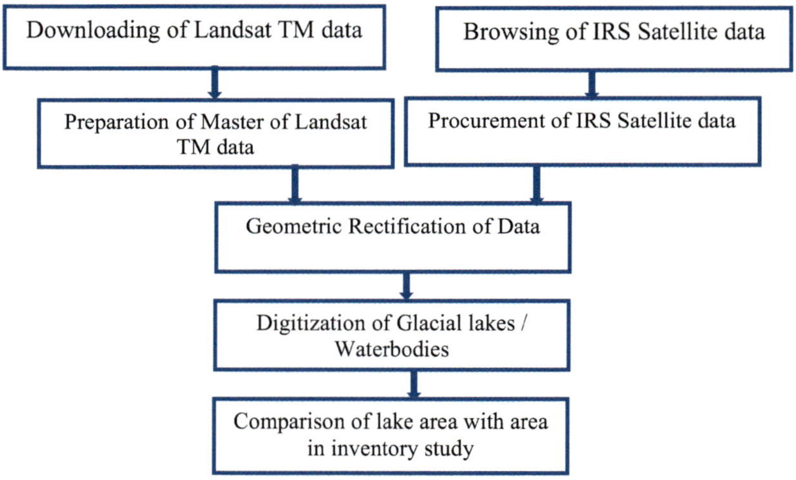

Fig. 8.2 Methodology implemented and its process flow in the interpretation of the satellite imageries (*Source* Authors)

Himalaya. All the delineated lakes have been assigned lake id's (Randhawa and Sharma 2013). The lakes have been further classified based on their aerial distribution such as lakes with area more than 10 hectares (ha), between 5–10 ha and less than 5 ha and all such lakes with area more than 5 ha have been taken into consideration for the tabulation purpose of the inventory of lakes in the present study.

8.5.2 Lakes Inventory of Chenab Basin

The Chenab basin instigates from the Baralacha region of Himachal Pradesh. Two major tributaries Chandra and Bhaga instigate from this region and flows in the reverse direction before meeting at Tandi near Keylong The Chandra, Bhaga and Miyar sub-basins has been analyzed through the satellite image of 12 July 2013, and a total of 116 lakes were identified. The number of lakes identified in the basin using satellite data for 2001 and 2003 was only 55 (Randhawa et al. 2005), suggesting an increase of 61 lakes during the last decade. The distribution of these 116 lakes further suggests that 19 lakes fall in the Chandra sub-basin, 14 in the Bhaga sub-basin, and a maximum of 83 falls in the Miyar sub-basin (Fig. 8.3).

Further analyses revealed that in the Chandra sub-basin, 2 lakes are having area more than 10 ha, another 2 lakes are within the aerial range of 5 ha to 10 ha and 15 lakes were <5 ha. Likewise, Bhaga sub-basin, the aerial distribution of 14 lakes suggests that only one lake is having area more than 10 ha, 3 lakes are within the aerial range of 5 to 10 ha, and the remaining 10 lakes are having an area less than 5 ha. Similarly, in the sub-basin having a maximum number of lakes, i.e. the Miyar Sub-basin, only 3 lakes are having an area between 5 to 10 ha, and the remaining 80 lakes are of area <5 ha. As far as the vulnerable lakes in the Chenab basin are concerned, the lakes with id 1and 3 (Table 8.2) having an area of 80.12 ha and 146.31 ha seems to be extremely susceptible, particularly the lake with ID 1 appears to be quite deep. This is attached with the snout of the glacier under which the aerial extent is gradually increasing, i.e. from 27.0 ha (1976) to 55.4 ha (2001) and 80.12 ha in 2013. Likewise, the other lake with

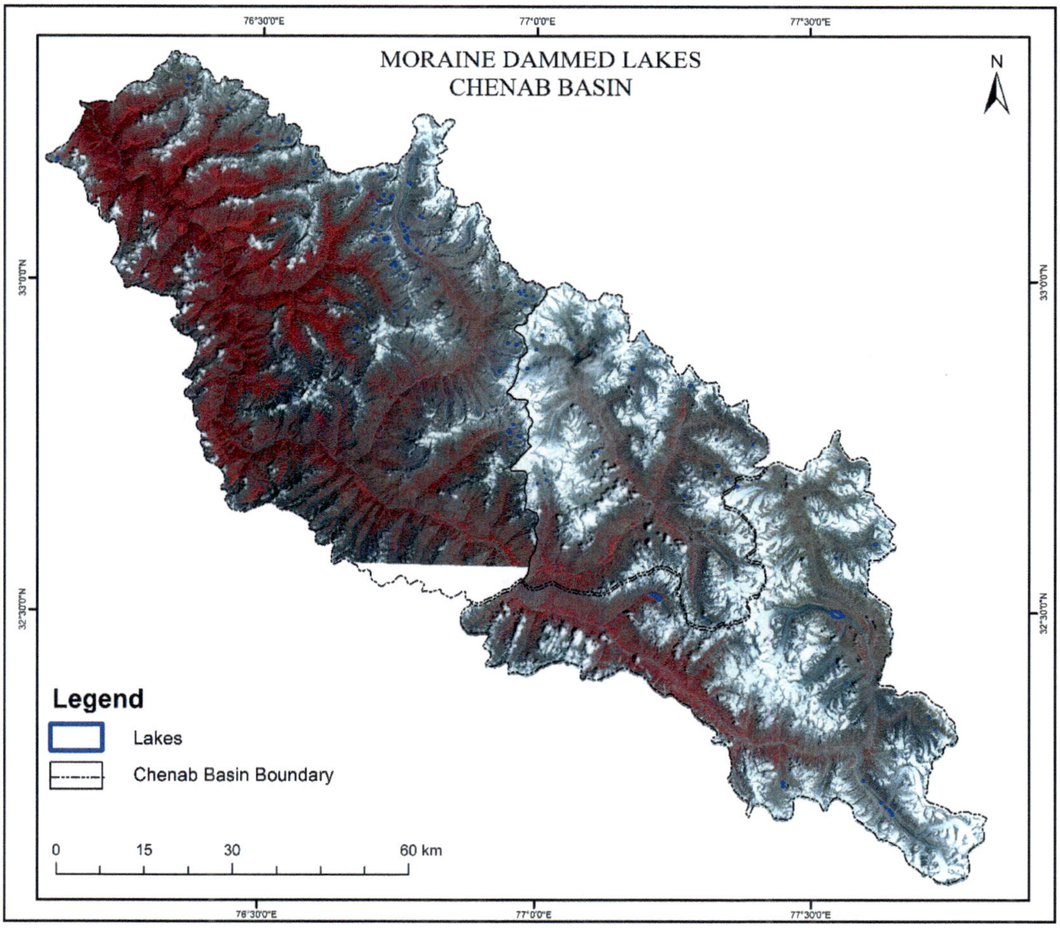

Fig. 8.3 Moraine dammed lakes in Chenab basin as delineated from IRS, LISS-3 image of 12 July 2013 (*Source* IRS, LISS-3)

ID 3 whose area has increased from 30 ha (1972) to 105.3 ha (2001) and 146.31 ha in 2013. Based on field evidence and paleo-history, a large moraine dammed lake at the snout Samundra Tapu glaciers in the Chenab basin has increased at a faster rate during the last three decades (Dhar et al. 2010). Thus, the whole Table 8.2 reflects that in the Chenab basin, 3 lakes having >10 ha area and 8 lakes with ranging from 5 to 10 ha area, due to that these warrant for regular monitoring.

8.5.3 Lakes Inventory in the Beas Basin

Beas basin instigates from the Rohtang region of Kullu district and the river emerging from this basin is named as Beas River. This basin has been studied in three parts as Jiwa, Parbati and the Beas. The Parbati River acts as the main tributary and joins the Beas River at Bhuntar in Kullu district. The analysis of IRS LISS III image dated 29 July 2013, showed 28 lakes in Parbati sub-basin and 39 lakes in the Jiwa sub-basin further downstream in the catchment. Because of the dense cloud cover, no information could be retrieved from the image in the upper catchment of the Beas basin (Fig. 8.4).

Further spreading of lakes in other sub-basins suggests that out of 28 lakes identified in the Parbati sub-basin, 2 lakes are having an area of more than 10 ha and the remaining 26 lakes are such which have an area less than 5 ha. Likewise, in the Jiwa sub-basin, 02 lakes show an area ranging from 5 to 10 ha, and the remaining 37 lakes are with an area <5 ha (Table 8.3).

8.5.4 Lakes Inventory of Ravi Basin

The LISS III data of 04 October 2013, reflected 22 lakes in the whole Ravi basin (Fig. 8.5).

A good quality cloud-free image of October has been used which showed 22 lakes in the Ravi basin and most of the lakes exist along the natural geomorphic depressions of the prominent Dhauladhar range. Further, the distribution of these 22 lakes suggests that there are 2 lakes which have an area of more than 10 ha and only 1 lake is with an aerial range between 5 to 10 ha, whereas the remaining lakes (19) in the Ravi basin are of lesser size with an area <5 ha (Table 8.4).

8.5.5 Lakes Inventory of Satluj Basin

Satluj basin having a large spatial extent has been studied using satellite data during the period July–September 2012/2013 in different path rows viz. 96–48, 96–49, 97–48, 97–49, 98–48, 99–49, respectively. The analysis reveals the presence of 391 moraine-dammed lakes in the entire catchment of the Satluj right from its origin in the Tibetan Himalaya (Fig. 8.6).

Table 8.2 Areal range of moraine-dammed lakes in Chenab basin (based on the analysis of IRS, LISS–3 data)

Sl.	Name of sub-basin	Lake ID (Number)	Longitude	Latitude	Area (ha)
1	Bhaga	11	77.306289	32.630207	10.20
2	Bhaga	3	77.398433	32.761989	6.71
3	Bhaga	6	77.329385	32.722573	7.73
4	Bhaga	8	77.347478	32.705244	5.13
5	Chandra	1	77.222620	32.524214	80.12
6	Chandra	3	77.546984	32.499906	146.31
7	Chandra	6	77.617488	32.604030	5.00
8	Chandra	13	77.447949	32.245131	7.44
9	Miyar	6	76.546224	33.216837	8.33
10	Miyar	43	76.671748	32.929764	5.91
11	Miyar	83	76.123668	33.182620	8.56

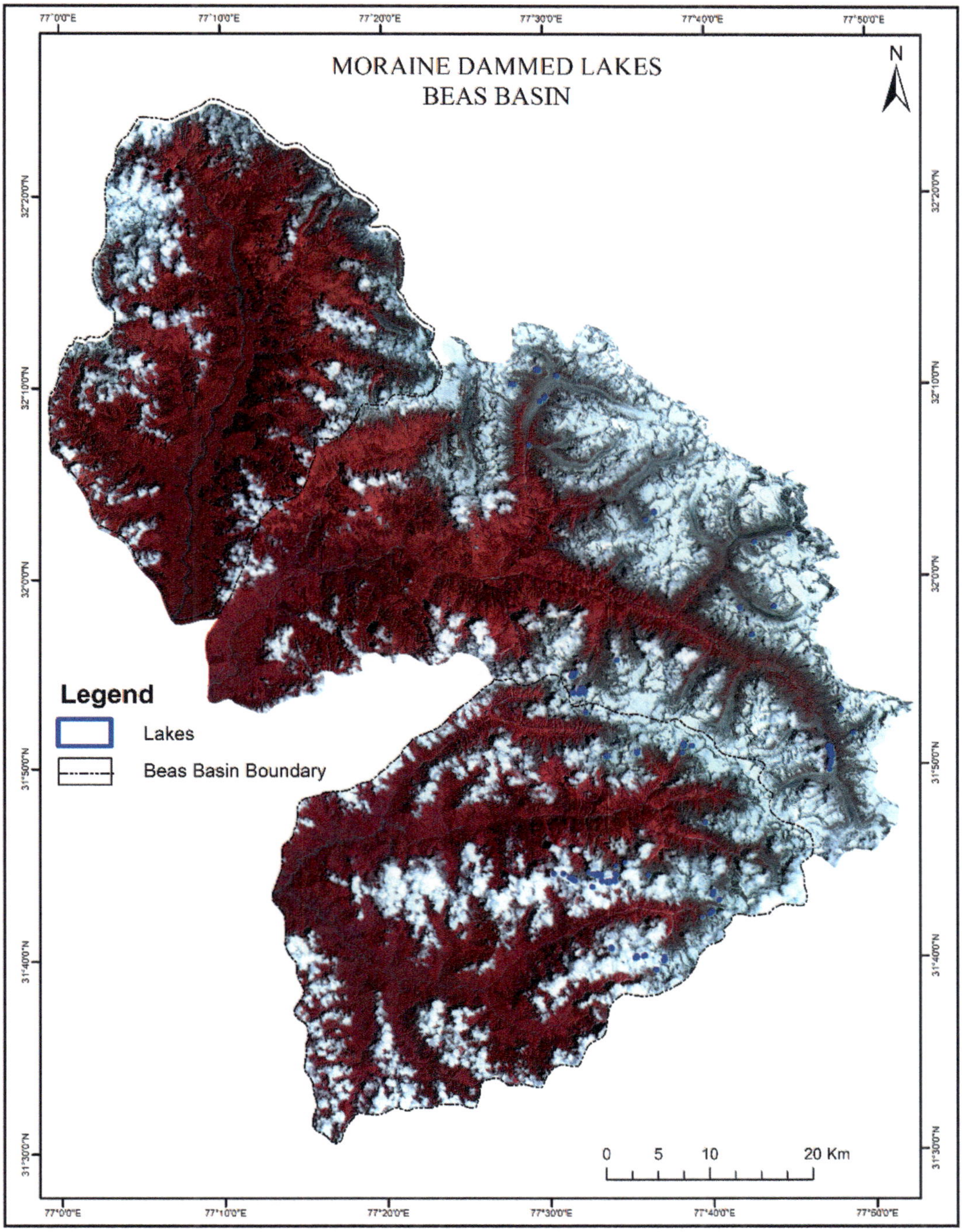

Fig. 8.4 Moraine dammed lakes in Beas basin as delineated from the image of 29 July 2013 (*Source* IRS, LISS-3 image)

Table 8.3 Areal spread of moraine dammed lake in Beas basin (*Source* IRS LISS–3 image)

Sl.	Basin name	Sub basin	Lake ID number	Longitude	Latitude	Area (ha)
1	Beas	Parbati	21	77.572690	31.926278	13.58
2	Beas	Parbati	26	77.526890	31.897405	13.08
3	Beas	Jiwa	9	77.668512	31.721823	5.07
4	Beas	Jiwa	34	77.618027	31.665046	6.07

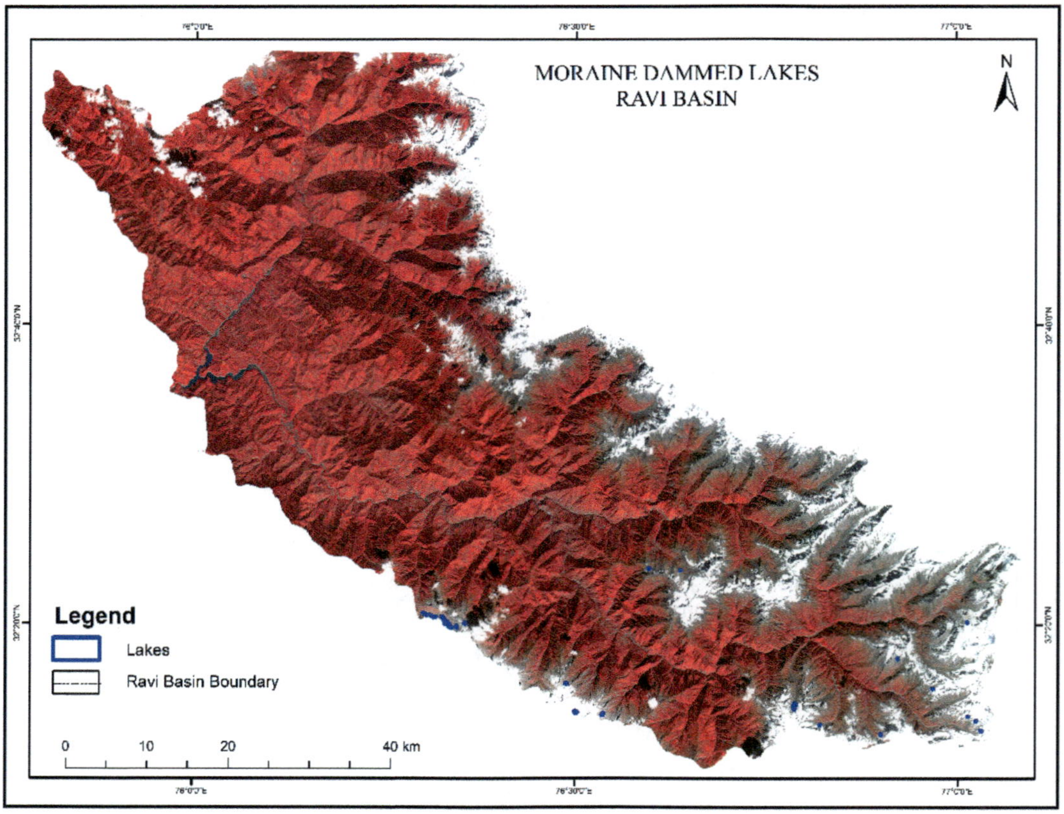

Fig. 8.5 Moraine dammed lakes in Ravi basin as delineated from the image of 4 October 2013 (*Source* IRS, LISS-3)

Table 8.4 Areal extent of moraine dammed lake in Ravi basin (*Source* IRS, LISS–3 image)

Sl.	Basin	Lake id number	Longitude	Latitude	Area (ha)
10	Ravi	10	76.787141	32.244530	14.87
12	Ravi	12	76.500869	32.237279	7.09
16	Ravi	14	76.330737	32.337624	33.51

The supplementary, analysis revealed that out of 391 lakes, 118 lakes are in sub-basin 1 (Spiti sub-basin), 15 lakes in sub-basin 2 (lower Satluj) and the 258 lakes in the sub-basin 3 (upper Satluj) in the Tibetan Himalayan region. As far as the aerial distribution of the lakes is concerned, 40 lakes in the entire Satluj basin having an area of more than 10 ha, 75 lakes are having

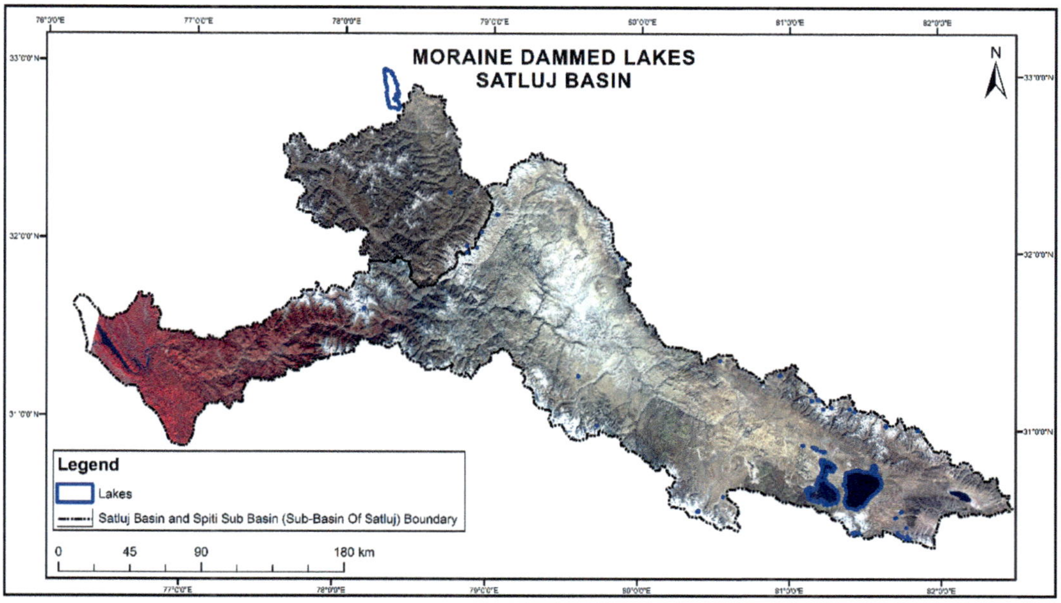

Fig. 8.6 Moraine dammed lakes as delineated in Satluj basin (*Source* IRS, LISS-3)

an area ranging from 5 to 10 ha and 276 lakes are of the lesser size of <5 ha (Table 8.5).

The results related to the lakes from all the basins have been summarized and the category wise distribution of lakes has been presented through Table 8.6. The aerial distribution of the lakes suggests that the Satluj basin possesses 391 lakes, from which 275 are of smaller size (less than 5 ha), 75 lakes range between 5 to 10 ha and 40 lakes are of size greater than 10 ha. Similarly, the Chenab basin (Chandra, Bhaga, Miyar) possesses 116 lakes which are approximately double than (55) lakes identified in 2001 satellite imageries (Randhawa et al. 2005). Out of the total 105 lakes are in the range of area <5 ha, 8 lakes are in the range of 5 to 10 ha and only 3 lakes are of area >10 ha. In the Beas basin (Jiwa, Parbati), out of 67 lakes 63 lakes of area <5 ha, 2 lakes of area 5 to10 ha and the remaining 2 lakes are of area >10 ha. The cloudy image of the upper part of the Beas basin put the constraints so no lake identification has been made. In Ravi basin, out of 22 lakes, 2 lakes are of area >10 ha, only 1 lake is with area ranging from 5 to 10 ha and remaining 19 lakes are with area <5 ha.

8.6 Conclusions

The status of lakes in the basins and sub-basins of Chenab (Bhaga, Chandra, Miyar), Beas (Jiwa, Parbati), Ravi, Satluj have been analyzed based on LISS III of the year 2013 and compared with the earlier lake inventories in these basins. The results disclosed a substantial increase in the numeral of moraine dammed lakes in all the basins mentioned above. The comparison with the earlier reported inventory of lakes in these basins based on the IRS IA & IB satellite imageries in the years 1993 disclosed about 38 lakes in Satluj basin (Kulkarni et al. 1999) having 14 lakes in the Himachal Pradesh while 24 lakes in the Tibetan region. The same basin in 2013 has been covered with 391 lakes including the Baspa and Spiti basins. Similarly, the basin Chenab has been reported to have 55 lakes based on the IRS IC/ID satellite imageries of the year 2003, which has been increased to 116 in LISS III satellite image of the year 2013.

This temporal variation of lakes in these basins is very clearly indicating the increasing

Table 8.5 Areal distribution of all moraine-dammed lakes in Satluj Basin

S. No.	Lake id	Basin ID	Longitude	Latitude	Area (ha)	S. No.	Lake id	Basin No	Longitude	Latitude	Area (ha)	S. No.	Lake id	Basin No	Longitude	Latitude	Area (ha)
1	37	1	78.70461	32.69416	5.4	41	560	3	79.12591	32.321	5.95	81	680	3	80.18395	31.44718	19.79
2	47	1	78.49398	32.44269	6.61	42	562	1	78.9954	32.23965	8.77	82	681	3	80.19622	31.44245	17.71
3	49	1	78.74175	32.31269	20.3	43	567	3	78.95518	32.10657	5.23	83	683	3	80.22435	31.43245	17.2
4	58	3	79.01097	32.32503	9.93	44	568	3	78.95625	32.09329	6.44	84	684	3	80.23391	31.43146	18.24
5	62	3	78.98791	32.30671	8.8	45	571	3	78.93541	32.001	9.59	85	685	3	80.29606	31.45177	8.56
6	63	3	78.99401	32.28709	9.57	46	573	1	78.84455	32.02254	7.67	86	686	3	80.26917	31.45958	6.28
7	67	3	79.05035	32.19611	23.74	47	576	1	78.76645	32.03093	7.85	87	687	3	80.18071	31.40144	7.49
8	68	3	79.03454	32.21077	8.53	48	580	3	78.85223	31.90085	22.32	95	98	3	81.0332	31.28473	14.95
9	74	1	78.98079	32.17743	5.31	49	581	3	78.79443	31.91341	6.59	96	107	3	81.0845	31.23876	5.85
10	77	3	78.9864	32.14782	9.17	50	585	1	78.71304	31.89844	7.92	97	109	3	81.13771	31.23316	11.01
11	82	3	78.95104	32.10916	5.26	51	587	1	78.73075	31.88669	5.84	98	112	3	81.14911	31.1978	12.15
12	86	3	78.94946	32.09378	10.12	52	590	2	78.74747	31.52819	6.03	99	113	3	81.19457	31.18094	52
13	87	3	78.9417	32.09407	11.46	53	595	2	78.14109	31.73377	6.97	93	114	3	81.15269	31.17821	23.25
14	88	3	78.95636	32.07898	5.02	54	597	2	78.20633	31.90931	6.49	94	116	3	81.38243	31.1381	9.21
15	89	3	78.93825	32.07883	8.14	55	601	1	78.45281	32.16677	5.64	95	134	3	81.85202	31.00371	24.36
16	93	3	78.95291	32.04592	7.59	56	605	1	78.50053	32.362	7	96	140	3	81.531	30.82868	6.57
17	94	3	78.92	32.00424	5.67	57	173	2	78.69657	31.22521	11.86	97	142HWL	3	81.56941	30.80125	364.14
18	97	1	78.88407	32.00381	8.33	58	177	2	78.3288	31.31149	24.84	98	145HWL	3	81.54085	30.7279	41066.37
19	99	1	78.85532	32.01603	17.36	59	180	2	78.25397	31.34068	5.94	99	148HWL	3	81.59198	30.7591	194.77
20	101	1	78.85464	31.97947	22.81	60	614	2	77.87832	31.69373	10.84	100	172	3	81.71995	30.44545	7.12
21	102	3	78.87972	31.95617	7	61	615	2	78.16767	31.66174	27.7	101	173	3	81.67561	30.44572	9.89
22	106	3	78.8476	31.96766	6.59	62	639	2	78.01247	31.40694	7.87	102	179	3	81.71341	30.42744	25.09
23	109	1	78.77321	32.02695	8.6	63	184	3	79.39264	32.37784	27.63	103	184	3	81.71968	30.41953	16.95
24	111	1	78.76805	32.03095	6.54	64	190	3	79.68327	32.25382	6.96	104	187	3	81.77671	30.40834	7.28
25	113	1	78.76372	32.02884	6.63	65	201	3	79.84579	31.99178	5.43	105	203	3	81.09009	31.26536	5.06
26	117	3	78.83234	31.91934	5.53	66	202	3	79.87211	31.97695	10.78	106	210	3	81.55487	30.77131	53
27	118	3	78.8365	31.91645	5.28	67	203	3	79.87563	31.96732	8.5	107	223	3	81.42486	31.1366	5.46
28	119	3	78.83168	31.91673	9.61	68	207	3	79.87748	31.95564	5.17	108	228	3	81.43578	31.11286	7.25
29	121	3	78.82885	31.91557	8.04	69	209	3	79.86544	31.92432	34.57	109	233	3	81.50399	31.09335	5.5

(continued)

Table 8.5 (continued)

S. No.	Lake id	Basin ID	Longitude	Latitude	Area (ha)	S. No.	Lake id	Basin No	Longitude	Latitude	Area (ha)	S. No.	Lake id	Basin No	Longitude	Latitude	Area (ha)
30	122	3	78.79484	31.90425	11.65	70	645	3	78.79411	31.44988	10.61	110	235	3	81.50304	31.11081	7.5
31	131	1	78.71485	31.90381	5.2	71	647	2	78.75113	31.55366	8.66	111	237	3	81.54287	31.116	11.88
32	133	2	78.76188	31.5392	7.32	72	652	3	79.59381	32.35287	6.2	112	242	3	81.54608	31.10174	6.81
33	135	3	78.8036	31.4354	7.48	73	656	3	79.78836	32.02724	10.35	113	253	3	81.72296	31.06934	8.6
34	154	1	78.81737	32.04471	10.5	74	662	3	79.85466	31.97757	7.21	114	607	3	81.16383	30.91277	14.7
35	155	1	78.81706	32.06681	8.21	75	664	3	79.85751	31.96684	5.37	115	608	3	81.09476	30.92422	15.97
36	551	1	78.70612	32.71675	8	76	96	3	79.60051	31.30347	24.34	116	692	3	81.1102	31.26913	5.24
37	553	1	78.72321	32.69407	7	77	171	3	79.514	31.13195	6.7	117	731	3	80.40077	30.55246	37.88
38	554	1	78.72233	32.69336	7.53	78	178	3	79.41423	31.05959	6.22	118	142HWL	3	81.56941	30.80125	364.14
39	556	1	78.84073	32.40785	7.9	79	399	3	79.72944	31.02881	6.04	119	145HWL	3	81.54085	30.7279	41066.37
40	557	1	78.90868	32.31144	8.02	80	669	3	79.4774	31.3577	11.92	120	148HWL	3	81.59198	30.7591	194.77

Table 8.6 Lakes spreading in the basins/sub-basins of Himachal Pradesh analyzed from the IRS, LISS-3 image (2013)

Sl.	Basin/sub-basin	No. of lakes with area >10 ha	No. of lakes with an area between 5–10 ha	No. of lakes with area <5 ha	Number of lakes in total
1	**Chenab**	**3**	**8**	**105**	**116**
	Bhaga	1	3	10	14
	Chandra	2	2	15	19
	Miyar	0	3	80	83
2	**Beas**	**2**	**2**	**63**	**67**
	Jiwa	0	2	37	39
	Parbati	2	0	26	28
3	**Ravi**	**2**	**1**	**19**	**22**
4	**Satluj**	**40**	**75**	**276**	**391**

trend of the lakes in the higher Himalaya of Himachal Pradesh. The results of 2013 indicated that the rise in the number of lakes are much higher for smaller sized lakes having an area less than 5 ha and formed mostly related to the snout retreat and moraines damming that may be an indication of a noticeable impact of changing climate in these regions.

One of the disastrous incidents of 2013, in the Kedarnath region of Uttrakhand, was also related to the bursting of a lake at the snout of the Chorabari in association with cloud burst causing a catastrophe in the downstream region (Dobhal et al. 2013). Hence, the importance of lake monitoring should not be ignored to have preparedness against the likely disaster due to GLOF. The lake formation and swelling of the lake are dynamic and its stability depends on the stability of the damming materials, as well as the variability of the impounded water quantity in lakes. The increased water volume in lakes may induce dam breaching and consequential situation of a flash flood in the downstream.

The reported lakes in the present study having an area >10 ha are highly potential, while the lakes with an area between 5–10 ha are moderately potential towards becoming GLOF in case of higher meltwater intrusion in the lakes and associated cloud burst disorder in the basins of Chenab, Beas, Ravi and Satluj. Thus, regular monitoring of existing and new formation of lakes are indispensable to avoid any possibility of catastrophe due to GLOF in the region.

References

Bahuguna IM (2003) Satellite stereo data analysis in snow and glaciated region. Training document: Course on remote sensing for glaciological studies. Manali, India, SAC/RESA/MWRG/ESHD/TR/13/2003, pp 85–99

Bhardwaj A, Sam L, Singh S, Kumar R (2016) Automated detection and temporal monitoring of crevasses using remote sensing and their implications for glacier dynamics. Annals Glaciol 57(71):81–91. https://doi.org/10.3189/2016AoG71A496

Bhat MS, Alam A, Ahmad B, Kotlia BS, Farooq H, Taloor AK, Ahmad S (2019) Flood frequency analysis of river Jhelum in Kashmir basin. Quat Int. 507:288–294

Bisht H, Arya PC, Kumar K (2018) Hydro-chemical analysis and ionic flux of meltwater runoff from Khangri Glacier, West Kameng, Arunachal Himalaya India. Environ Earth Sci. 77:1–16. https://doi.org/10.1007/s12665-018-7779-6

Bisht H, Kotlia BS, Kumar K, Joshi LM, Sah SK, Kukreti M (2020). Estimation of the recession rate of Gangotri glacier, Garhwal Himlaya (India) through kinematic GPS survey and satellite data. Environ Earth Sci 13(6):1–12. https://doi.org/10.1007/s12665-020-090780

Bisht H, Rani M, Kumar K, Sah S, Arya PC (2019) Retreating rate of Chaturangi glacier, Garhwal Himalaya, India derived from kinematic GPS survey and satellite data. Curr Sci 116:304–311. https://doi.org/10.18520/cs/v116/i2/304-311

Denton GH and Hughes TJ (1981) The last great ice sheets, John Wiley and Sons, New York USA, 7–10

Dhar S, Kulkarni AV, Rathore BP, Kalia R (2010) Reconstruction of the moraine dammed lake, based on field evidences and paleohistory, Samudra Tapu Glacier, Chandra Basin., Himachal Pradesh. J Indian Soc Remote Sens 38:133–144

Dobhal DP, Gupta AK, Mehta M, Khandelwal DD (2013) Kedarnath Disaster: facts and plausible causes. Current Sci 105:171–174

Haque S, Kannaujiya S, Taloor AK, Keshri D, Bhunia RK, Ray PKC, Chauhan P (2020) Identification of groundwater resource zone in the active tectonic region of Himalaya through earth observatory techniques. Groundwater Sust Develop 10: P100337. https://doi.org/10.1016/j.gsd.2020.100337

Khan A, Govil H, Taloor AK, Kumar G (2020) Identification of artificial groundwater recharge sites in parts of Yamuna river basin India based on remote sensing and geographical information system. Groundwater Sust Develop 11:P100415. https://doi.org/10.1016/j.gsd.2020.100415

Kulkarni AV, Philip V, Thakur VC, Sood RK, Randhawa SS, Chandra R (1999) Glacial inventory of the Satluj Basin using remote sensing technique. Himal Geol 20(2):45–52

Kumar R, Bahuguna IM, Ali SN, Singh R (2020a) Lake inventory and evolution of glacial lakes in the Nubra-Shyok Basin of Karakoram range. Earth Syst Environ 4:57–70. https://doi.org/10.1007/s41748-019-00129-6

Kumar R, Sharma RK, Pradhan P, Sharma N, Shrestha DG (2020b) Melt Runoff Characteristics and Hydro Meteorological Assessment of East Rathong Glacier in Sikkim Himalaya, India. Earth System and Environment. https://doi.org/10.1007/s41748-020-00168-4

Kumar R, Kumar R, Singh S, Singh A, Bhardwaj A, Kumari A, Randhawa SS, Saha A (2018) Dynamics of suspended sediment load with respect to summer discharge and temperatures in Shaune Garang glacierized catchment. Western Himalaya. Acta Geophysica 66(5):1109–1120. https://doi.org/10.1007/s11600-018-0184-4

Randhawa SS, Sharma A (2013) A technical report on the inventory of moraine dammed glacial lakes (GLOFs) in Satluj, Beas, Chenab and Ravi Basins in Himachal Pradesh using IRS LISS III satellite data. Unpublished report. http://www.hpccc.gov.in/documents/Inventory%20of%20Moraine%20Dammed%20Glacial%20Lakes%20%20in%20Himachal%20Pradesh.pdf

Randhawa SS, Sood RK, Rathore BP, Kulkarni AV (2005) Moraine dammed lakes study in the Chenab and Satluj river basins using IRS data. J Indian Soc Remote Sens 33(2):285–290

Sarkar T, Kannaujiya S, Taloor AK, Ray PKC, Chauhan P (2020) Integrated study of GRACE data derived interannual groundwater storage variability over water stressed Indian regions. Groundwater Sustain Develop 10:P100376. https://doi.org/10.1016/j.gsd.2020.100376

Singh AK, Jasrotia AS, Taloor AK, Kotlia BS, Kumar V, Roy S, Ray PKC, Singh KK, Singh AK, Sharma AK (2017) Estimation of quantitative measures of total water storage variation from GRACE and GLDAS-NOAH satellites using geospatial technology. Quat Int 444:191–200

Singh S, Sood V, Taloor AK, Prashar S, Kaur R (2020) Qualitative and quantitative analysis of topographically derived CVA algorithms using MODIS and Landsat-8 data over Western Himalayas. Quat Int, India. https://doi.org/10.1016/j.quaint.2020.04.048

Sood V, Gusain HS, Gupta S, Taloor AK, Singh S (2020a) Detection of snow/ice cover changes using subpixel-based change detection approach over Chhota-Shigri glacier. Quat Int, Western Himalaya, India. https://doi.org/10.1016/j.quaint.2020.05.016

Sood V, Singh S, Taloor AK, Prasher S, Kaur R (2020b) Monitoring and mapping of snow cover variability using topographically derived NDSI model over north Indian Himalayas during the period 2008-19. Applied Computing and Geosciences. https://doi.org/10.1016/j.acags.2020.100040

Taloor AK, Kotlia BS, Jasrotia AS, Kumar A, Alam A, Ali S, Kouser B, Garg PK, Kumar R, Singh AK, Singh B (2019) Tectono-climatic influence on landscape changes in the glaciated Durung Drung basin, Zanskar Himalaya, India: a geospatial approach. Quat Int 507(262):273

Dr. Surjeet Singh Randhawa has been associated with Himachal Pradesh Council for Science, Technology and Environment since 1993 and presently working as Principal Scientific Officer. He joined Honours School in Geology, Punjab University Chandigarh and did B.Sc. (Hons.), M.Sc (Hons) in Geology and Doctorate in Geology from Department of CAS in Geology, PU Chandigarh in the field of Applications of Remote Sensing in Applied Geomorphology. He has expertise in the mapping of natural resources in Himachal Pradesh using space data, exploration of ground water and has expertise in the digital image processing in snow and glacier studies, mapping and management of natural disasters in Himachal Pradesh. Presently, coordinating activities of the Himachal Pradesh State Centre on Climate Change established under the aegis of the State Council for Science Technology & Environment. He has also worked with SWISS Government and carried out a study under Indian Himalayan Climate Adaptation Programme (IHCAP) in Kullu District and undertook mapping of Permafrost in Indian Himalayas for the first time using space data. Member of the Indian Scientist team who visited Switzerland in 2015 on an exposure visit under ICHAP programme. He has published more than 35 research papers in various national and international journals and also co-authored in one of the book chapters along with Swiss Scientists on GLOFs (Glacial lake Outbursts Floods). Scientific Reports, Atlases and through Vedas portal of SAC

Prof. Sunil Dhar completed his Ph.D., M.Sc (Hon's Geology), B.Sc (Hon's Geology) from Panjab University, Chandigarh and Post-Doctoral Fellow, University of Bern, Switzerland. Subsequently, he joined Post-Graduate, Department of Geology (Himachal Pradesh, University) Dharamsala. Apart from heading the department there, he remained actively involved in research project (funded by UGC, DST, ISRO) on North Western Himalayas, MoU's and consultancies. Later, Prof. Dhar joined Central University of Jammu, J&K (UT) and he is at present heading the Department of Environment Science and Earth Sciences at Central University of Jammu. His areas of research interest include quaternary geology, glaciology, seismic hazards geochemical precursors/zonation and contemporary geomorphology. Being an ardent supporter of field geology, Prof. Dhar has travelled intensely in the Higher Himalayan terrain of Himachal Pradesh. Recently, he has also successfully completed the Chandrayan-I lunar geology project of ISRO and presently he is working on research project of DST on the glacio-meterological attributes of the glaciers of lower Chenab Basin of J&K (UT). Prof. Dhar has more than 25 years of teaching and research experience and has published about 65 international, national research papers, book chapters and supervised 11 Ph.D./M.Phil. students

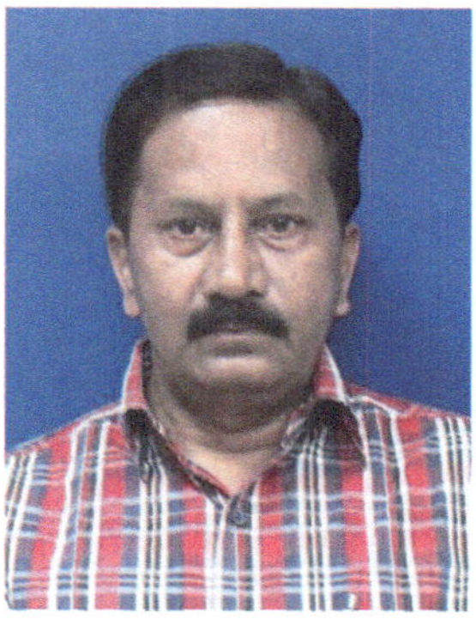

Er. Bhanu Prakash Rathore, a Civil Engineer with specialization in Water Resources Management is a Senior Scientist at Space Applications Centre ISRO, Ahmedabad. He has been involved in remote sensing and GIS-based analysis of satellite and ground data for snow and glacier monitoring in the perspective of hydrological and climatological-related applications of Himalayan Region. He has contributed in different capacities to many national-level projects for mapping; monitoring and long-term analysis of glaciers response to climate change; snow cover mapping and monitoring using WiFS, AWiFS and INSAT series sensors to develop a long-term database, proglacial lake monitoring in assessing GLOF scenario, developing snow and glacier melt runoff model, RISAT-1 SAR data utilization for glacier facies, wetland mapping and monitoring, etc. He is recipient of ISRO team award (Snow and Glacier) in year 2008. He has shared and showcased his research work in journals, conferences, book chapters, scientific reports, atlases and through Vedas portal of SAC

Dr. Rajesh Kumar has received Ph.D. and M.Sc. (Tech.) in Geophysics (Meteorology) from Banaras Hindu University (BHU), Varanasi. Presently, he is working as Professor in the Department of Environmental Science, School of Earth Sciences (SES), Central University of Rajasthan, Ajmer and holds the responsibility of Head, Dept. of Environmental Science and Dean, School of Earth Sciences, Central University of Rajasthan. The research area of Prof. Kumar is air pollution, climate change, glaciology, glacial water quality and quantity assessment in the snow and glacier-fed river systems of Himalaya, and application of remote sensing and GIS on cryosphere research. He has received eight research grants from National and international organizations like WWF-India; DST, Govt. of India; USAID (USA); South Asia Water Initiative (SAWI), ICIMOD, Nepal; MoES, Govt. of India and SAC (ISRO). He, with research and teaching experience of 21 years, has major work on the influence of changing climate and global warming on snow and glaciers that impact the availability of water in terms of quality and quantity in the Indian Himalayan Region and downstream. He has published 95 papers in international and national journals/book chapters/encyclopaedia presented in 38 papers in national and international conferences and supervised 10 Ph.D. scholars. He has been listed among the top 2% scientist of the world as per the list released in 2020 by the team of researchers of Stanford University, USA

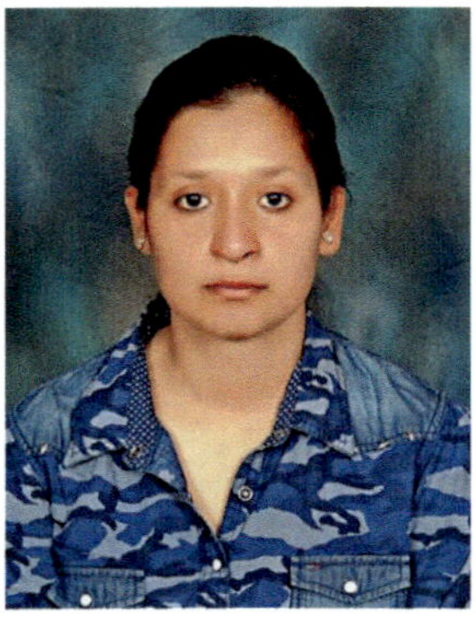

Ms. Neha Thakur did her M. Phil from Himachal Pradesh University, Shimla, entitled as 'Regional and Development Planning of Una District, Himachal Pradesh'. She has completed advanced 1-year Diploma in Remote Sensing and GIS and Master's Degree in Geography from Himachal Pradesh University, Shimla. She has completed 12-week regular course on 'Geospatial Technologies and Applications', NRSC, Hyderabad. She has joined Himachal Pradesh State Centre on Climate Change as Scientific Professional in 2018 and is currently working in project entitled as 'Geo-Resources and Impact Assessment of Geo (Exogenic) Processes in N. W. Himalayan Ecosystem" (WIHG Project). She has worked earlier in NGO—Institute for Spatial Planning and Environment Research, (ISPER), Panchkula. She works in area of remote sensing and Geographic Information System (GIS) in Integrated Watershed Management Programme, spatial planning, land use land cover, etc. Her expertise is in the field of digital image processing and geographic information system

Smt. Pooja Rana has done Master's Degree in Geology from Govt. PG College Dharamshala and completed her Diploma in Remote Sensing and GIS from Indian Institute of Remote Sensing, ISRO, Dehradun. Her research interest and area of expertise is in the field of remote sensing and its applications. She has joined Himachal Pradesh State Centre on Climate Change as Scientific Professionaland and has over 8 years of working in the above-said domain

Mr. Duni Chand Rana post-graduate in English literature and Disaster Mitigation is working in the Administrative Services for the last 21 years. He held various positions in the State Govt. The last one being Director, Environment, Science and Technology, Member Secretary HIMCOSTE and Director cum Special Secretary (Disaster Management). Presently, he is posted as Deputy Commissioner Chamba. His areas of interest include disaster management, climate change, solid waste management, etc.

Dr. Ajay Kumar Taloor has obtained his Doctorate in Remote Sensing and GIS applications in hydrogeology from University of Jammu, NAAC accredited A+ University of India. Thirteen years of research experience in the applications in geospatial technology for natural resources management of land and water resources. He has excelled twice with best paper presentation award in India. Being an expert of remote sensing applications in water science, cryosphere and climate change, tectonic and quaternary geomorphology, he is working on two major research projects on using space-based inputs for glacier mapping and climate change in Himalayas. He has also published many articles in tectonic and quaternary geomorphology in the recent years. He has high scientific temper and strong HR relations in science world, with high professional and managerial skills. He has edited many volumes in the top-rated journals in the Elsevier and Springer publishers, member of Editorial Board of the Quaternary Science Advances, and reviewer of the many top-rated international journals in science world

9 Late Quaternary Glacial Geomorphology of Kashmir Valley, NW Himalayas: A Case Study of the Sind Basin

Reyaz Ahmad Dar, Omar Jaan Paul, Khalid Omar Murtaza, and Shakil Ahmad Romshoo

Abstract

Glacial geomorphological mapping from satellite imagery supported with detailed field studies can form the basis for reconstructing the nature and extent of paleoglaciation of an area. In this paper, we present the glacial geomorphological setting of the Sind Basin Kashmir Valley, NW Himalayas, India. Various glacial landforms mapped in the area included cirques, glacial valleys, arêtes, moraine ridges, tarns and overdeepened tongue basins. The mapping of the erosional and depositional landforms revealed that the Sind Basin has predominantly experienced the alpine type of glaciation. The nature and extent of paleoglaciation determined on the basis of glacial geomorphological features suggests that the glaciers covered an area of ~914 km^2 during the last glacial maximum, whereas the present glacial cover in the area is 31.939 km^2. The average reconstructed cirque glacier thickness in the basin is 313 m with some of the cirque glaciers attaining a thickness of more than 500 m. It is believed that the accelerated glacial erosion concomitant with the glacial recession has resulted in the rock uplift as manifested by the straight mountain fronts and linear, deep stream incised valleys. The current research on the glacial landform studies and the outline of the maximum glacial advancement can help for precise paleo-glacial reconstruction in the Kashmir Himalaya.

Keywords

Cirque glacier · Deglaciation · Geomorphological map · Marginal moraine · Sind basin

9.1 Introduction

The nature and extent of the past glaciation is important for reconstructing the paleoclimate of a region and provides insights into the role played by glaciers in landscape evolution (Dubey et al. 2019). Numerous studies reveal that glacial geomorphic features provide the first-hand information regarding the nature, timing and extent of glaciers (Fu et al. 2013). Detailed glacial landform mapping focussing mainly on the reconstruction of the regional glacial extent has

R. A. Dar · O. J. Paul · K. O. Murtaza
S. A. Romshoo (✉)
Department of Earth Sciences, University of Kashmir, J&K, Srinagar, India
e-mail: shakilrom@kashmiruniversity.ac.in

R. A. Dar
e-mail: reyazsopore@gmail.com

O. J. Paul
e-mail: geopaul65@gmail.com

K. O. Murtaza
e-mail: komurtaza@gmail.com

A. K. Taloor et al. (eds.), *Water, Cryosphere, and Climate Change in the Himalayas*,
Geography of the Physical Environment, https://doi.org/10.1007/978-3-030-67932-3_9

been carried out in numerous glaciated Himalayan Ranges (Shi 2002; Thackray et al. 2008; Ali et al. 2013; Bali et al. 2013; Ali and Juyal 2013; Ali et al. 2017; Shukla and Yousuf 2017; Dubey et al.2019; Taloor et al. 2019; Sood et al. 2020; Singh et al. 2020; Kumar et al. 2020). In Kashmir Himalaya, most of the cryosphere studies have either documented the impact of climate change on glacier recession (Romshoo et al. 2015; Murtaza and Romshoo 2017), changing snowfall patterns (Dar et al. 2014a), snow and glacier melt contribution to streamflow (Alam et al. 2011; Dar and Romshoo 2013; Marazi and Romshoo 2018; Bhat et al. 2019) or the impact of black carbon on glacier melting (Bhat et al. 2017). However, only a few studies have documented the glacial geomorphology of the glaciated valleys of the Kashmir Himalayas including the Thajwas Valley (Dar et al. 2017) and the Lidder Valley (Rashid et al. 2017).

This chapter extends the previous glacial geomorphic work to the Sind Basin of the Kashmir Valley. The basin is poorly studied in terms of glacial geomorphology despite the fact that the landforms related to glacial advance and retreat are remarkably well preserved in the basin, especially in its upper reaches. The Sind Basin is largely glacier-free today with glaciers occurring mainly as remnants of the last glacial maximum (LGM) in the form of cirque and small valley glaciers. Thajwas is one of the main glaciers, occurring in the form of cirque glaciers and covering an area of ~ 3.60 km^2. However, the sediments and landforms associated with the glacier advance and retreat are commonly observed in the basin reflecting the past extent, nature and thickness of the glaciers. Additionally, tectonic-geomorphic features like linear truncation of moraine deposits and mountain aprons, linear and deep valley incisions, reflecting the landscape response to tectonics and postglacial rebound, are obviously present in the basin. Dar et al. (2017) have reported the major glacial advances and the response of the Thajwas glacier to the regional climate change. Nevertheless, a detailed and comprehensive mapping of the glacial geomorphology of the entire basin to strengthen the paleo-glaciological reconstructions is poorly defined. The present work would, therefore, improve our understanding of the nature and pattern of paleo-glaciation and the role played by the glaciers in sculpturing the landscape of the Sind Basin in particular and the Kashmir Himalayas in general. We particularly investigated the nature and extent of landform patterns in terms of paleo-glacial extent, thickness and major glacial advances. The research is largely based on the information derived from the mapping of landforms from the Google Earth Pro Imagery and extensive field observations in the basin. The paleo extents of the glaciers in the basin were determined on the basis of the glacial geomorphological features like end moraines, outwash plains, trim lines, etc.

9.2 Study Area

The Kashmir valley, an intermontane basin in NW Himalaya, is nestled between the Great Himalayan Range in NE, Pir-Panjal Range in SW, Kazinag Range in SE and Saribal Range in NW. All the mountain ranges in the Himalayan region are tectonically active which are due to the continued collision of the Indian and Eurasian lithospheric plates (Bisht et al. 2020). The mountains possess impressive glacial sediments and landforms that provide clear evidence of the Late Quaternary glaciation. The moisture sources for the NW Himalayan glaciers, particularly in the Kashmir Valley, are largely provided by the western disturbances originating from the Mediterranean Sea and the Atlantic Ocean (Dimri and Chevuturi 2016). The role of tectonics has been found important in controlling the distribution of the moisture and thus influencing the overall growth and distribution of the glaciers in various Himalayan basins including the Sind basin (Brozović et al. 1997), which has been chosen as a study area for the present study.

Spread over an area of 1590 km^2, Sind Basin forms one of the major sub-basins of the Jhelum Basin, situated on the north–eastern margin of the Kashmir Valley (Fig. 9.1). River Sind is the major tributary of the centrally flowing River

Jhelumof the Kashmir Valley. Having a marvellous course of 96 km, the river Sind is fed by various streams emanating from the lofty peaks of Zoji-La (3256 m) and other headstreams originating from snow and glacier resources near the Amaranth, Kolahoi, Gangbal and Panjtarni areas (Fig. 9.1). Preliminary investigations suggest that most of the glaciers in the basin are valley type delimited by topography and have played a major role in the evolution of the high mountain topography. Geologically the Sind Basin is very diverse consisting of Recent Alluvium, Late Quaternary Loess-paleosol sequences, Triassic-Jurassic Limestone, Zewan Limestone, Fenestella Shale, Panjal volcanic, Granites and Muth Quartzites (Dar et al. 2013; Stojanovic et al. 2016). The River Sind obliquely runs across the basin from Sonamarg to Ganderbal town (Fig. 9.1.)

9.3 Methodology

The glacial geomorphology of the basin was mapped using the visual interpretation technique from the Google Earth Pro imagery. All the glacial landforms were mapped using on-screen digitization at 1:1000 scale. The landforms were validated in a 3D environment using ASTER (Advanced Space Borne thermal emission and reflection radiometer) DEM draped onto the Landsat ETM + imagery (Jansson and Glasser 2005; Greenwood and Clark 2009a, b). Field expeditions, using GPS, were carried out at several accessible locations of the basin to validate the mapping results obtained from the remote sensing data. The extent and advance of the last glacial maximum was delineated using the location, shape and morphology of various glacial landforms, especially lateral and terminal

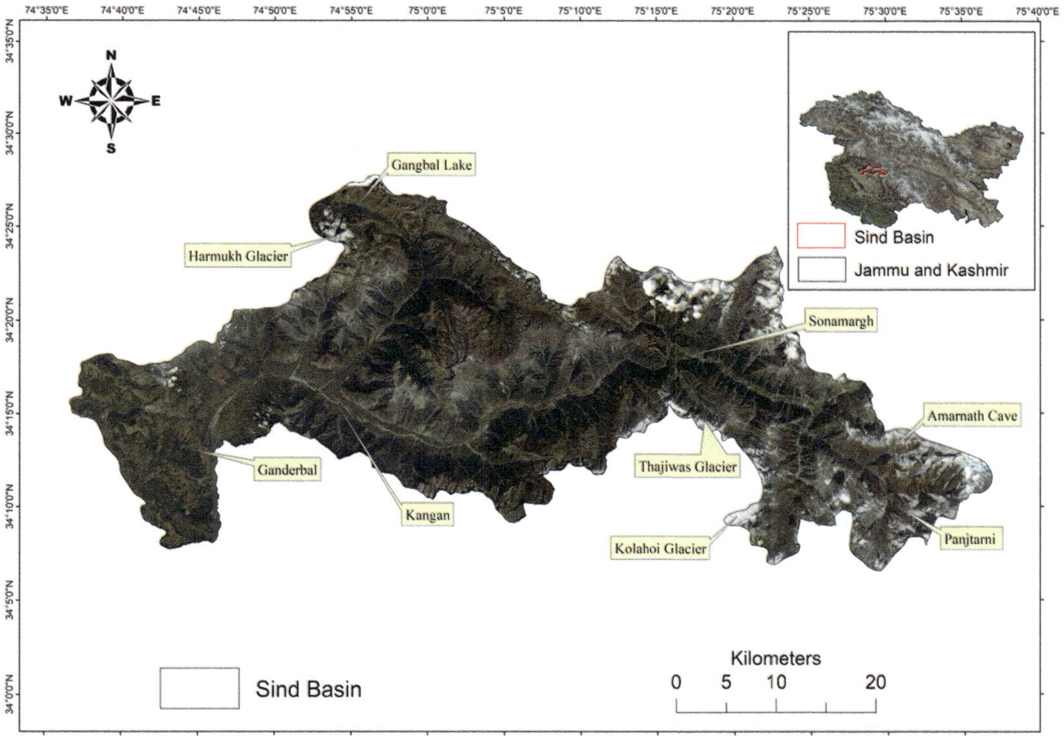

Fig. 9.1 Google earth imagery showing the location of the study area. Main glaciers and key landmarks are also shown in the image (*Source* Google image)

moraines. The top to bottom height of the de-glaciated cirques and glacial troughs was used to make inferences about the past thickness of the cirque and valley glaciers. Various geomorphic indices like mountain front sinuosity index (S_{mf}) and valley floor width to height ratio (V_f) were used to evaluate the response of the basin to tectonics and postglacial adjustments (Dar et al. 2014b; Alam et al. 2017, 2018; Taloor et al. 2020; Thakkar et al. 2020).

9.4 Results

The original form of mountainous regions is modified by glacial erosion and deposition to a greater extent with the development of various peculiar glacial landforms (Brozović et al. 1997). Glacial landforms in the Sind Basin, currently harbouring only vestiges of glaciers on the Great Himalayan Flank of the Kashmir Valley, indicated the extent and style of the former glaciers in the form of valley type covering an area of ~914 km^2(Fig. 9.2). The prominent glacial landforms related to the valley glaciation identified in the basin are; terminal, lateral and recessional moraines, cirques, glacial troughs, hanging valleys, glacial lakes, arêtes, etc. The description and paleo-glacial insights of the observed landforms in the basin are provided in the following sections.

9.4.1 Cirques

Cirques are bowl shaped depressions formed on the sides of the mountains by the scooping action of glaciers (Barr and Spagnolo 2013). Cirques form the second dominant glacial landforms after glacial moraines in the basin (Fig. 9.3a–g and 9.4a).

A large number of de-glaciated and glacier-covered cirques were mapped in the basin. Majority of the cirques in the basin occur in the eastern and the south-eastern part, in the upper reaches towards the western part near Naranag, Gangabal and upper southern mountainous areas of Mammer. Currently, the cirque glaciers cover more than 16 km^2 area in the basin. Cirque elevations and their aspects have resulted in the current existence and preservation of glaciers. The deglaciation observed in the basin is attributed to Quaternary climate changes (Dar et al. 2017). However, the other possible reasons are the high gradient of the glaciated valleys and the basal conditions of the ice. Faster retreat, formation of moraine dammed glacial lakes and the abundant inflow of meltwater at the ice base do explain this deglaciation. The moraine dammed lakes encourage the fast melting of the ice margins (Ehlers et al. 2011). The average bottom to a top height of the de-glaciated cirques is about 313 m which provides an insight into the nature and thickness of the past glaciers. Rock or debris-covered glaciers are also observed in a few cirques (Fig. 9.4b) in the basin. Some of the cirque and glacial troughs have produced overdeepened lakes (Fig. 9.4a), for example, Gangabal Lake having an area of 1.63 km^2, forming the famous tourist destinations in the Kashmir Valley.

The formation of these lakes is evidenced by the overdeepened tongue basins. These overdeepened valley basins are reported from many glaciated regions of the world and usually form in ablation zones where the ice velocity is high and the hydrostatic pressure of the basal meltwater drainage is higher. Since the tongue areas always develop more or less in the same position, the basins are, therefore, scoured repeatedly during the successive glaciations (Ehlers et al. 2011). The overdeepening also indicates the strong linear ice and meltwater drainage and weak bedrock lithology (Murtaza et al. 2020).

9.4.2 Arête

Arete is a saw shaped narrow ridge crest of a glaciated mountain which is formed due to the erosion of most of the material between two cirques, due to their progressive growth (Dar et al. 2017). The sharp edges are usually attributed to wedging by frost action and glacial erosion. Several arêtes were identified and mapped along the western and south-western side of the

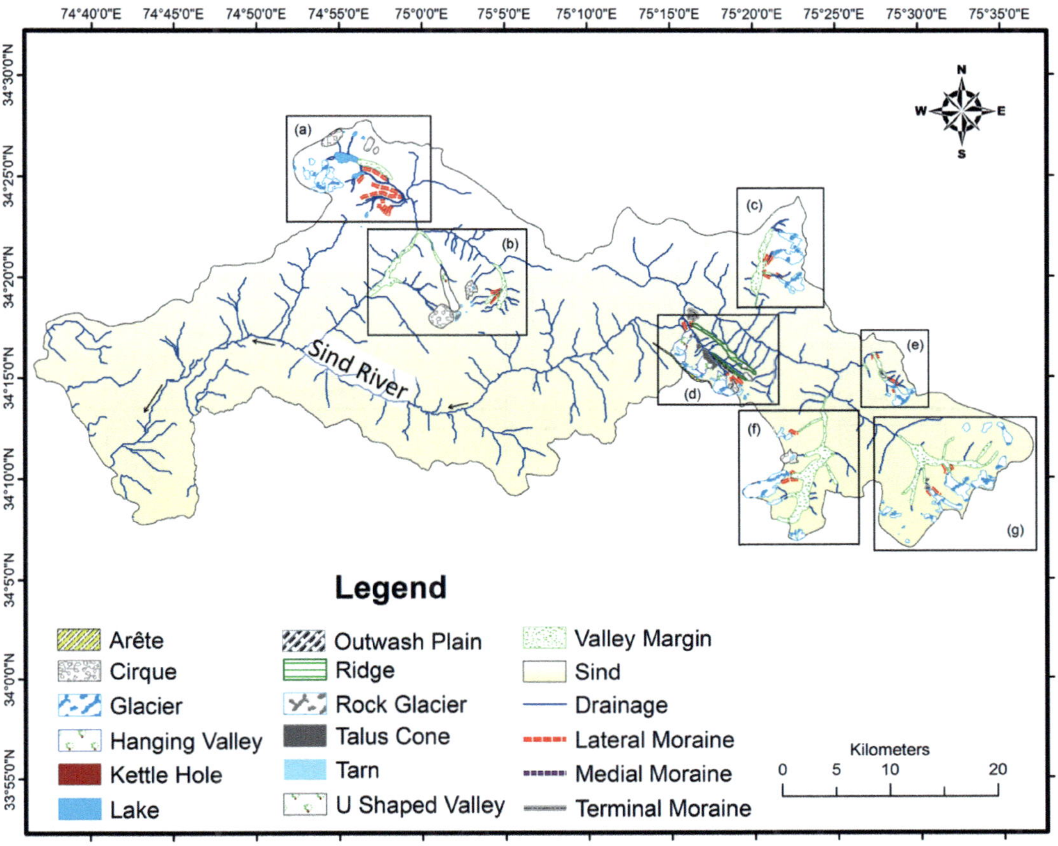

Fig. 9.2 Glacial Geomorphological map of the Sind Basin. The areas delineated using square boxes **a–g** show different glacial valleys, which are further elaborated in Fig. 9.3 (*Source* USGS)

basin (Fig. 9.4a). The arêtes indicate the potential glacial erosion and the subsequent gradual lowering of the ice level in the basin. Since most of the glaciers have already receded and the area is undergoing isostatic exhumation, therefore, most of the glacier features, including arêtes are prominent.

9.4.3 Moraines

Moraines are depositional landforms resulting from the accumulation of glaciogenic sediments at the bottom, end or margins of the glacier. In most cases, moraines are direct indicators of the extent and the thickness of the glaciers. Lateral moraines are the commonest landform features observed in the basin and are well preserved, especially in the upper reaches. The height of the moraines from the valley bottom ranges from 95 m to about 200 m with the longest marginal moraine observed in the basin measuring ~7 km long and attaining an average height of 150 m. A few end moraines and numerous recessional moraines were also identified in the basin (Fig. 9.3a–g). The glacial retreat has been mostly recessional with short-term re-advances (Dar et al. 2017).

9.4.4 U Shaped Valleys

These are sculptured when glaciers flow downhill, eroding the glacier valleys both laterally and vertically. The side walls are nearly vertical and the floor is broad and flat than in the fluvial

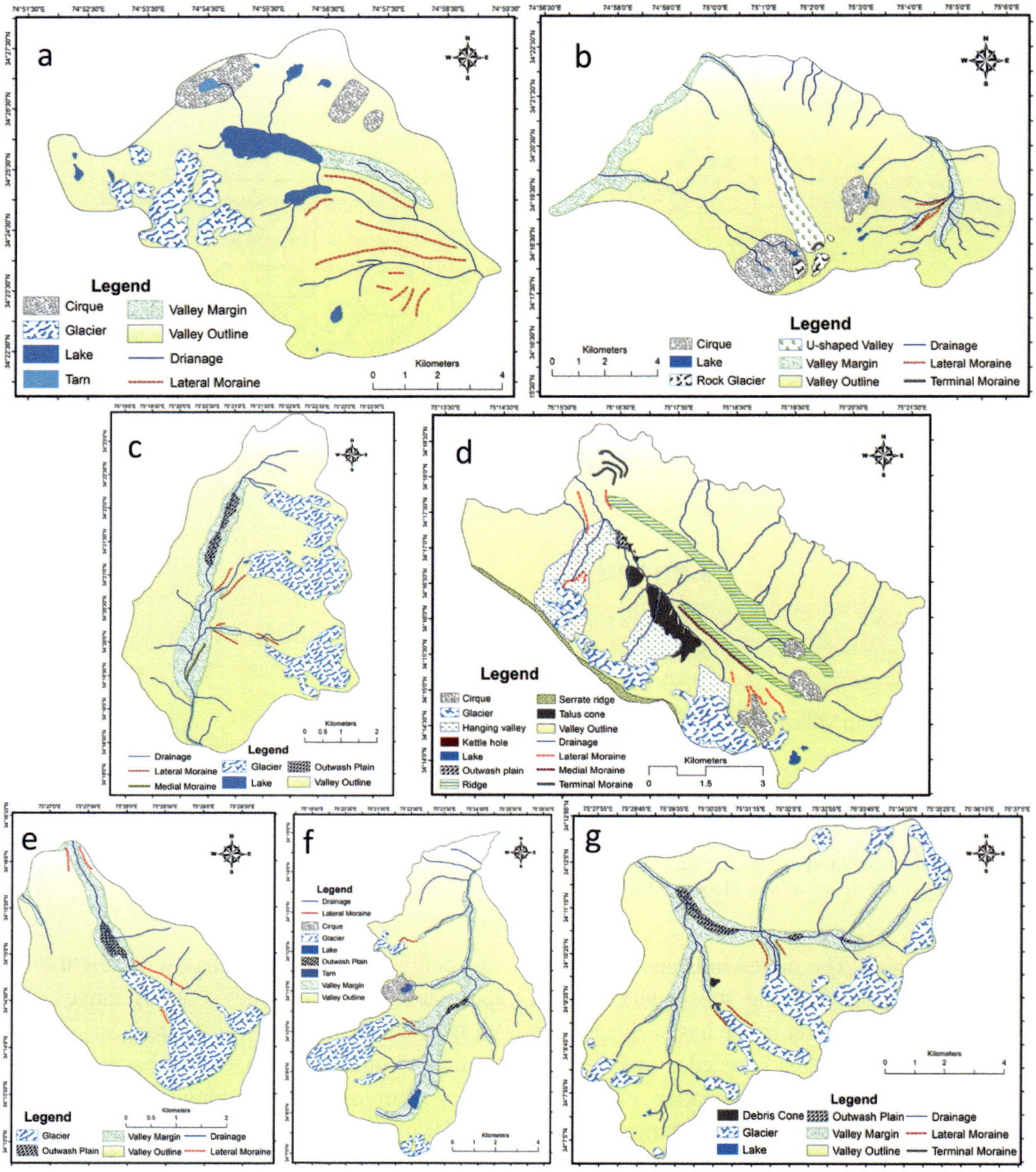

Fig. 9.3 Showing detailed glacier geomorphology of various sub-valleys of the Sind Basin (*Source* USGS)

valleys. Nevertheless, widening and deepening depends on the lithology and the erosive power of the glacier (Ehlers et al. 2011). Most of the U-shaped valleys identified and mapped in the basin are in the western and south-western side and a few are located in the eastern side and southern ranges of Kullan, Gund, Naranag Mountains (Figs. 9.2 and 9.3b). Some of the de-glaciated U-shaped valleys have attained a V-shape because of the subsequent fluvial erosion and the post-glacial rebound which helped the fluvial system to incise deeper the lower furrows of the valleys. Although it is possible that these valleys have tectonic or fluvial origin, but the presence of clear glacial erosional signatures (for example trimlines) and the existence of the relict glacier at their upper reaches points to their glacial origin. A large number of hanging valleys are also

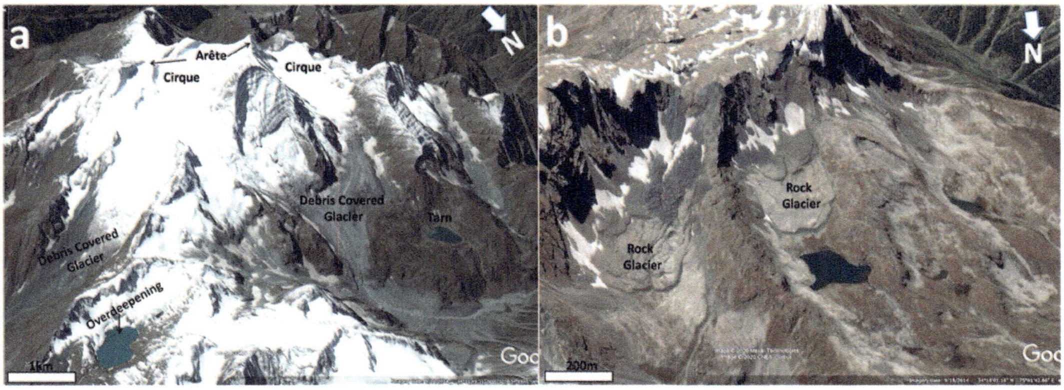

Fig. 9.4 Google Earth imagery **a** showing the presence of arête, cirque, debris-covered glaciers, overdeepening and tarn, **b** showing rock glaciers in the study area (*Source* Google earth images)

present in the basin and have played an important role in increasing the overall valley width.

9.4.5 Morphotectonic Inferences

The presence of numerous erosional features, especially cirques and deep glaciated valleys reflect accelerated glacial erosion which, in combination with the glacial recession, might have resulted in the isostatic uplift of the rocks and mountain ranges in the basin (Dar et al. 2014b). In order to test this assumption, we carried out the morphotectonic analysis of numerous representative deglaciated troughs and the mountain fronts using two geomorphic indices; Mountain Front Sinuosity Index (S_{mf}) and Valley Floor Width to Height Ratio Index (V_f). The geomorphic features, viz., mountain fronts and glacial troughs, were selected as the glaciers have not obscured the tectonic-geomorphic evidences in the selected areas. S_{mf} values of more than 100 mountain front segments in the basin range from 1.007 to 1.300 with an average value of 1.088. Since the values are less than the threshold value of 1.6 indicating straight and active mountain fronts, which reveals that the area is experiencing exhumation or postglacial uplift. The straight and linear truncation of the mountain fronts and moraine deposits also reflects the high tectonic activity of the basin (Bull and McFadden 1977; Jaan et al. 2019). Further, the V_f index was calculated at 39 locations across the basin and the value ranges between 0.011 and 1.603 with an average value of 0.243 which also suggests that the basin is tectonically very active. V-shaped valley formation is commonly associated with the active uplift and deep, linear stream incision (Pazzaglia et al. 1998). Large scale rock deformation and linear pattern of river incision and presence of knickpoints in the river course (Taloor et al. 2017) provides evidence for tectonic uplift and isostatic rebound at the end of glaciation.

9.5 Discussion

The Sind Basin harbours glacial geomorphic features over a significant portion of the basin. Glaciers currently occur as cirque and small valley glaciers, especially in the east and north-eastern side of the basin. However, the role played by glaciers in shaping the landscape is well established by the presence and preservation of glacial landforms, like moraine deposits, cirques, glacial troughs, hanging valleys, glacial lakes, tongue basins, erratic boulders, arêtes, etc. Mountainous areas of the basin are subjugated by erosional glacial landforms, while the piedmont areas and valleys record both the erosional and depositional glacial landforms (Fig. 9.2). In the upper mountain reaches of the basin, the typical alpine landforms like cirques, arêtes, overdeepened valleys, horns exist and the lower reaches are dominated by a glacial trough and glacial depositional features like moraine ridges, erratics. Much of the eroded material has been transported beyond the periphery of the current glaciated boundary most probably by the high energy fluvial activity. However, a large amount of the glacial deposits still exist on the sides, base and at the end of the glacial valleys. Deposits of the older glaciation are either eroded by the recent glacial activity or are covered by trees and younger sediments (Fig. 9.5a–b).

The locations of the lateral and recessional moraines indicate glacial advances and retreat due to Late Quaternary climate change. The distribution and location of cirques, glacial valleys and lateral moraines indicates that the basin has witnessed alpine-style glaciations in the past, with centres of glaciation on the higher mountain areas (Fig. 9.2).

On the basis of the pattern and number of lateral and terminal moraine ridges, Dar et al. (2017) suggested that Thajwas, one of the main glaciers of the basin has witnessed three major glacial advances during the glacial maximum. It is believed that the glacial advances in the other glacial valleys of the basin, like Harmukh valley, might be concomitant with the Thajwas glacier advance (Murtaza et al. 2020). Nevertheless, the application of accurate dating methods, such as carbon dating, cosmogenic radionuclide or OSL dating, and in-depth studies of glacial borne sediments would help to precisely determining the exact number of glacial advances in the basin. The current estimates worked out in this study suggest that the glaciers covered almost 914 km^2 of the basin in the past. The estimate does not correspond to glacier coverage for a particular point in time but rather refers to the collective outcome of the glaciations during the maxima of glaciation. Further, the current glacial expansion estimates may vary because some of the glaciers deposited landforms are not detected in the remote sensing data as they might lie outside the delineated boundaries. Therefore, more detailed field investigations are required to reveal the sediment-landform patterns that occur outside of the reconstructed ice margins.

Loess, windblown sediment deposited during the dry windy glacial paleoclimatic times, blankets large areas of the Sind Basin (Dar et al. 2016). The thickest deposit of the basin (7 m thick loess-paleosol section) occurs at the Woyil plateau, near Kangan Village consisting of four paleosol profiles (Gupta et al. 1991). It is good to point out here that paleosols represent soils of the past and develop during warm and humid climatic conditions (Rendell et al. 1989). However, the loess pattern in the basin is determined by the general topography with terraces mantled by thin loess sheets and plateau settings preserving thick loess deposits. The loess deposits have been reported as products of glacial conditions and are deposited due to the transportation of sediments

Fig. 9.5 Field photograph **a** showing moraines covered with vegetation and **b** shows left lateral (older) moraine of the Thajwas Glacier. Note the presence of erratic boulders and dense vegetation cover on the moraine ridges (*Source* Authors)

by strong winds (Dar et al. 2013). Per se, it seems that the periods of major loess accumulation should be synchronous with the major glacial advances (Youn et al. 2014). It is generally believed that when the glaciers expanded in the Kashmir Himalayas, loess was deposited in the adjacent regions especially in the Kashmir Valley. Therefore, detailed paleoclimatic studies of loess can also provide important evidences for the past climatic changes and the associated glacial dynamics. Previously some researchers have tried to deduce the number of glacial and interglacial periods documented in the loessic sediments of the Kashmir Valley (Singhvi et al. 1987; Agrawal et al. 1989; Dilli and Pant 1994). However, due to the paucity of high resolution chronology, the initiation of loess deposition and the exact number of glacial advances remains a debatable issue in the Kashmir Valley (Dar and Zeeden 2020). Kumar et al. (2018) on the basis of carbon dating of the loess paleosols sequences suggested that Kashmir Valley witnessed LGM during 18.6–22.3 ka. It has been felt that still a fair amount of work is needed to know the exact number of glacial advances witnessed by the Kashmir Himalayas. The future work, therefore, should focus on the high resolution carbon, TL, OSL and cosmogenic radionuclide dating of loess deposits and glacial moraines to reconstruct the timing and nature of the Kashmir Himalayan glaciations.

The observations made in this study show the importance of remote sensing and GIS techniques for mapping various geomorphic features in the mountainous Himalaya. Unless supported by the absolute dating, the inferences about the palaeoclimatic changes and glacial advances in the area, though informative, shall be treated indicative only. The dating techniques (e.g., ^{14}C, OSL, TL and CRN dating) need to be carried out in the glaciated areas following the remote sensing and GIS studies for knowing the glacier advances and retreats in the area. Further, the landform mapping needed to be improved by integrating the remotely sensed data with stratigraphic data, information on the internal structure of landforms and variations in sedimentological characteristics (Dunlop and Clark 2006; Napieralski et al. 2007). The integrated analysis of morphometry, internal structure, and larger-scale patterns when analyzed together shall provide additional insights into the glaciological processes and landscape evolution.

9.6 Conclusion

The results from this study demonstrate the usefulness of the high resolution remote sensing data for glacial landform mapping and delineating the extent of former glaciers in inaccessible and rugged Himalayan terrain. The glacial landforms were identified and the extents of the maximum glacial advance outlined which will form a guide and basis for detailed field investigations aimed at paleo-glacial reconstructions in Kashmir Himalaya. However, a consistent and accurate chronology of the glaciations within the basin, and in broader Kashmir Himalayas and its correlation with chronostratigraphic systems of other glaciated Himalayan regions have not yet been established. Some attempts have been made to provide insight into the questions of local glacial chronology using lichenometry and morphological characteristics of the glacial moraines, but this has proved insufficient. Future research in the region might be focused on the absolute geochronology of the glacial deposits to establish the timing and nature of the glacial advances in the region to improve our understanding of the deglaciation patterns and process in the basin. The other important aspect that needs the attention of the researchers would be to estimate the impact of tectonics on the glacier dynamics or the postglacial uplift on the preservation of glacial landforms, especially at the higher elevations in the Kashmir Himalayas.

Acknowledgments The research work was conducted as part of the Department of Science and Technology (DST), Government of India sponsored national project titled "Centre of Excellence for Glacial Studies in Western Himalaya" and the financial assistance received from the Department under the project to accomplish this research is thankfully acknowledged.

References

Agrawal DP, Dodia R, Kotlia BS, Razdan H, Sahni A (1989) The Plio-Pleistocene geologic and climatic record of the Kashmir valley, India: a review and new data. Palaeogeogr Palaeoclimatol Palaeoecol 73:267–286

Alam A, Romshoo SA, Bhat MS (2011) Estimation of Snowmelt Runoff using Snowmelt Runoff Model (SRM) in a Himalayan Watershed. World 1:37–42

Alam A, Bhat MS, Kotlia BS, Ahmad B, Ahmad S, Taloor AK, Ahmad HF (2017) Coexistent pre-existing extensional and subsequent compressional tectonic deformation in the Kashmir basin, NW Himalaya. Quat Int 444:201–208

Ali SN, Juyal N (2013) Chronology of late quaternary glaciations in Indian Himalaya: a critical review. J Geol Soc India 82:628–638

Ali SN, Biswas RH, Shukla AD, Juyal N (2013) Chronology and climatic implications of Late Quaternary glaciations in the Goriganga valley, central Himalaya, India. Quat Sci Rev 73:59–76

Ali I, Shukla A, Romshoo SA (2017) Assessing linkages between spatial facies changes and dimensional variations of glaciers in the upper Indus Basin, western Himalaya. Geomorphology 284:115–129

Bali R, Ali SN, Agarwal KK, Rastogi SK, Krishna K, Srivastava P (2013) Chronology of late Quaternary glaciation in the Pindar valley, Alaknanda basin, Central Himalaya (India). J Asian Earth Sci 66:224–233

Barr ID, Spagnolo M (2013) Palaeoglacial and palaeoclimatic conditions in the NW Pacific, as revealed by a morphometric analysis of cirques upon the Kamchatka Peninsula. Geomorphology 192:15–29

Alam A, Bhat, MS, Kotlia BS, Ahmad B, Ahmad S, Taloor AK, Ahmad HF (2018) Hybrid tectonic character of the Kashmir basin: response to comment on "Coexistent pre-existing extensional and subsequent compressional tectonic deformation in the Kashmir basin, NW Himalaya (Alam et al., 2017)" by Shah (2017) Quat Int 468:284–289

Bhat MA, Romshoo SA, Beig G (2017) Aerosol black carbon at an urban site-Srinagar, Northwestern Himalaya, India: seasonality, sources, meteorology and radiative forcing. Atmos Environ 165:336–348

Bhat MS, Alam A, Ahmad B, Kotlia BS, Farooq H, Taloor AK, Ahmad S (2019) Flood frequency analysis of river Jhelum in Kashmir basin. Quat Int 507:288–294

Bisht H, Kotlia BS, Kumar K, Dumka RK, Taloor AK, Upadhyay R (2020) GPS derived crustal velocity, tectonic deformation and strain in the Indian Himalayan arc. Quat Int. https://doi.org/10.1016/j.quaint.2020.04.028

Brozović N, Burbank DW, Meigs AJ (1997) Climatic limits on landscape development in the northwestern Himalaya. Science 276:571–574

Bull WB, McFadden L (1977) Tectonic geomorphology north and south of the Garlock fault, California. In: Geomorphology in arid regions, pp 115–128

Dar RA, Romshoo SA (2013) Estimating daily stream flow in the glacierized Mountainous Kashmir Himalayan Basin. J Res Dev 12

Dar RA, Zeeden C (2020) Loess-Palaeosol sequences in the Kashmir Valley, NW Himalayas: a review. Front Earth Sci 8:113. https://doi.org/10.3389/feart.2020.00113

Dar RA, Chandra R, Romshoo SA (2013) Morphotectonic and lithostratigraphic analysis of intermontaneKarewa Basin of Kashmir Himalayas, India. J Mount Sci 10:1–15

Dar RA, Rashid I, Romshoo SA, Marazi A (2014a) Sustainability of winter tourism in a changing climate over Kashmir Himalaya. Environ Monit Assess 186:2549–2562

Dar RA, Romshoo SA, Chandra R, Ahmad I (2014b) Tectono-geomorphic study of the KarewaBasin of Kashmir Valley. J Asian Earth Sci 92:143–156

Dar RA, Romshoo SA, Chandra R, Ahmad I (2016) Response to "No major active backthrust bounds the PirPanjal Range near Kashmir basin, NW Himalaya" by Shah. J Asian Earth Sci 123:58–60

Dar RA, Jaan O, Murtaza KO, Romshoo SA (2017) Glacial-geomorphic study of the Thajwas glacier valley, Kashmir Himalayas, India. Quat Int 444:157–171

Dilli K, Pant RK (1994) Clay-minerals as indicators of the provenance and paleoclimatic record of the Kashmir-loess. J Geol Soc India 44:563–574

Dimri AP, Chevuturi A (2016) Western disturbances—an Indian meteorological perspective. Springer, Switzerland

Dubey J, Ali SN, Sharma A, Morthekai P, Sing R, Sharma RK, Srivastava V (2019) Glacial geomorphology and landscape evolution of the Thangu Valley, North Sikkim Himalaya, India. J Indian Soc Remote Sens 47(5):821–883

Dunlop P, Clark CD (2006) The morphological characteristics of ribbed moraine. Quat Sci Rev 25:1668–1691

Ehlers J, Ehlers J, Gibbard PL, Hughes PD (2011) Quaternary glaciations-extent and chronology: a closer look, vol 15. Elsevier

Fu P, Harbor JM, Stroeven AP, Hättestrand C, Heyman J, Zhou L (2013) Glacial geomorphology and paleoglaciation patterns in Shaluli Shan, the southeastern Tibetan Plateau—evidence for polythermal ice cap glaciation. Geomorphology 182:66–78

Greenwood SL, Clark CD (2009a) Reconstructing the last Irish Ice Sheet 2: a geomorphologically-driven model of ice sheet growth, retreat and dynamics. Quat Sci Rev 28:3101–3123

Greenwood SL, Clark CD (2009b) Reconstructing the last Irish Ice Sheet 1: changing flow geometries and ice flow dynamics deciphered from the glacial landform record. Quat Sci Rev 28:3085–3100

Gupta SK, Sharma P, Juyal N, Agrawal DP (1991) Loess—palaeosol sequence in Kashmir: correlation of mineral magnetic stratigraphy with the marine palaeoclimatic record. Quat Sci 6:3–12

Jaan O, Lone S, Malik R, Lone A, Wasim M, Nissa A (2019) Morphotectonic and morphometric analysis of Vishav basin left bank tributary of Jhelum River SW Kashmir Valley India. Int J Eco Environ Geol 6(2):17–26

Jansson KN, Glasser NF (2005) Using Landsat 7 ETM + imagery and digital terrain models for mapping glacial lineaments on former ice sheet beds. Int J Remote Sens 26:3931–3941

Kumar P, Shrivastava JP, Chandra R, Chopra S, RoonwalGS, Sharma R (2018) High resolution 14C AMS ages (~ 50 ka) of organic matter associated with the loess palaeosol Holocene-Late Pleistocene (8–130 ka) sediments of Dilpur Formation, Karewa Group, Kashmir, India. Quat Geochron 47:170–179

Kumar D, Singh AK, Taloor AK, Singh DS (2020) Recessional pattern of Thelu and Swetvarn glaciers between 1968 and 2019, Bhagirathi basin, Garhwal Himalaya, India. Quat Int. https://doi.org/10.1016/j.quaint.2020.04.049

Marazi A, Romshoo SA (2018) Streamflow response to shrinking glaciers under changing climate in the Lidder Valley, Kashmir Himalayas. J Mount Sci 15:1241–1253

Murtaza KO, Romshoo SA (2017) Recent glacier changes in the Kashmir alpine Himalayas, India. Geocarto Int 32:188–205

Murtaza KO, Dar RA, Paul OJ, Bhat NA, Romshoo SA (2020) Glacial geomorphology and recent glacial recession of the Harmukh Range. QuatInt, NW Himalaya. https://doi.org/10.1016/j.quaint.2020.08.044

Napieralski J, Harbor J, Li Y (2007) Glacial geomorphology and geographic information systems. Earth-Sci Rev 85:1–22

Pazzaglia FJ, Gardner TW, Merritts DJ (1998) Bedrock fluvial incision and longitudinal profile development over geologic time scales determined by fluvial terraces. Geophys monograph, vol 107, pp 207–236. American Geophysical Union

Rashid I, Romshoo SA, Abdullah T (2017) The recent deglaciation of Kolahoi valley in Kashmir Himalaya, India in response to the changing climate. J Asian Earth Sci 138:38–50

Rendell HM, Gardner RAM, Agrawal DP, Juyal N (1989) Chronology and stratigraphy of Kashmir Loess. Zeitschrift fur GeomorphologieSuppliment 76:213–223

Romshoo SA, Dar RA, Rashid I, Marazi A, Ali N, Zaz SN (2015) Implications of shrinking cryosphere under changing climate on the streamflows in the Lidder catchment in the Upper Indus Basin, India. Arctic Antarctic Alpine Res 47:627–644

Shi Y (2002) Characteristics of late Quaternary monsoonal glaciation on the Tibetan Plateau and in East Asia. Quat Int 97:79–91

Shukla A, Yousuf B (2017) Evaluation of multisource data for glacier terrain mapping: a neural net approach. Geocarto Int 32:569–587

Singh S, Sood V, Taloor AK, Prashar S, Kaur R (2020) Qualitative and quantitative analysis of topographically derived CVA algorithms using MODIS and Landsat-8 data over Western Himalayas, India. Quat Int. https://doi.org/10.1016/j.quaint.2020.04.048

Singhvi AK, Bronger A, Pant RK, Sauer W (1987) Thermoluminescence dating and its implications for the chronostratigraphy of loess-paleosol sequences in the Kashmir Valley (India). Chem Geol: Isotope Geosci Sect 65:45–56

Sood V, Gusain HS, Gupta S, Taloor AK, Singh S (2020) Detection of snow/ice cover changes using subpixel-based change detection approach over Chhota-Shigri glacier, Western Himalaya, India. Quat Int. https://doi.org/10.1016/j.quaint.2020.05.016

Stojanovic D, Aitchison JC, Ali JR, Ahmad T, Dar RA (2016) Paleomagnetic investigation of the early Permian Panjal Traps of NW India; regional tectonic implications. J Asian Earth Sci 115:114–123

Taloor AK, Ray PK, Jasrotia AS, Kotlia BS, Alam A, Kumar SG, Kumar R, Kumar V, Roy S (2017). Active tectonic deformation along reactivated faults in Binta basin in Kumaun Himalaya of north India: Inferences from tectono-geomorphic evaluation. Z Geomorphol 61(2):159–80

Taloor AK, Kotlia BS, Jasrotia AS, Kumar A, Alam A, Ali S, Kouser B, Garg PK, Kumar R, Singh AK, Singh B (2019) Tectono-climatic influence on landscape changes in the glaciated Durung Drung basin, Zanskar Himalaya, India: a geospatial approach. Quat Int 507:262–273

Taloor AK, Joshi LM, Kotlia BS, Alam A, Kothyari GC, Kandregula RS, Singh AK, Dumka RK (2020) Tectonic imprints of landscape evolution in the Bhilangana and Mandakini basin, Garhwal Himalaya, India: a geospatial approach. Quat Int. https://doi.org/10.1016/j.quaint.2020.07.021

Thackray GD, Owen LA, Yi C (2008) Timing and nature of late Quaternary mountain glaciation. J Quat Sci 23:503–508

Thakkar MG, Kothyari GC, Jani C, Chauhan G, Lakhote A, Taloor AK (2020) Time assessment of tectonic and climatic forcing on the formation of Khari bedrock gorge, Kachchh, western India: a mathematical approach. Quat Int. https://doi.org/10.1016/j.quaint.2020.06.035

Youn JH, Seong YB, Choi JH, Abdrakhmatov K, Ormukov C (2014) Loess deposits in the northern Kyrgyz Tien Shan: implications for the paleoclimate reconstruction during the Late Quaternary. CATENA 117:81–93

Dr. Reyaz Ahmad Dar works as an Assistant Professor in the Department of Earth Sciences, University of Kashmir, J & K, India. He did his Ph.D. in Geology from the University of Kashmir in 2014. His areas of expertise include Quaternary Geology and Paleoclimate. In the last 5 years, he has published more than 15 research papers in reputed journals. The notable journals in which his research has been published include Earth Science Reviews, Frontiers in Earth Sciences, Quaternary International, Journal of Asian Earth Science, Arctic Antarctic and Alpine Research. He is making significant contributions to Quaternary Himalayan Geology through a sustained and comprehensive study of the Kashmir Himalayas. These include glacial geomorphology, isotopic studies, micromorphologic studies, phytolith analysis, rock magnetic studies of Late Quaternary loess-paleosol sequences of the Karewa Group of sediments leading to the reconstruction of the paleoclimate of the Kashmir Valley. He has organized many national and international field expeditions to the Himalayas. In the year 2016, he was one among the four persons selected for the 'Teach the Teachers workshop' on glaciology and related areas at the Universities of Fribourg and Zurich, Switzerland

Mr. Omar Jaan Paul is pursuing Doctorate at the Department of Earth Sciences, University of Kashmir, Srinagar. He received his Master's from the same university in the subject of Geology. His research interests include glacial geomorphology, remote sensing, cosmogenic radionuclide dating and paleoclimate reconstructions. He has a research experience of more than 6 years. Initially, he worked for a period of 2 years in an Indo-Australian project and then for a period of 3 years on Cosmogenic Radionuclide Dating of glacial moraines from Great Himalayan Range. Presently, he is Working as a Project Associate in 'Center of Excellence for glacial studies in western Himalayas programme' sponsored by Department of Science and Technology, New Delhi, Govt. of India at the University of Kashmir. He is a life member of Indian Society of Geomatics and Indian Metrological society

Dr. Khalid Omar Murtaza has received his Ph.D. in Geoinformatics from the University of Kashmir and is currently working as Contractual Lecturer in Department of Geoinformatics, University of Kashmir. His research interests include glaciology, glacial-geomorphology and geoinformatics. He is presently working on glacial dynamics of Kashmir Himalaya using state-of-the-art Earth observation data and extensive ground observations with special emphasis on glacier mass balance, glacier-geomorphology and ice-surface velocity estimations. He has published 12 research papers in peer-reviewed and refereed journals. He has been teaching Remote Sensing and Glaciology from 2017 at the Department of Geoinformatics, University of Kashmir and has supervised seven Master's dissertations on research related to the applications of geospatial technology in various fields of Earth and Environmental Sciences

Prof. Shakil Ahmad Romshoo has Master's degree in remote sensing and GIS from the Asian Institute of Technology, Bangkok and doctoral degree from the University of Tokyo, Japan and is currently serving as Dean for Research at the University of Kashmir, India. He has 32 years of multidisciplinary academic experience researching issues related to geomatics, hydrology, glaciology and climate change with the geographic focus in the Himalaya. He has headed and taught both in the Department of Earth Sciences and Department of Geoinformatics at the University of Kashmir. He has served in the past as Fellow, The Energy Research Institute (TERI), New Delhi and Scientist at the Japan Aerospace Exploration Agency, Japan. He has written more than 150 articles published in refereed journals or edited collections and is an Elected Fellow of the Indian Academy of Sciences, Indian Society of Remote Sensing and Indian Society of Geomatics. He is serving on the Editorial Board of PLOS ONE

Snow Cover and Land Surface Temperature Assessment of Mana Basin Uttarakhand India Using MODIS Satellite Data

10

Manish Rawat, Sateesh Karwariya, Ritik Raushan, Shruti Kanga, Ajay Kumar Taloor, and Asha Thapliyal

Abstract

Climate change has a significant impact on the Himalayan glaciers in the last three decades. Several methods and techniques have been used for the control of spatial or temporal trends in the snow cover and temperature of the Himalayas. Understanding the relationship between climate change and glacial dynamics is essential for better considerate the interactions among various drivers, including land cover change, climate change, local ecological assessments and adaptation research. In this study, daytime LST retrieval from the MODIS (MOD11A2) data was analyzed against the snow cover retrieval trend using MODIS (MOD10A2) for the Mana basin of Chamoli district of Uttarakhand. The results from the trend analysis of surface temperature and snow cover from 2001 to 2018 showed that the maximum and mean surface temperature has been rising in all four zones of the basin with different growth rates. Also, it was found that the percentage of snow cover has been slightly declining from the year 2001 to 2018 due to the rise in the surface temperature. Season-wise trends during 2001–2018 in all the temperature series also showed significant positive warming trends. The trend analysis of the MODIS satellite products for LST and snow cover showed promising results in understanding the climate change phenomenon operating in the basin affecting downstream requirements of the water for energy and drinking water supplies.

Keywords

Climate change · Drinking water · Mana basin · MODIS LST · MODIS snow cover · Trend analysis

M. Rawat
Water Resource Department, NIH, Roorkee, India
e-mail: rawat1992manish@gmail.com

S. Karwariya
SAC-ISRO, Ahmadabad, India
e-mail: sateesh.karwariya@gmail.com

R. Raushan
Department of Water Engineering and Management, Central University of Jharkhand, Ranchi, India
e-mail: ritikraushan333@gmail.com

S. Kanga (✉)
Centre for Climate Change and Water Research, Suresh Gyan Vihar University Jagatpura, Jaipur, India
e-mail: shruti.kanga@mygyanvihar.com

A. K. Taloor
Department of Remote Sensing and GIS, University of Jammu, Jammu 180006, India
e-mail: ajaytaloor@gmail.com

A. Thapliyal
Uttarakhand Space Application Centre (USAC), Dehradun, Uttarakhand 248001, India
e-mail: ashath08@gmail.com

A. K. Taloor et al. (eds.), *Water, Cryosphere, and Climate Change in the Himalayas*, Geography of the Physical Environment, https://doi.org/10.1007/978-3-030-67932-3_10

10.1 Introduction

Glaciers are much susceptible to change of climate and act as a good global warming predictor. Globally the average air temperature of the earth's surface has been risen by 0.6 °C during the twentieth century (IPCC 2007), and several small glaciers retreated sharply. In the last few decades, the glaciers of the Indian Himalaya have been widely affected by changing climate (Kothyari and Singh 1996; Singh and Sontakke 2002; Kothawale et al. 2010; Bolch et al. 2012; Bisht et al. 2019; Taloor et al. 2020a; Sood et al. 2020a, b). The regional impacts of warming climate in the Indian sub-continent have been examined in numerous studies (Dash and Hunt 2007, Yadav et al. 2004; Kulkarni 2007; Bhutiyani et al. 2007; Taloor et al. 2019; Bisht et al. 2020; Kumar et al. 2020; Singh et al. 2020). Several methods of remote sensing were employed to determine spatio-temporal temperature and snow cover patterns in the Himalayas region. For mapping of snow-covered area and surface temperature estimation, several sensors are available, such as the Advanced Wide Field Sensor (AWiFS), Landsat 5 Thematic Mapper (TM) and MODIS. MODIS derived SCA is more advantageous compared to other sensors because the MODIS data correctly classifies the snow from the other features and has been enhanced with regard to special resolution and geolocation accuracy (Tekeli et al. 2005). The MODIS data recorded since 2000 is publicly accessible and has many ground, ocean, and atmospheric applications. Several products and their various MODIS variants are currently widely available. Several studies (Neteler 2010) have been performed for validating MODIS surface temperature and snow materials. The results of these surveys were exceptionally compared to the ground measurements or satellite snow products of high-resolution in various regions worldwide, including the Himalayas (Negi et al. 2007; Wang et al. 2008; Singh et al. 2017; Kannaujiya et al. 2020; Khan et al. 2020; Haque et al. 2020; Sood et al. 2020b). MODIS ground surface temperature measurements experiments have provided good results in many parts of the Himalayas and are consistent with different studies (Negi et al. 2007; Wang et al. 2007; Hachem et al. 2012; Pall et al. 2019; Romshoo et al. 2020). The surface terrain temperature (LST) is an integrated parameter for monitoring rational and latent heat flows in the energy exchange in atmosphere and land surface (Houghton et al. 2001; Tang et al. 2008; Sarkar et al. 2020; Taloor et al. 2020b). The LST from the MODIS sensor is currently applied in different fields because of its high spatial and temporal availability, but observations in cloudy conditions are very difficult to achieve. The main aim of the study is to establish snowfall and temperature trends for seasons in the four altitude zones of Mana basin, Uttrakhand, India.

10.2 Area of Study

Mana basin situated in the Chamoli district of Uttarakhand covering an area around 3503.17 km^2 with 3000–7000 m elevation. In this study, the basin separated into four elevation zones (Zone II to Zone V). Zone II lies between 3580 m and 4163 m amsl, considered as a terminus region, also called as end moraines. Zone III covers the area between 4060 and 4675 m amsl, which is considered as a snout region and influenced by seasonal snow cover. Zone IV covers the area between 4554 m and 5163 m, which cover the area under snow accumulation and ablation region. Zone V lies above the 5163 m amsl and is considered as a glacier region, which is covered by snow all through the year (Fig. 10.1).

10.3 Methodology

10.3.1 Objective

The purpose/objective of the study is to observe the temporal and spatial patterns of Mana basin's temperature and snowfall trends in winter and pre-monsoon from 2001 to 2018.

10.3.2 Data Sets Used

The MODIS LST (MOD11A2 L3 V5) time series product which provides 1 km of spatial resolution per-pixel temperature has been used in

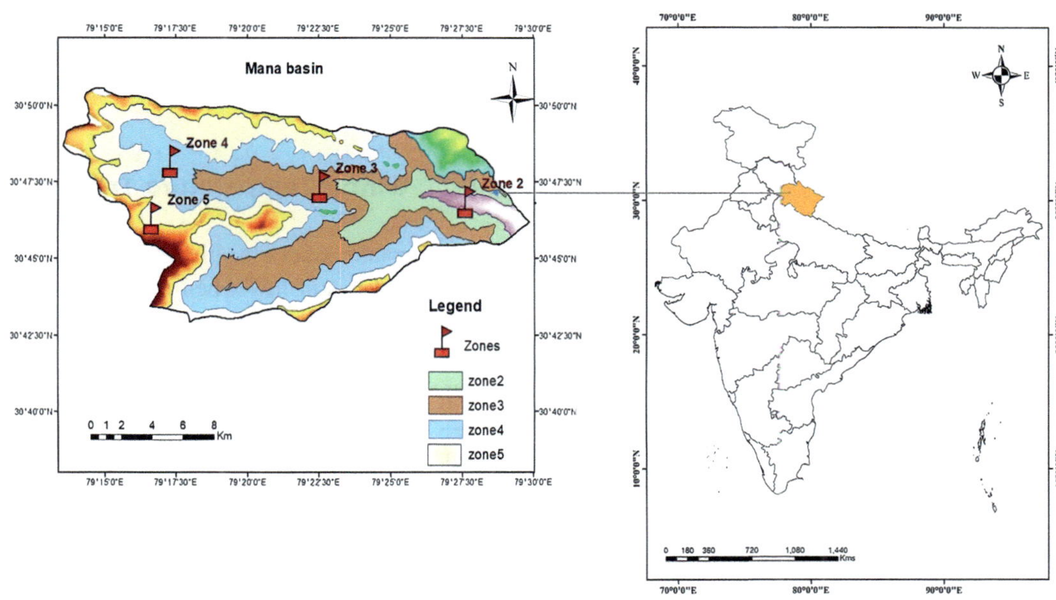

Fig. 10.1 Study area map for location (*Source* Authors generated map)

this study. The daytime LST and 8-day L3 Global 500 m Grid MODIS (MOD09A1) products on the basis of the algorithm used in snow mapping Normalized Difference Snow Index (NDSI) were also used to demonstrate the effect of snow on LST. In addition, study area elevation data was acquired from the ASTER 30-m DEM was resampled to a 1 km resolution to fit the MODIS data. The DEM was used to create the elevation zones at 500 m elevation intervals, and different elevation zones were calculated to have an average temperature.

10.3.3 Estimation Methods

The MODIS LST product is available under the Earth Observation System (HDF-EOS) Hierarchical Data Format. Nine scientific data sets (SDSs) are included in the LST product: LST, QC, Error LST, Emissivity 31, Emissivity 32, View angle, View time, Latitude and Longitude. To reproject the original sinusoidal image to the WGS 1984 UTM ZONE 44 N projection method, a MODIS Conversion Toolkit (MCT) was used in the ENVI programme. The LST images of the study area were extracted after reprojection and then the effective calibration formula for temperature estimation in degrees Celsius was applied.

To estimate the percentage of snow cover MODIS (MOD09A1), snow product was selected. Using the NDSI produced by MODIS data, the spectral signature of snow can be obtained. It utilizes the low and high snow reflectance of the electromagnetic spectrum in the visible (green) and short-wave infrared (SWIR) areas. To determine NDSI from MODIS images following equation were used:

$$\mathrm{NDSI} = \frac{\text{Band } 4 - \text{Band } 6}{\text{Band } 4 + \text{Band } 6} \tag{10.1}$$

One of the key benefits of the NDSI is the accurate snow cover estimation under the shadow area. Both photos from the time of research (2001–2018) were grouped into the categories of snow and without snow. Thus, maps of SCA for each year were generated between the month of January and April.

The 21 ground stations data were validated in the Himalayas of the north-west (over 5300 m), the Himalayas of the Middle (4000–5300 m) and the Himalayas in the Lower (2000–4000 m) in

Negi et al. (2007). The surface temperature product MODIS (MOD11A2) from multiple studies was comparatively high as well as reliable. So we have explicitly, without validation, implemented this LST product in this report. In the present study, LST products were superimposed on ASTER DEM and pixels present in each elevation zone were selected. Moreover, for the temperature trend estimation, the MOD11A2 temperature products were considered. The distribution of spatial temperature trend analysis for the selected four elevation zones is carried out on a seasonal basis using ground surface temperature.

10.4 Results and Discussion

10.4.1 Spatial Variability of Climatic Variables Over Four Different Zones

The mean monthly LST for each grid was calculated from the available data. The rate of increase or decrease in monthly LST values is calculated and the values are further used for the decade-based estimation of the rate. The two seasons have been considered for the current study, the period when the snow-covered surface is present, so that the period between January and April is taken into account. For the seasonal data of maximum temperature and snowfall for all four zones, graphs were plotted. For 18 years of MOD11A2 land surface temperature product, the linear trend in the graphs has been carried out. In 2001–2018, patterns in all temperatures indicate a big warming trend during both seasons. The designs are positive in both seasons. In these seasons, there has also been a major negative trend in the snow cover region, with growing trends in the winter and pre-monsoon weather. The findings suggest that global warming can have a potential effect on the snow-covered area in the mountains. The maximum and mean temperature trends in all four zones increased over January, February, March, April and the snow cover trend decreased in the same months, ranging from 3500 to 5600 m in height. Figure 10.2 indicates the LST spatial variability over the study region at different points of time.

10.4.2 Winter Temperature Trends

The temperature rises steadily from January to February, but is still low when the precipitation becomes adequate to accelerate snow cover formation and sublimation. The mean annual temperature in January indicates a warming trend of 0.0145, 0.0314, 0.029 and 0.027 °C for Zone II to Zone V, respectively, during the 2001–2018 period. This represents a significant temperature acceleration during the last two decades. Due to higher net radiation and greater absorption of long-wave radiation that warms the debris cover region, a slightly higher temperature trend in zone III is 0.0314 °C/18 yr, and similarly, for January, the max temperature follows an increasing tendency in all climatic zones of zone II to zone V that was found to be approximately 0.020 °C, 0.027 °C, 0.0086 °C, 0.0012 °C, respectively. Zone V has a freezing condition, it receives dry snow and is highly glaciated, and its altitude is more than 5000 m. The mean monthly temperature trend in zone V is 0.0002 °C for January. Overall, the mean and maximum temperature at all the zones has been consistently on the rise since 2001 (Table 10.1).

10.4.3 Pre-monsoon Temperature Trends

The mean temperature for Zone II to Zone V in April indicates a strong warming trend of 0.05, 0.025, 0.038, and 0.10 °C over the 2001–2018 period. This reflects a small temperature rise over the last two decades. We discovered that 0.10 °C/18 yr is a slightly higher rising temperature trend in zone V. Similarly, the maximum temperature for the month of April followed a growing pattern in all climatic zones.

Furthermore, Table 10.1 shows the rising pattern in surface temperature for four months in

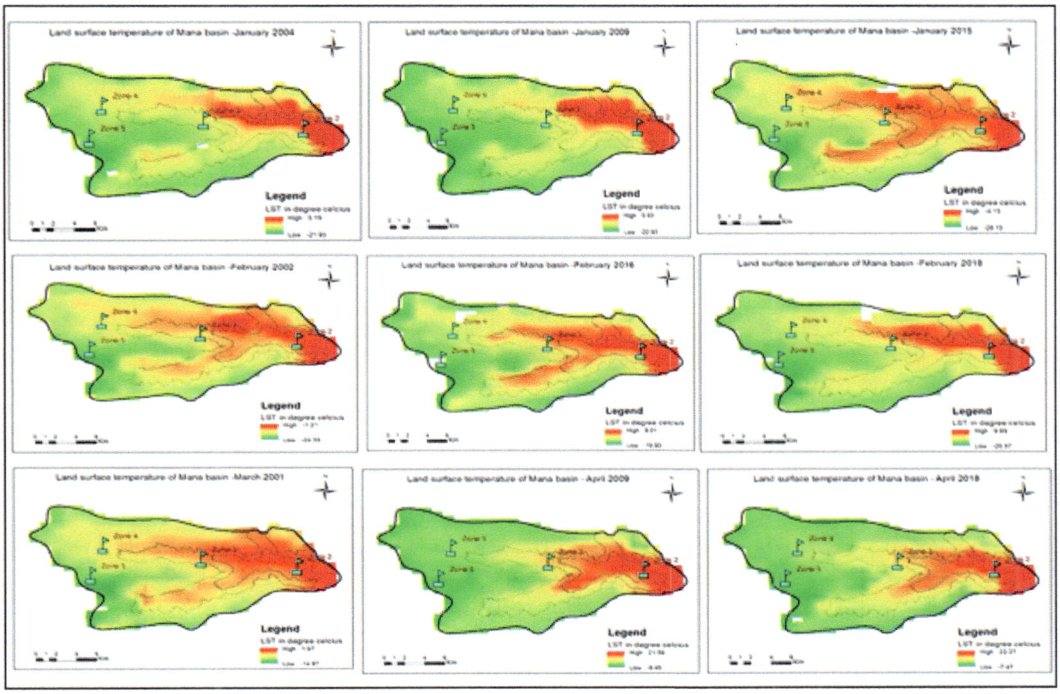

Fig. 10.2 LST time series maps of Mana basin derived from MODIS data at different points of time (*Source* Authors generated map)

Table 10.1 Linear Trend (°C/18 yr) in mean and maximum temperatures for all four different zones

2001–2018						
Zones	Elevation range	Temperature	January (°C)	February (°C)	March (°C)	April (°C)
Zone II	3581–4163	Mean temp	+0.0145	+0.262	+0.187	+0.05
		Max temp	+0.020	+0.160	+0.098	+0.084
Zone III	4060–4675	Mean temp	+0.0314	+0.228	+0.142	+0.025
		Max temp	+0.0274	+0.102	+0.223C	+0.040
Zone IV	4554–5163	Mean temp	+0.029	+0.140	+0.108	+0.038
		Max temp	+0.0086	+0.093	+0.216	+0.046
Zone V	5050–5666	Mean temp	+0.027	+0.104	+0.067	+0.10
		Max temp	+0.0002	+0.130	+0.076	+0.09

(*Source* Authors data)

all zones. It is found that +0.015 °C/year for February, the maximum mean temperature over zone II increased significantly by 0.262 °C. The surface temperatures mark the beginning of a cycle of the overcast situation with low net radiation and daily precipitation over the last few years with a consequent drop in LST fluctuation have been observed. The temperature trend shows that there is a lot of variance, but the trend is up overall. Out of Fig. 10.3a–h, LST values have been found to be high in recent years because they are characterized by low cloud cover, high net radiation and scanty precipitation.

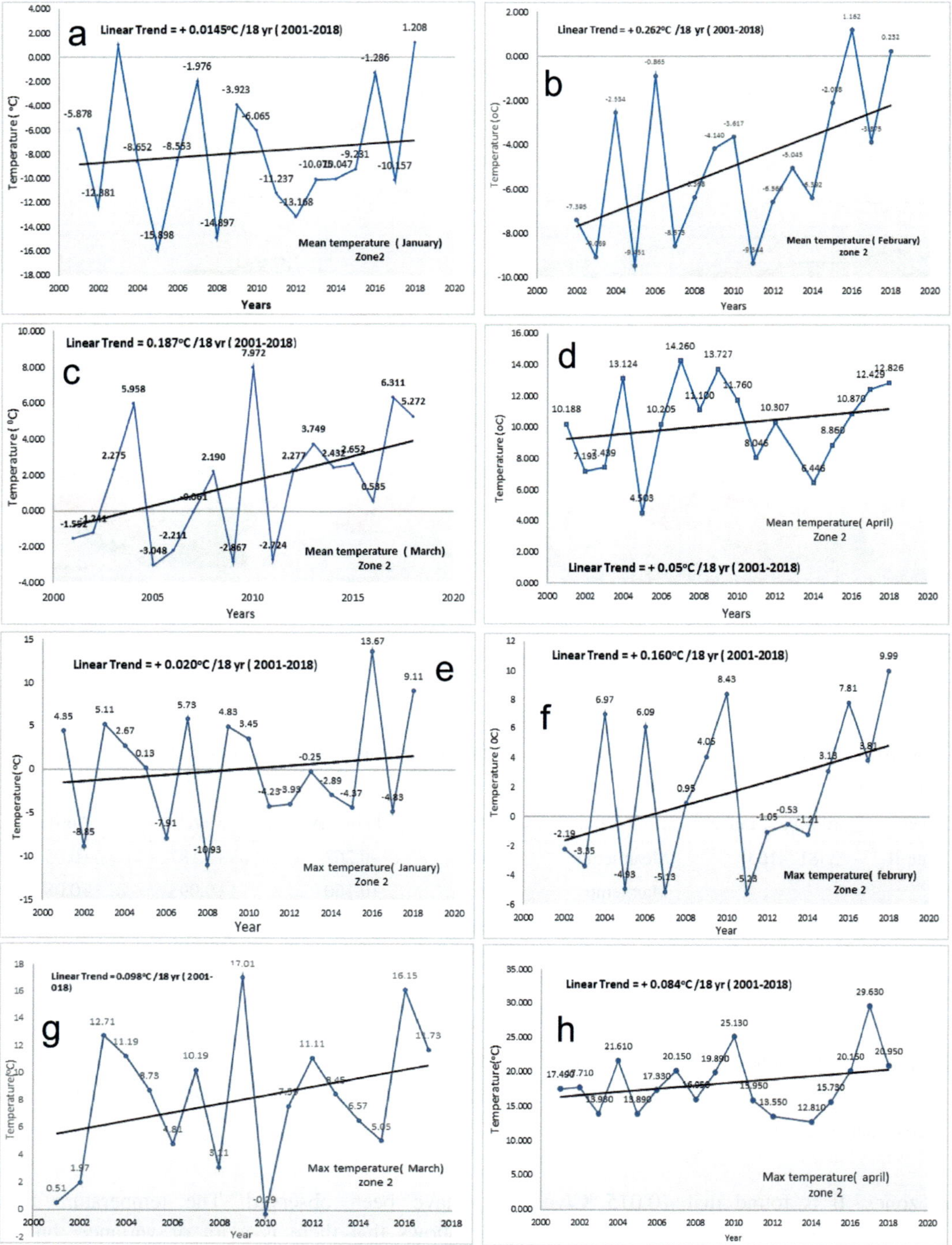

Fig. 10.3 **a–d** Variation of mean temperature and **e–h** variation of maximum temperature during 2001–2018 in zone II (*Source* Authors data)

10.4.4 Trend Analysis of Monthly LST (Zone II)

For the location of the Mana basin in Zone II, the monthly LST values of the last two decades were evaluated. The elevation ranges of Zone II are 3581–4163 m AMSL, which is considered to be a terminus region or a lower area (end of the moraine). The difference between the mean (a–d) and maximum (e–h) monthly LST values for zone II in the Mana range during the 2001–18 period is shown in Fig. 10.3. During the months of January, February, March and April, mean monthly LST values showed a growing pattern to 0.0145, 0.262, 0.187 and 0.05 °C in the altitude range of 3500–4163 m amsl for Zone II during the months of January, February, March and April. The trend of warming in this area is shown in the LST data by this anomaly. Similarly, maximum temperature trends have also shown a statistical significance for January, February, March and April of 0.020, 0.160, 0.098 and 0.084 °C over the last 18 years (2001–2018). The mean and maximum LST in zone II have increased according to the study, with a higher rate of 0.262 and 0.160 °C during the month of February.

10.4.5 Trend Analysis of Monthly LST Values (Zone III)

The monthly surface temperature values of the last two decades have been evaluated for the positions of the Mana basin in zone III. The elevation ranges of zone III are 4060–4675 m, which is known to be a zone filled with debris. The variance of the monthly mean (a–d) and (e–h) LST values for zone III during the period 2001–2018 is shown in Fig. 10.4. The elevation ranges of zone III are 4060–4675 m amsl, and during the months, the average monthly surface temperature shows a rising pattern. For the years 2000 to 2018, January, February, March and April were 0.0314 °C, 0.228 °C, 0.142 °C and 0.025 °C, respectively. The warming trend in this region is indicated by this anomaly in the LST data. Similarly, during January to April, the highest patterns of temperature over the last 18 years also showed statistically significant values of 0.0274 °C, 0.102 °C, 0.223 °C and 0.040 °C, respectively. An increasing trend with a higher rate during March at the rate of 0.0131 °C/year was the highest surface temperature in Zone III. The rate of temperature change in all the zones during a different month is shown in Table 10.1. The rate +0.0134 °C/year for February, the max mean temperature above zone III increased significantly by 0.2228 °C. Zone III is known as a debris-covered zone that decreases the latent heat flux and increases the temperature of the surface. In addition, the SCA percentage fluctuates with the annual time of year and clearly indicates that the pattern of snow cover variability decreases with an increase in the surface temperature, suggesting that the snow cover is negatively associated with LST values.

10.4.6 Trend Analysis of Monthly LST Values (Zone IV)

For the zone IV position of the Mana Basin, it is analysed that the estimated monthly mean LST values in the last two decades in Zone IV's elevation range is 4554–4675 m, which is known to be the accumulation field and part of the ablation. The variation of the monthly mean (a–d) and maximum (e–h) LST values for zone 2 in the Mana range during the 2001–18 period is shown in Fig. 10.5. During January, February, March and April, the mean monthly LST values showed a growing trend of 0.029, 0.140, 0.108 and 0.038 °C. The warming trend in this region is

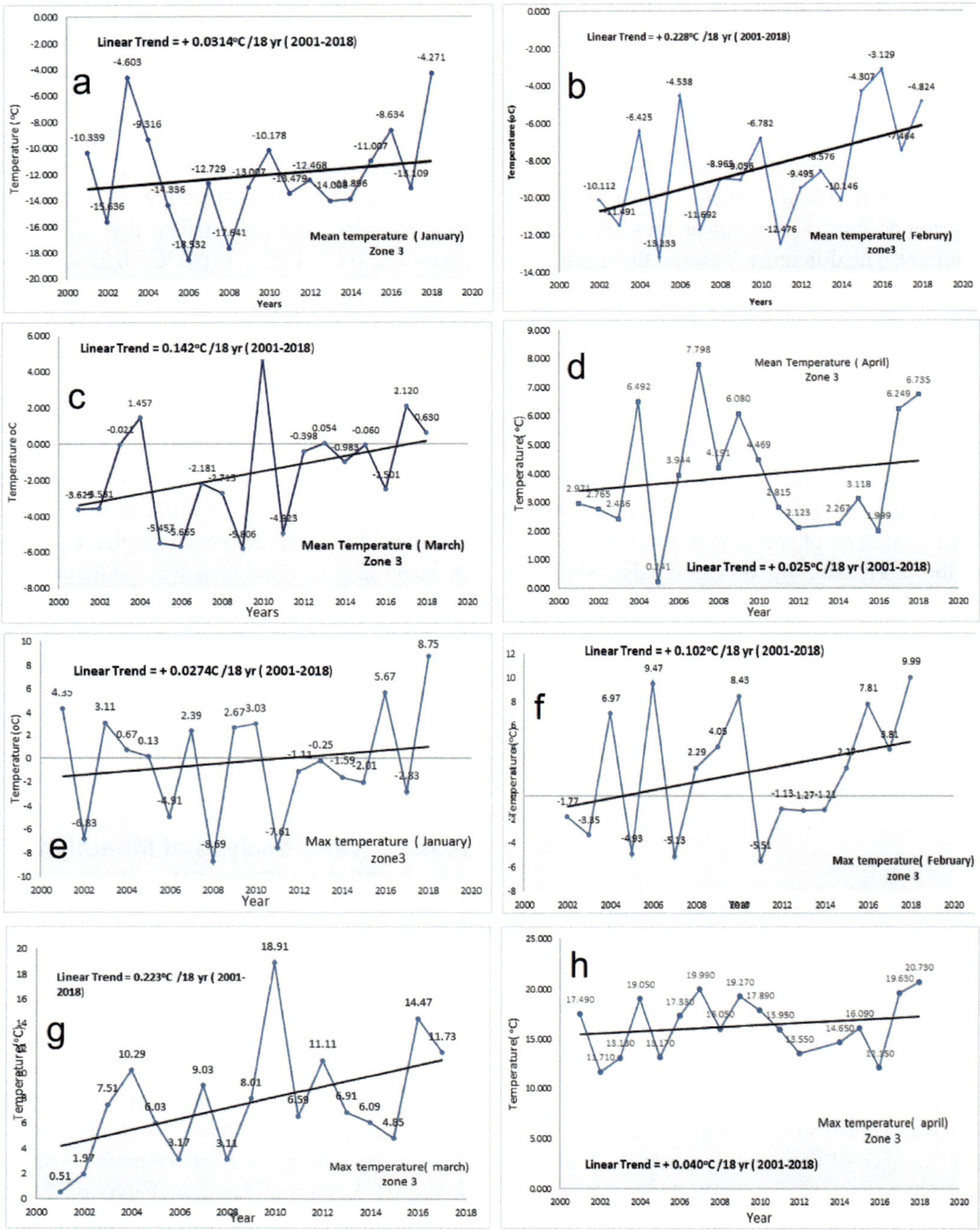

Fig. 10.4 **a–d** Variation of mean temperature and **e–h** variation of Maximum temperature during 2001–2018 in zone III (*Source* Authors data)

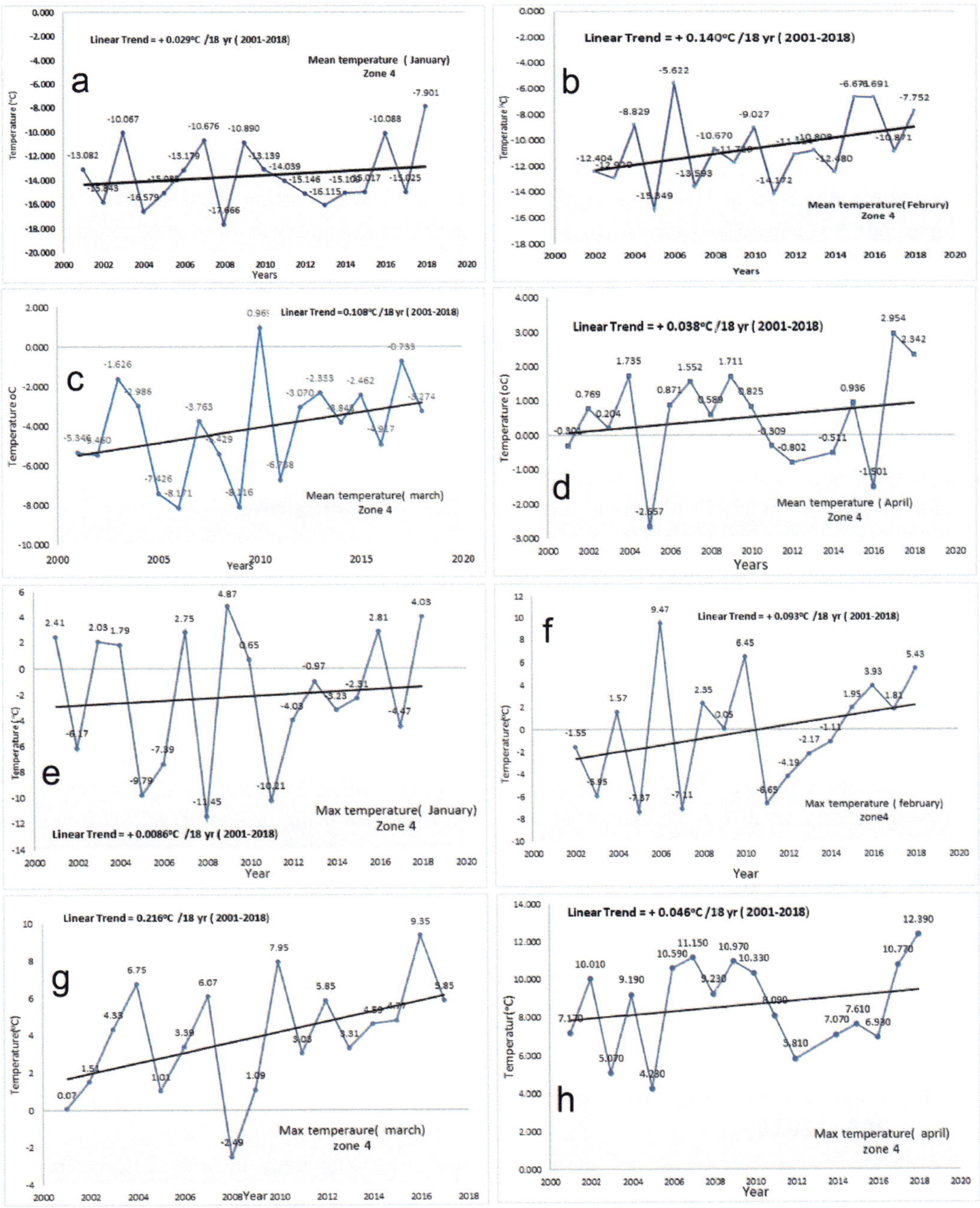

Fig. 10.5 **a–d** Variation of mean temperature and **e–h** variation of maximum temperature during 2001–2018 in zone IV (*Source* Authors data)

indicated by this anomaly in the LST data. Similarly, over the last 18 years (2001–2018), maximum temperature trends have also shown statistical significance for January, February, March and April as 0.0002 °C, 0.093 °C, 0.216 °C and 0.046 °C, respectively. The rate of temperature change in all the zones during a different month is shown in Table 10.1.

10.4.7 Trend Analysis of Monthly Mean LST Values (Zone V)

The monthly mean LST values of the last two decades have been analyzed for the Mana basin location of zone V. Zone V's elevation range is 5050–5666 m and is considered to be a glacier-accumulated area. The variance of the monthly mean (a–d) and (e–h) LST values for zone V in the Mana range during the 2001–18 period is shown in Fig. 10.6. From January to April, the mean monthly LST values showed a growing trend of 0.027, 0.104, 0.067 and 0.10 °C. Similarly, over the last 18 years (2001–2018), maximum temperature patterns have also shown a slightly increasing trend for the month of January, February, March, April as indicated from the study of 0.0002, 0.130, 0.076, and 0.09 °C respectively. This anomaly in the LST data shows the warming trend in this area and during the month of February, the mean and maximum LST in zone 2 increased at a higher rate.

10.4.8 Snow Cover Variability During 2001–2018

In Himalaya SCA, studies by Menon et al. (2010) indicate a decline of SCA from 1990 and 2001, whereas Gurung et al. (2011) recorded a slightly less decreasing trend in SCA of NWH for 2002–2010. The overall increasing trend in mean temperature over the basin has led to a decreasing trend for the overall long-term glacier and snow cover. All four zones in the mana basin exhibited an increasing trend of maximum temperature in January, February, March and April and also an increasing trend of mean temperatures during January to April at altitudes between 3500 and 5500 m amsl. A major negative trend in SCA was also established during these seasons, with growing trends in winter and spring temperatures. Results show potential effects of global warming in the high mountain region for precipitation and snow cover Fig. 10.7a–h.

10.5 Conclusions

One of the essential variables defining the climatological and hydrological processes operating over a glaciated environment is land surface temperature. In this report, using satellite-based temperature and snow cover products, trend analysis of surface temperature and snow cover from 2001 to 2018 was carried out. This study proposed MODIS (MOD11A2) daytime LST retrieval data using a mono-band algorithm and snow cover pattern analysis retrieval from the basin using MODIS snow product (MOD10A2) data. This study concluded that the maximum and mean surface temperature had risen with different growth rates in all four zones of the basin, and at the same time, it was also observed that the percentage of snow cover had decreased slightly between 2001 and 2018 due to surface temperature increase. The patterns in all the temperature series during 2001–2018 indicate a strong warming trend in both seasons. In addition, a major negative trend of SCA was also observed during these seasons, with growing temperature patterns in the winter and spring seasons.

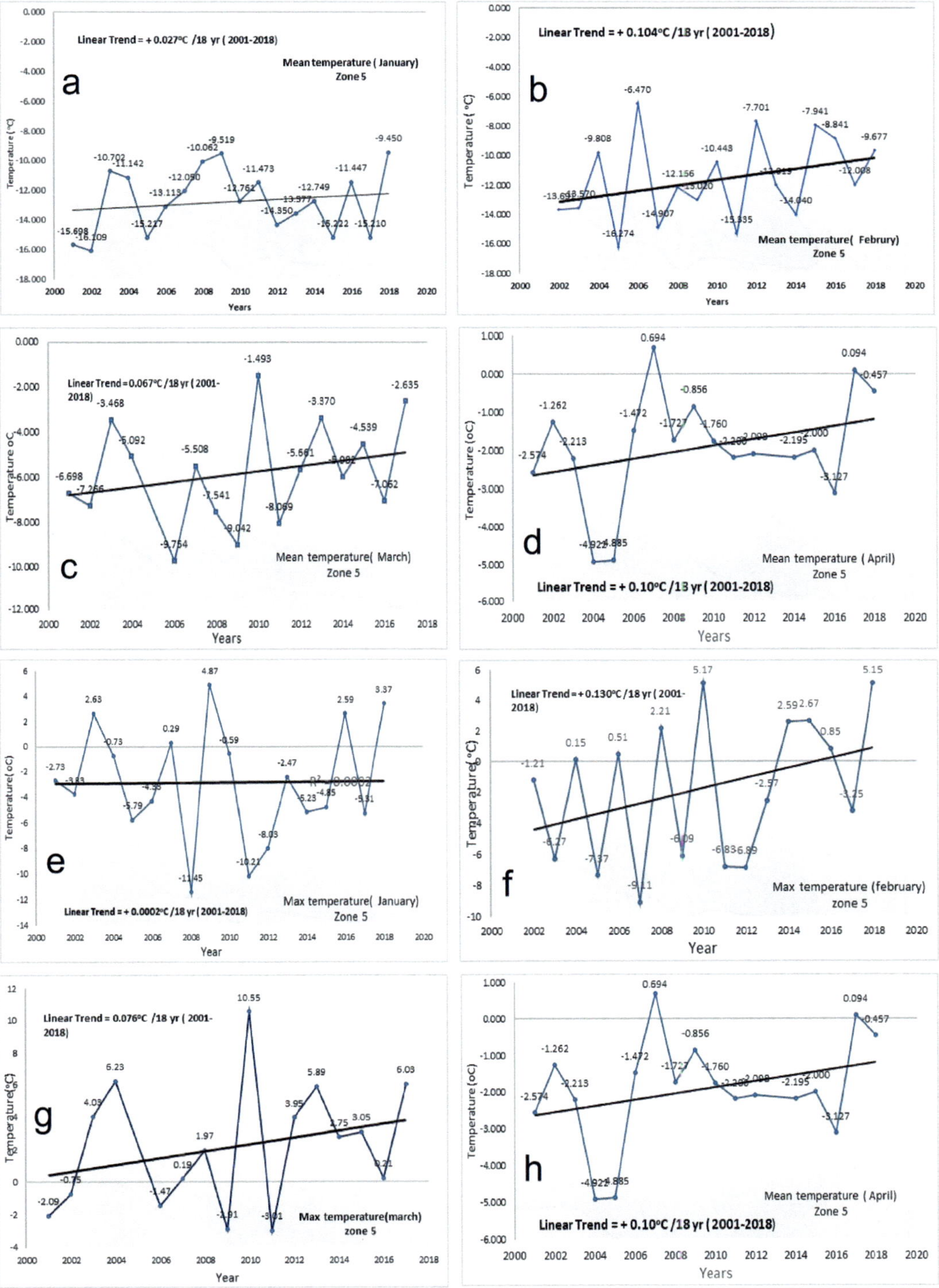

Fig. 10.6 **a–d** Variation of Mean temperature and **e–h** variation of Maximum temperature during 2001–2018 in zone V (*Source* Authors data)

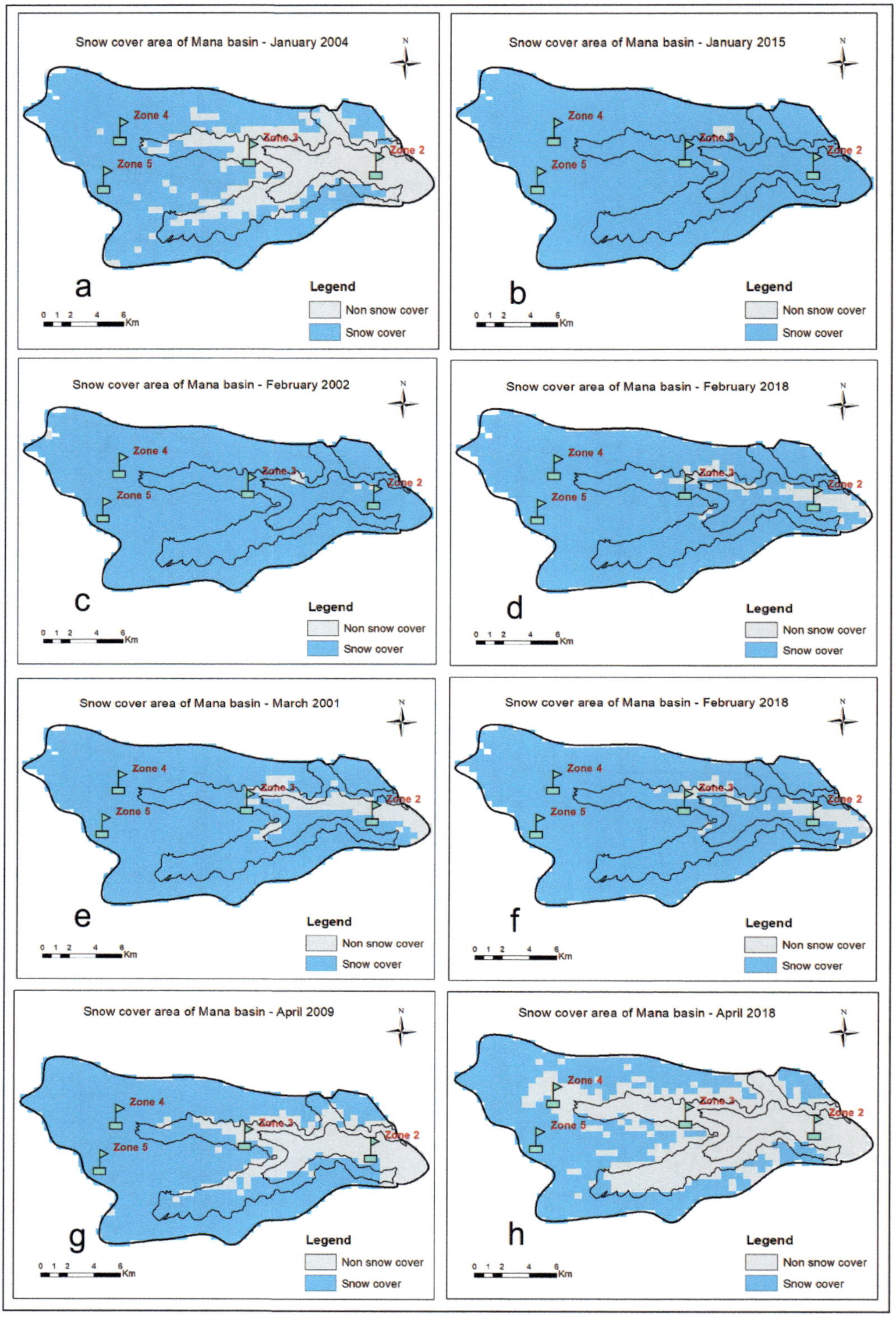

Fig. 10.7 a–h Variations of snow cover in the study area at different points of time (*Source* Authors data)

References

Bhutiyani MR, Kale VS, Pawar NJ (2007) Long-term trends in maximum, minimum and mean annual air temperatures across the North-western Himalaya during the twentieth century. Clim Change 85:159–177

Bisht H, Rani M, Kumar K, Sah S, Arya PC (2019) Retreating rate of Chaturangi glacier, Garhwal Himalaya, India derived from kinematic GPS survey and satellite data. Curr Sci 116:304–311. https://doi.org/10.18520/cs/v116/i2/304-311

Bisht H, Kotlia BS, Kumar K, Joshi LM, Sah SK, Kukreti M (2020) Estimation of the recession rate of Gangotri glacier, Garhwal Himlaya (India) through kinematic GPS survey and satellite data. Environ Earth Sci 13(6):1–12. https://doi.org/10.1007/s12665-020-090780

Bolch T, Kulkarni A, Kaab A, Huggel C, Paul F, Cogley JG, Frey H, Kargel JS, Fujita K, Scheel M, Bajracharya S, Stoffel M (2012) The state and fate of Himalayan Glaciers. Review Sci 336:310–314

Dash SK, Hunt JCR (2007) Variability of climate change in India. Curr Sci 93(6):782–788

Gurung DR, Kulkarni AV, Giriraj A, Aung KS, Shrestha B. Srinivasan J (2011) Changes in seasonal snow cover in Hindu Kush-Himalayan region. Cryosphere Discuss 5:755–777. https://doi.org/10.5194/tcd-5-755-2011

Hachem S, Duguay CR, Allard M (2012) Comparison of MODIS-derived land surface temperatures with the ground surface and air temperature measurements in continuous permafrost terrain. Cryosphere 6:51–69

Haque S, Kannaujiya S, Taloor AK, Keshri D, Bhunia RK, Ray PKC, Chauhan P (2020) Identification of groundwater resource zone in the active tectonic region of Himalaya through earth observatory techniques. Groundwater Sust Develop 10:P100337.https://doi.org/10.1016/j.gsd.2020.100337

Houghton JT, Ding Y, Griggs DJ, Noguer M, Vander Linden PJ, Dai X, Maskell K, Johnson CA (2001) Climate change 2001: the scientific basis—Contribution of working group I to the third assessment report of the intergovernmental panel on climate change. Cambridge University Press, New York, p 881

IPCC (2007) Climate change 2007: the physical science basis. contribution of working group I to the fourth assessment report of the intergovernmental panel on climate change. In: Solomon S, Qin D, Manning M, Chen Z, Marquis M, Averyt KB, Ignore M, Miller HL (eds). Cambridge University Press, Cambridge, UK, p 996

Kannaujiya S, Gautam PKR, Chauhan P, Roy PNS, Pal SK, Taloor AK (2020) Contribution of seasonal hydrological loading in the variation of seismicity and geodetic deformation in Garhwal region of Northwest Himalaya. Quat Int. https://doi.org/10.1016/j.quaint.2020.04.049

Khan A, Govil H, Taloor AK, Kumar G (2020) Identification of artificial groundwater recharge sites in parts of Yamuna river basin India based on remote sensing and geographical information system. Groundwater Sust Develop 11:P100415. https://doi.org/10.1016/j.gsd.2020.100415

Kothawale DR, Munot AA, Krishna KK (2010) Surface air temperature variability over India during 1901–2007, and its association with ENSO. Climate Res 42:89–104

Kothyari UC, Singh VP (1996) Rainfall and temperature trends in India. Hydrol Process 10:357–372

Kulkarni AV (2007) Effect of global warming on the Himalayan cryosphere. Jalvigyan Sameeksha 22:93–108

Kumar D, Singh AK, Taloor AK, Singh DS (2020) Recessional pattern of Thelu and Swetvarn glaciers between 1968 and 2019, Bhagirathi basin, Garhwal Himalaya, India. Quat Int. https://doi.org/10.1016/j.quaint.2020.05.017

Menon S, Koch D, Beig G, Sahu S, Fasullo J, Orlikowski D (2010) Black carbon aerosols and the third polar ice cap. Atmos Chem Phys 10:4559–4571. https://doi.org/10.5194/acp-10-45590

Negi HS, Thakur NK, Mishra VD (2007) Estimation and validation of snow surface temperature using modis data for snow-avalanche studies in NW-Himalaya. J Indian Soc Remote Sens 35:287. https://doi.org/10.1007/BF02990785

Neteler M (2010) Estimating daily land surface temperatures in mountainous environments by reconstructed MODIS LST data. Remote Sens 2:333–351

Pall IA, Meraj G, Romshoo SA (2019) Applying integrated remote sensing and field-based approach to map glacial landform features of the Machoi Glacier valley, NW Himalaya. SN Appl Sci 1:488

Romshoo SA, Fayaz M, Meraj G, Bahuguna IM (2020) Satellite-observed glacier recession in the Kashmir Himalaya, India, from 1980 to 2018. Environ Monit Assess 192:1–7

Sarkar T, Kannaujiya S, Taloor AK, Champati Ray PK, Chauhan P (2020) Integrated study of GRACE data derived interannual groundwater storage variability over water stressed Indian regions. Groundwater Sustain Devlop 10:100376. https://doi.org/10.1016/j.gsd.2020.100376

Singh S, Sood V, Taloor AK, Prashar S, Kaur R. Qualitative and quantitative analysis of topographically derived CVA algorithms using MODIS and Landsat-8 data over Western Himalayas, India. Quat Int. https://doi.org/10.1016/j.quaint.2020.04.048

Singh N, Sontakke NA (2002) On climatic fluctuations and environmental changes of the Indo-Gangetic plains, India. Clim Change 52:287–313

Singh AK, Jasrotia AS, Taloor AK, Kotlia BS, Kumar V, Roy S, Ray PKC, Singh KK, Singh AK, Sharma AK (2017) Estimation of quantitative measures of total water storage variation from GRACE and GLDAS-

NOAH satellites using geospatial technology. Quat Int 444:191–200
Singh SK, Kanga S, Mishra VK (2020) Spatial information science for natural resource management. IGI Global 1:355. https://doi.org/10.4018/978-1-7998-5027-4
Sood V, Gusain HS, Gupta S, Taloor AK, Singh S (2020a) Detection of snow/ice cover changes using subpixel-based change detection approach over Chhota-Shigri glacier, Western Himalaya, India. Quat Int. https://doi.org/10.1016/j.quaint.2020.05.016
Sood V, Singh S, Taloor AK, Prashar S, Kaur R (2020b) Monitoring and mapping of snow cover variability using topographically derived NDSI model over north Indian Himalayas during the period 2008–19. Appl Comput Geosci. https://doi.org/10.1016/j.acags.2020.100040
Taloor AK, Kotlia BS, Jasrotia AS, Kumar A, Alam A, Ali S, Kouser B, Garg PK, Kumar R, Singh AK, Singh B (2019) Tectono-climatic influence on landscape changes in the glaciated Durung Drung basin, Zanskar Himalaya, India: a geospatial approach. Quat Int 507:262–273. https://doi.org/10.1016/j.quaint.2018.09.030
Taloor AK, Kumar V, Singh VK, Singh AK, Kale RV, Sharma R, Khajuria V, Raina G, Kouser B, Chowdhary NH (2020a). Land use land cover dynamics using remote sensing and GIS techniques in Western Doon Valley, Uttarakhand, India. In: Geoecology of landscape dynamics 2020. Springer, Singapore, pp 37–51. https://doi.org/10.1007/978-981-15-2097-6_4
Taloor AK, Pir RA, AdimallaN AS, Manhas DS, Roy S, Singh AK (2020b) Spring water quality and discharge assessment in the Basantar watershed of Jammu Himalaya using geographic information system (GIS) and water quality Index (WQI). Groundwater Sust Develop 10:P100364. https://doi.org/10.1016/j.gsd.2020.100364
Tang BH, Bi Y, Li ZL, Xia J (2008) Generalized split-window algorithm for an estimate of land surface temperature from Chinese geostationary Fengyun meteorological satellite (FY-2C) data. Sensors 8:933–951
Tekeli AE, Akyurek Z, Sorman Arda A, Sensoy A, Sorman Unal A (2005) Using MODIS snow cover maps in modeling snowmelt runoff process in the eastern part of Turkey. Remote Sens Environ 97:216–230
Wang X, Xie H, Liang T (2008) Evaluation of MODIS snow cover and cloud mask and its application in Northern Xinjiang, China. Remote Sens Environ 112:1497–1513
Wang KZ, Wan PW, Sparrow M, Liu J, Haginoya S (2007) Evaluation and improvement of the MODIS land surface temperature/emissivity products using ground-based measurements at a semi-desert site on the western Tibetan Plateau. Int J Remote Sens 28 (11):2549–2565
Yadav RR, Park WK, Singh J, Dubey B (2004) Do the western Himalayas defy global warming. Geophys Res Lett 31:17201. https://doi.org/10.1029/2004GL020201

Mr. Manish Rawat had done his M. Tech. in Land and Water Resources Engineering from Indian Institute of Technology, Kharagpur and B. Tech. in Agricultural Engineering from Sam Higginbottom Institute of Agriculture Technology and Sciences, Allahabad. Currently, he is working as a Junior Research Fellow at National Institute of Hydrology (NIH), Roorkee on a project titled Flood modelling in western Himalayan region using HEC-RAS model. He has more than 3 years of research experience and has sound knowledge of hydrological modelling, remote sensing and GIS softwares like HEC-RAS, HEC-HMS, HEC-GeoRAS, ArcGIS, ERDAS Imagine, ENVI, SDSM statistical downscaling model and also good knowledge of handling of hyperspectral data

Dr. Sateesh Karwariya had done his Ph.D. in Remote Sensing and GIS from Mahatma Gandhi Chitrakoot Gramodaya University and Master's from Jiwaji University, Gwalior. He is presently associated with SAC- Indian Space Research Organization, Ahmedabad as a Research Associate in different in-house projects. Prior to this, he has more than 10 years of working experience in different reputed research organizations in India. He has published more than 15 national and international research papers in different scientific journals. His areas of interest are remote sensing applications on agriculture and hydrological research

Er. Ritik Raushan is presently working as Field Monitor at NABARD Consultancy Services (NABCONS) which is a wholly owned subsidiary promoted by National Bank for Agriculture and Rural Development (NABARD) and is engaged in providing consultancy in all spheres of agriculture, rural development and allied areas. Nabcons leverages on the core competence of the NABARD. He has completed M.Tech. from Central University of Jharkhand and B.Tech. from Sam Higginbottom University of Agriculture, Technology and Sciences. He was awarded the National Talent Scholarship of Indian Council of Agricultural Research

Dr. Shruti Kanga is an Associate Professor and Coordinator at the Centre for Climate Change and Water Research, Suresh Gyan Vihar University, Jaipur, India. She has experience teaching at several universities and working in industry in the areas of land management and resource development. She has published several articles in scientific journals and a few books to her credit as an author. She is presently on the reviewer panel for several research journals, and she is supervising several Ph.D. students on their dissertations

Dr. Ajay Kumar Taloor has obtained his Doctorate in Remote Sensing and GIS applications in hydrogeology from University of Jammu, NAAC accredited A+ University of India. Thirteen years of research experience in the applications in geospatial technology for natural resources management of land and water resources. He has excelled twice with best paper presentation award in India. Being an expert of remote sensing applications in water science, cryosphere and climate change, tectonic and quaternary geomorphology, he is working on two major research projects on using space-based inputs for glacier mapping and climate change in Himalayas. He has also published many articles in tectonic and quaternary geomorphology in the recent years. He has high scientific temper and strong HR relations in science world, with high professional and managerial skills. He has edited many volumes in the top-rated journals in the Elsevier and Springer publishers, member of Editorial Board of the Quaternary Science Advances, and reviewer of the many top-rated international journals in science world

Dr. Asha Thapliyal working as Scientist/Engineer in Uttarakhand Space Application Centre, Department of Information and Science Technology, Government of Uttarakhand, India. She has obtained her doctorate in Botany and M. Phil. in High Altitude Plant Physiology from HNB Garhwal (Central University), NAAC accredited A+ University of India. She has more than 13 years experience in the field of RS and GIS and involved in remote sensing and GIS-based analysis of satellite and ground data for snow, glacier and water quality monitoring in the perspective of hydrological and climatological-related applications in Himalayan Region. She has contributed in different capacities to many national-level projects (viz. mapping, monitoring and long-term analysis of glaciers response to climate change, snow cover analysis using AWiFS and MODIS, wetlands and ground water quality mapping and analysis, etc.). She was the member and environment expert of all India Mountaineering Foundation (IMF) for women expedition in the Bhagirathi peak-2 during year 2013. She has showcased her research work in various national, international journals, conferences, book chapters, scientific reports and atlases. She is the life time member of Indian Society of Geomatics and Indian Meteorological society (IMS). Currently, she is also organizing member of IMS-Dehradun chapter

Seasonal Ground Water Fluctuation Monitoring Using GRACE Satellite Technology Over Punjab and Haryana During 2005–2015

11

Anil Kumar Singh, Jayant Nath Tripathi, Ajay Kumar Taloor, Bahadur Singh Kotlia, Kamalesh Kumar Singh, and Shiv Dass Attri

Abstract

Optimum management of natural resources is critical for sustainable growth and development. Under rapidly increasing population and industrialization, the groundwater depletion rate is way more than that of the groundwater recharge rate in India. The situation is more alarming in North-West India, where the amount of precipitation is quite low for irrigation purpose. In the present study, groundwater fluctuation in Haryana and Punjab has been monitored during 2005–2015 using GRACE satellite data. Since 2002, Gravity Recovery and Climate Experiment (GRACE) satellite provided an estimation of various components of Earth's gravity field as it provides gravity data at 1° × 1° resolution for the estimation of terrestrial water storage change, i.e. surface water and groundwater. The land surface variable has been used to infer how Terrestrial Water Storage (TWS) is contributing to canopy water and soil moisture. In the present study, the groundwater storage change of the Punjab and Haryana was monitored by computing storage changes in GRACE TWS, GLDAS land surface state variables with the terrestrial water balance approach. The results indicate that the Groundwater fluctuation follows the cyclic yearly pattern with highs corresponding to the monsoon. Computed groundwater mean depletion thickness over Haryana was found 1.13 cm and for Punjab is 0.92 cm during 2005–2015 in the study area. There are clear signals of yearly and seasonal variation in the groundwater as well as the impact of the extreme event on the groundwater change. The impact of cumulative water loss through Evapotranspiration (ET) on the groundwater has also been analyzed, which shows a positive correlation with the groundwater fluctuation.

A. K. Singh · J. N. Tripathi
Department of Earth and Planetary Sciences, University of Allahabad, Allahabad 211002, India
e-mail: singhanil854@gmail.com

J. N. Tripathi
e-mail: jntripathi@gmail.com

A. K. Taloor (✉)
Department of Remote Sensing and GIS, University of Jammu, Jammu 180006, India
e-mail: ajaytaloor@gmail.com

B. S. Kotlia
Centre of Advanced Study in Geology, Kumaun University, Nainital 263001, India
e-mail: bahadur.kotlia@gmail.com

K. K. Singh · S. D. Attri
India Meteorological Department, New Delhi 110003, India
e-mail: kksingh2022@gmail.com

S. D. Attri
e-mail: sd.attri@imd.gov.in

A. K. Taloor et al. (eds.), *Water, Cryosphere, and Climate Change in the Himalayas*, Geography of the Physical Environment, https://doi.org/10.1007/978-3-030-67932-3_11

Keywords

GRACE · TWS · Noah model · Groundwater recharge · Remote sensing

11.1 Introduction

Because of overexploitation, diminishing freshwater accessibility is because of informal extraction and inappropriate administration of water resources. In the coming decades, India is going towards a significant freshwater emergency in the period of a quickly evolving atmosphere. The groundwater resources are depleting at an alarming rate globally and particular in north-west India over the last two decades (Georgakakos and Graham 2008; Xu et al. 2012; Bhat et al. 2019; Taloor et al. 2019; Haque et al. 2020; Kumar et al. 2020; Sood et al. 2020a).

India is the seventh-biggest nation on the planet in terms of the zone and second as far as populace with 1200 mm/y normal precipitation. Due to irregularity in rainfall and groundwater overexploitation, the depletion of groundwater level surges abruptly (Joshi and Tyagi 1994; Briscoe and Malik 2006; Kumar et al. 2005; Moore and Fisher 2012; Rodell et al. 2009; Jasrotia and Kumar 2014; Gautam et al. 2017; Jasrotia et al. 2019; Sood et al. 2020b; Khan et al. 2020). In states like Haryana and Punjab water level depletion is comparatively faster than any other states of India (Waters et al. 1990; Engman 1991; Gontia and Patil 2012; Kumar et al. 2008; Meijerink 1996; Sander et al. 1996; Siebert et al. 2010; Adimalla and Taloor 2020; Adimalla et al. 2020; Kannaujiya et al. 2020; Sarkar et al. 2020; Taloor et al. 2020; Singh et al. 2020). There is a need for appropriate scientific methods for sustainable water resource utilization and geospatial technology is rapidly gaining its applications in the monitoring and mapping of water resources in the last few decades. GRACE data is a very useful tool for identifying the impact caused by extreme climate events like drought and floods (Jin and Feng 2013; Tapley 2004; Andersen and Hinderer 2005; Longuevergne et al. 2013; Phillips et al. 2012). GRACE data are capable of identifying seasonal and long-term variation in TWS and quite useful for hydrological model development. The information of TWS variations of recent decades is quite important for the study of water and its temporal changes. In this study, GRACE RL05 products and Global Land Data Assimilation System (GLDAS) Noah LSM (Betts et al. 1997; Chen et al. 1996; Koren et al. 1999; Ek 2003; Rodell et al. 2004) products, combined with a data record of the Central Ground Water Board (CGWB) India, were used to determine the long-term TWS over the state of Haryana and Punjab. This study insights useful guidance for sustainable management of water resources and futuristic planning and research to improve groundwater storage for the future.

11.2 Study Area

The present study has been carried out in Haryana, Delhi and Punjab, India (Fig. 11.1). lies between 27° 39′ and 30° 35′ N scope and somewhere in the range of 74° 28′ and 77° 36′ E longitude. It has four primary topographical highlights viz. (i) The Yamuna-Ghaggar plain shaping the largest (piece of the state is likewise called Delhi doab comprising of Sutlej-Ghaggar doab (between Sutlej in the north in Punjab and Ghaggar stream coursing through northern Haryana). (ii) Ghaggar-Hakra doab (between Ghaggar waterway and Hakra or Drishadvati stream which is the paleochannel of the sacred Sarasvati River) and Hakra-Yamuna doab (between Hakra waterway and the Yamuna). (iii) The Shivalik hills towards the upper east the Bagar tract semi-desert dry sandy plain toward the southwest. (iv) The Aravalli Range in the south. The Haryana is very sweltering in summer at around 45 ° C and mellows in winter. The most sweltering months are May and June and the coldest December and January. The atmosphere is dry to semi-dry with normal precipitation of 354.5–530 mm. The dirt qualities are impacted to a restricted degree by the geography, vegetation and parent rock. The Punjab is separated into three particular areas based on soil types viz.

southwestern, focal and eastern. The most extreme temperatures, for the most part, happen in mid-May and June. The temperature stays over 40 °C in the whole locale during this period. Punjab encounters its base temperature from December to February and the average yearly precipitation of Punjab is 500 mm.

11.3 Data and Methodology

The GRACE satellite data downloaded from http://grace.jpl.nasa.gov/data/get-data/ to study the TWS changes from 2005 to 2015 in the state of Punjab, Haryana and Delhi. The monthly soil moisture anomalies (at a spatial resolution of

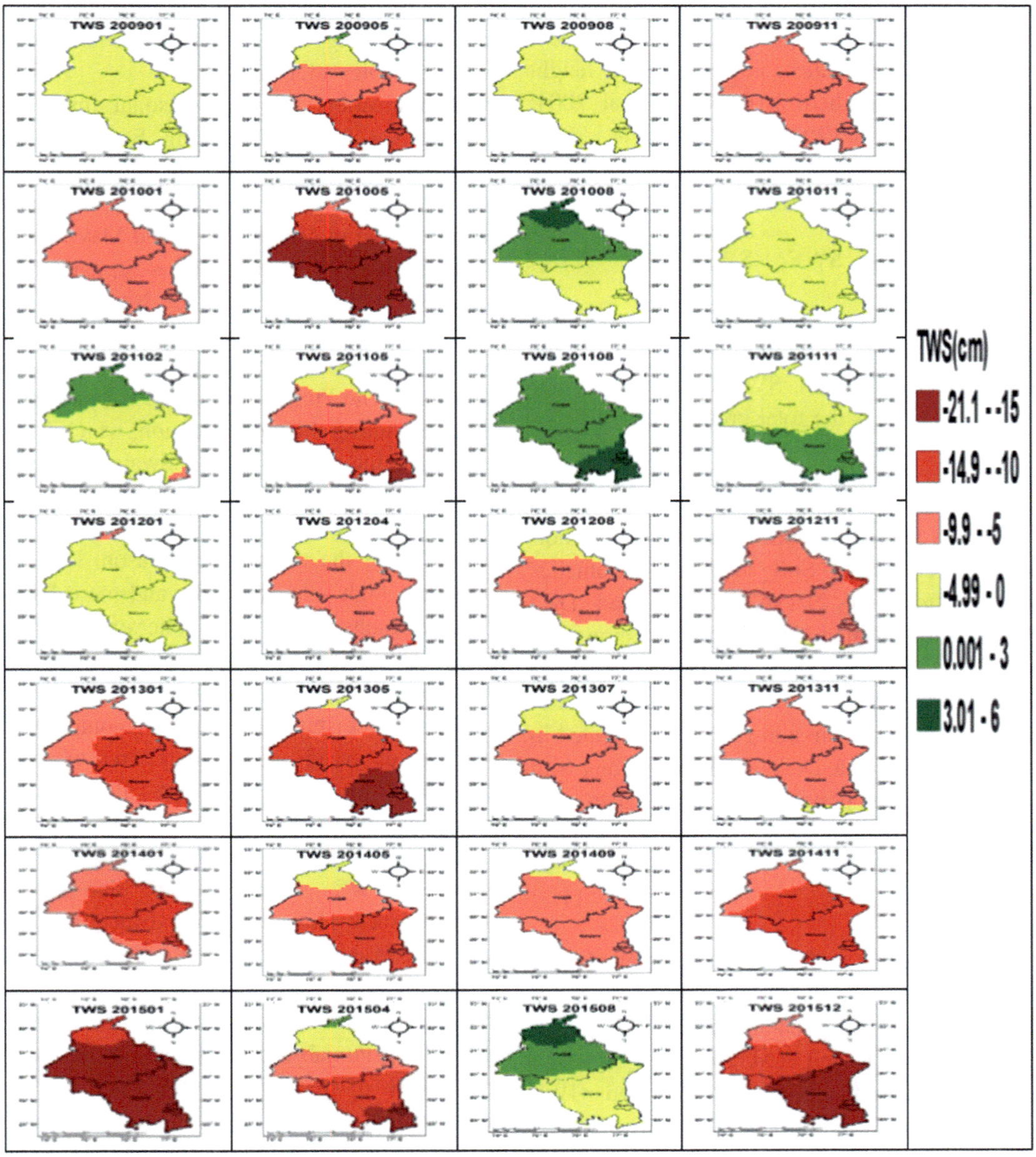

Fig. 11.1 Seasonal TWS spatial map over Haryana and Punjab from 2009 to 2015 (for winter January, May for pre-monsoon, August for monsoon and November for post-monsoon). **when satellite data is missing, we take adjacent month** (*Source* GRACE data)

1° × 1°) calculate the soil moisture. To independently evaluate groundwater storage change, there is a need to measure surface water storage change and expel it from GRACE perceptions. The GLDAS gauges used in the present study are from the Noah LSM (Ek 2003). The GRACE data provides the gravity mass anomalies to estimate TWS changes. These mass anomalies obtained by calculating the temporal variation in gravity which is expressed by monthly mean terrestrial water storage variation, equivalent water storage anomalies, as well as water height (Rodell and Famiglietti 2002; Rodell et al. 2004; Famiglietti et al. 2011; Rodell et al. 2009; Scanlon et al. 2012; Sun et al. 2012; Richey et al. 2015; Singh et al. 2017, 2019).

$$\mathrm{TWS}_t = \mathrm{SM}_t + \mathrm{SWE}_t + \mathrm{SW}_t + \mathrm{GW}_t, \quad (11.1)$$

Here TWS_t is the total terrestrial water storage, SM_t which is total soil moisture, SME_t is the snow water estimation, SW_t is the total surface water and GW_t is the total groundwater.

$$\Delta\mathrm{AW}_t = \mathrm{TWSA}_t - \Delta\mathrm{SWE}_t - \Delta\mathrm{SM}_t, \quad (11.2)$$

Here Δ is the time-mean variation of an individual parameter. Soil moisture anomalies derived from the NASA Global Land Data Assimilation System (GLDAS) as shown in Eq. (11.2). It isolates the contribution of groundwater storage changes to changes in total water storage. The reservoir storage changes applied in the state of Haryana, Delhi and Punjab along with soil moisture and previously described data

$$\Delta\mathrm{GW}_t = \mathrm{TWSA}_t - \Delta\mathrm{SWE}_t - \Delta\mathrm{SM}_t - \Delta\mathrm{SW}_t, \quad (11.3)$$

Here $\Delta\mathrm{SW}_\mathrm{t}$ is surface water anomaly for an individual month, Whereas errors in the GWS calculated using the parameters; TWSA, SM, SWE and SW (Rodell et al. 2004).

We compared our GRACE derived GWS variations with groundwater level (observed by monitoring dug wells and tube wells; data obtained from CGWB website).

11.4 Results and Discussions

The analysis of satellite data showed a continuous water deficit in Haryana and Punjab from 2009 to 2015 (Fig. 11.1). From winter to premonsoon, the depletion rate of TWS was 8–10 cm, while it was 5–8 cm after the monsoon. However, the rate of recharge during the monsoon was 10–13 cm. In the year 2010, 2011 and 2015, the month of August showed good recharge during monsoon because of heavy rainfall (Tables 11.1 and 11.2).

Quantification of the seasonal mean of groundwater from TWS using Eq. (11.3) has been depicted in Fig. 11.2. In this study, the mean of January and February for the winter session, mean of March, April, and May for premonsoon season, mean of June–September for monsoon season and mean of November and December for the post-monsoon season have been considered. The same pattern for both the States viz. The Punjab, Delhi and Haryana were observed during the winter session, which is due to less rainfall and more groundwater extraction for cropland irrigation. The ET values were higher due to more moisture present in soil and crop. During the premonsoon season, less precipitation, high solar radiation, and more ET on the field results in the higher water extraction for domestic and irrigation uses. During the monsoon season, large amount of rainfall resulted in higher soil moisture, and higher amount of ET was also observed over cropland area where as the TWS from satellite data also showed increasing trends. However, estimated groundwater (GW) from Eq. (11.3) was low as compared to TWS, which may be due to soil characteristics. However, groundwater recharge was more in the post-monsoon season due to the lag of soil moisture percolation. The results show that due to continuous water depletion over the state of the Punjab and Haryana and decreasing trend were of the order of 0.92 and 1.3 cm, respectively (Fig. 11.3). The above trends are in agreement with Central Ground Water Board results (Fig. 11.4).

Table 11.1 Monthly rainfall (cm) over Punjab from 2009 to 2015

Year	2009	2010	2011	2012	2013	2014	2015
January	1.04	0.27	0.36	3.58	0.93	2.18	1.77
February	1.71	1.23	3.3	0.29	5.01	2.01	3.13
March	1.07	0.05	0.67	0.19	1.16	3.03	6.85
April	2.13	0.03	1.22	2.02	0.34	2.45	2.98
May	0.44	0.35	1.45	0.08	0.36	2.08	1.67
June	1.02	3.45	9.88	0.96	12.03	2.06	4.87
July	16.67	20.86	8.5	6.77	11.79	7.63	13.16
August	8.24	12.13	16.13	0.51	21.71	4.19	8.86
September	7.18	9.46	11.43	8.35	2.44	10.58	6.92
October	0.27	0.58	0.06	0.28	1.62	0.6	0.92
November	0.61	0.04	0.02	0.04	0.61	0.07	0.08
December	0.2	1.66	0.34	0.82	0.66	1.41	0.07

Source India Meteorological Department

Table 11.2 Monthly rainfall (cm) over Haryana and Delhi from 2009 to 2015

Year	2009	2010	2011	2012	2013	2014	2015
January	2009	2010	2011	2012	2013	2014	2015
February	1.04	0.27	0.36	3.58	0.93	2.18	1.24
March	1. 71	1.23	3.3	0.29	5.01	2.01	0.65
April	1.07	0.05	0.67	0.19	1.16	3.03	7.16
May	2.13	0.09	1. 22	2.02	0.34	2.45	3.47
June	0.44	0.35	1.45	0.08	0.36	2.08	0.84
July	1.02	3.45	9.88	0.96	12.03	2.06	4.41
August	16.67	20.86	8.5	6.77	11.79	7.63	13.25
September	8.24	12.13	16.13	10.51	21.71	4.19	8.94
October	7.18	9.46	11.43	8.35	2.44	10.58	3.2
November	0.27	0.58	0.06	0.28	1.62	0.6	0.37
December	0.61	0.04	0.02	0.04	0.61	0.07	0.23

Source India Meteorological Department

11.5 Conclusion

Annual average groundwater losses over Haryana and Punjab were of the order of 1.13 cm/yr and 0.92 cm/yr, respectively. The vast majority of the groundwater withdrawal from the study area because of an expansion in irrigation and evapotranspiration as these areas are thickly populated and widely inundated. The groundwater assets are experiencing critical pressure as they are not being energized at a similar rate as they are found on the earth surface. Compelling administration is urgently needed to draw harmony among discharge and recharge in the study area. Moreover, the monthly satellite information can be used for ideal water management purposes.

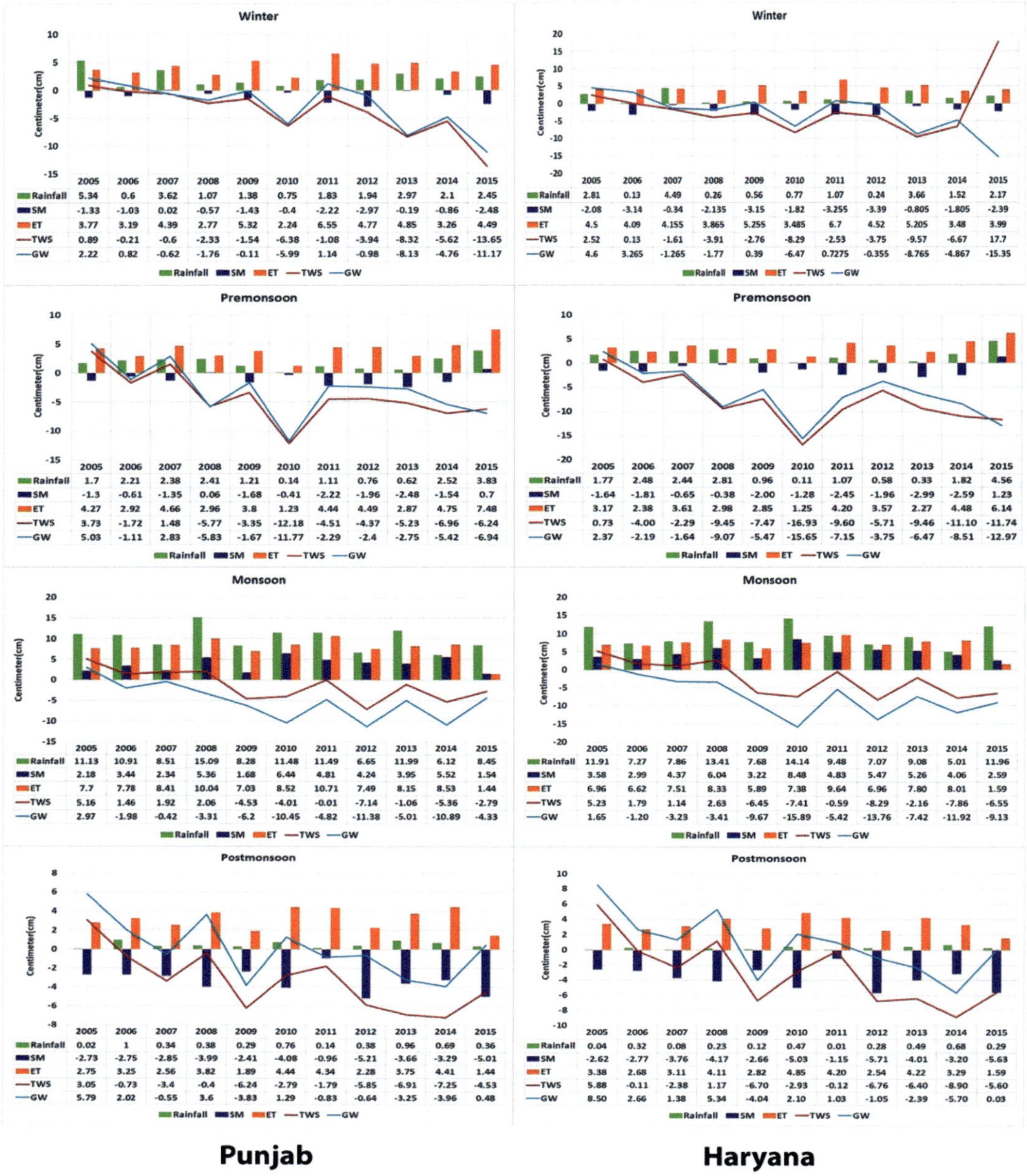

Fig. 11.2 Seasonal-mean calculated groundwater, TWS, SM and ET fluctuation over Punjab and Haryana State during 2005–2015 (*Source* India Meteorological Department)

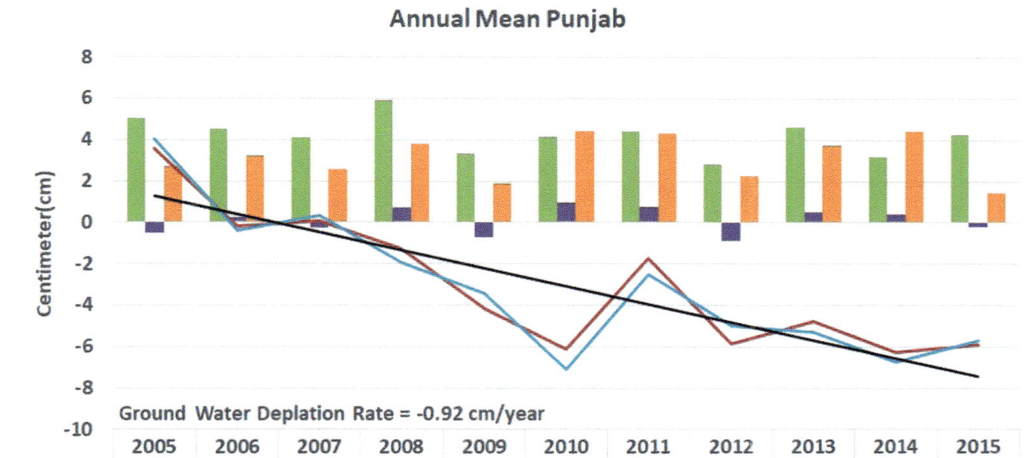

	2005	2006	2007	2008	2009	2010	2011	2012	2013	2014	2015
Rainfall	5.03	4.54	4.12	5.91	3.37	4.18	4.45	2.82	4.65	3.19	4.27
SM	-0.50	0.24	-0.27	0.71	-0.70	0.96	0.77	-0.87	0.53	0.43	-0.21
ET	2.75	3.25	2.56	3.82	1.89	4.44	4.34	2.28	3.75	4.41	1.44
TWS	3.56	-0.16	0.06	-1.25	-4.16	-6.14	-1.76	-5.85	-4.78	-6.28	-5.90
GW	4.06	-0.40	0.33	-1.95	-3.46	-7.10	-2.53	-4.98	-5.31	-6.71	-5.68

Rainfall SM ET TWS GW Linear (GW)

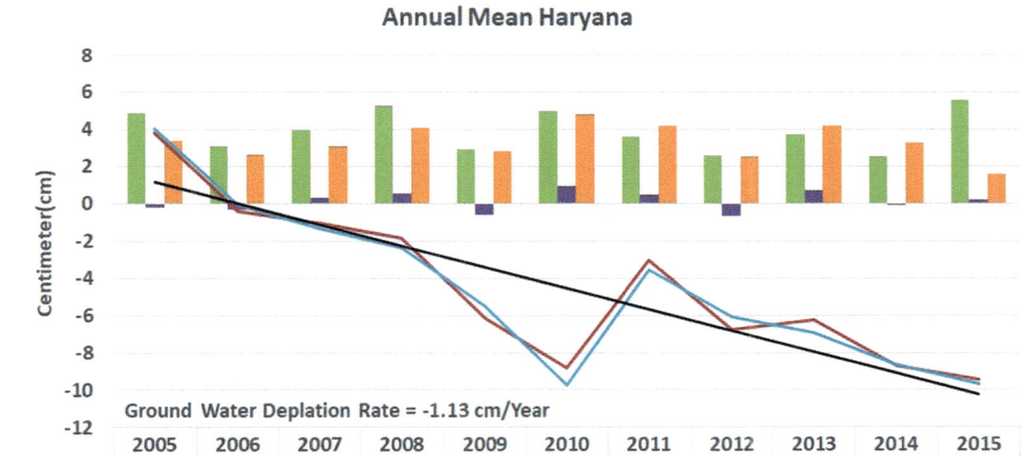

	2005	2006	2007	2008	2009	2010	2011	2012	2013	2014	2015
Rainfall	4.89	3.14	4.00	5.28	2.92	4.98	3.61	2.61	3.72	2.55	5.56
SM	-0.22	-0.34	0.30	0.52	-0.62	0.95	0.49	-0.66	0.68	-0.07	0.22
ET	3.38	2.68	3.11	4.11	2.82	4.85	4.20	2.54	4.22	3.29	1.59
TWS	3.82	-0.41	-1.06	-1.84	-6.15	-8.82	-3.05	-6.76	-6.28	-8.73	-9.47
GW	4.04	-0.07	-1.35	-2.37	-5.53	-9.76	-3.54	-6.10	-6.96	-8.66	-9.69

Rainfall SM ET TWS GW Linear (GW)

Fig. 11.3 Annual mean groundwater depletions observed by satellite over Punjab and Haryana (*Source* GRACE data)

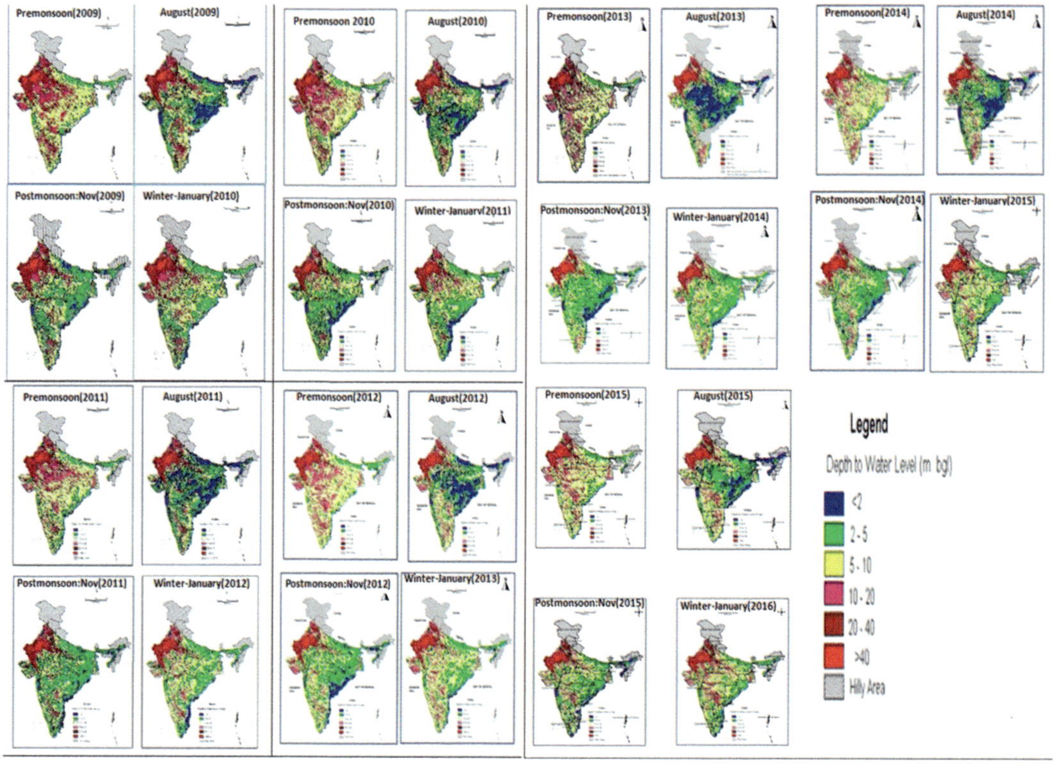

Fig. 11.4 Seasonal groundwater depth level observed by CGWB during 2009–2015 respectively (*Source* http://www.cgwb.gov.in/GW-Year-Book.html)

Acknowledgments The authors are highly thankful to the University of Allahabad, Prayagraj. India Meterology Department, Delhi; Head Department of Remote Sensing and GIS, University of Jammu, Jammu for encouragement and providing facilitating the present research work and especially thanks to NASA for providing GRACE & GLDAS Data set.

References

Adimalla N, Taloor AK (2020) Hydrogeochemical investigation of groundwater quality in the hard rock terrain of South India using Geographic Information System (GIS) and groundwater quality index (GWQI) techniques. Groundwater Sust Dev 10:P100288. https://doi.org/10.1016/j.gsd.2019.100288

Adimalla N, Dhakate R, Kasarla A, Taloor AK (2020) Appraisal of groundwater quality for drinking and irrigation purposes in Central Telangana, India. Groundwater Sust Dev 10:P100334. https://doi.org/10.1016/j.gsd.2020.100334

Andersen OB, Hinderer J (2005) Global inter-annual gravity changes from GRACE: early results. Geophys Res Lett 32:L01402. https://doi.org/10.1029/2004GL020948

Betts AK, Chen F, Mitchell KE, Janjić ZI (1997) Assessment of the land surface and boundary layer models in two operational versions of the NCEP eta model using FIFE data. Mon Weather Rev 125:2896–2916. https://doi.org/10.1175/1520-0493(1997)125%3c2896:AOTLSA%3e2.0.CO;2

Bhat MS, Alam A, Ahmad B, Kotlia BS, Farooq H, Taloor AK, Ahmad S (2019) Flood frequency analysis of river Jhelum in Kashmir basin. Quat Int 507:288–294. https://doi.org/10.1016/j.quaint.2018.09.039

Briscoe J, Malik RP (2006) India's water economy: bracing for a turbulent future. Oxford University Press, New Delhi. © World Bank. https://openknowledge.worldbank.org/handle/10986/7238. License: CC BY 3.0 IGO

Chen F, Mitchell K, Schaake J, Xue Y, Pan HL, Koren V, Duan QY, Ek M Betts A (1996) Modeling of land surface evaporation by four schemes and comparison with FIFE observations. J Geophys Res Atmos 101:7251–7268. https://doi.org/10.1029/95JD02165

Ek MB (2003) Implementation of Noah land surface model advances in the national centers for environmental prediction operational mesoscale Eta model.

J Geophys Res 108:8851. https://doi.org/10.1029/2002JD003296
Engman ET (1991) Applications of microwave remote sensing of soil moisture for water resources and agriculture. Remote Sens Environ 35:213–226. https://doi.org/10.1016/0034-4257(91)90013-V
Famiglietti JS, Lo M, Ho SL, Bethune J, Anderson KJ, Syed TH, Swenson SC, de Linage CR, Rodell M (2011) Satellites measure recent rates of groundwater depletion in California's Central Valley. Geophys Res Lett 38:2–5. https://doi.org/10.1029/2010GL046442
Gautam PK, Arora S, Kannaujiya S, Singh A, Goswami A, Champati PK (2017) A comparative appraisal of groundwater resources using GRACE-GPS data in highly urbanized regions of Uttar Pradesh, India. Springer International Publishing, Switzerland. https://doi.org/10.1007/s40899-017-0109-4
Georgakakos KP, Graham NE (2008) Potential benefits of seasonal inflow prediction uncertainty for reservoir release decisions. J Appl Meteorol Climatol 47:1297–1321. https://doi.org/10.1175/2007JAMC1671.1
Gontia NK, Patil PY (2012) Assessment of groundwater recharge through rainfall and water harvesting structures in Jamka Microwatershed using remote sensing and GIS. J Indian Soc Remote Sens 40:639–648. https://doi.org/10.1007/s12524-011-0176-1
Haque S, Kannaujiya S, Taloor AK, Keshri D, Bhunia RK, Ray PKC, Chauhan P (2020) Identification of groundwater resource zone in the active tectonic region of Himalaya through earth observatory techniques. Groundwater Sust Dev 10:P100337. https://doi.org/10.1016/j.gsd.2020.100337
Jasrotia AS, Kumar A (2014) Estimation of replenishable groundwater resources and their status of utilization in Jammu Himalaya, J&K, India. Eur Water 48:17–27
Jasrotia AS, Kumar A, Taloor AK, Saraf AK (2019) Artificial recharge to groundwater using geospatial and groundwater modelling techniques in North Western Himalaya, India. Arabian J Geosci 12:774. https://doi.org/10.1007/s12517-019-4855-5
Jin S, Feng G (2013) Large-scale variations of global groundwater from satellite gravimetry and hydrological models, 2002–2012. Glob Planet Change 106:20–30. https://doi.org/10.1016/j.gloplacha.2013.02.008
Joshi PK, Tyagi NK (1994) Salt affected and water logged soils in India: a review. In: Svendsen M, Gulati A (eds) Strategic change in Indian irrigation. ICAR and IFPRI, New Delhi, India and Washington, DC, USA, pp. 237–252
Kannaujiya S, Gautam PKR, Chauhan P, Roy PNS, Pal SK, Taloor AK (2020) Contribution of seasonal hydrological loading in the variation of seismicity and geodetic deformation in Garhwal region of Northwest Himalaya. Quat Int. https://doi.org/10.1016/j.quaint.2020.04.049
Khan A, Govil H, Taloor AK, Kumar G (2020) Identification of artificial groundwater recharge sites in parts of Yamuna river basin India based on remote sensing and geographical information system. Groundwater Sust Develop 11:P100415. https://doi.org/10.1016/j.gsd.2020.100415
Koren VI, Finnert BD, Schaake JC, Smith MB, Seo DJ, Duan QY (1999) Scale dependencies of hydrologic models to spatial variability of precipitation. J Hydrol 217:285–302. https://doi.org/10.1016/S0022-1694(98)00231-5
Kumar R, Singh RD, Sharma KD (2005) Water resources of India. Curr Sci 89:794–811. https://doi.org/10.1002/047147844X.wr243
Kumar MGM, Agarwal AKA, Bali R (2008) Delineation of potential sites for water harvesting structures using remote sensing and GIS. J Indian Soc Remote Sens 36:323–334. https://doi.org/10.1007/s12524-008-0033-z
Kumar D, Singh AK, Taloor AK, Singh DS (2020) Recessional pattern of Thelu and Swetvarn glaciers between 1968 and 2019, Bhagirathi basin, Garhwal Himalaya, India. Quat Int. https://doi.org/10.1016/j.quaint.2020.05.017
Longuevergne L, Wilson CR, Scanlon BR, Crétaux JF (2013) GRACE water storage estimates for the middle east and other regions with significant reservoir and lake storage. Hydrol Earth Syst Sci 17:4817–4830. https://doi.org/10.5194/hess-17-4817-2013
Meijerink AMJ (1996) Remote sensing applications to hydrology: groundwater. Hydrol Sci J 41:549–561. https://doi.org/10.1080/02626669609491525
Moore S, Fisher JB (2012) Challenges and opportunities in GRACE-based groundwater storage assessment and management: an example from Yemen. Water Resour Manag 26:1425–1453. https://doi.org/10.1007/s11269-011-9966-z
Phillips T, Nerem RS, Fox-Kemper B, Famiglietti JS, Rajagopalan B (2012) The influence of ENSO on global terrestrial water storage using GRACE. Geophys Res Lett 39. https://doi.org/10.1029/2012GL052495
Taloor AK, Pir, RA, Adimalla N, Ali S, Manhas DS, Roy S, Singh AK (2020) Spring water quality and discharge assessment in the Basantar watershed of Jammu Himalaya using geographic information system (GIS) and water quality Index(WQI). Groundwater Sust Develop 10:P100364. https://doi.org/10.1016/j.gsd.2020.100364
Richey AS, Thomas BF, Lo MH, Reager JT, Famiglietti JS, Voss K, Swenson S, Rodell M (2015) Quantifying renewable groundwater stress with GRACE. Water Resour Res 51:5217–5237. https://doi.org/10.1002/2015WR017349
Rodell M, Famiglietti JS (2002) The potential for satellite-based monitoring of groundwater storage changes using GRACE: the High Plains aquifer. Central US J Hydrol 263:245–256. https://doi.org/10.1016/S0022-1694(02)00060-4
Rodell M, Famiglietti JS, Chen J, Seneviratne SI, Viterbo P, Holl S, Wilson CR (2004) Basin scale estimates of evapotranspiration using GRACE and other observations. Geophys Res Lett 31:L20504. https://doi.org/10.1029/2004GL020873

Rodell M, Velicogna I, Famiglietti JS (2009) Satellite-based estimates of groundwater depletion in India. Nature 460:999–1002. https://doi.org/10.1038/nature08238

Sander P, Chesley MM, Minor TB (1996) Ground water assessment using remote sensing and GIS in a rural groundwater project in ghana: lessons learned. Hydrogeol J 4:40–49

Sarkar T, Kannaujiya S, Taloor AK, Ray PKC, Chauhan P (2020) Integrated study of GRACE data derived interannual groundwater storage variability over water stressed Indian regions. Groundwater Sust Develop 10:P100376. https://doi.org/10.1016/j.gsd.2020.100376

Scanlon BR, Faunt CC, Longuevergne L, Reedy RC, Alley WM, McGuire VL, McMahon PB (2012) Groundwater depletion and sustainability of irrigation in the US high plains and Central Valley. Proc Natl Acad Sci 109:9320–9325. https://doi.org/10.1073/pnas.1200311109

Siebert S, Burke J, Faures JM, Frenken K, Hoogeveen J, Döll P, Portmann FT (2010) Groundwater use for irrigation—a global inventory. Hydrol Earth Syst Sci 14:1863–1880. https://doi.org/10.5194/hess-14-1863-2010

Singh S, Sood V, Taloor AK, Prashar S, Kaur R (2020) Qualitative and quantitative analysis of topographically derived CVA algorithms using MODIS and Landsat-8 data over Western Himalayas, India. Quat Int. https://doi.org/10.1016/j.quaint.2020.04.048

Singh AK, Jasrotia AS, Taloor AK, Kotlia BS, Kumar V, Roy S, Ray PKC, Singh KK, Singh AK, Sharma AK (2017) Estimation of quantitative measures of total water storage variation from GRACE and GLDAS-NOAH satellites using geospatial technology. Quat Int 444:191–200. https://doi.org/10.1016/j.quaint.2017.04.014

Singh AK, Tripathi JN, Kotlia BS, Singh KK, Kumar A (2019) Monitoring groundwater fluctuations over India during Indian Summer Monsoon (ISM) and Northeast monsoon using GRACE satellite: Impact on agriculture. Quat Int 507:342–351. https://doi.org/10.1016/j.quaint.2018.10.036

Sood V, Gusain HS, Gupta S, Taloor AK, Singh S (2020a) Detection of snow/ice cover changes using subpixel-based change detection approach over Chhota-Shigri glacier. Quat Int, Western Himalaya, India. https://doi.org/10.1016/j.quaint.2020.05.016

Sood V, Singh S, Taloor AK, Prasher S, Kaur R (2020b) Monitoring and mapping of snow cover variability using topographically derived NDSI model over north Indian Himalayas during the period 2008–19. Applied Computing and Geosciences. https://doi.org/10.1016/j.acags.2020.100040

Sun AY, Green R, Swenson S, Rodell M (2012) Toward calibration of regional groundwater models using GRACE data. J Hydrol 422:1–9. https://doi.org/10.1016/j.jhydrol.2011.10.025

Taloor AK, Kotlia BS, Jasrotia AS, Kumar A, Alam A, Ali S, Kouser B, Garg PK, Kumar R, Singh AK, Singh B (2019) Tectono-climatic influence on landscape changes in the glaciated Durung Drung basin, Zanskar Himalaya, India: a geospatial approach. Quat Int 507:262–273. https://doi.org/10.1016/j.quaint.2018.09.030

Tapley, BD, Bettadpur S, Ries JC Thompson PF Watkins MM (2004) GRACE measurements of mass variability in the earth system. Science 305(80–):503–505. https://doi.org/10.1126/science.1099192

Waters P, Greenbaum D, Smart PL, Osmaston H (1990) Applications of remote sensing to groundwater hydrology. Remote Sens Rev 4:223–264. https://doi.org/10.1080/02757259009532107

Xu X, Huang G, Zhan H, Qu Z, Huang Q (2012) Integration of SWAP and MODFLOW-2000 for modeling groundwater dynamics in shallow water table areas. J Hydrol 412–413:170–181

Dr. Anil Kumar Singh received the B.E. degree in Information Technology from University of Rajasthan, Jaipur, India and M.Tech. degree in Remote Sensing and GIS from SHUATS University Prayagraj Uttar Pradesh, India, in 2010 and 2012, respectively. Subsequently, he received PG Diploma in Remote Sensing and GIS from IIRS (ISRO), Dehradun, India in 2013. He is pursuing his Ph.D. degree in Climate Risk Management System for Agriculture from University of Allahabad, Prayagraj, India and parallelly he worked as a Senior Research Fellow at India Meteorology Department, India from 2016 to 2020. Currently, he has been working in Dehaat Company, Gurugram as a Remote Sensing GIS Research Lead. His research interest includes the remote sensing, satellite meteorology, agrometeorology, hydrology and watershed management. He has been involved in studying remote sensing application in agriculture. He has published more than 10 publications, including peer-reviewed papers and book chapters. He is a peer reviewer for several international journals such as Journal of Hydrologic Engineering, International Journal of Water Resources and Environmental Engineering, Journal of Hydrology and Royal Meteorological Society

Prof. Jayant Nath Tripathi received his B.Sc.(Hons.), M.Sc. Tech. (Exploration Geophysics) and Ph.D. (Geophysics) from Banaras Hindu University (B.H.U.) in 1983, 1987 and 1994, respectively. He also obained his M.E. (Software Engineering) degree from MNREC (University of Allahabad) (present MMNIT) Prayagraj in 2000. He joined the University of Allahabad as a Lecturer in 1992 and is serving as Professor since January 2009. He has been a recepient of the Department of Science and Technology (DST), India—International Centre for Theoretical Physics (ICTP), Italy. He received Fellowship to visit ICTP, Italy during September–October, 2000. He was awarded Post-Doctoral Fellowship of the Spanish Government Fellowship for Foreign Scientists and worked as a residential scientist at Observatori de l'Ebre, Spain from November 2002 to March 2004. He visited University of Tohoku, Japan under Visiting Professor scheme of JSPS during December 2007 to August 2008. He was awarded under Leadership Program for Academicians (LeAP) scheme of PMMMNMTT of MHRD, Government of India and visited Ohio State University, USA in September 2019. He is Life Member of American Geophysical Union, Indian Society of Remote Sensing and many other academic societies. He has more than 40 published research articles to his credit, published in international and national journals/periodicals in the field of theoretical seismology, hazard assessment, hydrology and applications of remote sensing in meteorology and agriculture. Till date, he has supervised 2 Ph.D. and 44 Master's students

Dr. Ajay Kumar Taloor has obtained his Doctorate in Remote Sensing and GIS applications in hydrogeology from University of Jammu, NAAC accredited A+ University of India. Thirteen years of research experience in the applications in geospatial technology for natural resources management of land and water resources. He has excelled twice with best paper presentation award in India. Being an expert of remote sensing applications in water science, cryosphere and climate change, tectonic and quaternary geomorphology, he is working on two major research projects on using space-based inputs for glacier mapping and climate change in Himalayas. He has also published many articles in tectonic and quaternary geomorphology in the recent years. He has high scientific temper and strong HR relations in science world, with high professional and managerial skills. He has edited many volumes in the top-rated journals in the Elsevier and Springer publishers, member of Editorial Board of the Quaternary Science Advances, and reviewer of the many top-rated international journals in science world

Dr. Bahadur Singh Kotlia is Research Scientist (Professor) at Centre of Advanced Study, Department of Geology, Kumaun University, Nainital. He received his Ph.D. degree in Geology from Panjab University, Chandigarh. He is a recipient of Alexander von Humboldt Fellowship. He received his Post-Doctoral degree from University of Bonn, West Germany, Institute of Palaeontology. He has executed various projects in the field of Earth Sciences. He has published over 80 research papers in international journals and 50 research papers in national journals

Dr. Kamalesh Kumar Singh is currently working in India Meteorological Department as Additional DG and Head, Agromet Division. He did post-graduation in Geophysics in 1983 and Ph.D. on Crop yield forecasting in 1988 from Banaras Hindu University, Varanasi, India. He joined the National Centre for Medium Range Weather Forecasting, Department of Science and Technology, New Delhi in 1990 and has immensely contributed in the establishment and development of operational agriculture weather forecasting and Agrometeorological Advisory Service (AAS) system for benefits of farmers in the country. As Project Director for Gramin Krishi Mausam Sewa (GKMS), a flagship multi-institutional scheme of Govt. of India, he is working to enhance climate services for agriculture risk management by setting up Agromet units at district level in Krishi Vigyan Kendras to provide agrometeteorological advisories at block/panchayat level. He worked in the field of application of state-of-the-art tools like dynamic crop simulation models, decision support system for resilient farming, remote sensing, GIS, etc. to address the impact of changing climate and disaster at farm level. He made a synergistic effort to evolve capacity building for HRD in these domains including development and support of the Climate Risk Management and Disaster Risk Reduction (DRR) Tool and Technology Transfer at district level in the entire country. He has supervised M.Sc. and Ph.D. scholars from different universities, institutes and IITs. He has more than 150 research papers to his credit published in peer-reviewed national and international journals.He is associated with a number of international and Govt. of India funded project in the field of seasonal climate forecast, product development using satellite data, operational crop yield forecasting, climate resilient agriculture, crop insurance scheme, viz. PMFBY, drought assessment and management leading to food security and sustainable rural livelihood security. He is also an expert member of many high-profile committees in Ministry of Science and Technology, Water Resources, Agriculture and Farmers Welfare, to guide the flagship government programmes

Dr. Shiv Dass Attri is presently Scientist-G/Additional Director General of Meteorology, India Meteorological Department, Ministry of Earth Sciences, New Delhi. He has been Member of Commission for Atmospheric Sciences Management Group of World Meteorological Organization (WMO), Geneva Switzerland; Primary Contact of Global Atmosphere Watch in the country and Expert Member Task forces of Global Framework for Climate Services of WMO, United Nations. He is presently Member of the Standing Committee on Services to Agriculture—Expert Team on Agromet Risk Management of WMO. He has been felicitated and awarded Commendation Certificate by the **Hon'ble Prime Minister of India**, in November, 2007 for his contribution to the work of the **Intergovernmental Panel on Climate Change**, which is the Joint Winner of the **Nobel Peace Prize 2007**. He has been bestowed upon '**Best IMD Officer**' in 2010 and '**Pride of University**' in 2018 by GJ University of Science and Technology. He has served as Member of the Environmental Appraisal Committees (Thermal Power, Industries Mining and Coal Mining) of the Ministry of Environment, Forests and Climate Change for environmental clearance. He participated and presented papers in many national/international meetings/workshops/ conferences on Agromet, environment, climate change, data management, etc. including **Congress and Executive Council of WMO**, Session of Regional Association, Technical Commission, WMO/IAEA Meeting of Experts, and has also served as **Chairman/Rapporteur of Working Groups**. He published more than 100 research papers, Met. Monogrpah, books and reports in national and international journals. He served as External Expert on the Board of Post-Graduate Studies and Research in Environmental Sciences and Engineering at I.P. University, New Delhi , GJ University Hisar and MDU Hisar. Presently, he is the Executive Editor of International Journal—Mausam and referee of many national and international journals

12 Importance of Regulating Transboundary Aquifers in the World with Special Reference to Indian Subcontinent: A Review

Ashima Awasthi, Madhuri S. Rishi, and Ashu Khosla

Abstract

A significant number of the groundwater aquifers, rivers and lakes are jointly shared by two or more countries, which cross man-made political borders. The absence of legitimate systems and the hidden nature of transboundary groundwater regimes lead to misunderstanding and confusion on the part of many policymakers. Transboundary aquifers (TBAs) or internationally shared aquifers have performed a decisive part in efficiently utilizing water supply especially for drinking and irrigation. The increasing depletion of groundwater cause considerable unpredictability on provincial farmers, food security of multiple regions and food commodities imported from TBAs with descending groundwater extent. Effective management of groundwater is imminent in order to curtail the overexploitation of TBAs and to maximize the valuable usage of groundwater resources through regional and international efforts. Transboundary aquifers have been constantly under environmental risks due to climate change, increasing population, urbanization and human-induced water pollution. The main objective of the chapter is to describe a general idea regarding present situation of TBAs worldwide with special reference to the Indian subcontinent in terms of major studies, research undertaken, focused problems, management efforts and legal aspect including recommendations.

Keywords

Effective management · Food security · Overexploitation · Transboundary aquifers · Water pollution

12.1 Introduction

Out of 60% of all the freshwater resources that lie within the internationally shared boundaries throughout the globe, only an estimated 40% are administered by basin agreements (Wolf 2010; Rivera 2015; Golovina 2018; Haque et al. 2020). Water resources play a vital role with reference to sustainable development and the range of services provided by water support growth of economy, environmental sustainability and poverty reduction (Skoulikaris and Zafirakou 2019). Transboundary aquifer is a body of groundwater, which is lying across an internationally shared border

A. Awasthi · M. S. Rishi
Department of Environment Sciences, Panjab University, Sector-14, Chandigarh 160014, India
e-mail: aaashima.15@gmail.com

M. S. Rishi
e-mail: madhuririshi@gmail.com

A. Khosla (✉)
Department of Geology, Panjab University, Sector-14, Chandigarh 160014, India
e-mail: khosla100@yahoo.co.in

A. K. Taloor et al. (eds.), *Water, Cryosphere, and Climate Change in the Himalayas*, Geography of the Physical Environment, https://doi.org/10.1007/978-3-030-67932-3_12

with the possible associated risk of clash over a shared resource (Cobbing et al. 2008). Groundwater is widely embraced for production of food, irrigation of crops, industrial and domestic usage in the communities of urban and rural areas, accounting for around 25% of total water usage in Asian continent (FAO 2016; Lee et al. 2018). Groundwater resource is extremely important so as to encounter the need of the rapid rise in population, urban development and competition for development in the economy (Nwankwoala 2011; Oni and Aizebeokhai 2017; Adimalla and Taloor 2020). When an aquifer is present under the territory of two or more states/countries, then it is called a transboundary aquifer (Bourne 1992) as depicted in Fig. 12.1. As the groundwater travels from one country to another, TBAs also flow across the international political boundaries (Sanchez et al. 2018). There is a strong possibility that the groundwater recharge might happen in one country, whereas abstraction might occur in the other countries (Puri 2003; Wada and Heinrich 2013). Internationally shared resources have significantly contributed towards the human water requirements, e.g. agriculture and ecosystems of the nature (Bittinger 1972; Hayton and Utton 1989; Foster and Chilton 2003; Ahmad et al. 2005; Llamas and Martínez-Santos 2005; Puri and Aureli 2005; Davies et al. 2013).

Various authors (Eckstein and Eckstein 2003, 2005; Eckstein 2007; Davies et al. 2013) defined six types of transboundary aquifers (Fig. 12.2).

Type 1: Hydraulically linked unconfined aquifer with a river, both of which flow along an international border between two countries.

Type 2: Hydraulically linked unconfined aquifer with a river, both being intersected by the same international border.

Type 3: Hydraulically linked unconfined aquifer with a river that flows across an international border, whereas the river flows completely within the territory of one state.

Type 4: Hydraulically linked unconfined aquifer with a river that is completely within the territory of one state, whereas the river flows across an international border.

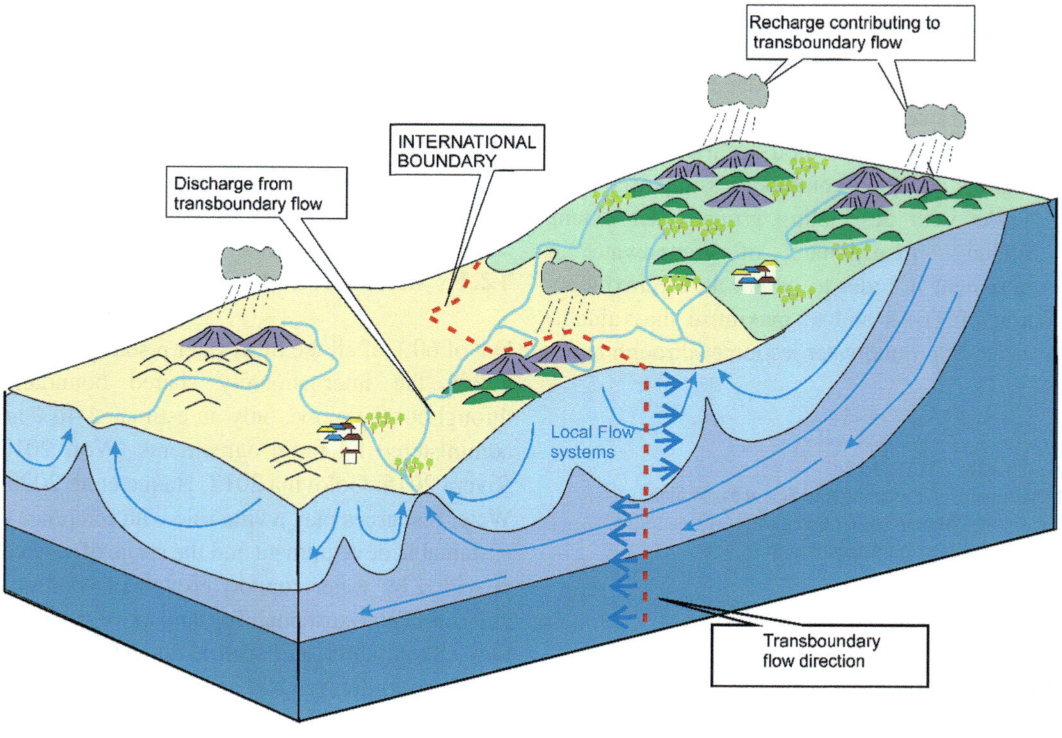

Fig. 12.1 Schematic drawing of a transboundary aquifer. *Source* Puri and Naser (2002)

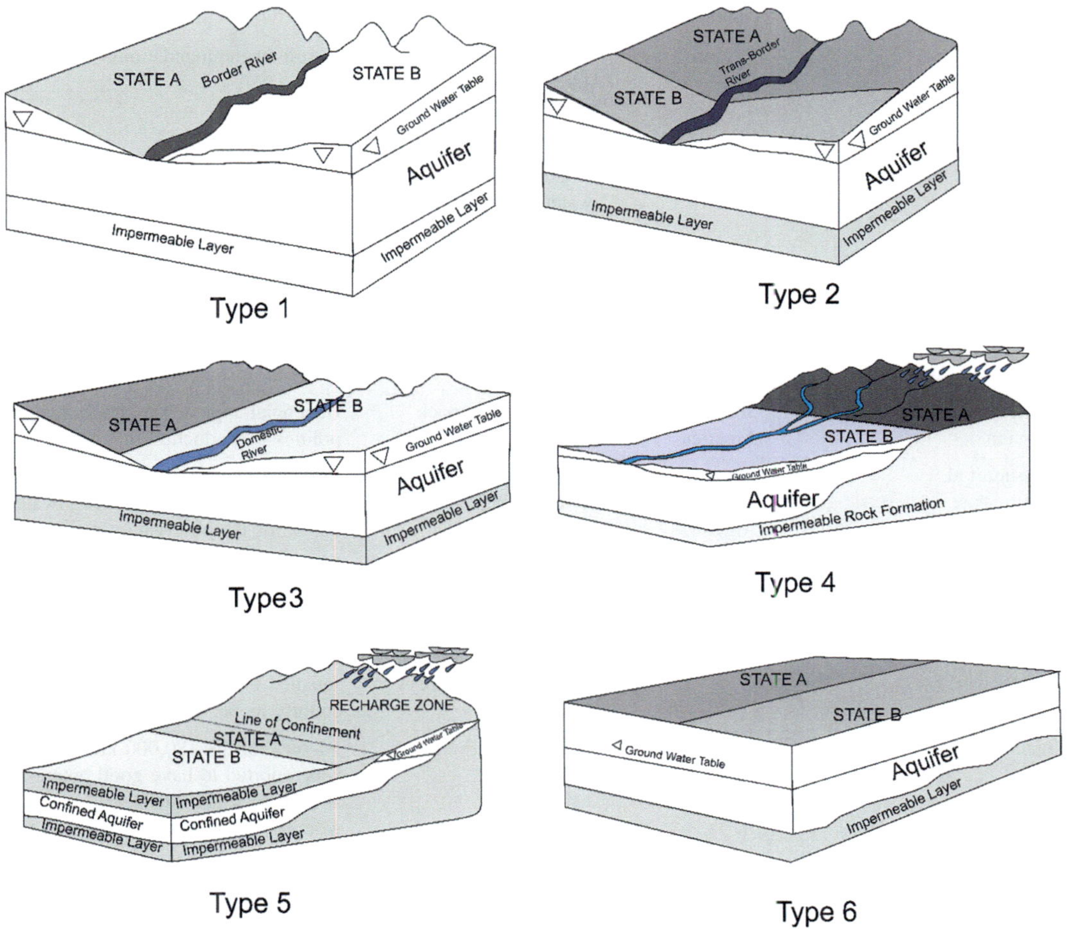

Fig. 12.2 Types of transboundary aquifers. *Source* Eckstein and Eckstein (2003, 2005), Eckstein (2007), Davies et al. (2013)

Type 5: Hydraulically unconnected confined aquifer with any surface water body, with a zone of recharge that traverses an international boundary or that is located completely in another country.

Type 6: A TBA unrelated to any surface water body and devoid of any recharge.

12.2 Review of Literature

Summary of some of the studies related to transboundary aquifers are given in Table 12.1.

12.3 Transboundary Aquifers of the World

Deeper aquifers are still unidentified, and so far, are only sparingly exploited around the world for the abstraction of freshwater (UNESCO-IHP and UNEP 2016). Approximately 40% of the people of this planet reside in river basins and aquifer systems that are shared by the international borders of two or more countries (Munia et al. 2016). As per International Groundwater Resources Assessment Centre (2014), there are

Table 12.1 Studies related to transboundary aquifers

Name of the researcher/institute that conducted the study	Year	Region	Methodology adopted/developed	Results/recommendations
Skoulikaris and Zafirakou	2019	European Union	Coupling of River water bodies ecological status with their chemical status	Issues related to water quality and quantity could be fixed by using constant and consistent datasets
Zeng et al.	2019	North China	Hybrid Game Theory and Mathematical programming Model (HGT-MPM)	Indistinct preliminary rights of water can be equitably demarcated through mutual bilateral consultations between upstream and downstream
Dasilva and Hussein	2019	South America	Historical approach	Beneficial approach to assess hydro-politics is production of scales
Hassen et al.	2019	Central Tunisia and Algeria	Three-dimensional hydrogeological modelling	Rational connection between observed and calculated groundwater levels for the observation wells
Alkhatib et al.	2019	Jordan	Ground Water Model	"Safe yield alternative" is the most appropriate for basin
Nijsten et al.	2018	Africa	Global water-use model and questionnaire surveys	Among 72 TBAs identified till date values of groundwater quality were satisfied in only 38 segments of the country as per standards
Sanchez et al.	2018	Mexico and Texas	Transboundariness approach	50–60% of the 180,000 km^2 area was reported to have good potential as well as water quality whereas ~25% were considered to have poor potential and water quality
Vizintin et al.	2018	Slovenia/Italy	Numerical hydrological model	Spread of contamination occurred quickly through karst aquifers and steadily through intergranular aquifers
Kattan	2018	Syria	Hydrochemistry and environmental isotopes	Current age and renewability of groundwater in aquifer were specified by ^{13}C and ^{14}C radio isotopes
Rahman et al.	2016	South Asia	SWAT	Scarcity of water in the wetland was due to the less seepage to the local groundwater table
Central Ground Water Board	2016	Delhi, Haryana, U. P	GIS	Potential zone of aquifer has been recognized along the western Yamuna canal passing through the northwest of the district
Sanchez et al.	2016	Mexico-US	Research analysis methods	Projected an outline for the assessment of these aquifers as a means for refining groundwater management in the border region
Halder et al.	2013	Indo-Bangladesh Ganges basin	Water governance plan	Inadequate quantity of surface water was present so as to fulfil the irrigational demands

(continued)

Table 12.1 (continued)

Name of the researcher/institute that conducted the study	Year	Region	Methodology adopted/developed	Results/recommendations
Wada and Heinrich	2013	World	AQUASTAT	Overexploitation by manmade activities leaded to the stress conditions in the 8% of the world's aquifer
Altchenko and Villholth	2013	Africa	Harmonised approach and aquifer Mapping	TBAs identified ~42% of the continental area and 30% of the population
Alley	2013	US (Mexico)	Hydrological modelling	Main source of recharge of aquifer systems is stream flow which is contaminated in urban area due to surface drainage
Gunawardhana et al.	2011	Japan	Hadley centre's climate model	Predicted that the expected groundwater recharge in 2080 would reduce by 1–26%
Kourakos and Mantoglou	2011	Semi-arid region	SEAWAT	Trade-off curves indicated rise in the cost of economy as the recharge of the groundwater was reduced due to change in climate
Megdal and Scott	2011	United States-Mexico border	Binational cooperative framework	Intense efforts are vital to identify and accommodate unevenness in the legal and regulatory frameworks
Cooley et al.	2009	World	Integrated approach	Summarized the risks to transboundary water agreements caused by climate change
Feitelson and Fischhendler	2009	Israel	Conceptual model	Initial assistance with positive outcomes can be vital for the successive development
Chadha	2008	Indus Basin	Lithologs	Intensification in the water strained situations in the arid zones of Indus basin
Chermak et al.	2005	World	Cooperative, noncooperative and myopic approach	For all parts of aquifer, cooperative solution was implemented followed by noncooperative and myopic
Puri	2002	World	Multidisciplinary approach	Economic, environmental, legal and institutional inputs are required for the better understanding of the hydrogeology
Wolf	1999	World	Generalized principles for delineating water allocations	Explained the practice of sharing water resources and compared the principles and policy of water equality
Utton	1978	United States	Legal regimes and management machinery	Recommended probable legal rules and administration machinery to accomplish the groundwater reserves

608 recognized transboundary aquifers throughout the world underlying all the nations except most of islands (ContiKI 2014). Some of the well-known global TBAs are listed in Table 12.2. The controversy over TBAs arises from the shortage of precise knowledge related to the subject in dispute. In light of the fact that groundwater must be characterized, inventoried and undergo a socio-technical evaluation to provide information to the states that permit them to acquire a policy according to the nature of globally or internationally shared groundwater (Kuri 2018). Diffuse and dispersed locations might be the source for the entry of polluting substances in the aquifer system. In such cases, so as to monitor and detect the exact cause and effect relationship, and assessing the related share of pollution from all the sources is a very laborious work (Scheumann and Alker 2009).

12.4 Transboundary Aquifers in the Indian Subcontinent

India, having about ~23% of the population at global level within only ~3% of the area of the world's land, encompasses some of the world's largest fluvial systems (River Brahmaputra, Ganges and Indus basins), which host some of the highest yielding aquifers on the earth (Mukherjee et al. 2015). Ganges basin is the largest river basin in the country hosting catchment area of ~ 86.1 million ha (CWC 2010). Indus-Ganges–Brahmaputra is one of the most productive systems worldwide that jointly drain the northern Indian plains and give rise to the regional alluvial aquifer system (MacDonald et al. 2016). Depletion of groundwater is considerably diverse over TBAs including the Indian River Plain (India and Pakistan) (Wada and Heinrich 2013). Along its international border, India shares its groundwater resources and there are minimum of eight aquifers that have been recognized as TBAs in the adjoining countries (Sharma 2009; Dhiman and Jain 2010) as shown in Fig. 12.3. Focused studies related to TBAs are extremely important as the concerned issues like water and food security are associated with them. There is a requirement for the formulation of efficient plans of management for TBAs in the Indian subcontinent so as to address the different groundwater issues related to the concerned countries. In the state of Punjab in India, the transboundary aquifers lie in the Indus basin (Chadha 2008). There is a need to measure the quality and quantity of groundwater. In recent scenario understanding of cooperation on transboundary aquifers than that of Transboundary Rivers is more important (Eckstein 2007).

Studies carried out by Dhiman and Jain (2010) revealed that 3D geometry for the aquifers at greater depth is yet to be entrenched and the parameters like water quality, recharge areas, flow of groundwater are yet to be established. There is a need to assess the cross-border impacts within TBAs so as to establish the international cooperation and management of aquifer system (Davies et al. 2013) as aquifers know no boundaries, they only obey hydraulic heads (UNESCO-IHP and UNEP 2016). Examples of Indian transboundary aquifers are given in (Table 12.3).

12.5 Problems Related to Transboundary Aquifers

Internationally shared aquifers have received negligible consideration from policymakers (Bourne 1992; Vandam and Wessel 1993; Puri and Aureli 2005; Rivera 2015). The grapple for water is intensifying legislative pressures and aggravating influence on ecosystems (Vorosmarty et al. 2000; Swyngedouw 2009). There is a rapid decline in the quality of surface as well as groundwater due to the overdose of agrochemicals and the overexploitation of groundwater resources (Ballabh 2008). Agricultural production and economic development would slow down gradually without the appropriate quantity and quality of water (Rai et al. 2019; Jasrotia and Kumar 2014). Groundwater is perpetual supply of water and is a vital reserve for the fundamental requirements and growth of the society as it acts as a safeguard in times of drought at reasonable cost (Masiyandima and Giordano 2007; Nijsten

Table 12.2 List of TBAs across the world.

S.no	Name of aquifers	Aquifer sharing nations
1	Guarani aquifer	Brazil, Argentina, Paraguay, Urugua
2	Nubion Sandstone aquifer	Chad, Egypt, Libya & Sudan
3	Kalahari/Karoo aquifers	Botswana, Namibia & South Africa
4	Mekong River plain aquifer	Thailand, Laos, Cambodia & Vietnam
5	Ili River plain aquifer	China & Kazakhstan
6	Himalayan foothill aquifer	Nepal & India
7	West Altai	Russia & Kazakhstan
8	New Guinea Island	Indonesia, Papua New Guinea
9	South Burma	Burma & Thailand
10	Ertix River plain	Russia, Kazakhstan
11	Anti-Lebanon	Lebanon, Syria
12	Pretashkent aquifer	Kazakhstan, Uzbekistan
13	Syrt	Kazakhstan, Russia
14	Indus River plain aquifer	Pakistan, India
15	Khorat Plateau aquifer	Laos, Thailand
16	Saq-Ram aquifer system	Jordan, Saudi Arabia
17	Salween River aquifer	Myanmar, Thailand
18	Tacheng basin/Alakol	Kazakhstan, China
19	Yenisei Upstream	Mongolia, Russia
20	East Ganges River plain	Bangladesh, India
21	Abbotsford-Sumas	British Columbia, Washington
22	Okanagan-Osoyoos	British Columbia, Washington
23	Grand Forks	British Columbia, Washington
24	Poplar	Saskatchewan, Montana
25	Estevan	Saskatchewan, North Dakota
26	Milk River	Alberta, Montana
27	Judith River	Saskatchewan, Alberta, Montana
28	Northern Great plains	Manitoba, Saskatchewan, N. Dakota, S. Dakota
29	Wajid Aquifer system	Saudi Arabia, Yemen
30	Basalt Aquifer system (South)	Jordan, Syria

Source CGWB (2010), UNESCO (2011), Rivera (2015), Lee et al. (2018)

et al. 2018; Olago 2018). The pollution caused to the aquifers is almost irreversible as they are located in the subsurface and are visible only through the eyes of hydrogeology. As the wider percentage of freshwater in liquid state lies underground, the rising requirement of freshwater often leads to overutilization of groundwater (Feitelson and Fischhendler 2009). The overexploitation of groundwater helped in considerable fiscal and welfare gains and also lead to decline in its quality and quantity that affect the sustainability of current processes and practices (Llamas and Custodio 2003; Balooni and Venkatachalam 2016; Jasrotia et al. 2018). Futile planning of water resources as well as lack of making people aware regarding the root cause behind this problem and unsuccessful policies are the other reasons that contribute towards the

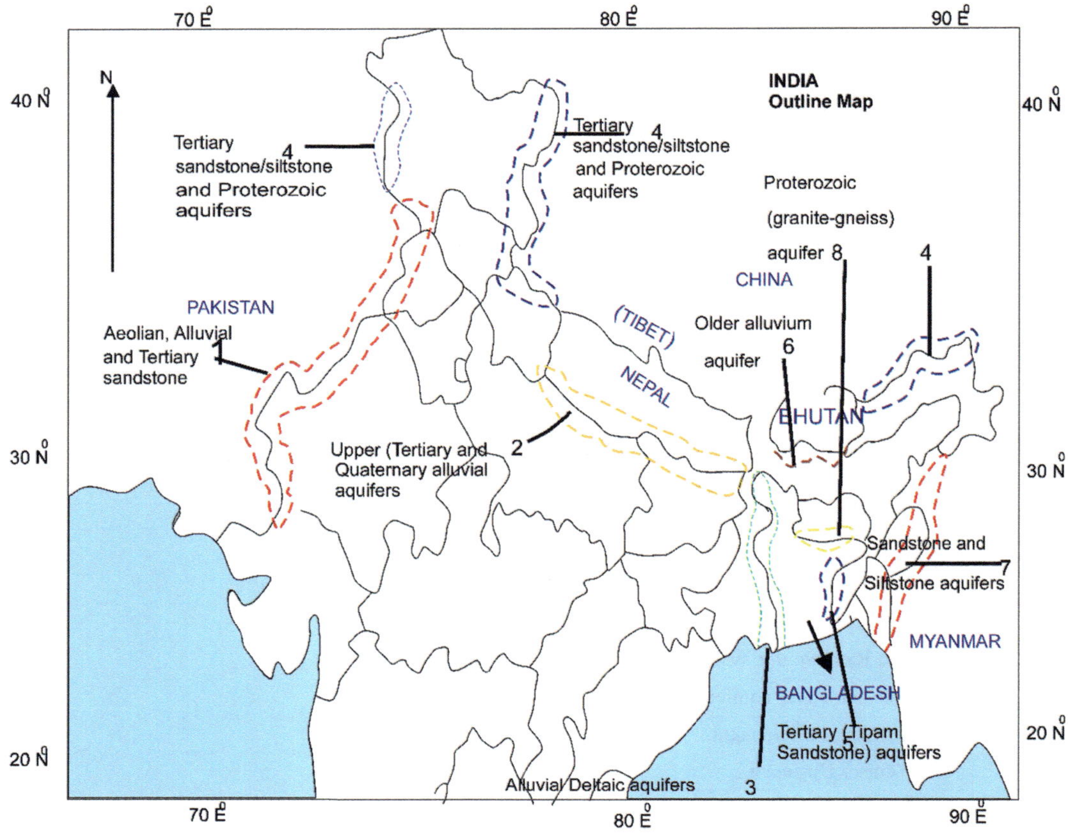

Fig. 12.3 TBAs of Indian subcontinent. *Source* Sharma (2009), Dhiman and Jain (2010)

Table 12.3 TBAs of Indian subcontinent

S. no	Name of transboundary aquifer system	Countries sharing aquifer system
1	Aeolian, alluvial and Tertiary sandstone aquifers	India, Pakistan
2	Upper Tertiary and Quaternary alluvial aquifers; Bhabbar Terai aquifers	India, Nepal
3	Alluvial/Deltaic aquifers	India, Bangladesh
4	Tertiary Sandstone/Siltstone and Proterozoic (Granite, Phyllite, Quartzite) aquifers	India, China, Pakistan
5	Tertiary (Tipam Sandstone) aquifers	India, Bangladesh
6	Older Alluvium aquifers	India, Bhutan
7	Sandstone and Siltstone aquifers	India, Myanmar
8	Proterozoic (Granite-Gneiss) aquifers	India, Bangladesh

Source Sharma (2009), Dhiman and Jain (2010)

scarcity of freshwater in the country (Kumar and Pramod 2019). Even if the scarcity of water is not a contemporary issue but for the future reference, the quantity as well as the quality of water needs to be forewarned at this phase (Sinisi 2003). Integrated management of shared resources requires some effective agreements and laws (Brooks et al. 2013).

Globally or internationally shared transboundary groundwater resources remain indeterminate due to unavailability of the required data, differences in the aquifer methodologies and also due to the lack of collaboration and cooperativeness among the government and local bodies of the related countries so as to deal with the challenges of groundwater from a binational/multinational prospect (Price 2016; Sanchez et al. 2018; Jasrotia et al. 2019). The majority (72%) of the withdrawal of groundwater in Asia is caused by rigorous agricultural activities and abrupt growth of population over the area including India, Bangladesh, Pakistan, China and Iran (Shah 2005; Gleeson et al. 2012; FAO 2016; Lee et al. 2018). Bilateral understanding of transboundary groundwater resources is hampered by the inadequacy of training, cooperation and sharing of the data between the riparian countries. Predetermined issue is that the transboundary water resource not administered by united and comprehensive method by one nation may be over-utilized or exploited by other nations (Godfrey and Dyk 2002; Jarvis et al. 2005; Cobbing et al. 2008). Some of the government schemes that addressed transboundary waters are manifested by their patchy geographic division, reach and legislative hostile sign (Matsumoto 2002; Jarvis et al. 2005; Walter 2010). Debates on rising concepts on water reliability and water energy food connections are the good examples that connect policymaking and scientific knowledge (Mirumachi and Chan 2014; Artioli et al. 2017). In recent scenario, shared water resources are turning into a tool of power, encouraging rivalry inside and among the countries (Rivera 2015). The countries like Africa are vigorously dependent upon groundwater resources, including transboundary groundwater, with an approximately 75% of its people reliant on this resource for essential water supplies (Braune and Xu 2008; Altchenko and Villholth 2013; Fraser et al. 2018). TBAs play an imperative role in assisting water requirements of Asia and water resource system for the globe (Zaisheng et al. 2008; Rivera 2015; Sanchez et al. 2018).

Regional studies carried out by Rodell et al. (2009); Tiwari et al. (2009); Singh et al. (2017); Sarkar et al. (2020) using the Gravity Recovery and Climate Experiment (GRACE) exposed a substantial quantity of depletion in groundwater, i.e. the constant ejection of groundwater from aquifer storage as a result of groundwater abstraction in excess of groundwater recharge, from the aquifer underlying the countries like India, Pakistan and Bangladesh, most of which is used for irrigation for the production of food. Shortage of water repeatedly leads to water-related battles. Simultaneously, globally shared aquifers with their huge depository of water resources often deliver other possible elucidation to avert conflicts and retain peace (Naff and Matson 1984; Tignino 2010; Khan et al. 2020). Tensions at national or international levels over transboundary waters remain a hotly debated matter. The role and implications of TBAs in the development of human and environment have received very less consideration. Bounding lines of transboundary aquifers remain badly known or most of the aquifers remain poorly known due to the differences in approaches of geological litho-stratigraphy, irregular accessibility of data and communication gap between the concerned countries (UNESCO 2011). As a political context, interaction of transboundary water may provide strategic motives, for example, to bring exterior funding, or to share the load of cleaning costs (Bernauer 2002; Zeitoun and Mirumachi 2008).

12.6 Management of Transboundary Aquifers

In the twenty-first century, management of water is a principal component on the global agenda (United Nations 2015; Yasuda et al. 2017). Usage of aquifer in the last half-century has added towards the mitigation of poverty and the advancement of conditions associated with public health (Nwankwoala 2015; Oni and Aizebeokhai 2017). The aquifer system contains excellent quality of water, which can be affected

Table 12.4 Indicator components for the management of transboundary aquifers

Scientific	Institutional	Legal	Socioeconomic	Environmental
Provide direction for the growth of conceptual models	Provide direction related to the legal frameworks	Provide direction related to the authority and powers for the joint management	Provide guidance for the demographics, land use, current and forecast needs for agricultural environment	Provide directions for hydrology, biodiversity, climate change and ethical issues

Source Puri and Aureli (2005), Davies et al. (2013), Sanchez et al. (2018)

Table 12.5 Approaches with the issues associated with overexploitation of aquifers.

Preventive approach	Remedial approach
The chief purpose of the preventive approach is to obstruct overutilization and overexploitation by executing and accomplishing suitable or pertinent groundwater legislation	It is applicable for the instances where the issue of overexploitation has already occurred and generally involves artificial recharge of the aquifers

Source Zaisheng et al. (2006)

by the poor management practices. In the recent future, shared aquifers are important for the food security and drinking water supply (CGWB 2010). When basins enclose diverse sovereign states, the preeminent affair is how to plan and assist institutions for the equitable sharing of global water resources (Sneddon and Fox 2006). Assessment of aquifer is a prerequisite to aquifer management. Effective management plans are needed in order to address sundry aquifers associated matter of concern to the countries involved. Comprehensive data is required about these aquifers so as to articulate the effective management strategy across the international border. The main components that might help the management perspective of TBAs are given in Table 12.4.

Management of water is practically carried out within hydrological basins. In general, there are two approaches for the management of the problems related to the overexploitation of aquifers (Zaisheng et al. 2006, 2008) and are summarized in Table 12.5.

A further feasible way to solve the overutilization and overexploitation of groundwater is to increase the supply of surface water and to limit or reduce the supply of groundwater. Therefore, it is the need of the hour to have consolidated management of surface as well as groundwater resources (Huntington and Niswonger 2012; National Water Commission 2014).

In the discussion about the crises of water in the world, the aspect of the Dublin Conference (1992) has gained attention that safeguards market interference in the management of water, in contrary to those who assert water as a public good (Kuri 2018). In case of climate change occurring due to the retrieving glaciers of Himalayas, management of transboundary aquifers is urgently required (Dhiman and Jain 2010). Thus, the management of transboundary aquifers leads to the direct and indirect contribution towards the development of economy, trade at international level, water and political security and alleviation of poverty. The rudiments for successful management of TBAs are political will and obligation from all government, at all stages. There is no ubiquitous way out but the knowledge behind the hydrology of the TBAs, Law and administration, monetary and reasonable skills are the important points that need to be acknowledged for everlasting, reliable and sustainable transboundary water management (Haldar et al. 2013).

12.7 Conclusion

It is unquestionably required to protect the water resources in the aspect of the risk posed by overexploitation. Economic and human expansion, and aquifer problem alertness and management are strongly correlated. There is absence of awareness in most of the decision-makers and users regarding the harmful effects of overexploitation and contamination of the aquifers. The absence of this basic knowledge leads to the poor management of inadequate water resources throughout the world. Proper management practices are required at local level as transboundary water problems are geographically specific. Also, there is a lack of financial resources for scientific research due to which there is scarcity of excellent geological and hydrogeological knowledge and modern technical data. There is a need to delineate the main areas of transboundary aquifers, which are at higher risk of stress and degradation. Effective water management policies and full understanding of social, environmental, legal issues would help to dramatically protect the transboundary resources prudently. There should be a strict and clear rule governing the usage, conservation and overall management of TBAs across the borders as the rules governing these resources are indistinct. Formal treaties or agreements for the management of TBAs should be initiated. There should be transnational governance agreement between scientists so as to safeguard the sustainability and safety of future global water resources.

12.8 Recommendations

- The gap in the basic data and knowledge at local, regional and international level indicates that special attention is needed so as to carry out further research in this area.
- There should be a consolidated management of both surface as well as groundwater resources.
- Major recommendation is to establish collaborative awareness and monitoring programmes so as to improve the cooperation at national and international levels.
- It is also recommended that the historical understanding of transboundary aquifers needs to be modernized.

Acknowledgements Ashu Khosla (corresponding author) is thankful to book Editors for inviting him to contribute the article. The first author is grateful to PhD supervisors and mentors (Madhuri S. Rishi and Ashu Khosla) for helping in many ways.

References

Adimalla N, Taloor AK (2020) Hydrogeochemical investigation of groundwater quality in the hard rock terrain of South India using Geographic Information System (GIS) and ground water quality index (GWQI) techniques. Groundwater Sust Develop 10:P100288. https://doi.org/10.1016/j.gsd.2019.100288

Ahmad M, Bastiaanssen WGM, Feddes RA (2005) A new technique to estimate net groundwater use across large irrigated areas by combining remote sensing and water balance approaches, Rechna Doab, Pakistan. Hydrogeol J 13:653–664

Albrecht TR, Varady RG, Zuniga-Teran AA, Gerlak AK, Staddon C (2017) Governing a shared hidden resource: a review of governance mechanisms for transboundary groundwater security. Water Security 2:43–56

Alkhatib J, Engelhardt I, Ribbe L, Sauter M (2019) An integrated approach for choosing suitable pumping strategies for a semi-arid region in Jordan using a groundwater model coupled with analytical hierarchy techniques. Hydrogeol J, 1–15. https://doi.org/10.1007/s10040-019-01925-0

Alley WM (2013) Five-year interim report of the United States-Mexico transboundary aquifer assessment program, 2007–2012. US Department of the Interior, US Geological Survey, pp 1–31

Altchenko Y, Villholth KG (2013) Transboundary aquifer mapping and management in Africa: a harmonised approach. Hydrogeol J 21(7):1497–1517

Artioli F, Acuto M, Mcarthur J (2017) The water-energy-food nexus: an integration agenda and implications for urban governance. Polit Geogr 6:215–223

Ballabh V (2008) Governance of water: institutional alternatives and political economy. Sage, p 396

Balooni K, Venkatachalam L (2016) Managing water for sustainable development: An Indian perspective. IIM Kozhikode Soc Manage Rev 5(1):7–13. https://doi.org/10.1177/2277975215625500

Bernauer T (2002) Explaining success and failure in international river management. Aquatic Sci 64 (1):1–19

Bittinger MW (1972) Survey of interstate and international aquifer problems. Groundwater 10(2):44–54

Bourne CB (1992) The international law commission's draft articles on the law of international watercourses: principles and planned measures. Colorado J Int Environ Law and Pol 3:65–92

Braune E, Xu Y (2008) Groundwater management issues in Southern Africa-An IWRM perspective. Water SA 34(6):699–706
Brooks DB, Trottier J, Doliner L (2013) Changing the nature of transboundary water agreements: the Israeli-Palestinian case. Water Int 38(6):671–686
Central Water Commission (CWC), Ministry of Water Resources (2010) Water and related statistics, information systems organization, water planning and project wing. Central Water Commission, New Delhi, India
Central Ground Water Board (2010) Special issue on transboundary aquifer system. CGWB, Ministry of water Resources
Central Ground Water Board (2016) Aquifer mapping and groundwater management plan of NCT Delhi, Ground Water Year Book, p 162
Chadha DK (2008) Development, management and impact of climate change on transboundary aquifers of Indus Basin. In 4th International Symposium on Transboundary Water Management, Thessaloniki, Greece, p 4
Chermak JM, Patrick RH, Brookshire DS (2005) Economics of transboundary aquifer management. Groundwater 43(5):731–736
Cobbing JE, Hobbs PJ, Meyer R, Davies J (2008) A critical overview of transboundary aquifers shared by South Africa. Hydrogeol J 16(6):1207–1214
ContiKI (2014) Factors enabling transboundary aquifer cooperation: a global analysis. Delft, The Netherlands
Cooley H, Christian-Smith J, Gleick PH, Allen L, Cohen M (2009) Understanding and reducing the risks of climate change for transboundary waters. Oakland: Pacific Institute, Preservation Park Oakland, California, p 96
Dasilva LPB, Hussein H (2019) Production of scale in regional hydropolitics: an analysis of La Plata River Basin and the Guarani Aquifer System in South America. Geoforum 99:42–53
Davies J, Robins NS, Farr J, Sorensen J, Beetlestone P, Cobbing JE (2013) Identifying transboundary aquifers in need of international resource management in the Southern African Development Community region. Hydrogeo J 21(2):321–330
Dhiman SC, Jain SK (2010) An overview of transboundary aquifers of the Indian Sub-continent. Editorial Board 25(1):1
Eckstein GE, Eckstein Y (2003) A hydrogeological approach to transboundary ground water resources and international law. AmerUnivInt Law Rev 19:201
Eckstein Y, Eckstein GE (2005) Transboundary aquifers: Conceptual models for development of international law. Groundwater 43(5):679–690
Eckstein GE (2007) Commentary on the UN international law commission's draft articles on the law of transboundary aquifers. Colorado J Int Environ Law and Policy 18:537
European Water Framework Directive EU-WFD (2000) Directive 2000/60/EC of the European Parliament and of the Council of 23 October 2000. Establishing a framework for Community action in the field of water policy, Article 175, OJL 327, 22 December 2000, pp. 1–73
FAO (2016) AQUASTAT main database, Food and Agriculture Organization of the United Nations (FAO). https://www.fao.org/nr/water/aquastat/water_use
Feitelson E, Fischhendler I (2009) Spaces of water governance: the case of Israel and its neighbors. Ann Assoc Amer Geogr 99(4):728–745
Foster SSD, Chilton PJ (2003) Groundwater: the processes and global significance of aquifer degradation. Phil Trans Roy Soc London. Sr B: Bio Sci 358 (1440): 1957–1972
Fraser CM, Kalin RM, Rivett MO, Nkhata M, Kanjaye M (2018) A national approach to systematic transboundary aquifer assessment and conceptualisation at relevant scales: a Malawi case study. J Hydr Reg Stud 20:35–48
Giordano M, Villholth KG (2007) The agricultural groundwater revolution: opportunities and threats to development, vol 3. CAB International 13:978
Gleeson T, Wada Y, Bierkens MF, van Beek LP (2012) Water balance of global aquifers revealed by groundwater footprint. Nature 488(7410):197
Godfrey L, van Dyk G (2002) Reserve determination for the Pomfret-Vergelegen dolomitic aquifer, North West Province, part of catchments D41C, D. CSIR Report ENV-PC, E and F, p 31
Golovina EI (2018) Problems of groundwater extraction from Transboundary aquifers and complexes. In IOP conference series. Ear Environ Sci 151(1)
Gunawardhana LN, Kazama S, Kawagoe S (2011) Impact of urbanization and climate change on aquifer thermal regimes. Wat Res Manage 25(13):247–276
Haldar S, Sadhukhan D, Naha N (2013) Transboundary water governance in Indo-Bangladesh lower Ganges basin. In International on Water Resources of South Asia, Dhaka, Bangladesh, pp 1–13
Haque S, Kannaujiya S, Taloor AK, Keshri D, Bhunia RK, Ray PKC, Chauhan P (2020) Identification of groundwater resource zone in the active tectonic region of Himalayathroughearth observatory techniques. Groundwater Sust Develop 10: P100337. https://doi.org/10.1016/j.gsd.2020.100337
Hassen I, Milnes E, Gibson H, Bouhlila R (2019) Impact of groundwater flow across tectonic aquifer compartments in a Miocene Sandstone aquifer: three-dimensional hydrogeological modeling of the Kasserine aquifer system in central Tunisia and northeastern Algeria. Hydrogeol J 1:1–17
Hayton RD, Utton AE (1989) Transboundary groundwaters: the Bellagio draft treaty. Nat Res J 1:663–722
Huntington JL, Niswonger RG (2012) Role of surface water and groundwater interactions on projected summertime stream flow in snow dominated regions: An integrated modeling approach. Wat Resour Res 48 (11). https://doi.org/10.1029/2012WR012319
International Lake Environment Committee (2011) Methodology for the GEF transboundary waters

assessment programme Volume 2: methodology for the assessment of transboundary aquifers, UNEP, p113
ISARM (2004) International Shared Aquifer Resource Management Initiative, UNESCO, IHP-VI, Paris. https://www.isarm.nitg.tno.nl/
Jarvis T, Giordano M, Puri S, Matsumoto K, Wolf A (2005) International borders, ground water flow and hydroschizophrenia. Groundwater 43(5):764–770
Jasrotia AS, Kumar A (2014) Estimation of replenishable groundwater resources and their status of utilization in Jammu Himalaya, J&K, India. Eur Water 48:17–27
Jasrotia AS, Taloor AK, Andotra U, Bhagat BD (2018) Geoinformatics based groundwater quality assessment for domestic and irrigation uses of the Western Doon valley, Uttarakhand, India. Groundwater Sustain Dev 6:200–212
Jasrotia AS, Taloor AK, Andotra U, Kumar R (2019) Monitoring and assessment of groundwater quality and its suitability for domestic and agricultural use in the Cenozoic rocks of Jammu Himalaya, India: a geospatial technology-basedapproach. Groundwater Sustain Dev 8:554–566
Kattan Z (2018) Using hydrochemistry and environmental isotopes in the assessment of groundwater quality in the Euphrates alluvial aquifer, Syria. Environ Ear Sci 77:45. https://doi.org/10.1007/s12665-017-7197-1
Khan A, Govil H, Taloor AK, Kumar (2020) Identification of artificial groundwater recharge sites in parts of Yamuna river basin India based on remote sensing and geographical information system. Groundwater Sustain Dev. https://doi.org/10.1016/j.gsd.2020.100415
Kourakos G, Mantoglou A (2011) Simulation and multi-objective management of coastal aquifers in semi-arid regions. Water Res Manage 25(4):1063–1074
Kumar A, Pramod KA (2019) Groundwater vulnerability and contamination risk assessment using GIS-based modified DRASTIC-LU model in hard rock aquifer system in India. Geocarto Int 1:1–30
Kuri GH (2018) A joint management of transboundary aquifers: from asymmetries to environmental protection. Frontera Norte 30(59):129–154
Lee E, Jayakumar R, Shrestha S, Han Z (2018) Assessment of transboundary aquifer resources in Asia: status and progress towards sustainable groundwater management. J Hydol Reg Stud 20:103–115
Llamas MR, Custodio E (2003) Intensive use of groundwater: a new situation which demands proactive action. Intensive use of groundwater: challenges and opportunities, pp 13–31
Llamas MR, Martínez-Santos P (2005) Intensive groundwater use: a silent revolution that cannot be ignored. Water Sci Tech 51(8):167–174
MacDonald AM, Bonsor HC, Ahmed KM, Burgess WG, Basharat M, Calow RC, Dixit A, Foster SSD, Gopal K, Lapworth DJ, Lark RM (2016) Groundwater quality and depletion in the Indo-Gangetic Basin mapped from in situ observations. Nature Geosci 9 (10):762
Masiyandima M, Giordano M (2007) Sub-Saharan Africa: opportunistic exploitation. The agricultural groundwater revolution: opportunities and threats to development, pp 79–99
Matsumoto K (2002) Transboundary groundwater and international law: past practices and current implications. Master's Thesis. Department of Geosciences, Oregon State University, p 67
Megdal SB, Scott CA (2011) The importance of institutional asymmetries to the development of binational aquifer assessment programs: the Arizona-Sonora experience. Water 3(3):949–963
Mirumachi N, Chan K (2014) Anthropocentric hydro politics? Key developments in the analysis of international transboundary water politics and some suggestions for moving forward. Aquatic Procedia 2:9–15
Mukherjee A, Saha D, Harvey CF, Taylor RG, Ahmed KM, Bhanja SN (2015) Groundwater systems of the Indian sub-continent. J Hydr Regl Stud 4:1–14
Munia H, Guillaume JHA, Mirumachi N, Porkka M, Wada Y, Kummu M (2016) Water stress in global transboundary river basins: significance of upstream water uses on downstream stress. Environ Res Lett 11 (1):014002
Naff T, Matson RC (1984) Water in the Middle East: conflict or cooperation? Boulder CO, Westview, AGRIS
National Water Commission (2014) Integrating groundwater and surface water management in Australia. NWC, Canberra, pp 6–11
Nijsten GJ, Christelis G, VillholthKG BE, Gaye CB (2018) Transboundary aquifers of Africa: review of the current state of knowledge and progress towards sustainable development and management. J Hydr Regl Stud 20:21–34
Nwankwoala HO (2011) An integrated approach to sustainable groundwater development and management in Nigeria. J Geo Min Res 3(5):123–130
Nwankwoala HO (2015) Hydrogeology and groundwater resources of Nigeria. NY Sci J 8(1):89–100
Olago DO (2018) Constraints and solutions for groundwater development, supply and governance in urban areas in Kenya. Hydrogeol J 1:1–20
Oni AA, Aizebeokhai AP (2017) Impacts of groundwater development on poverty reduction and alleviation in Nigeria. J Inf Matham Sci 9(2):455–484
Price G (2016) Rethinking water climate cooperation in south Asia. ORF Issue brief. Embassy of the Federal Republic of Germany and New Delhi
Puri S, Appelgren B, ArnoldG AA, Burchi S, Burke J, MargatJ PP (2001) Internationally shared (transboundary) aquifer resources management: their significance and sustainable management. United Nations Educational, Scientific and Cultural Organization (UNESCO)
Puri S, El Naser H (2002) Intensive use of groundwater in transboundary aquifers. Challenges and opportunities, intensive use of groundwater, p 415
Puri S (2003) Transboundary aquifer resources: international water law and hydrogeological uncertainty. Wat Int 28(2):276–279

Puri S, Aureli A (2005) Transboundary aquifers: a global program to assess, evaluate, and develop policy. Groundwater 43(5):661–668

Rahman MM, Thompson JR, Flower RJ (2016) An enhanced SWAT wetland module to quantify hydraulic interactions between riparian depressional wetlands, rivers and aquifers. Environ Mod Softw 84:263–289

Rai SP, Sharma N, Lohani AK (2019) Novel approach for issues identification in transboundary water management using fuzzy c-means clustering. Appl Wat Sci 9 (1):11

Rivera A (2015) Transboundary aquifers along the Canada–USA border: science, policy and social issues. J Hydrol Reg Stud 4:623–643

Rodell M, Velicogna I, Famiglietti JS (2009) Satellite-based estimates of groundwater depletion in India. Nature 460(7258):999

Sarkar T, Kannaujiya S, Taloor AK, Ray PK, Chauhan P (2020) Integrated study of GRACE data derived interannual groundwater storage variability over water stressed Indian regions. Groundwater Sustain Develop. P100376. https://doi.org/10.1016/j.gsd.2020.100376

Sanchez R, Lopez V, Eckstein G (2016) Identifying and characterizing transboundary aquifers along the Mexico–US border: an initial assessment. J Hydrol 535:101–119

Sanchez R, Rodriguez L, Tortajada C (2018) The transboundariness approach and prioritization of transboundary aquifers between Mexico and Texas. Ambio 47(7):760–770

Scheumann W, Alker M (2009) Cooperation on Africa's transboundary aquifers-conceptual ideas. Hydr Sci J 54(4):793–802

Shah T (2005) Groundwater and human development: challenges and opportunities in livelihoods and environment. Water Sci Tech 51(8):27–37

Sharma SK (2009) Transboundary aquifer system of Indian subcontinent, technical paper. Workshop on transboundary aquifer systems by CGWB, MOWR, New Delhi

Skoulikaris C, Zafirakou A (2019) River Basin Management Plans as a tool for sustainable transboundary river basins management. Environ Sci and Pollu Res 1–14

Singh AK, Jasrotia AS, Taloor AK, Kotlia BS, Kumar V, Roy S, Ray PKC, Singh KK, Singh AK, Sharma AK (2017) Estimation of quantitative measures of total water storage variation from GRACE and GLDAS-NOAH satellites using geospatial technology. Quat Int 444:191–200

Sinisi L (2003) Impact assessment of aquifer recharge. Health risks in aquifer recharge using reclaimed Water-State of the art report, pp. 154–170

Sneddon C, Fox C (2006) Rethinking transboundary waters: a critical hydropolitics of the Mekong basin. Political Geogr 25(2):181–202

Swyngedouw E (2009) The political economy and political ecology of the hydrosocialcycle. J Contemp Wat Res Edu 142(1):56–60

Tignino M (2010) Water international peace and security. Int Rev Red Cross 92(879):647–674

Tiwari VM, Wahr J, Swenson S (2009) Dwindling groundwater resources in northern India, from satellite gravity observations. Geophy Res Lett 36(18):1–5. https://doi.org/10.1029/2009GL039401

Turton AR, Earle A, Malzbender D, Ashton PJ (2005) Hydropolitical vulnerability and resilience along Africa's international waters. Nairobi, UNEP. Hydropolitical vulnerability and resilience along international waters. UNEP, p 148

UNESCO (2011) Second Assessment of Transboundary Rivers, lakes and groundwater: a United Nations Publication, Economic commission for Europe. https://www.unece.org/fileadmin/DAM/env/water/publications/assessment/English/ECE_Second_Assessment_En.pdf

UNESCO-IHP, UNEP (2016) Transboundary aquifers and groundwater systems of small island developing states: status and trends, summary for policy makers. TWAP, United Nations Environment Programme (UNEP), Nairobi

United Nations (2015) Transforming our world: the 2030 agenda for sustainable development. United Nations, Department of Economic and Social Affairs, New York

Utton AE (1978) International groundwater management: the case of the US-Mexican frontier. Neb L Rev 57:633

Vandam JC, Wessel J (1993) Transboundary river basin management and sustainable development, vol ii. Technical documents in hydrology, UNESCO, Paris

Vizintin G, Ravbar N, Janez J, Koren E, JanezN ZL, Treu F, Petric M (2018) Integration of models of various types of aquifers for water quality management in the transboundary area of the Soca/Isonzo river basin (Slovenia/Italy). Sci Total Environ 619:1214–1225

Vorosmarty CJ, Green P, Salisbury J, Lammers RB (2000) Global water resources: vulnerability from climate change and population growth. Science 289 (5477):284–288

Wada Y, Heinrich L (2013) Assessment of transboundary aquifers of the world-vulnerability arising from human water use. Environ Res Lett 8(2):024003

Waithaka EL (2018) Role of shared resource management in enhancing inter-state cooperation in the horn of Africa: a case of ilemi triangle (2011–2016). Doctoral dissertation, institute of diplomacy and international studies, university of Nairobi

Walter M (2010) Managing transboundary aquifers: Lessons from the field. In: Research paper collection, international conference on Transboundary aquifers: challenges and new directions (ISARM), pp. 1–7

Wolf AT (1999) Criteria for equitable allocations: the heart of international water conflict. Nat Resour Forum 23(1):3–30

Wolf AT (2010) Sharing water, sharing benefits: working towards effective transboundarywater resources management. UNESCO, France

Yasuda Y, Aich D, Hill D, Huntjens P, Swain A (2017) Transboundary water cooperation over the Brahmaputra River: legal political economy analysis of current and future potential cooperation.The Hague Institute for Global Justice, p 149
Zaisheng H, Hao W, Rui C (2006) Transboundary aquifers in Asia with special emphasis to China. United Nations Educational, Scientific and Cultural Organization 10–18
Zaisheng H, Jayakumar R, Ke L, Hao W, Rui C (2008) Review on transboundary aquifers in People's Republic of China with case study of Heilongjiang-Amur River Basin. Environ Geol 54(7):1411–1422
Zeitoun M, Mirumachi N (2008) Transboundary water interaction I: reconsidering conflict and cooperation. Int Environ Agreements. Politics 8(4):267–297. https://doi.org/10.1007/s10784-008-9083-5
Zeng Y, Li J, Cai Y, Tan Q, Dai C (2019) A hybrid game theory and mathematical programming model for solving trans-boundary water conflicts. J Hydrol 570:666–681

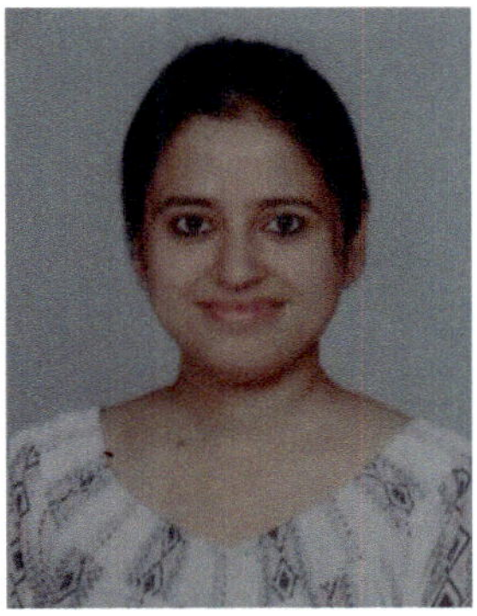

Ms. Ashima Awasthi holds Master's degree in Environmental Sciences from Central University of Himachal Pradesh. Currently, she is Research Scholar in the Department of Environment Studies, Panjab University, Chandigarh and doing research on transboundary aquifers. She has experience in remote sensing and GIS applications in groundwater quality. She was trainee in National Remote Sensing Centre (ISRO, Department of Space, Government of India), Hyderabad for 12 weeks under geospatial technologies and applications course. She was placed First in Panjab University Ph.D. entrance exam and has qualified UGC-NET exam. She has acquired acute insights from seminars and has presented papers in conferences

Dr. Madhuri S. Rishi is presently Assistant Professor in the Department of Environment Studies, Panjab University, Chandigarh, India and has successfully completed her tenure as Chairperson of the Department from September 2016 to September 2019. She has been very sincere in her approach to education and science in general—in fact, a very devoted person in pursuing projects and gaining knowledge, which is supplemented by her academic achievements, being first class and gold medalist at graduate and post-graduate levels. She has teaching and research experience for more than 23 years and has supervised 20 Ph.D. scholars working in the fields of groundwater quality and quantity issues, artificial recharge of groundwater, mining and its impacts on environment, environmental impacts of hydroelectric power projects, uranium contamination in SW Punjab and related issues of societal significance. She is having a good knowledge of computer skills and software like Minitab, Map Info, Aqua Chem, SPSS which have helped in bringing out more than 80 quality research publications in national and international journals. She was successfully associated with 'Scheme for Rain Water Harvesting and Artificial Recharge to Groundwater at Panjab University Campus, Chandigarh' amounting Rupees 7.76 Crore. She was conferred upon with the National Water Award, 2019 by National Water Mission in association with MoWR, GR and RD, for following Best Campus Practices on Panjab University Campus. She has also been a resource person to Centre for IAS and other competitive examinations, Panjab University, Chandigarh and State Institute of Family Health and Welfare, Mohali

Prof. Ashu Khosla accomplished Ph.D. in Geology from Panjab University in 1997 and later undertook a Post-Doctorate from Montpellier University, France in 1997–1998. Presently, he is a faculty (Professor) in the Department of Geology, Panjab University, Chandigarh. His work has been acknowledged worldwide by the palaeontologists and palaeobiogeographers, as it covers diverse issues such as evolution, diversity and biogeography of vertebrates and microbiota associated with the Cretaceous fragmentation and drift of the Indian plate. He has to his credit several exciting fossil discoveries from the Late Cretaceous of India. He has published 61 research papers in peer-reviewed national/international journals, apart from few in press, in high-impact factor journals, including two papers in illustrious American journal *SCIENCE AAAS* and other journals, for instance, Earth and Planetary Science Letters, Global and Planetary Change, Palaeo 3, Journal of Asian Earth Sciences, Journal of Vertebrate Palaeontology, Geological Journal, Cretaceous Research and Historical Biology. He has already successfully completed six research projects funded by the Department of Science and Technology (Government of India), New Delhi. He has published two important books. First on the global Cretaceous (Cretaceous Period: Biotic Diversity and Biogeography). The volume was published in the New Mexico Museum of Natural History and Science Bulletin in 2016 and second on the Indian Late Cretaceous dinosaur eggs of peninsular India in October, 2020 (Springer Nature, Switzerland). Presently, he is working on groundwater problems of Punjab and Riverbed mining its geoenvironmental and social impacts in Himalayan terrain of Himachal Pradesh, India

Chemical Weathering in Jhelum River and its Tributaries, Kashmir Basin, Western Himalaya

13

Riyaz Ahmad Mir, Farooq Ahmad Dar, and Ghulam Jeelani

Abstract

The study describes the chemical analysis of water and sediment to understand the hydro-geochemical processes, chemical weathering rates and its intensities in the upper river Jhelum and its major tributaries in Kashmir Basin, Western Himalaya. A total of 50 water samples and 15 riverbed sediment samples were analyzed. It was found that Ca and Mg contribute 82% of the major cations and HCO_3^- contributes 92% of the anion budget with only 4% and 3% contributions from SO_4 and Cl, respectively. The Chemical Weathering Rates (CWR) varied spatially among tributaries with Sukhnag showing the lowest and Sindh, the highest CWR during High Flow Period (HFP). However, during the Low Flow Period (LFP), Sukhnag recorded the lowest and Romush recorded the highest CWR thereby reflecting the influence of varied precipitation, discharge, basin lithology, topography and probably active tectonics. The CWR in the main Jhelum river ranged from 31.4×10^2 t/km^2/month (LFP) to 107.0×10^2 t/km^2/month (HFP) with an annual CWR of 11.1×10^2 t/km^2/year. The Left Bank Tributaries (LBT) draining the tectonically active Pir-Panjal range showed higher CWR than the Right Bank Tributaries (RBT) draining the Great Himalayan range (except Sindh). However, in comparison to other Himalayan rivers and World averages (0.36×10^2 t/km^2/year), the observed CWR in the main Jhelum river and its tributaries were also found to be higher. Furthermore, the main Jhelum river showed an average annual Silicate Weathering Rate (SWR) of 0.10 t/km^2/year and Carbonate Weathering Rate (CrWR) of 0.35 t/km^2/year. The Pohru tributary showed the lowest average SWR and CrWR of 0.04 t/km^2/year, whereas the Sukhnag showed the highest SWR and CrWR of 0.41 t/km^2/year and 0.48 t/km^2/year respectively. The Chemical Index of Alteration (CIA) and Chemical Index of Weathering (CIW) support these findings and revealed a moderate weathering in the basin quite usual of cold regions.

Keywords

Chemical weathering rate · Chemical index of alteration · Chemical index of weathering · Resistant index of maturity · Kashmir himalaya

R. A. Mir (✉)
Department of Earth Sciences, Indian Institute of Technology Roorkee, Roorkee 247667, India
e-mail: riyazgsi@gmail.com

F. A. Dar
Department of Geology, University of Kashmir, Leh Campus, Leh 194101, India
e-mail: farooq.dar1@gmail.com

G. Jeelani
Department of Earth Sciences, University of Kashmir, Srinagar 190006, India
e-mail: geojeelani@gmail.com

A. K. Taloor et al. (eds.), *Water, Cryosphere, and Climate Change in the Himalayas*, Geography of the Physical Environment, https://doi.org/10.1007/978-3-030-67932-3_13

13.1 Introduction

River basins, occupying about 69% of the total land area, transport an estimated 19 billion tonnes of eroded material annually out of which about 20% is in solution form (Nanson and Gibling 2004; Bailey et al. 2014). Generally, mountainous rivers originate from the surface runoff, snow/ice melt, interflow, and baseflow. Consequently, the chemical load associated with these input parameters controls the chemistry of the river water. However, the concentration of the chemical constituents in the precipitation and snow/ice melt is generally less as compared to waters contributed by interflow and baseflow. Therefore, the general chemical composition of the river water is acquired by the dissolution of minerals of the rocks and soil during weathering and erosion. The proportion of this river-transported matter is a complex function and depends on the relative elemental composition of minerals, mode and rate of weathering, catchment lithology, geochemical behaviour of elements, vegetation, and biogenic activities (Lyu et al. 2009; Khan et al. 2020; Haque et al. 2020; Sarkar et al. 2020). Besides this, the climatic parameters (temperature and precipitation), runoff, tectonics, landscape age, and bedrock exhumation, as well as processes that affect transportation, sedimentation, and diagenesis, and intense anthropogenic activities exert a strong control on the quantity of this chemical load (Lyu et al. 2009; Tipper et al. 2012). The human-induced factors also depend on land use, population density, and pollution (Bailey et al 2014; Taloor et al. 2020). The whole process is a dynamic multi-component system that varies as a function of time and distance from the source area because of continuous adjustments in the catchment areas.

The quantitative analysis of chemical load associated with the water of the riverine system and streambed sediments along with the information of sediment-texture and catchment parameters is a reliable proxy in understanding the weathering and erosional processes, the provenance of geochemical elements, tectonic setting of the catchment, source of transported matter, the impact of human activities (e.g., Singh et al. 2005; Bailey et al 2014; Gaillardet et al. 1999; Mortatti and Probst 2004; Lyu et al. 2009; Taloor et al. 2020; Adimalla et al. 2020; Adimalla and Taloor 2020). The signature of these processes is imprinted on the dissolved and particulate material derived from the catchments of the rivers, which transport and recycle nearly 90% of the global lithogenic and anthropogenic load (Pattanaik et al. 2013). Hence, the rivers are fundamental geochemical tools to investigate how the chemical weathering processes govern the elemental cycling on the earth (Berner and Berner 1996; Drever 1997) because each cycle is controlled by the weathering processes (Probst et al. 1994). Hence, the estimation of chemical weathering rates of river basins is very significant because it helps in the management of the catchment resources. The study also helps to estimate the long-time impact that the weathering process has on the CO_2 budget of the atmosphere (Lyu et al. 2009; Wu et al. 2008).

Numerous studies have assessed the chemical weathering rates in the worldwide river catchments keeping in view one or more of the above factors. There are a number of studies on the Himalayan rivers (e.g., Chakrapani and Veizer 2005; Soumya et al. 2011) and Indian Himalayan rivers (e.g., Chakrapani and Subramanian 1996; Ramesh et al. 2000; Mir and Jeelani 2015a). No such study has been carried out in the upper river Jhelum Basin located in Kashmir Basin, western Himalaya therefore; the present work focuses on the hydrochemistry and sediment geochemistry of the river and its major tributaries to assess the chemical flux variability, chemical weathering rates, weathering intensity, and anthropogenic pollution in its catchment areas.

13.2 Study Area

13.2.1 Geography and Climate

The river Jhelum Basin drains the whole Kashmir Valley located in the north of India between 34°17′ N to 37°6′ N and 73°6′ E to 80°30′ E coordinates. The NW–SE elongated boat-shaped valley is about 140 km long and 50 km wide surrounded by the Greater Himalaya in the north-east and the Pir-Panjal in the south-west (Mir and Jeelani 2015a; Mir et al. 2016). The river Jhelum is one of the major Himalayan river and important tributary of the upper Indus River Basin. It normally originates in the south-east of the valley and is fed by the snow and glacier meltwater and springs. The river is fed by several left bank tributaries (LBT) and right bank tributaries (RBT) originating from the Great Himalaya and the Pir-Panjal ranges. For instance, Sandrin, Bringi, Kuthar, Lidder, Arapal, Sindh, and Pohru tributaries join on the right bank and the Vishav, Dudganga, Sukhnag, Rambiara, Romush, and Ningal on the left bank. The location map of the study area is shown in (Fig. 13.1). The altitude of the valley ranges between 1100 and >5400 m amsl. The average height of 1850 m (6070 ft) is around low land areas, but the surrounding Pir-Panjal and Greater Himalayan ranges have an average elevation of more than 3000 m (Mir et al. 2016).

The climate of the Jhelum Basin is normally temperate where the month of January is the coldest (average minimum temperature is -5 °C) and July the hottest (average maximum temperature is 30 °C). The mean annual rainfall is about 1100 mm (Mir and Jeelani 2015b; Bhat et al. 2019) with March receiving maximum precipitation and October the least. The river is a perennial source of irrigation and domestic uses and generates huge hydropower as well. The maximum water discharge occurs in May and June coinciding with the melting of the snow and glaciers, whereas the lowest discharge occurs between October and February (Mir and Gani 2019).

13.2.2 Geomorphology and Geology

The geomorphology of the upper river Jhelum Basin is characterized by high structural hills, small mounds of Karewas, colluvial fans below hill slopes, and the alluvial filled valley (Mir et al. 2016). The valley contains one of the finest developments of the stratigraphic successions right from Precambrian up to Recent times (Ganju and Khar 1984; Alam et al. 2017, 2018). The geological formations present in the area include Syringothyris Limestone, Fenetella Shales, Muth-Quartzites, Agglomeratic Slates, Nishatbagh beds, Panjal Volcanics, Mammal beds, Zewan formation, Triassic Limestone, Jurrasics, Karewa Group of sediments, and recent alluvium. The metasedimentary rocks of Cambrio-Silurian age are reported at many places while the Dogra Slates, Zewan, and Muth-Quartzites are of lesser distribution. Panjal Volcanics and Triassic Limestone are the two main geological formations constituting the valley bedrock and the surrounding mountain ranges. The Triassic formation consists of a thick series of compact blue limestone, slates, and dolomites (Wadia 1975). Karewas are the deposits of Plio-Pleistocene age and are composed of fine silty clays with sand and boulder gravel (Bhat 1989). The stratigraphic sequence of the geological formations of Kashmir Valley is shown in (Fig. 13.2).

13.3 Materials and Methods

Fifty water samples, 25 each in high flow (June 2008) period (HFP), and low flow (January 2009) period (LFP) were sampled from RBT and LBT (Fig. 13.1). Two samples were taken at each confluence point; one from the main channel at upstream and another from the tributary at upstream from the joining point. The details of the sampling locations are given in (Table 13.1). The procedure of water sampling and chemical analysis is given in (Mir et al. 2016). Besides, 15-bed load sediments at some feasible locations were collected in polythene bags using a scoop from the shallower depth (<1 m) and an Ekman dredge from deeper depths (>1 m) for the major elemental oxide analysis. Four samples were collected at each site at different places and combined to form a single composite sample. Grain size analysis was carried out by sieving

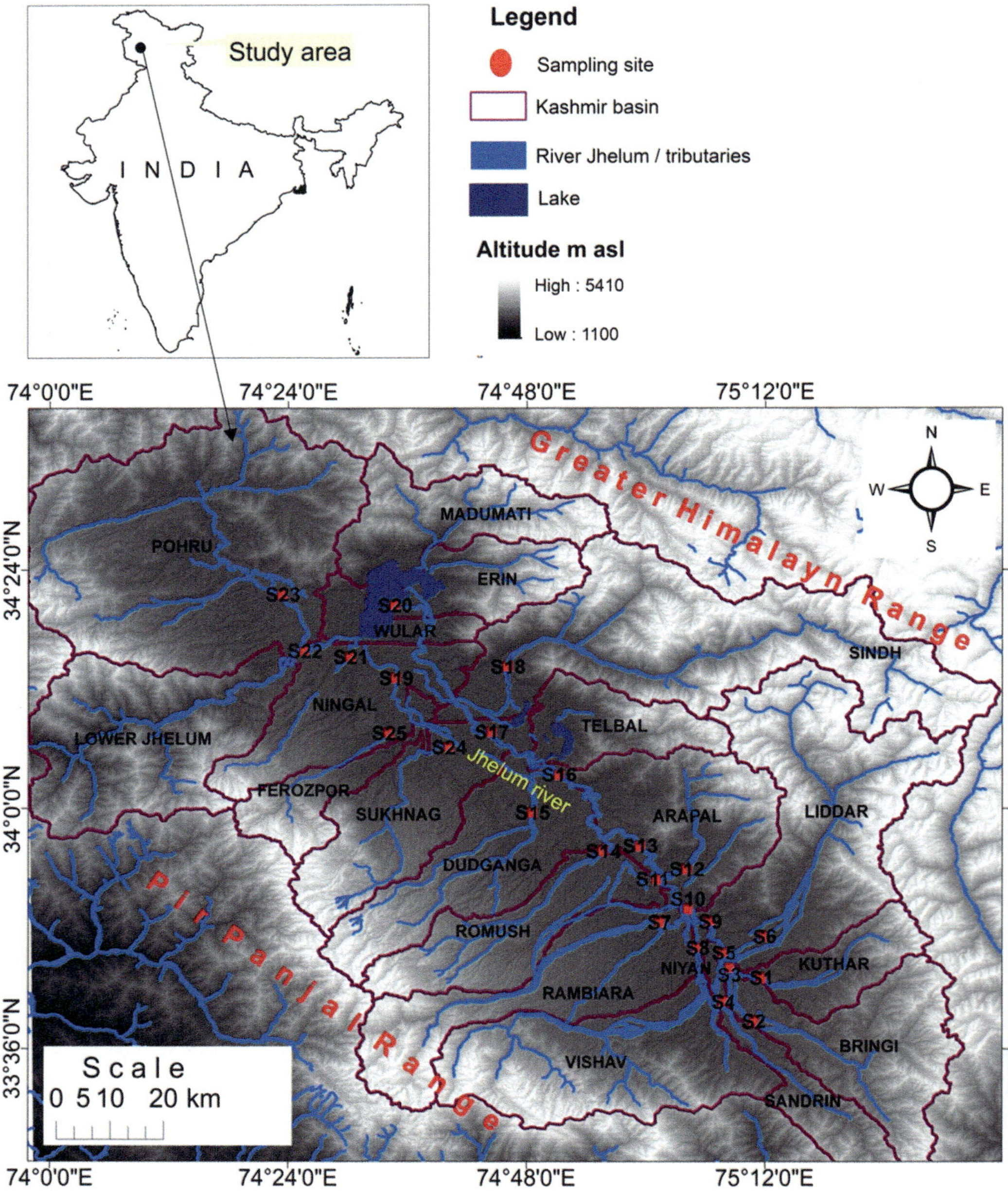

Fig. 13.1 Location map of the study area. The details of sampling sites shown on the map with corresponding sample IDs are given in Table 13.1. *Source* Authors collected data

(Lindholm 1987) and major oxides following the methodology of Ghosh and Mathur (1996). The sediments were pulverized in a mortar by rubber-tipped pestle and samples were decomposed by HNO_3, HCl, H2SO_4, and HF mixture to form solutions. The solution was prepared by digesting and subsequent fusing of 0.5 gm sample with NaOH for 5 minutes in a nickel crucible. The mixture was cooled to room temperature, bleached with distilled water, covered with a lid, and allowed to stand overnight. The solution was then acidified with 20 ml of 1:1 HCl and boiled for 10 min on a hot plate. After cooling, the solution was diluted to 1000 ml and stored in

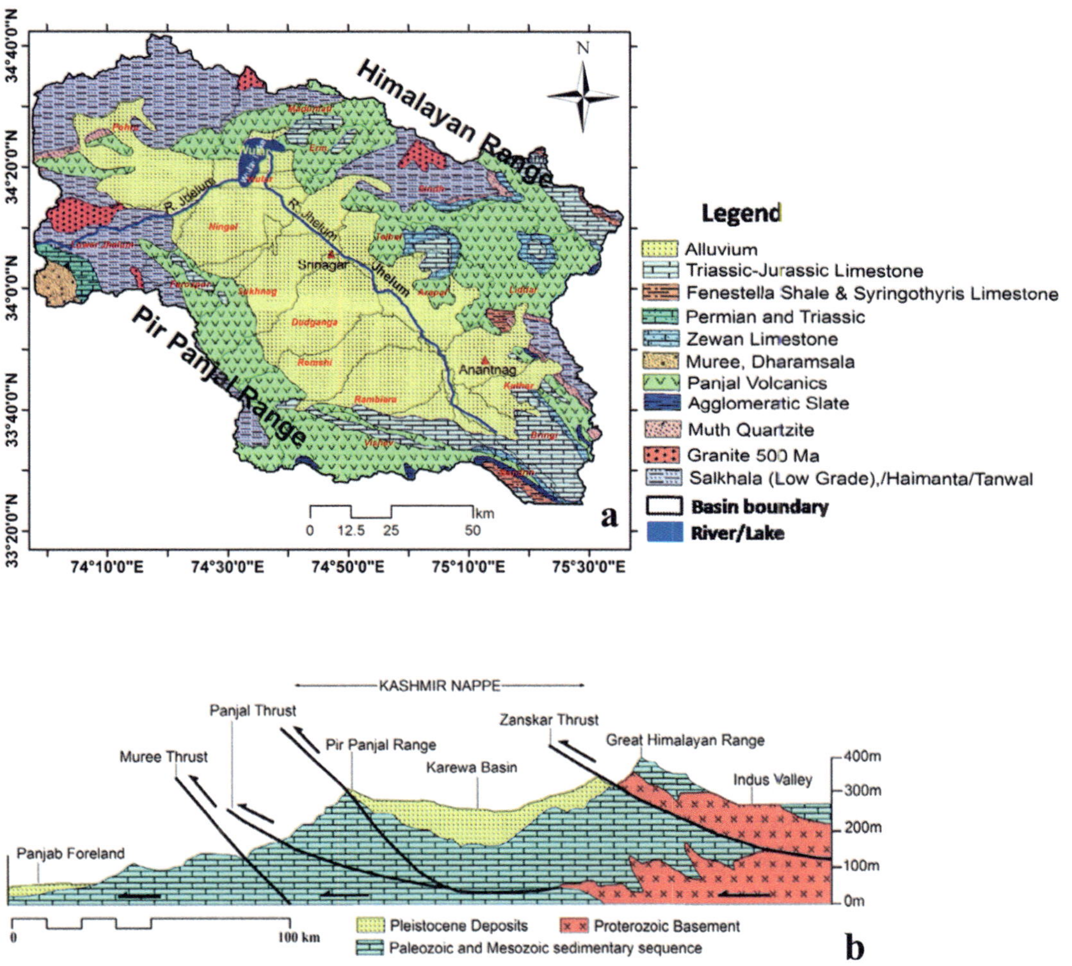

Fig. 13.2 **a** Geological map of Kashmir Valley (after Thakur and Rawat 1992). **b** Geological cross-section across the Kashmir Himalaya showing Kashmir Nappe Zone and rock formations of Kashmir Basin (after Wadia 1975)

polythene bottles. From the sample solution, CaO was determined by EDTA titration using Pattens and Readers reagents as an indicator, MgO using thymolphthalein as an indicator, Fe_2O_3 using sulphosalsalic acid indicator, and Al_2O_3 using xylenol orange indicator. The SiO_2 was estimated by the baking method and TiO_2, P_2O_5, MnO, Na_2O, K_2O were determined by Flame Photometry. Furthermore, the water discharge at each location was measured by the velocity area method during the sample collection. However, a daily discharge of each location for the whole study period was also collected from P and D, Flood Control Department, Srinagar. The surface area of each sub-basin of the Jhelum catchment was calculated using the ASTER DEM in the GIS platform.

13.4 Results and Discussion

13.4.1 Stream Water Chemistry

The detailed spatiotemporal overview of hydrogeochemistry of the main Jhelum river, its RBT, and LBT for two periods has been discussed in Mir et al. (2016), the statistics of which are presented in (Table 13.2). As observed, the

Table 13.1 Summary of the sampling sites. *Note* The water samples were collected at 25 sites, whereas the sediment samples were collected at 15 sites only represented by bold font text

S. No	River/Tributary	Sample ID	Site	Latitude (N)	Longitude (E)	Altitude (m amsl)
1	**Kuthar(Arapat)**	**S1**	**Bangidar (Khanabal)**	33°44′037”	75°08′257”	1582
2	Bringi	S2	Bangidar	33°44′042”	75°08′574”	1586
3	**Jhelum**	**S3**	**Kursherpur (Khanabal)**	33°44′268”	75°07′692”	1604
4	Sandrin	S4	Kureshpur	33°44′343”	75°07′626”	1599
5	**Jhelum**	**S5**	**Gur (Khanabal)**	33°44′564”	75°07′937”	1606
6	**Liddar**	**S6**	**Gur**	33°45′018”	75°07′859”	1613
7	**Rambiara**	**S7**	**Naiyan (Sangam)**	33°49′069”	75°07′778”	1611
8	**Vishav**	**S8**	**Naiyan**	33°49′060”	75°03′998”	1611
9	Jhelum	S9	Sangam	33°44′829”	75°04′092”	1590
10	Vishav	S10	Sangham	33°49′838”	75°04′062”	1597
11	**Jhelum**	**S11**	**Chursu (Awantipora)**	33°57′099”	75°55′806”	1616
12	**Arapal**	**S12**	**Chursu**	33°57′099”	75°55′806”	1616
13	**Jhelum**	**S13**	**Kakapora**	33°57′108”	75°55′806”	1583
14	**Romush**	**S14**	**Kakapora**	33°57′074”	75°55′077”	1573
15	Dudganga	S15	Shaltang (Srinagar)	34°03′288”	74°47′309”	1604
16	Jhelum	S16	Ram Munshibagh	34°04′142”	74°49′494”	1555
17	Jhelum	S17	Shadipora	34°10′990”	74°40′747”	1522
18	**Sindh**	**S18**	**Shadipora**	33°59′375”	74°55′498”	1523
19	**Haritar**	**S19**	**Haritar (Sopore)**	34°19′275”	74°33′528”	1579
20	**Jhelum**	**S20**	**Ningal (Sopore)**	34°17′253”	74°30′499”	1584
21	**Ningal**	**S21**	**Ningal**	34°17′253”	74°30′499”	1584
22	Jhelum	S22	Daubgau (Sopore)	34°15′921”	74°25′269”	1574
23	**Pohru**	**S23**	**Daubgam**	34°15′935”	74°25′935”	1575
24	Sukhnag	S24	Singpora (Patan)	34°06′835”	74°50′335”	1590
25	Ferozpor	S25	Palhalan (Patan)	34°08′875”	74°52′365”	1585

(*Source* Authors collected data)

catchment area (A) of the RBT is relatively large (sum 6409 km^2 and mean per stream 1068 km^2) than the LBT (sum 4867 km^2 and mean per stream 608 km^2). Among all the streams, the RBT such as S23, S18, and S6 have higher catchment areas and generally follow the descending order as S23 > S18 > S6 > S2 > S12 > S1, whereas the LBT follow an order of S7 > S15 > S24 > S21 > S14 > S25 > S4. Similarly, an analysis of the stream discharge (Q) revealed that the LBT consists of a higher discharge always as compared to the RBT annually as well as during both the seasons. The average annual Q of RBT was estimated to be is 44 L/m while as, for LBT, the Q is 55 l/m. Among the tributaries, the LBT, i.e., S18 has the highest Q with a general order followed as S15 > S7 > S4 > S14 > S21 > S8 > S24 > S25.

The mean values of chemical variables were calculated as an arithmetic mean of data of two periods, i.e., high and low flow periods. After analysis, it was observed that the Ca^{2+}, Mg^{2+}, and HCO_3^- are the major ions followed by Na^+, K^+, Cl^-, SO_4^{2-} and F-. Besides, the

Table 13.2 Statistics of the physico-chemical parameters of waters of the main channel of river Jhelum, RBT, and LBT for high flow and low flow periods, some Himalayan rivers, and precipitation collected at Kolahoi Glacier in Pahalgam area (upstream of the Lidder stream, a tributary of river Jhelum)

High flow period															
		T	pH	EC	TDS	Hardness	Ca^{2+}	Mg^{+2}	Na^-	K^+	Cl^-	HCO_3-	SO_4^{2-}	SiO_2	NO_3^-	F^-
RBT	Min	15.1	7.6	121.0	77.0	102.0	25.0	4.4	6.6	0.2	5.0	130.0	1.6	3.6	6.9	0.8
	Max	21.2	8.0	264.0	169.0	160.0	44.0	22.8	13.4	0.5	8.5	195.0	16.8	7.4	9.2	1.1
	Mean	18.4	7.7	195.3	125.0	136.7	33.7	10.8	9.8	0.3	6.9	162.5	6.6	5.4	8.5	1.0
	SD	2.0	0.2	60.6	39.0	27.2	7.1	6.7	2.4	0.1	1.4	27.0	7.4	1.4	0.8	0.1
	CV	11.2	2.0	31.0	31.2	19.9	21.1	62.1	24.5	30.8	20.7	16.6	110.8	26.0	9.7	10.4
LBT	Min	13.8	7.5	122.0	78.0	90.0	23.0	4.4	6.3	0.3	4.4	115.0	1.0	1.8	8.4	0.7
	Max	22.2	7.9	291.0	187.0	196.0	45.0	19.9	12.4	1.1	8.2	200.0	14.2	6.2	10.1	1.1
	Mean	18.4	7.7	205.1	138.8	134.0	32.3	10.9	8.6	0.5	6.0	157.2	6.7	4.9	9.1	0.9
	SD	2.8	0.1	51.8	37.5	34.3	6.2	5.4	1.8	0.3	1.2	25.9	4.6	1.5	0.6	0.1
	CV	15.0	1.6	25.3	27.0	25.6	19.1	49.8	20.8	55.2	20.6	16.5	69.0	30.0	6.1	14.8
Main Jhelum channel	Min	17.4	7.5	100.0	114.0	178.0	24.0	4.9	6.6	0.3	3.6	120.0	2.4	6.8	3.8	0.6
	Max	23.2	7.7	144.0	187.0	240.0	38.0	17.5	10.4	1.0	8.0	175.0	12.0	10.3	7.4	1.1
	Mean	19.9	7.6	120.9	134.8	199.8	32.0	10.4	8.6	0.4	5.4	145.2	7.3	9.0	6.1	0.9
	SD	2.2	0.1	15.8	22.7	20.4	4.5	4.0	1.3	0.2	1.4	18.7	3.1	1.1	1.4	0.1
	CV	10.9	1.0	13.1	16.8	10.2	14.0	38.1	14.6	53.6	25.9	12.9	43.3	12.0	23.2	17.6

(continued)

Table 13.2 (continued)

High flow period																
		T	pH	EC	TDS	Hardness	Ca^{2+}	Mg^{+2}	Na^{-}	K^{+}	Cl^{-}	HCO_3-	SO_4^{2-}	SiO_2	NO_3^{-}	F^{-}
Low flow period																
		T	pH	EC	TDS	Hardness	Ca^{2+}	Mg^{2+}	Na^{+}	K^{+}	Cl^{-}	HCO_3^{-}	SO_4^{2-}	SiO_2	NO^{2-}	F^{-}
RBT	Min	4.7	7.7	149.1	95.4	75.0	29.0	4.1	8.3	0.1	1.8	150.0	4.0	16.2	3.8	0.7
	Max	9.8	8.4	364.0	233.0	135.0	44.0	23.9	11.1	0.7	6.2	225.0	50.3	19.2	5.9	0.9
	Mean	7.8	8.1	291.8	186.8	105.0	37.8	12.5	9.9	0.4	4.4	181.0	17.0	17.2	5.0	0.8
	SD	1.9	0.3	84.9	54.4	26.2	6.1	7.7	1.2	0.2	1.7	27.7	19.2	1.4	0.8	0.1
	CV	24.1	3.6	29.1	29.1	25.0	16.0	61.8	12.3	44.5	39.4	15.3	113.0	8.0	15.6	13.6
LBT	Min	3.7	7.5	182.8	117.0	85.0	26.0	4.5	6.4	0.3	3.7	120.0	2.3	14.5	3.5	0.7
	Max	8.9	8.4	430.0	275.2	158.0	40.0	20.0	15.0	0.9	10.7	220.0	15.4	18.0	6.5	0.9
	Mean	6.1	8.0	262.0	167.6	113.3	33.1	11.8	10.3	0.5	5.2	168.0	7.0	16.5	5.2	0.8
	SD	1.5	0.3	77.7	49.8	20.6	4.5	4.7	2.7	0.2	2.1	28.6	4.5	1.0	1.0	0.1
	CV	23.7	3.7	29.7	29.7	18.2	13.5	40.0	26.3	46.3	40.1	17.0	64.5	6.2	18.7	8.7
Main Jhelum channel	Min	4.6	7.8	95.0	131.8	206.0	25.0	7.4	8.0	0.3	3.2	145.0	1.1	3.9	15.6	0.6
	Max	9.9	8.3	115.0	210.6	329.0	43.5	18.9	11.1	0.8	6.2	195.0	20.4	6.0	17.8	0.9
	Mean	7.3	8.0	103.0	160.0	250.6	33.6	11.9	9.6	0.5	4.9	163.8	7.6	4.9	16.7	0.8
	SD	1.6	0.2	5.6	25.9	40.2	4.9	3.6	1.1	0.2	1.2	15.6	7.2	0.7	0.8	0.1
	CV	22.3	2.2	5.4	16.2	16.0	14.6	30.0	11.4	38.6	24.6	9.6	94.3	13.4	4.8	11.6
Precipitation at Kolahoi Glacier	–	10.2	8.9	64	64	–	4.01	0.97	0.69	0.02	0.35	21.31	2.12	1.67	–	–
Indus river	Mean	–	–	–	122	–	27	1	1	2	9	64	15	15	–	–
Brahmaputra river	Mean	–	–	–	Na	–	14	5	7	3	11	56	4	4	–	–
Ganga river	Mean	–	–	–	214	–	25	8	11	3	10	128	11	11	–	–

(*Source* Authors collected data)

hydrochemistry exhibited a clear and significant variation among the catchments. Overall, the contribution of major ions like Na^+, K^+, Cl^- from atmospheric sources is almost negligible as observed by the chemical composition of rainwater collected during July-2008 in the upper reaches of the Lidder Valley (S6) near Kolahoi Glacier (Table 13.2). The analysis showed that the rainwater is normally alkaline and has a very low concentration of various ions indicating non-polluted rains in the area. The general order of cations was found as Ca^{2+} > Mg^{2+} > Na^+ > K^+ and anions as HCO_3^{2-} > SO_4^{2-} > SiO_2 > Cl^- with an overall order of HCO_3^- > Ca^{2+} > SO_4^{2-} > Mg^{2+} > Na^+ > K^+ > Cl^- > F^-. Among the tributaries, the RBT have higher pH, Ca^{2+}, Mg^{2+}, HCO_3^-, Na^+, SO_4^{2-}, SiO_2, and F^-; whereas there was almost a similar distribution pattern during low and high flow periods (Fig. 13.3). As discussed in Mir et al. (2016) most of these elevated values are usually the by-products of carbonate and sulphide dissolution and indicate chemical weathering as the governing process of hydrochemistry in RBT/ catchments. Moreover, the widely distributed limestones (Fig. 13.2) possessing gypsum and pyrite minerals, Ca^{2+} and Mg^{2+} rich rock minerals of Panjal traps, granites, quartzites, slates, shales, and Karewa sediments, few sulphur springs in the area, and alkaline pH are reported to be responsible for the release of higher values of these ions from rock-water interaction (geogenic) with a little contribution from anthropogenic sources (Mir et al. 2016). Besides, the higher values of Na^+, F^-, and SiO_2 also reflected silicate weathering in these catchments. On the contrary, LBT waters have higher T, EC, TDS, TH, K^+, Cl^-, and NO_3^- with almost similar behaviour when comparing the data of low and high flow periods, respectively (Fig. 13.3). Nevertheless, the prominent variability in the case of Na^+, K^+, Cl^-, and NO_3^- have been generally related to the anthropogenic activities in these catchments (Mir et al. 2016). Human activities affecting these waters (e.g., S15, S21, S24, S25) are ascribed to the domestic sewage, fertilizers, organic matter, and influx from contaminated water bodies that result in an overall increase in EC of water. Cultivation is also reported to be more intense in these catchments. However, the presence of hard waters, high K^+, Cl^- and NO_3^- also reflect weathering of Ca-Mg bearing minerals, silicates, and evaporites of Karewas in these catchments (Mir et al. 2016).

13.4.2 Streambed Sediment Texture

The results of the textural analysis of 15 streambed sediment samples of the river Jhelum and its tributaries are presented in (Table 13.3).

13.4.2.1 Graphic Mean Size (Mz)

The mean size of sediments varied from 0.28Φ at Pohru (S23) to the highest of 2.53Φ at Lidder (S6) with an average of 1.25Φ corresponding to sand and mud size particles (Mir and Jeelani 2015b). Most of the samples consist of very-coarse to a very-fine sand fraction which on average constitutes >95% of sediment amount, while mud contributes >2%. Out of sand fraction, 53% of samples possess medium sand while 40% possess coarse-grained sand and 8% possess fine sand. As per the sediment size, the decreasing order of sites is as S6 > S2 > S7 > S8 > S15 > S14 > S4 > S1 > S12 > S18 > S23 > S21. Except for S21, all the LBT sediments are larger (average 1.36Φ) than RBT sediments (average 1.15Φ). However, S6 and S2 have higher sediment sizes than all tributaries. The large grain size revealed a higher transporting velocity of water of LBT due to the observed higher Q values than the RBT. This is also observed from gravel size sediments at few LBT sites (e.g., Rambiara, S7 and Vishav, S8) that contribute >3% to the particle size.

13.4.2.2 Graphic Standard Deviation (σI)

The standard Deviation of sediment distribution varied from the lowest 0.7Φ to a highest of 1.4Φ at Kuthar (S1) with an average of 0.95Φ. This sediment parameter indicates moderately well sorted (low σI) to poorly sorted (higher σI) sand (Mir and Jeelani 2015b). Except for S6, all the RBT sediments have higher graphic standard

Table 13.3 Textural parameters and major oxide chemistry of the sediments of the river Jhelum and its tributaries. *Index* Mz-mean size, σI-standard deviation, SkI-skewness, KG-kurtosis, CIA-Chemical Index of Alteration, Rm-Resistant Index of Maturity

Sampling station	Sample ID	Mz	σI	SkI	KG	SiO_2	Al_2O_3	CaO	Fe_2O_3	MgO	Na_2O	K_2O	MnO	TiO_2	P_2O_5	CIA	Rm
Kuthar	S1	0.8	1.4	0.0	1.2	63.8	13.2	7.6	7.5	2.8	0.8	1.0	0.3	0.9	0.4	66	5
Bringi	S3	1.9	1.1	-0.1	1.2	60.8	12.8	8.9	6.1	4.8	0.9	0.7	0.5	1.5	0.4	63	4
Lidder	S6	2.5	0.8	0.1	1.2	49.7	15.2	7.8	12.1	8.4	1.3	1.1	0.2	0.8	0.5	67	3
Arapal	S12	0.8	1.0	0.3	1.4	63.3	13.7	7.6	6.1	3.4	0.7	0.9	0.4	1.2	0.5	68	5
Sindh	S18	0.6	1.1	0.0	1.0	52.3	10.0	23.4	5.2	1.6	1.4	0.6	0.4	1.7	0.4	37	2
Pohru	S23	0.3	1.2	0.1	1.2	69.6	16.6	2.5	4.4	1.2	1.4	0.6	0.4	0.9	0.7	82	12
Sandrin	S4	1.0	0.7	0.0	1.1	58.3	12.1	13.7	8.5	0.8	0.9	0.8	0.4	1.5	0.4	53	4
Vishav	S8	1.5	0.9	0.1	1.3	62.5	11.4	11.2	7.0	2.4	0.9	0.8	0.3	1.2	0.6	56	4
Rambiara	S7	1.8	0.7	0.0	1.6	70.5	12.1	6.2	4.1	1.6	1.5	0.9	0.5	0.9	0.4	65	7
Romush	S14	1.1	1.0	0.1	1.1	65.1	14.5	6.1	6.4	1.8	1.2	0.8	0.3	1.3	0.4	71	7
Dudganga	S19	1.4	0.7	0.0	1.2	73.7	12.0	2.0	4.5	1.0	1.4	0.8	0.4	0.8	0.4	78	14
Ningal	S21	0.3	1.0	0.0	1.1	71.0	11.9	4.2	5.9	0.6	0.7	0.9	0.6	0.8	0.6	73	11
Jhelum	S22	1.4	0.9	0.0	1.2	72.3	12.7	5.1	3.6	0.6	1.5	0.8	0.4	0.7	0.4	69	9

(*Source* Authors collected data)

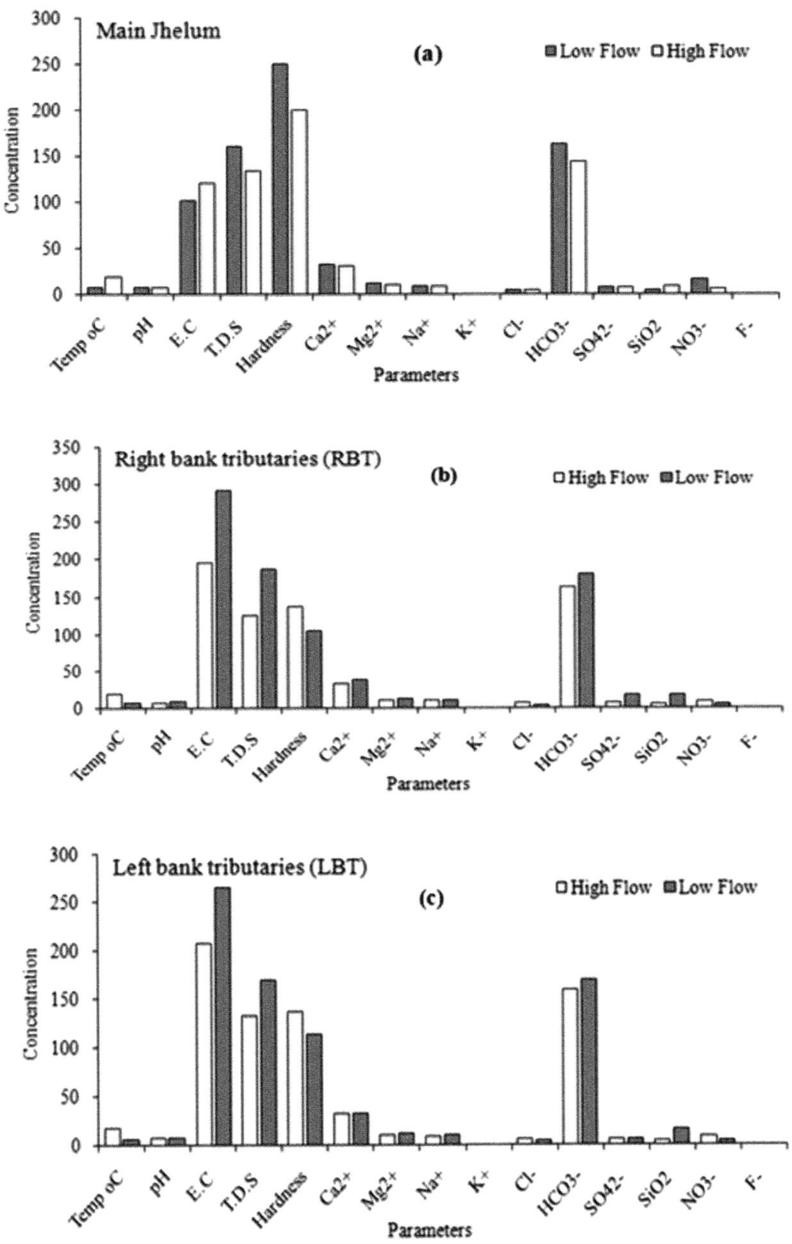

Fig. 13.3 Temporal variation of physico-chemical parameters of the main Jhelum river, RBT, and LBT. (*Source* Authors collected data)

deviation (average 1.1Φ) than LBT sediments (average 0.8Φ) which indicate that sediments of RBT/catchments are a mixture of sand and thus have large transported distance and high weathering intensity. This is also shown by the higher amount of mud particles in most of the RBT. The decreasing order of standard deviation follows an order as S1 > S23 > S2 > S18 > S12 > S14 > S21 > S8 > S6 > S4 > S7 > S15.

13.4.2.3 Skewness and Kurtosis

A similar observation is drawn from the skewness of sediments which showed that on average the RBT sediments are highly skewed and have high kurtosis. About 20% of the samples were poorly sorted followed by 66% moderately sorted and remaining moderately well sorted. About 14% of samples are highly skewed and majority of sediemnts are near-symmetrical. In the upstream of main river Jhelum, both RBT as well as LBT consists an abundance of coarse fraction thereby possessing coarse skewness. In the downstream, the fine mode generally increases and the skewness becomes fine in both side tributaries. Kurtosis of the sediments of river Jhelum and its tributaries varied from 1.13Φ to 2Φ with an average of 1.205Φ, suggesting leptokurtic to very leptokurtic sediments. 20% of the sediments were leptokurtic and 60% were mesokurtic. The parameter indicated that the sediments were deposited under moderate to low energy conditions associated with fluctuations in the velocity of the depositing medium. The leptokurtic nature indicates a high deposition of fine particles of sediments. The kurtosis does not show any specific trend downstream.

13.4.3 Sediment Chemistry

All the 15 streambed sediment samples were analyzed for major oxide composition, the results of which are given in (Table 13.3). The results indicated that the SiO_2, Al_2O_3, CaO, Fe_2O_3, and MgO were the most abundant oxides constituting > 65.5%, 13.2%, 8.4%, 6.4%, and 2.4 to the total oxide sum of the sediments, respectively. The SiO_2 varied from 49.7% at Lidder (S6) to the highest 73.7% at Dudganga (S15). Most of the LBT sediments have higher silica content (average SiO_2 = 66.9%) than RBT sediments (average SiO_2 = 59.9%) which reflected fast release or intense weathering of silicates from rocks of RBT catchments. The descending order of silicate content is as S15 > S21 > S7 > S23 > S14 > S1 > S12 > S8 > S2 > S4 > S18 > S6. Similarly, Al_2O_3content ranged from 10% at Sind (S18) to the highest of 16.61% at Pohru (S23). The average Al_2O_3 in the RBT samples (13.58%) is higher than the LBT sediments (12.33%). The descending order of Al_2O_3 content is as S23 > S6 > S14 > S12 > S1 > S2 > S4 > S7 > S15 > S21 > S8 > S18. The CaO also exhibited high spatial variation and ranged from 1.68% at Dudganga (S15) to a distinctively highest 23.37% at Sind (S18) reflecting control of short transportation and presence of carbonate lithology in their catchments. The carbonate weathering dominantly controls the concentration of CaO in RBT with values of 9.66% than LBT values of 7.2%. The order of decreasing CaO content is S18 > S4 > S8 > S2 > S6 > S1 > S12 > S7 > S14 > S21 > S23 > S15. This order is contrary to the silica concentration order, thereby reflecting the dominant control of carbonate weathering in RBT and silicate weathering in the LBT.

Fe_2O_3 percentage varied from 3.6% at (S7) to the highest of 12.11% at (S6). For RBT samples, the average value is 6.9% and for LBT, it is 6.1%. The higher values of RBT reflected the weathering of carbonate and Fe-bearing minerals. The decreasing order of Fe_2O_3 content is S6 > S4 > S1 > S8 > S14 > S2 > S12 > S21 > S18 > S15 > S23 > S7. However, the overall mixed behaviour in all the samples indicated that the Fe oxides are present in the rocks of all catchments. The MgO percentage showed a wide variation among the streams. The lowest value of 0.6% was found at Ningal (S21), whereas the highest value of 8.4% observed at Lidder (S6) was ascribed to the presence of limestones particularly dolomitic limestones in this catchment. The average value (3.7%) is higher in the RBT sediments than the LBT (1.4%). The decreasing order of MgO content is S6 > S2 > S12 > S1 > S8 > S14 > S18 > S7 > S23 > S15 > S4 > S21. The Na_2O is widely distributed in the samples and ranges from 0.7% at Ningal (S21) to 1.5% at Rambiara (S7). No significant variability was observed between LBT (average = 1.1%) and RBT samples (average = 1.08%) which reflected that Na is present in both catchment rocks. However, comparatively higher values of Na_2O along with higher SiO_2 of LBT may indicate more silicate weathering. The descending order of Na_2O content is S7 > S18 > S23 > S15 > S6 > S14 > S2 > S4 > S8 > S1 > S12 > S21. The K_2O percentage was lowest (0.6%) at Pohru (S23) and highest (1.1%) at Lidder (S6). All the

LBT samples showed intermediate values, while as RBT have two groups; S6, S1, and S12 have higher amount and S2, S18 and S23 have lower. The decreasing order of K_2O content is shown in (Table 13.3). Overall, K_2O is higher in LBT (average = 0.42%) than RBT (average = 0.3%) which reflect the dominance of K-silicate bearing rocks in their catchments. The MnO, TiO_2 and P_2O_5 percentage also showed variation with the corresponding values ranging from 0.2 to 0.6%, 0.7 to 1.7%, and 0.36 to 0.7%, respectively. The decreasing order of oxide concentration is shown in (Table 13.3). On average, RBT samples have a higher amount of these oxides. The elements are generally associated with rocks of varied compositions such as dolomites, and mafic igneous rocks which indicates that these oxides are released from diverse lithology of the catchments. In general, the results reveal that the average values of SiO_2, Na_2O, and K_2O are found to be higher in LBT sediments, whereas the average values of Al_2O_3, CaO, Fe_2O_3, MgO, MnO, TiO, and P_2O_5 are more in RBT sediments. The overall order of the oxides followed the pattern as $SiO_2 > Al_2O_3 > CaO > Fe_2O_3 > MgO > Na_2O$ in 40% of the samples followed by $SiO_2 > Al_2O_3 > Fe_2O_3 > CaO > Na_2O > MgO$ in 26% samples and $SiO_2 > Al_2O_3 > Fe_2O_3 > CaO > MgO > Na_2O$ in 20% of samples. 7% of the samples showed $SiO_2 > Al_2O_3 > Fe_2O_3 > CaO > MgO > Na_2O$ and $SiO_2 > Al_2O_3 > Fe_2O_3 > CaO > MgO > Na_2O$ each.

13.4.4 Chemical Weathering in the Area

13.4.4.1 Total Dissolved Load

The chemical composition of rainwater collected in the upper reaches of Lidder Valley was used to carry out the atmospheric corrections of chemical composition of the river Jhelum water and its tributaries irrespective of the fact that there exists negligible atmospheric fallout. Considering this, the amount of dissolved load/elements ($X = Na^+$, K^+, Ca^{2+}, Mg^{2+}, and SO_4^{2-}) supplied via chemical weathering were calculated by subtracting the weighted mean contribution of rainwater as

$$X^* = X_{river} - X_{rain} \quad (13.1)$$

The solute transport of the tributaries was calculated from the discharge weighted average annual loads. The dissolved load was calculated using major ions (Ca^{2+}, Mg^{2+}, Na^+, K^+, HCO_3^-, Cl^-, SO_4^{2-} and SiO_2) for 14 tributaries and the main outlet of the Jhelum river at Doubgau, Baramullah (S22). About 99% of the dissolved load is constituted by the above major ions, out of which alkalinity alone constitutes 67%, Ca 14%, silica 5%, and Mg, Na, and SO_4 4% each to the dissolved load. The detail results are given in (Table 13.4).

13.4.4.2 Total Dissolved Load in Water

The total dissolved load in the water (TDS_w) was calculated by adding the constituents of chemical weathering and dissolved silica and estimating the monthly averages for high and low flow periods as

$$TDS_w = \sum X^* + SiO_2 \quad (13.2)$$

The annual loads were calculated by averaging the total dissolved loads for two flow periods (June 2008 and January 2009), respectively. The average TDS_w during HFP was estimated at 0.61 t/mth in RBT and 0.69 t/mth in LBT/streams. During LFP, the average TDS_w was 1.11 and 1.12 t/mth, respectively. Similarly, the monthly average over the annual period was found to be 0.86 and 0.91 t/mth for RBT and LBT. The decreasing order of TDS_w is as S4 > S12 > S15 > S2 > S18 > S21 > S1 > S7 > S14 > S24 > S25 > S8 > S23 > S6. The initial 7 streams having higher TDS_w, 57% belong to RBT/ catchments. Thus, it is observed that the overall higher dissolved load of RBT is significantly dominat in comparison to the LBT. Moreover, the TDS_w was lowest at S23 and S6 which have overall a lower sum of Ca + Mg + HCO_3 + Na + K and SO_4 + NO_3 + Cl. The mean monthly TDS_w for RBT follows the order as S12 > S2 > S18 > S1 > S23 > S6 and S4 > S15 > S21 > S7 > S14 > S24 > S25 > S8 for LBT.

Table 13.4 Average seasonal fluxes of TDS and chemical Weathering rates (CWR) of the river Jhelum and its tributaries. (*Note* S22* represents the TDS flux and CWR for the main river Jhelum which is the average of TDS flux and CWR estimated at nine sites from upstream to downstream of the river and mth- represent month)

S. No	Site Id	Area Km^2	High flow period (HFP)				Low flow period (LFP)				Annual period			
			TDS × 10^{-7} t	Discharge l mth^{-1} × 10^{12}	Flux × 10^6 t	CWR × 10^2 t /Km^2 / mth	TDS × 10^{-7} t	Discharge l mth^{-1} × 10^{12}	Flux × 10^6 t	CWR × 10^2 t /Km^2 / mth	TDS × 10–7 t	Discharge l mth^{-1} × 10^{12}	Flux × 10^6 t	CWR × 10^2 t /Km^2 /mth
1	SI	390	0.61	17	0.07	17.0	1.17	0.67	0.08	20.2	0.89	1	0.13	3.22
2	S2	665	0.89	17	0.11	16.7	1.45	0.57	0.08	12.5	1.17	1	0.16	2.43
3	S6	1243	0.13	11	0.13	10.8	0.31	7.78	0.24	19.7	0.22	10	0.21	1.71
4	S12	658	1.05	7	0.05	7.4	1.69	1.71	0.29	43.9	1.37	2	0.22	3.34
7	S18	1526	0.83	208	3.18	208.1	1.51	6.64	1.00	65.7	1.17	32	3.73	24.44
10	S23	1927	0.15	1.63	0.02	1.3	0.53	2.46	0.13	6.8	0.34	3	0.12	0.61
12	S4	291	1.12	80	0.23	79.8	2.11	0.36	0.08	26.3	1.62	2	0.27	9.22
5	S8	985	0.14	12	0.12	12.1	0.64	7.70	0.49	50.0	0.87	10	0.83	11.07
6	S7	751	0.72	90	0.67	89.7	1.02	3.50	0.36	47.7	0.39	11	0.42	4.31
13	S14	524	0.67	70	0.36	69.6	0.93	6.01	0.56	107.2	0.80	8	0.62	11.78
14	S15	700	1.22	149	1.04	148.6	1.32	2.41	0.32	45.6	1.27	7	0.87	12.46
11	S24	648	0.52	10.19	0.53	81.7	0.75	8.45	0.63	97.8	0.64	9	0.58	8.96
8	S25	355	0.44	10.11	0.44	125.3	0.63	4.20	0.26	74.5	0.54	8	0.41	11.42
9	S21	613	0.67	19	0.11	18.7	1.57	0.67	0.11	17.2	1.12	1	0.14	2.22
15	**S22***	**12,372**	**1.23**	**107.62**	**13.24**	**107.0**	**0.85**	**45.57**	**3.88**	**31.4**	**1.04**	**132**	**13.74**	**11.11**
16	TN	145	–	–	–	–	-	–	–	–	2.5	2.65	5.9	406.89
17	AN	11,800	–	–	–	–	–	–	–	–	1.12	14.1	1.58	1.34
18	BH	7800	–	–	–	–	–	–	–	–	0.89	8.3	0.74	0.95
19	YN	140,000	–	–	–	–	–	–	–	–	2.22	93	21	1.50
20	GG	975,000	–	–	–	–	–	–	–	–	1.78	393	70	0.72
21	BP	580,000	–	–	–	–	–	–	–	–	1.00	609	60	1.05
22	GA	101,000,000	–	–	–	–	–	–	–	–	1.15	31,400	3 611	0.36

Note * = Main Jhelum River, TN = Telbal Nala (Jeelani and Shah 2006), AN = Alaknanda (Singh 1998), BH = Bhagirathi (Pandey et al. 1999), YN = Yamuna, GG = Ganga, BP = Brahamputra (Sarin et al. 1989), GA = Global Average (Hu et al. 1982) (*Source* Authors collected data)

13.4.4.3 Dissolved Flux

Flux or total dissolved load in tonnes per month and year (Q × TDS_w) was also estimated in this study. The month and year will be read as per 'mth' and per 'yr' here after. Average flux (Q × TDS_w) during HFP was determined to be 0.59 × 10^6 t/mth for RBT and 0.44 × 10^6 t/mth for LBT. During LFP, the values were 0.3 × 10^6 and 0.35 × 10^6 t/mth, respectively. The monthly average of the two periods was 0.45 × 10^6 and 0.39 × 10^6 t/mth for RBT and LBT, whereas the monthly average over the annual period was 0.76 × 10^6 and 0.52 × 10^6 t/mth for RBT and LBT, respectively. This data shows that RBT have in general higher flux than the LBT. The flux averaged over two flow periods is plotted and shown in as (Fig. 13.4a–g). As seen, the dissolved load flux was found to be relatively higher in LBT as compared to the RBT except S18 (right bank stream). Furthermore, it is notable that the overall flux was highest at S18 probably due to its higher catchment area and long transportation. The decreasing order of mean monthly flux for RBT is S18 > S6 > S12 > S2 > S1 > S23, whereas for LBT, it is S15 > S24 > S7 > S14 > S25 > S8 > S4 > S21. Similarly, the annual flux order is S18 > S12 > S6 > S2 > S1 > S23 for RBT and S15 > S8 > S14 > S24 > S7 > S25 > S4 > S21 for LBT. Overall order of monthly average flux is S18 > S15 > S24 > S7 > S14 > S25 > S8 > S6 > S12 > S4 > S21 > S2 > S1 > S23.

13.4.4.4 Chemical Weathering Rate

From the estimated dissolved load, chemical weathering rates (CWR) were determined for each sub-catchment of the Jhelum river as

$$\mathrm{CWR} = \mathrm{Q} \times \frac{\mathrm{TDS_w}}{\mathrm{Area}} \text{ or Flux/Area} \qquad (13.3)$$

The average value of CWR during high flow (HFP) was determined to be 43.55 × 10^2 t/km^2/mth for RBT and 78.2 × 10^2 t/km^2/mth for LBT. During LFP, the CWR was 28.13 × 10^2 and 58.3 × 10^2 t/km^2/mth, respectively. The monthly average of two periods was found as 35.8 × 10^2 and 68 × 10^2 t/km^2/mth for RBT and LBT and similarly, the monthly average over the annual period was 5.9 × 10^2 and 8.9 × 10^2 t/km^2/mth for right and left streams respectively. Overall, the average CWRs were found low in RBT than LBT. The CWR averaged over two flow periods is plotted and shown in (Fig. 13.4a, b). CWR was highest at S18 (right stream) than all tributaries followed by higher in most of the LBT. The decreasing order of mean monthly flux for RBT was S18 > S12 > S1 > S6 > S2 > S23 and S15 > S15 > S24 > S14 > S7 > S4 > S8 > S21 for LBT. Overall order of average monthly CWR was S18 > S25 > S15 > S24 > S14 > S7 > S4 > S8 > S12 > S1 > S21 > S6 > S2 > S23. Moreover, to understand the CWRs comprehensively, the CWRs of Jhelum Basin sub-catchments were classified into different categories as given in (Table 13.5) and as shown in (Fig. 13.5). From the results, it is observed that the LBT is characterized by higher CWR's than its RBT/ counterparts, both seasonally as well as annually. Furthermore, the CWR of river Jhelum and its tributaries are also much higher than other Himalayan rivers (Table 13.4). However, the CWR of the river Jhelum and its tributaries are much lower than the CWR (406.9 × 10^2 t/Km^2/yr) of Telbal Nala (Jeelani and Shah 2006). Overall, from the above observations it can be inferred that the varied topography has also an effect on rates of the chemical weathering processes in the area. Previously, its has been established and reported in many of the watersheds that the factors such as temperature, runoff, and topographic relief are very significant and important in controlling the weathering processes (White and Blum 1995; Moon et al. 2007). Addtionally, the CWR during the HFP was higher most probably due to higher water flow which exceeded the water flow in observed in the LFP. The chemical weathering rates observed higher during the HFP reflected a significant and major control of higher discharge and precipitation produced as a result of higher rainfall during this season. The LBT such as Sandrin, Rambiara, Vishav, Romush, Dudganga, Sukhnag, and Ferozpor showed higher CWR than the RBT. The higher CWR in LBT draining the Pir-Panjal range was also attributed to the higher erosion

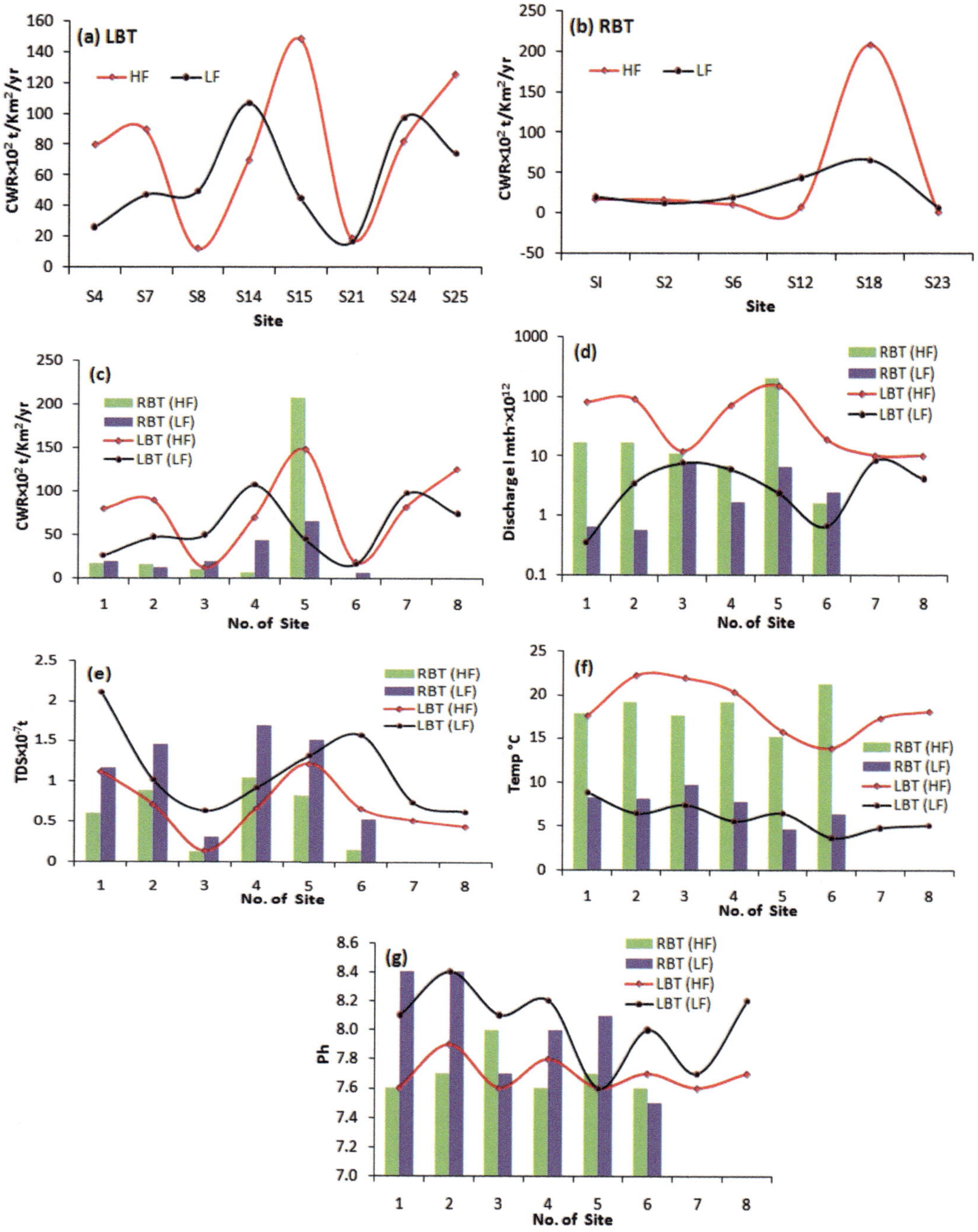

Fig. 13.4 Comparison of the CWR, discharge, and TDS concentrations of the LBT and the RBT during HFP and LFP. (*Source* Authors collected data)

Table 13.5 Chemical weathering classes estimated for each catchment of the river Jhelum Basin (*Note* yr-represents year)

S. No	CWR × 10^2 t/Km^2/yr	Classes	RBT	LBT
1	<5	Low	Bringi, Kuthar, Lidder, Arapal, Pohru	Ningal, Vishav,
2	6–10	Moderate	–	Sandrin, Sukhnag
3	11–15	High	–	Rambiara, Romush, Dudganga, Ferozpor
4	16–20	Highest	–	–
5	>20	Extreme Highest	Sind	–

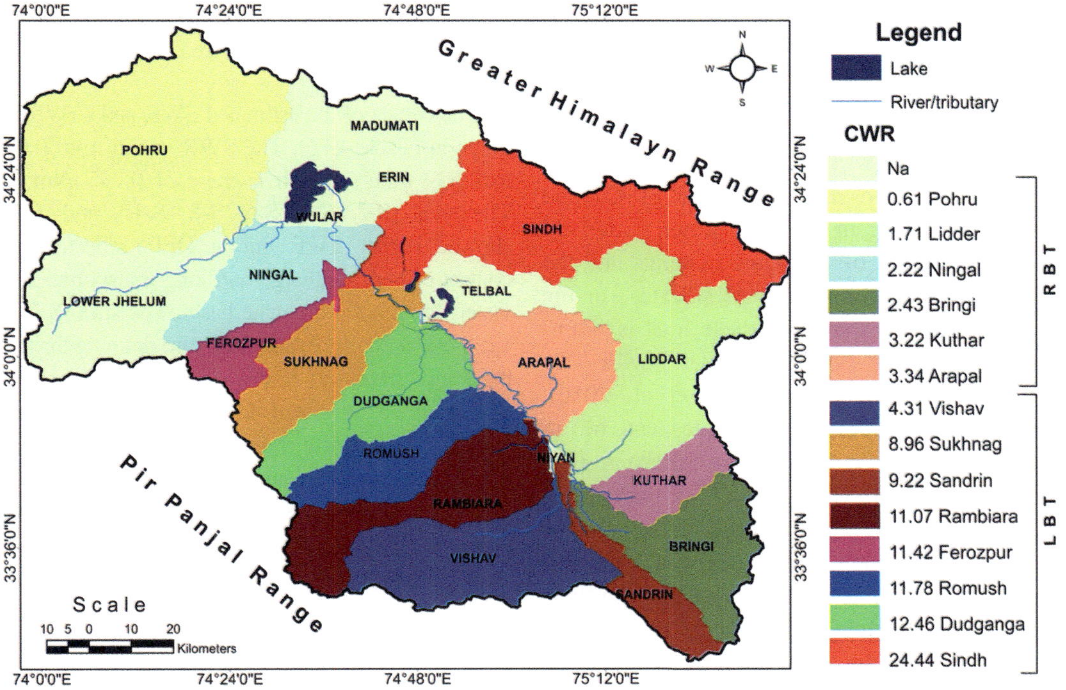

Fig. 13.5 Catchment wise Chemical weathering rate (CWR) map of the Jhelum Basin (ref: Table 13.4). *Note* LBT and TBT means left right bank tributary (*Source* Authors collected data)

and physical weathering of the rock formations caused due to the active tectonics and deformation in this region (Zaz and Romshoo 2012; Shabir et al. 2013; Dar et al. 2014). The higher deformation of the rocks increases and exposes higher surface area of the rock fragments and creates the chances of higher rock-water interaction vis a vis higher CWR. Moreover, the colder climate, higher relief and topographic gradient, and complimentary tectonics in the source region of rivers favours higher physical weathering and erosion rates. The weathering rates in the river Jhelum Basin, being tectonically active, is observed to be higher than other zones in this region because the tectonic activities are reported to induce high weathering and sedimentation rates (Fernex et al. 2001; Dar et al. 2014).

13.4.4.5 Silicate Weathering Rate and Carbonate Weathering Rate

After the correction of major ions for the atmospheric input, the silicate components of the Ca and Mg were determined from the following equation (Krishnamurthy et al. 1986) as

$(HCO_3)_{car.} = 0.74\ (Ca)_{tot.} + 0.4\ (Mg)_{tot.}$

And $(HCO_3)_{sil.} = (HCO_3)_{tot.} - (HCO_3)_{car.}$

Since the total bicarbonate in the river water has two fractions, namely, (i) bicarbonate derived from primary carbonates as well as calcareous cement, (ii) bicarbonate derived from silicates. Similarly, the total Ca and Mg also have the same two sources. In summary.

$(HCO_3)_{tot.} = (HCO_3)_{\ car.} + (HCO_3)_{sil.}$

$(HCO_3)_{\ car.} = (Ca)_{\ car.} + (Mg)_{\ car.}$

$(Ca)_{tot.} = (Ca)_{\ car.} + (Ca)_{sil.}$

$(Mg)_{tot.} = (Mg)_{\ car.} + (Ca)_{sil.}$

Holland opined that $74 \pm 10\%$ of Ca and $40 \pm 20\%$ of Mg in the river water is derived from the solution of carbonate minerals and the remaining largely comes from silicates. In the present area, the source of these ions is dominantly reported to be the carbonates with little contribution from silicates (Mir et al. 2016). Also, the dissolution of evaporites can be an important source for Na and K in the rivers. But, such conditions are very rare and not likely to be met in the present area therefore, the sources of these ions from such type of deposits is considered to be negligible in this study area. Furthermore, the estimation of the silicate components of Na and K in the source waters of the study area were estimated following the equation as follows (Krishnaswamy et al. 1998):

$(Na_{sil.} + K_{sil.}) = (Na_r - Cl_r) + 0.85K_r$

where the subscript ($_{sil}$) and ($_r$) refer to silicate and river. In this approach, the Na contribution to the rivers from atmospheric deposition is assumed to equal the Cl concentration (Sarin et al. 1989; Krishnaswamy et al. 1998). But, the validity of these calculations requires that there are no alkaline/saline deposits containing carbonates/bicarbonates and sulphates of sodium along the drainage basins of the source waters. In the present river basin, such conditions are not likely to be met and it was considered to provide a negligible contribution of Na and K. Therefore, following the Wu et al. (2008) the silicate and carbonate weathering rates were determined from the water discharge, drainage basin area and the calculated silicate component of total cations and carbonate component of Ca and Mg. The Silicate Weathering Rate (SWR) and Carbonate Weathering Rate (CrWR) were determined by the following equations as

$$\mathrm{SWR} = \frac{(\mathrm{Casil} + \mathrm{Mgsil} + \mathrm{Nasil} + \mathrm{Ksil})}{\mathrm{Drainage\ Area}} \times \mathrm{Q} \tag{13.4}$$

$$\mathrm{CrWR} = \frac{(\mathrm{Cacar} + \mathrm{Mgcar})}{\mathrm{Drainage\ Area}} \times \mathrm{Q} \tag{13.5}$$

The details of the estimated SWR and CrWR are given in (Table 13.6). The SWR varied from 0.030 to 0.539 t/km^2/yr with average of 0.285 t/km^2/yr. The sites (S-2, S-4, S-6, S-8, S-15, and S-17) showed higher SWR during HFP, reflecting the control of high discharge and long transportation of water through the drainage basin. The sites (S-7, S-12, S-14, S-20, S-22, and S-23) are draining through the Panjal traps and Karewas showed higher SWR that may be attributed to higher release of ions because of its low resistance to the weathering processes (Krishnaswami and Singh 2005). The varied topography, temperature, precipitation pattern and runoff are also expect to have an effect on the variable rates of the chemical weathering in the area. Previously, a number of studies have also reported that the factors such as temperature, runoff, and topographic relief are very important in controlling the weathering rates of the river catchments in other regions (White and Blum 1995; Moon et al. 2007). Similarly, the river Jelum basin and its all sub-catchments possess a highly variable relief, topography, geomorphology, temperature and precipitation pattern as well as a highly variable production of runoff from each sub-catchment (Dar et al. 2014; Shafiq et al. 2020). The mainstream of river Jhelum showed SWR varying from 0.098 t/km^2/yr to 0.105 t/km^2/yr with an average 0.101 t/km^2/yr. The high SWR during the LFP was ascribed to the higher concentration ions resulting due to low discharge in this season. The CrWR

Table 13.6 Silicate and Carbonate Weathering Rates of the mainstream of river Jhelum and its tributaries during high and low flow periods (*Note* MS-represents main stream)

S. no	River/ tributary	Station ID	Average Discharge l $yr^{-1} \times 10^{12}$	High flow		Low flow		Avg. SWR $t/km^2/yr$	Avg. CrWR $t/km^2/yr$	Source
				SWR $t/km^2/yr$	**CrWR $t/km^2/yr$**	**SWR $t/km^2/yr$**	**CrWR $t/km^2/yr$**			
1	Kuthar	S-1	0.80	0.05	0.06	0.05	0.07	0.05	0.07	Present study
2	Bringi	S-2	1.35	0.08	0.09	0.03	0.04	0.06	0.07	
3	Sandrin	S-4	1.66	0.33	0.32	0.06	0.06	0.20	0.19	
4	Lidder	S-6	8.90	0.15	0.23	0.13	0.19	0.14	0.21	
5	Rambiara	S-7	3.44	0.08	0.12	0.11	0.15	0.10	0.14	
6	Vishav(i)	S-8	9.63	0.2	0.25	0.13	0.2	0.17	0.23	
7	Arapal	S-12	1.47	0.07	0.06	0.09	0.09	0.08	0.08	
8	Romush	S-14	4.89	0.22	0.32	0.44	0.61	0.33	0.47	
9	Dudganga	S-15	3.69	0.29	0.34	0.15	0.15	0.22	0.25	
10	Sindh	S-17	20.69	0.49	0.07	0.27	0.35	0.38	0.21	
11	Ningal	S-20	0.62	0.09	0.1	0.17	0.18	0.13	0.14	
12	Pohru	S-22	2.95	0.04	0.03	0.04	0.04	0.04	0.04	
13	Sukhnag	S-23	7.25	0.27	0.34	0.54	0.61	0.41	0.48	
14	Ferozpor	S-24	3.68	0.17	0.23	0.27	0.34	0.22	0.29	
15	**Jhelum**	**(MS)**	**66.41**	**0.0976**	**0.121**	**0.1049**	**0.5716**	**0.10**	**0.35**	
1	World Average	–	–	–	–	–	–	2	–	Singh (2005)
2	Amasom	–	–	–	–	–	–	2.2	11.1	Gaillardet et al. (1999)
3	Indus	–	–	–	–	–	–	1.8	–	Singh (2005)
4	Ganga	–	–	–	–	–	–	7.9	–	Singh (2005)
5	Brahmaputra	–	–	–	–	–	–	11.8	–	Singh (2005)
6	Brahmaputra upper reaches	–	–	–	–	–	–	0.5–2.5	3.9–14.1	Hren et al. (2007)
7	Congo	–	–	–	–	–	–	0.84	1.6	Gaillardet et al. (1999)
8	Chumaer He	–	–	–	–	–	–	0.18	0.38	Wu et al. (2008)

(*Source* Authors collected data)

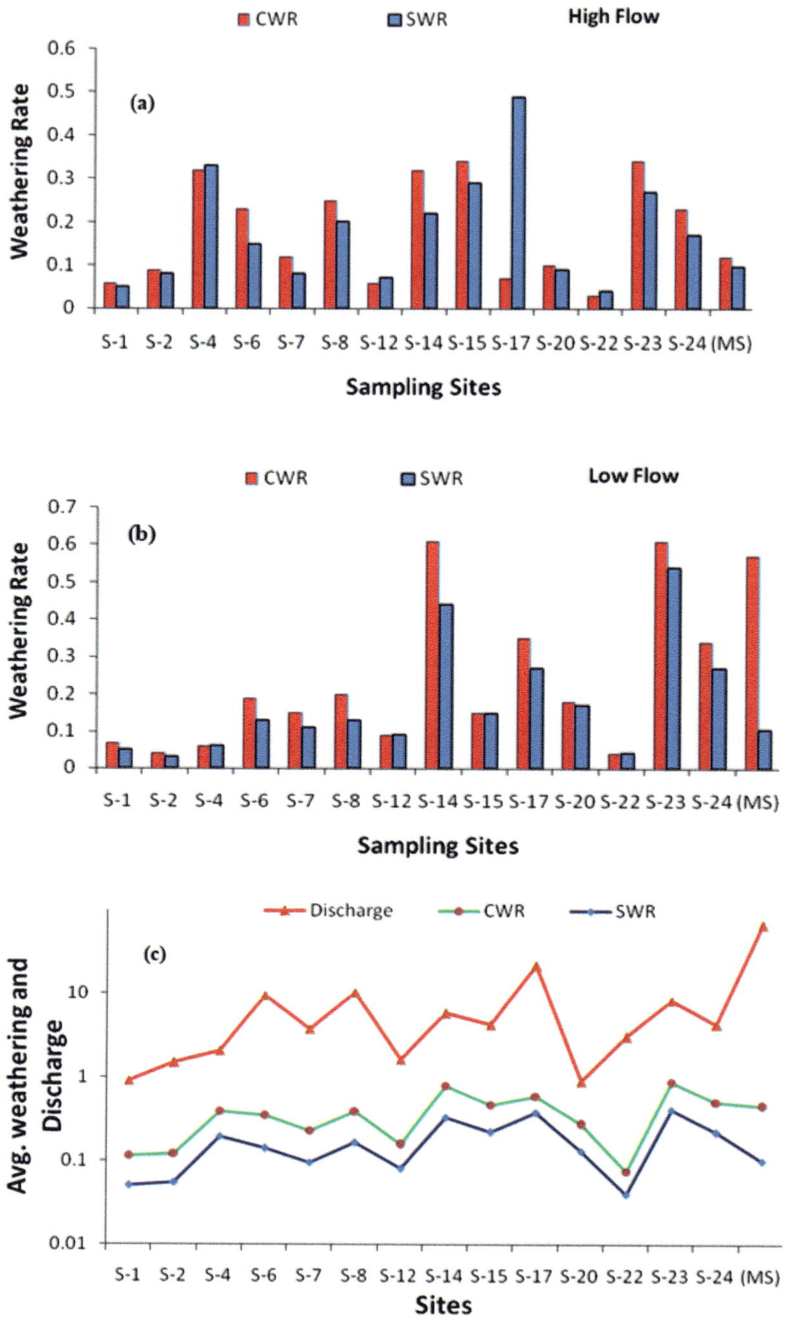

Fig. 13.6 Comparision of SWR and CrWR (ref: Table 13.6) during **a** high flow period (HFP), **b** low flow period (LFP) and **c** also with mean river discharge (*Source* Authors collected data)

varied from 0.037t/km^2/yr to 0.724 t/km^2/yr with an average of 0.381t/km^2/yr. The mainstream of the river Jhelum showed the CrWR varying from 0.121 t/km^2/yr to 0.572 t/km^2/yr with an average of 0.346 t/km^2/yr. Like SWR, CrWR exhibited the same trend during both the periods (Fig. 13.6a, b). Overall, it is observed that the higher Q may be probably causing more weathering processes and acting as a significant controlling factors in the area. However, it is also notable that faster dilution leads to faster transport of dissolved matter which might lead to lower weathering rates in certain cases. A comparison of the CrWR and SWR of the mainstream of river Jhelum and its tributaries

Table 13.7 The influence of chemical weathering rate and anthropogenic activities downstream. The chemical weathering rates show a decrease downstream, whereas the anthropogenic effects are higher towards downstream areas. (*Note* Normal fonted represent RBT whereas, the bold fonted represent LBT)

Tributary/ streams	ID	High value		Low value		Remarks
		In water	In sediments	In water	In sediments	
Dudganga	**S15**	Q, HCO_3, Ca, Mg, Na, K, SO_4, NO_3	SiO_2, Na_2O	T, pH, F, Cl	Al_2O_3, CaO, Fe_2O_3, MgO,K_2O,TiO_2, P_2O_5, σI	Major anthropogenic, acidic water, little silicate weathering
Ningal	**S21**	HCO_3, Mg, K, SO_4, Cl	SiO_2, MnO, P_2O_5	A T, pH, NO_3	Al_2O_3, CaO, MgO, Na_2O, TiO_2, Mz	Anthropogenic, weathering (carbonate), acidic pH favours weathering
Arapal	S12	HCO_3, Mg, Na, F, Cl, NO_3	Al_2O_3, MgO, K_2O, MnO, TiO_2	Q, SiO_2, pH, Ca, SO_4	Na_2O, Mz	Dolomitic weathering, acidic water, little ferromagnesian silicate weathering
Rambiara	**S7**	A, Q, SiO_2, T, pH, Ca, SO_4	SiO_2, Na_2O, K_2O, MnO, Mz	HCO_3, Mg, Na, Cl, NO_3	Fe_2O_3, TiO_2, P_2O_5, σI	High silicate and little carbonate weathering, alkaline water, almost absent anthropogenic
Pohru	S23	A, T, Na, F, SO_4	CaO, Fe_2O_3, MgO, K_2O, σI	Q, pH, HCO_3, Ca, Cl, NO_3	SiO_2, Al_2O_3, Na_2O, P_2O_5, Mz	Dominant silicate weathering, acidic, anthropogenic
Sindh	S18	A, Q, SiO_2, HCO_3, Ca, Mg, K, F, SO_4	CaO, Na_2O, TiO_2, σI	T, Na, Cl, NO_3	SiO_2, Al_2O_3, Fe_2O_3, K_2O, Mz	Carbonate weathering dominant, higher ions/oxides in a dissolved state in water than sediment oxides, low anthropogenic sources
Kuthar	S1	pH, HCO_3, Ca, Na, Cl	Fe_2O_3, MgO, K_2O, σI	A SiO_2, Mg, F	Na_2O, MnO	Alkaline water, more silicate (shale, slate, volcanic, quartzite) and little carbonate weathering, negligible anthropogenic
Bringi	S2	SiO_2, T, pH, Cl	CaO,MgO, MnO, TiO2, σI, Mz	K	SiO2, K2O	High carbonate and Na-silicate weathering, alkaline water, high temperature
Romush	**S14**	Q, SiO_2, pH, Ca, K, Cl	Al_2O_3, TiO_2	A Mg, F	CaO, MnO, P2O5	Alkaline water, carbonate, and silicate weathering, anthropogenic
Lidder	S6	A, T, F, NO3	Al_2O_3, Fe_2O_3, MgO, K_2O, P_2O_5, Mz	Q, HCO_3, Mg, Na, K, SO_4, NO_3	SiO_2, MnO, TiO_2, σI	Various members (mixed water), carbonate and silicate weathering, anthropogenic, fast flushing of chemicals
Sandrin	**S4**	Q, SiO_2, Mg, Na	CaO, Fe_2O_3, TiO_2	A, Ca, K	SiO_2, MgO, P_2O_5, σI	Carbonate (dolomitic) weathering, little silicate and anthropogenic

(continued)

Table 13.7 (continued)

Tributary/ streams	ID	High value		Low value		Remarks
		In water	In sediments	In water	In sediments	
Vishav	**S8**	A T, F, NO_3	CaO, Fe_2O_3, P_2O_5, Mz	SiO_2, HCO_3, Ca, Mg, Na, K, SO_4, Cl	Al_2O_3, Na_2O, MnO	Dominant silicate weathering, good anthropogenic
Sukhnag	**S24**	K, NO_3	–	Q, SiO_2, T, pH, HCO_3, Ca, F, SO_4	–	Highly anthropogenic, fertilizers, waste
Ferozpor	**S25**	pH, NO_3	–	A, Q, SiO_2, T, Na, K, F, SO_4	–	Dominant anthropogenic, industries, negligible weathering

(*Source* Authors collected data)

showed CWR to be higher than SWR during both the periods except the sites (S-4, S-12, and S-22). In comparison to the World average and other river basins of the world and India, the SWR and CrWR rates are very low (Table 13.6). The average silicate and carbonate weathering rates showed a positive correlation with the average discharge in the mainstream of river Jhelum and its tributaries except for the sites (S-6 and S-22), reflecting that the weathering rates in these streams are governed by the other factors also (Fig. 13.6c). Overall, in this study, the discharge has been suggested to be a major factor in controlling chemical weathering rates.

13.4.4.6 Chemical Index of Alteration and Weathering

Also, the Chemical Index of Alteration (CIA; Nesbitt and Young 1982) and Chemical Index of Weathering (Harnois 1988) were calculated to evaluate the extent of weathering in the area of the provenance of the river Jhelum and its tributaries. The equations of the above indices are

$$\mathrm{CIA} = \left(\frac{\mathrm{Al_2O_3}}{\mathrm{Al_2O_3} + \mathrm{CaO}^* + \mathrm{Na_2O} + \mathrm{K_2O}}\right) \times 100 \quad (13.6)$$

$$\mathrm{CIW} = \left(\frac{\mathrm{Al_2O_3}}{\mathrm{Al_2O_3} + \mathrm{CaO}^* + \mathrm{Na_2O}}\right) \times 100 \quad (13.7)$$

where CaO* is the amount incorporated in silica fraction of rocks which is corrected following the McLennan (1993) method.

The CIA of river sediments varied from a lower value of 64.5 at S18 to the highest value of 80.1 at S12 with an average value of 73.5 (Table 13.3). In general, a moderate weathering was inferred in the provenance of most of the sites. However, the highest CIA at S12 indicated extreme weathering in its provenance which is ascribed to the active tectonic upliftment and higher deformation of rocks in the area (Singh 2005; Riebe et al. 2001; Shabir et al. 2013; Dar et al. 2014). The average CIA of RBT sediments is higher (74.4) than LBT sediments (72.5) which indicated relatively higher weathering in these areas. Similarly, CIA values of the study area were also compared with other World standards and it was observed that the average CIA values were higher than the CIA values of Post-Archean Shale (PASS; Taylor and McLennan 1985, average 69), Upper Continental Crust (Taylor and McLennan 1985 average 47), and Brahmaputra river (Singh et al. 2005, range 58–65), thereby, reflecting comparatively a higher weathering in this region.

The Al_2O_3-(CaO^* + Na_2O)-K_2O (A-CN-K) ternary diagram, which is widely used to reflect silicate weathering trends (Nesbitt and Young 1982; 1984), distinctly indicated that sediment oxides approach more towards Al_2O_3 apex (Fig. 13.7). This suggests the removal of Ca and Na-bearing silicate minerals from the source rocks. Accordingly, the river sediments belong to the intermediate stage of silicate weathering. The progressive weathering of an igneous rock tends to drive the composition parallel to A-CN towards the right corner losing K so that more extremely weathered rocks will have more aluminous composition. The plot shows most of the right bank stream sediments (S18, S6, S23) higher on the diagram. This is the reason that most of the right streams have a higher Al_2O_3 percentage (average 9.66%) indicating high weathering intensity. Extreme weathering produces residuals completely depleted in alkali and alkaline earth that plot at A = 100 (Kaolinite, Gibbsite, and Bauxite). Thus, S18 indicates more weathering intensity than all other sites.

Overall, the weathering trend in studied samples is parallel to the A-CN line, reflecting the removal of Ca and Na (for example, plagioclase dissolution dominates) via silicate weathering in the drainage basins and K-bearing minerals are less altered. It is observed that most of the river sediments seem to be weathered from similar parent rocks, having close to the average composition of the mafic, despite the variable weathering intensities. The trend clearly showed that sediments have followed a basalt pattern of weathering. Further, all samples emanate from the Panjal volcanic or trap sediments that have acted as a major source (Taylor and McLennan 1985) as shown in (Fig. 13.7). The river sediments showed a marked loss in Na and Ca when compared to the UCC, as they tend to plot towards A-K edge close to Illite, Muscovite

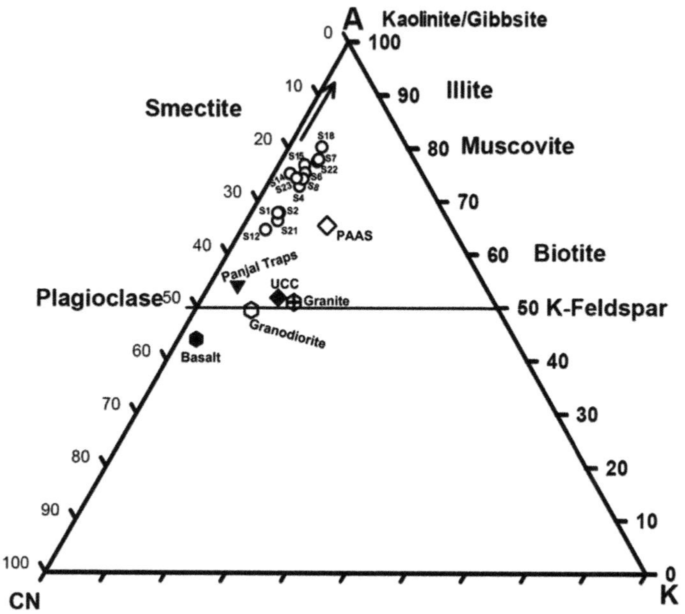

Fig. 13.7 A-CN-K plots (Nesbit and Young 1984) for studied samples. Plotted for reference are the UCC, PASS, Granite, Granodiorite Basalt, and Panjal Trap. Note the increasing order of weathering as indicated by the increasing shift of plot towards the A apex. (*Source* Authors collected data)

minerals above PAAS tending towards high weathering condition.

Furthermore, the comparison of major elemental oxides of sediments with an average UCC, PAS, and NASC in the form of normalized multi-element patterns was also carried out to understand the weathering processes (Fig. 13.8). The oxides such as SiO_2, TiO_2, MnO, MgO, and CaO revealed positive anomalies and therefore, suggested moderate weathering at the source. This also infers that the mafic minerals have concentrated in finer fraction after weathering. However, the negative anomaly of Al_2O_3, Fe_2O_3, Na_2O, K_2O, P_2O_5 than the limits could be attributed to the dilution effects of $CaCO_3$ as envisaged from the positive CaO anomaly and dilution effect of quartz (Das and Dhiman 2003). The reason is that Na and K are mainly contained in minerals which weather rapidly, specifically plagioclase.

13.4.4.7 Resistant Index of Maturity

RM is used to understand the degree of weathering in the provenance (Wakatsuki et al. 1977). In this study, the RM was estimated quantitatively by the concentration of the resistant component (SiO_2) relative to the more mobile alkaline ($Na_2O + K_2O$) and alkaline earth ($CaO + MgO$) components in the sediments using equation;

$$RM = SiO_2/(Na_2O + K_2O + CaO + MgO) \tag{13.8}$$

The RM for the river Jhelum sediments varied from 2.7 at Lidder S6 to 14.2 at Dudganga S15 with an average value of 6.66 (Table 13.3). The low RM value (average 5.6) of RBT sediments indicated the immaturity of sediments and higher RM in the LBT (average 7.7) indicated matured sediments. This reflected that the streams with higher weathering and lower RM (RBT in general) generate huge weathered material which gets little time to become mature while as streams with lower weathering rate and high anthropogenic sediments (LBT in general) have matured sediments.

13.4.5 Weathering and Anthropogenic Impact

The weathering processes in any catchment are also affected by anthropogenic factors. However, this impact was not accounted for in the calculation of weathering rates. For example, Romush (S14), Dudganga (S15), Sukhnag (S24),

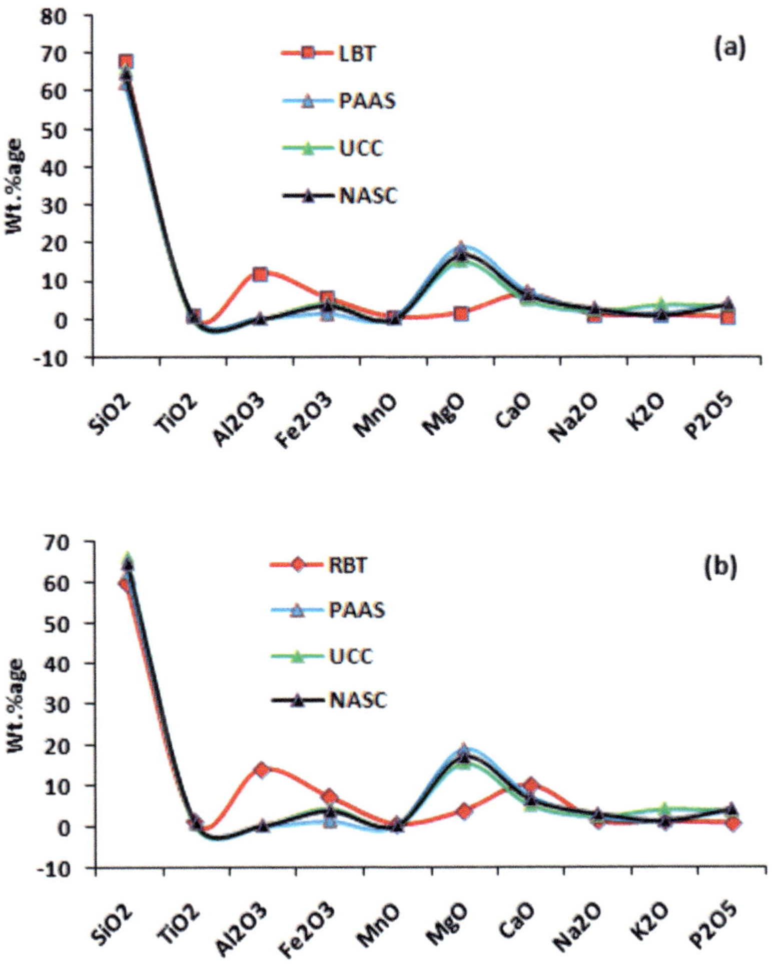

Fig. 13.8 Comparison of major oxide composition of sediments of the **a** LBT and **b** RBT against PAS, UCC, and NASC. (*Source* Authors collected data)

Ferozpor (S25), etc streams show higher chemical weathering rates (CWR); whereas these catchments have more anthropogenic influence. The calculated CWR thus considers only a few parameters while neglecting the role of important factors such as pH, T, and F-, on the rate of weathering. Furthermore, streambed sediment size and sediment chemistry were not considered in this calculation. Thus, combining all the water chemistry parameters, sediment characteristics, and their chemical composition, a more plausible assessment of weathering processes and anthropogenic intensity in the catchments was given (Table 13.7). It is construed from this analysis that the right bank catchments have a higher tendency in general for a higher rate of weathering with a small exception than left ones where anthropogenic forces also affect the processes. Therefore, Rambiara (S7), Romush (S14), and Dudganga (S15) are LBT and revealed high weathering rate which is ascribed to the ongoing neo-tectonic activities which assist in the creation and exposure of related structures (fractures, faults, etc) and physical disintegration of rocks in the Panjal Traps. These streams have also higher flow velocities (Table 13.7), which erode and transport the material very quickly along the densely created fracture surfaces. The larger catchment area and high water temperature might also be a probable reason for more weathering processes in these catchments. This observation is inferred from the large catchment area and high surface water temperature of S7. Moreover, the alkaline pH of the water and high dissolved

SiO_2 is the indication of more silicate weathering than carbonate dissolution (low value of Ca^{2+}, Mg^{2+}, and HCO_3^-). This is also supported by the higher SWR and CrWR rates in the LBT/ catchments.

The river Sindh (S18) originated from the highly elevated greater Himalayan range with varied rock formations varying from shale, slates, phyllite, granites, quartzites, volcanic rocks, limestone/dolomites, etc) in its catchment. Besides, its catchment area is very large and results in the generation of the highest discharge. All these factors are responsible for higher weathering rates in this sub-catchment area of the Jhelum Basin. Moreover, the area has higher carbonate weathering, a higher concentration of ions/oxides in a dissolved state in water, and lower concentrations in sediment oxides and low anthropogenic effects due to less population. A similar role of varied geological and tectonic setup is responsible for higher weathering in Bringi (S2), Kuthar (S1), and Arapal (S12) catchments. However, the Lidder (S6), which is also located in the same area as above three streams, is dominated by volcanic (Panjal Traps) rocks and very less of carbonate and other rocks. Pohru (S23) has lower rates in all the right streams mainly because of the generation of less stream discharge and the role of anthropogenic factors is dominating than weathering parameters. The Sandrin (S4) is a small watershed that produces very low weathered material but is located in a densely populated area of Anantnag district which generates more anthropogenic pollutants. A similar interpretation can be drawn for Vishav (S8), which has very high NO_3^- content and intermediate SO_4^{2-} and Cl^-. Likewise, Sukhnag (S24), Ningal (S21), and Ferozpor (S25) catchments are located in densely populated areas of Kashmir Valley, i.e., Srinagar and adjoining areas, and are fed by many highly polluted local drains of these areas. Thus, the rate of pollution is very high and these streams are small in area, have low discharge, and colder water with acidic pH (an indication of sewage water). Overall, the streams with higher weathering rates have usually alkaline water, i.e., low acidic. The streams with lower pH have a more anthropogenic nature and have lower weathering rates in general. Furthermore, ions/oxides, which are higher in water, are found to be generally lower in sediments, which indicate their release from the minerals and remain in dissolved forms. This observation is more clearly visible in the streams with a higher weathering rate (WR).

13.5 Conclusion

The geochemical load brought by the rivers in different forms, such as bed load, suspended load and dissolved load, from the catchment areas to the low lying areas also carries with it the signature of several processes that operate and generate this matter. The evaluation of the water chemistry and streambed sediments of the river Jhelum in the Kashmir Himalayas indicated that the carbonate rock weathering is the dominant source of the water composition in the river and its tributaries. The CWR showed variation among different tributaries and reflected the control of variable basin lithology, higher precipitation, and discharge. The annual CWR of the mainstream of river Jhelum is 11.1×10^2 t/km^2/yr. The LBT draining Pir-Panjal range showed higher CWR probably due to the active tectonic activities in this mountain range than the RBT. However, the left streams are additionally affected by the human-induced forces, which require consideration in further studies. The Sindh which is a right bank tributary exhibited higher CWR of 24.4×10^2 t/km^2/yr than the main river Jhelum and other tributaries which is attributed to its high relief, longer transportation, steep gradient, higher sediment and glacial erosion in its upper catchment. The CWR was also higher during the high flow period. In general, the river Jhelum and its tributaries show higher CWR as compared to other Himalayan rivers and World averages (0.36×10^2 t/km^2/yr). A similar pattern is almost exhibited by the Silicate weathering and Carbonate weathering rates. The Jhelum showed an average 0.10 t/km^2/yr of silicate weathering rate (SWR) and 0.35 t/km^2/yr of carbonate weathering rate (CrWR). The Pohru stream showed the lowest average SWR and CrWR of

0.04 t/km^2/yr whereas, the Sukhnag showed the highest SWR of 0.41 t/km^2/yr and CrWR of 0.48 t/km^2/yr, respectively. The derived CIA values in conjunction with low contents of Na suggested moderately weathered provenance and moderately humid climate characteristic of cold climate regions. The Resistant Index of Maturity (R_M) suggested immaturity of the river sediments. Furthermore, the present study attempts to quantify the chemical load and chemical erosion rates in the upper Jhelum catchment and its tributaries. These studies can therefore further be extended to understand the carbon dioxide sequestration processes involved in the region for better understanding the climate change occurring in this mountainous area.

Acknowledgements The authors are highly thankful to Head, Department of Earth Sciences, University of Kashmir Srinagar for providing laboratory and other facilities during Master of Philosophy (M. Phil) program. Sincere thanks are also to Mr. Ali Mohammad (Chemist) at Khyber Cement Industry Ltd, Khonmoh, Srinagar and P&D, Flood Control Department, Srinagar for their support during the work. The authors also recognize Dr. Malik Zubair Ahmad, Assistant Professor Department of Geology, University of Kashmir, Leh Campus for his valuable suggestions while interpreting the chemical data.

References

Adimalla N, Dhakate R, Kasarla A, Taloor AK (2020) Appraisal of groundwater quality fordrinking and irrigation purposes in Central Telangana. India. Groundwater Sust Develop 10:P100334. https://doi.org/10.1016/j.gsd.2020.100334

Adimalla N, Taloor AK (2020) Hydrogeochemical investigation of groundwater quality in thehard rock terrain of South India using Geographic Information System (GIS) andgroundwaterquality index (GWQI) techniques. Groundwater Sust Develop 10:P100288. https://doi.org/10.1016/j.gsd.2019.100288

Alam A, Bhat MS, Kotlia BS, Ahmad B, Ahmad S, Taloor AK, Ahmad HF (2017) Coexistent pre-existing extensional and subsequent compressional tectonic deformation in the Kashmir basin, NW Himalaya. Quat Int 444:201–208

Alam A, Bhat MS, Kotlia BS, Ahmad B, Ahmad S, Taloor AK, Ahmad HF (2018) Hybrid tectonic character of the Kashmir basin: Response to comment on "Coexistent pre-existing extensional and subsequent compressional tectonic deformation in the Kashmir basin, NW Himalaya (Alam et al., 2017)" by Shah (2017). Quat Int 468:284–289

Bailey SW, Brousseau PA, McGuire KJ, Ross DS (2014) Influence of landscape position and transient water table on soil development and carbon distribution in a steep, headwater catchment. Geoderma 226:279–289

Berner EK, Berner RA (1996) Global environment: Water. Air and Geochemical Cycles, Prentice Hall. New Jersey., p 376

Bhat DK (1989) Lithostratigraphy of Karewa Group. Kashmir Valley India, and a Critical Review of Its Fossil Record Memoirs GSI. 122:1–85

Bhat MS, Alam A, Ahmad B, Kotlia BS, Farooq H, Taloor AK, Ahmad S (2019) Flood frequency analysis of river Jhelum in Kashmir basin. Quat Int 507:288–294

Chakrapani GJ, Subramanian V (1996) Fractionation of heavy metals and phosphorus in suspended sediments of the Yamuna river, India. Environ Monit Assess 43:117–124

Chakrapani GJ, Veizer J (2005) Dissolved inorganic carbon isotopic compositions in the Upstream Ganga River in the Himalaya. Curr Sci 83:10–25

Dar RA, Romshoo SA, Chandra R, Ahmad I (2014) Tectono-geomorphic study of the Karewa Basin of Kashmir Valley. J Asian Earth Sci 1.92:143–156

Das BK, Dhiman SC (2003) Water and sediment chemistry of Higher Himalayan lakes in the Spiti valley control on weathering, provenance and tectonic setting of the basin. J Environ Geol 44:717–730

Drever JI (1997) The geochemistry of natural waters: Surface and Ground Water Environments. Prentice Hall, New Jersey., p 436

Fernex F, Valle PZ, Sanchez RH, Michand F, Parron C, Dalmasso J, Genevieve BF, Manuel GA (2001) Sedimentation rates in Lake Chapala (Western Mexico): possible active tectonic control. Chem Geol 177:213–228

Gaillardet J, Dupre B, Louvat P, Allegre CJ (1999) Global silicate weathering and CO_2 consumption rates deduced from the chemistry of large rivers. Chem Geol 159:3–30

Ganju JL, Khar BM (1984) Tectonics and hydrocarbon prospects of Kashmir valley- possible exploration targets. Petrol Asian J 207–216

Ghosh SN, Mathur VK (1996) Testing and Quality. Control. 3:457–466

Harnois L (1988) The CIW index: a new chemical index of weathering. Sedimen Geol 55:319–322

Haque S, Kannaujiya S, Taloor AK, Keshri D, Bhunia RK, Ray PKC, Chauhan P (2020) Identification of groundwater resource zone in the active tectonic region of Himalaya throughearth observatory techniques. Groundwater Sust Develop 10:P100337. https://doi.org/10.1016/j.gsd.2020.100337

Hren MT, Chamberlain CP, Hilley GE, Blisniuk PM, Bookhagen B (2007) Major ion chemistry of the Yarlung Tsangpo–Brahmaputra river: chemical weathering, erosion, and CO2 consumption in the southern

Tibetan plateau and eastern syntaxis of the Himalaya. Geochim Cosmochim Acta 71(12):2907–2935
Hu M, Stallard RF, Edmond JM (1982) Major ion chemistry of some large Chinese rivers. Nature 298:550–553
Jeelani G, Shah AQ (2006) Geochemical characteristics of water and sediment from the Dal Lake, Kashmir Himalaya: constraints on weathering and anthropogenic activity. Environ Geol 50:12–23
Khan A, Govil H, Taloor AK, Kumar G (2020) Identification of artificial groundwater recharge sites in parts of Yamuna river basin India based on remote sensing andgeographical information system. Groundwater Sust Develop 11:P100415. https://doi.org/10.1016/j.gsd.2020.100415
Krishnamurthy RV, Bhattacharya SK, Kusumgar S (1986) Palaeoclimatic changes deduced from 13C/12C and C/N ratios of Karewa Lake sediments, India. Nature 323:150–152
Krishnaswami S, Singh SK (2005) Chemical weathering in the river basins of the Himalaya. India. C Sci 89 (5):841–849
Krishnaswami S, Singh SK, Dalal TK (1998) Silicate weathering in Himalaya: Role in contributing to major ions and radiogenic Sr to Bay of Bengal; In Proc. INSA Symposium on 'Present trends and future directions in ocean sciences, pp 23–51. Indian National Science Academy and Akademia International
Lindholm RC (1987) A practical Approach to Sedimentology. Allen and Unwin, London, p 276
Lyu X, Tao Z, Gao Q, Peng H, Zhou M (2009) Chemical weathering and riverine carbonate system driven by human activities in a subtropical Karst Basin. South China. Water 10:1524
McLennan SM (1993) Weathering and global denudation. J Geol 101:295–303
Mir RA, Gani MK (2019) Evaluation of water quality of upper part of river Jhelum basin using multivariate statistical techniques. Arab J Geosci 12:445
Mir RA, Jeelani G (2015a) Hydrogeochemical assessment of river Jhelum and its tributaries for domestic and irrigation purposes, Kashmir valley. India. Curr Sci 109(2):311–322
Mir RA, Jeelani G (2015b) Textural characteristics of sediments and weathering in river Jhelum basin located in Kashmir valley western Himalaya. J Geol Soc India 86(4):445–458
Mir RA, Jeelani G, Dar FA (2016) Spatio-temporal variations and factors controlling the hydrogeochemistry of river Jhelum located in Kashmir valley, western Himalayas. Environ Mont Assess 188:438
Moon S, Huh Y, Qin J, van Pho N (2007) Chemical weathering in the Hong (Red) River basin: rates of silicate weathering and their controlling factors. Geochim Cosmochim Acta 71:1411–1430
Mortatti J, Probst JL (2004) Silicate rock weathering and atmospheric/soil CO_2 uptake in the Amazon basin estimated from river water geochemistry: seasonal and spatial variations. Chem Geol 197:177–196
Nanson GC, Gibling MR (2004) Anabranching and anastoming rivers. In Encylopedia of Geomorphology (A. S. Goudie, Ed.). International Conference of Geomorphologists. Routeledge. London. 21:213–225
Nesbitt HW, Young GM (1982) Early Proterozoic climates and plate motions inferred from major element chemistry of lutites. Nature 299.5885:715–717
Nesbitt HW, Young GM (1984) Prediction of some weathering trends of plutonic and volcanic rocks based on thermodynamics and kinetic considerations. Geochim Cosmochim Acta 54:1523–1534
Pandey SK, Singh AK, Hasnain SI (1999) Weathering and geochemical processes controlling solute acquisition in Ganga headwater–Bhagirathi river, Garhwal Himalaya, India. Aqua Geoche 5(4):357–379
Pattanaik J, Balakrishnan S, Bhutani R, Singh P (2013) Estimation of weathering rates and CO_2 drawdown based on solute load: Significance of granulites and gneisses dominated weathering in the Kaveri River basin, Southern India. Geochim Cosmochim Acta 121:611–636
Ramesh R, Ramanathan AL, Ramesh S, Purvaja R, Subramanian V (2000) Distribution of rare earth elements and heavy metals in the surficial sediments of the Himalayan river system. J Geochem 34:295–319
Riebe CS, Kirchner JW, Granger DE, Finkel RC (2001) Strong tectonic and weak climatic control of long-term chemical weathering rates. Geol 26:511–514
Sarin MM, Krishnaswami S, Dilli K, Somayajulu BLK, Moore WS (1989) Major ion chemistry of Ganga-Brahmaputra river system: Weathering processes and fluxes to the Bay of Bengal; Geochim. Cosmochin Acta 53:997–1009
Sarkar T, Kannaujiya S, Taloor AK, Ray PKC, Chauhan P (2020) Integrated study of GRACE data derived interannual groundwater storage variability over water stressed Indian regions. Groundwater Sustain Develop 10:P100376. https://doi.org/10.1016/j.gsd.2020.100376
Shabir A, Bhat MI, Madden C, Bali BS (2013) Geomorphic analysis reveals active tectonic deformation on the eastern flank of the Pir-Panjal Range. Arab J Geosci, Kashmir Valley, India. https://doi.org/10.1007/s12517-013-0900-y
Shafiq ul M, Ashraf I, ul Islam Z, Ahmed P, Dimri AP (2020) Response of streamflow to climate variability in the source region of Jhelum River Basin in Kashmir valley, India. Nat Haz 104(1):611–637
Singh SK, Sarin MM, Lenord C (2005) Chemical erosion in the eastern Himalayas, major ion composition of the Brahmaputra and $\sigma^{13}C$ of inorganic carbon. Geochim Cosmochim Acta 69:3573–3588
Soumya BS, Sekhar M, Riotte JJ, Audry S, Lagane C, Braun JJ (2011) Inverse models to analyze the spatiotemporal variations of chemical weathering fluxes in a granito-gneissic watershed: Mule Hole. South India. Geoderma 165(1):12–24

Taloor AK, Pir RA, Adimalla N, Ali S, Manhas DS, Roy S, Singh AK (2020) Spring water quality and discharge assessment in the Basantar watershed of Jammu Himalaya using geographic information system (GIS) and water quality Index(WQI). Groundwater Sust Develop 10:P100364. https://doi.org/10.1016/j.gsd.2020.100364

Taylor SR, McLennan SM (1985) The continental crust: its composition and evolution: Oxford, Black well's, pp. 312

Thakur VC, Rawat BS (1992) Geological Map of the Western Himalaya. Published Under the Authority of the Surveyor General of India. Printing Group of Survey of India, p. 101 (HLO)

Tipper ET, Lemarchand E, Hindshaw RS, Reynolds BC, Bourdon B (2012) Seasonal sensitivity of weathering processes: hints from magnesium isotopes in a glacial stream. Chem Geol 312–313:80–92

Wadia DN (1975) Geology of India. New Delhi, Tata McGraw Hill, p 344

Wakatsuki T, Furukawa H, Kyuma K (1977) Geochemical study of the redistribution of elements in soil—I. Evaluation of degree of weathering of transported soil materials by distribution of major elements among the particle size fractions and soil extract. Geochimica et Cosmochimica Acta 41(7):891–902

White AF, Blum AE (1995) Effects of climate on chemical weathering in watersheds. Geochim. Cosmochim. Acta 59:1729–1747

Wu W, XuS YangJ, Yin H (2008) Silicate weathering and CO_2 consumption deduced from the seven Chinese rivers originating in the Qinghai-Tibet Plateau. Chem Geol 249(3–4):307–320

Zaz SN, Romshoo SA (2012) Assessing the geoindicators of land degradation in the Kashmir Himalayan region, India. Nat haz 164(2):1219–1245

Dr. Riyaz Ahmad Mir is working as a Sr. Geoscientist in Geological Survey of India for the last 8 years. He has been honored with high ranks in joint CSIR-UGC (NET+JRF) examination and UPSC Geologist/Geoscientists Examination. After receiving his Master's degree in Geology from Bundelkhand University, Jhansi, he completed his M.Phil. in Hydrogeochemistry from University of Kashmir, Srinagar and his Ph.D. in Geoinformatics from Indian Institute of Technology (IITR), Roorkee. He has also worked with the National Institute of Hydrology (NIH), Roorkee for his Doctorate degree. He has been awarded with best Paper Presentation (SiD-2018) at Cluster University, Srinagar. He has a good working experience in the areas of hydro-and-sediment geochemistry, remote sensing and GIS and its applications in glacier, Glacier Lake Outburst Flood (GLOF), landslide studies and its modelling, climate change studies and its effect of water resources. He has also a vast field expertise in Geochemical Mapping (GCM), Specialized Geological Thematic Mapping (STM) and landslide zonation and susceptibility mapping, etc. He has published about 20 research articles in refereed international/national journals, about 10 extended abstracts in national and international conferences/symposiums and 1 book chapter (Springer publication). He has also completed about 7 Geological Reports of different research projects at GSI. He has published two articles in Hindi Magazine 'Tawi' published by GSI, J&K and is also writing to daily locals regarding geoscientific issues. He has delivered several guest lectures at Degree College and Intermediate level pertaining to the role of Earth Sciences in society

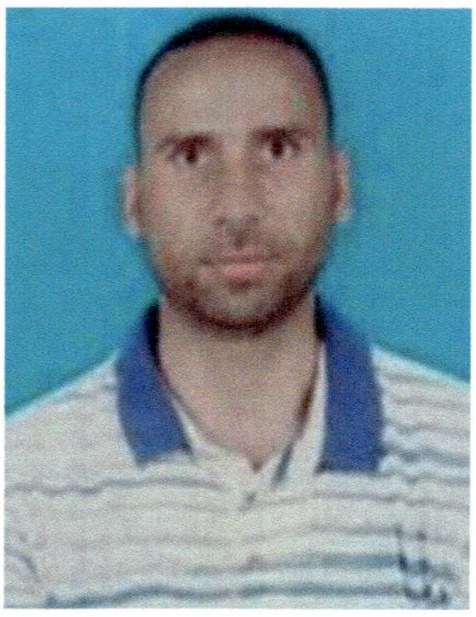

Dr. Farooq Ahmad Dar completed his Master's degree in Applied Geology from University of Kashmir and qualified CSIR-JRF. He did his Ph.D. from CSIR-National Geophysical Research Institute, Hyderabad and University of Hyderabad in 2014 on Karst Hydrogeology. His research interests include hydrogeology, karst geomorphology, hydrochemistry, stable isotopes, groundwater management, geophysics and climate change. The author has collaborated with many researchers from India and abroad and has published more than 15 research/review papers and book chapters in reputed SCI national and international journals on broader themes such as groundwater, hydrochemistry, environmental studies, remote sensing, GIS, etc. Currently working as Assistant Professor at University of Kashmir, Srinagar, Dr. Dar is teaching PG Geology programme and is working on a UGC sponsored research project

Prof. Ghulam Jeelani is presently serving as HOD, Department of Earth Science, University of Kashmir, Srinagar. He has carried out Master in Geology from Aligarh Muslim University, UP, India. Later, he has also carried out his Ph.D. from the Aligarh Muslim University in Hydrogeology. He has also carried out his Post-Doc from University of Kansas, Lawrence, USA. He has several awards to his credit like NET, CSIR/UGC, Excellent Grade Teacher Award by IQAC (University of Kashmir), INSA-Visiting Fellow and Fulbright-Nehru Senior Research Fellow and is a Visiting Scientist PRL, Ahmedabad, India. He has been Organizing Secretary/Convener of several workshops and is also acting as an Organizing Committee Member/Chairman of Committee/of several committees related to hydrogeology, climate change studies and sustainable development. He has completed several research projects and collaborations with many National Institutions. He has also supervised several candidates for their M.Phil. and Ph.D. degrees and has completed three Ph. D. and six M.Phils. He has authored about 50 articles published in International/national referred journals, about 40 abstracts presented in national/international, conferences and symposiums and six book chapters (Springer and Elsevier Publications). He has vast experience of more than 15 years in teaching and attended several national/international training programmes. He is a specialized worker in Hydrogeology, Hydrogeochemistry, Environmental Geology, Isotope Geology, Climate Change and Glaciology

Groundwater Storage Assessment Using Effective Downscaling GRACE Data in Water-Stressed Regions of India

14

Anuradha A. Karunakalage, Suresh Kannaujiya, Rajat S. Chatterjee, Ajay Kumar Taloor, Pranshu Pranjal, Prakash Chauhan, Prashant Kumar Champati Ray, and Senthil Kumar

Abstract

Globally, the groundwater is the most favourable and demandable freshwater resource. The threat to surface water resources and subsurface aquifer systems increased with climate change as well as surplus usage of groundwater in highly populated regions. Thus, in present day, groundwater is the primary resource for the sustainability of agriculture, industries and domestic activities in arid and semi-arid areas of the world. The overexploitation of subsurface water initiates land subduction. As water is the source of life on Earth, so it is essential to monitor and predict the capability of groundwater for secure sustainable management of subsurface water with the extreme climate conditions and population growth. The traditional way of keeping a check on groundwater level change is considering in situ or point measurements using the local network of well data. But these measurements are insufficient as hydrological models depend on the spatial data referring over large areas. Global Positioning System (GPS) and Gravity Recovery and Climate Experiment (GRACE) mission are perfect tools to overcome the drawbacks of the traditional groundwater monitoring. It measures the change in ice sheets and glaciers, near-surface and subsurface GWS changes, as well as sea-level changes by GRACE 1 mission and GRACE, Follow on (FO) mission. Most of the researches are based on GRACE satellite data to monitor GWS changes over a large-scale area as continental or regional achieved successful consequences. Although the past decade GRACE studies exhibited that

A. A. Karunakalage · P. Chauhan · P. K. Champati Ray · S. Kumar
Centre for Space Science and Technology Education in Asia and the Pacific, Indian Institute of Remote Sensing, Indian Space Research Organisation, Dehradun, Uttarakhand 248001, India
e-mail: anushikaworld@gmail.com

P. Chauhan
e-mail: prakash4140@gmail.com

P. K. Champati Ray
e-mail: champati_ray@rediffmail.com

S. Kumar
e-mail: askumar.nrsc@gmail.com

S. Kannaujiya (✉) · R. S. Chatterjee · P. Pranjal · P. Chauhan
Indian Institute of Remote Sensing, Indian Space Research Organisation, Dehradun, Uttarakhand 248001, India
e-mail: skannaujiya@iirs.gov.in

R. S. Chatterjee
e-mail: rsciirs@gmail.com

P. Pranjal
e-mail: pranshu.iirs@gmail.com

A. K. Taloor
Department of Remote Sensing and GIS, University of Jammu, Jammu, Jammu and Kashmir 180006, India
e-mail: ajaytaloor@gmail.com

A. K. Taloor et al. (eds.), *Water, Cryosphere, and Climate Change in the Himalayas*, Geography of the Physical Environment, https://doi.org/10.1007/978-3-030-67932-3_14

GRACE solution is capable of developing accurate quantitative estimations for GWS scenarios with the high temporal resolution. Still, it restricts only to continental or regional scale studies. Therefore, most of the recent studies took the step for effective downscaling of GRACE data.

Keywords

GRACE · GPS · Groundwater · Total water storage

14.1 Introduction

Groundwater is the most favourable and demandable freshwater resource in the world. The threat to surface water resources and subsurface aquifer systems increased with climate change (Wada et al. 2010; Jasrotia et al. 2019a; Bisht et al. 2020; Haque et al. 2020; Khan et al. 2020) as well as surplus usage of groundwater in highly populated regions. Thus in present day, groundwater is the primary resource for the sustainability of agriculture, industries and domestic activities in arid and semi-arid areas of the world like North-West India (Taloor et al. 2020; Jasrotia et al. 2018, 2019b; Adimalla and Taloor 2020b), Northern China, California in the USA, and regions without surrogate water resources like the Middle East and North Africa (Chen et al. 2014). The overexploitation of subsurface water initiates land subduction like in San Joaquin valley in California (Sneed and Brandt 2015) as well as some regions in North-West India as Chandigarh and some portions of Rajasthan (Singh et al. 2017; Kim et al. 2018). As water is the source of life on Earth, so it is essential to monitor and predict the capability of groundwater for secure sustainable management of subsurface water with the extreme climate conditions and population growth (Adimalla and Taloor 2020a; Adimalla et al. 2020). The traditional way of keeping a check on groundwater level change is considering in situ or point measurements using the local network of well data (Miro and Famiglietti 2018; Verma and Katpatal 2020). But these measurements are insufficient as hydrological models depends on the spatial data referring over large areas (Western and Blöschl 1998). Even though the in situ well data measurements have given high spatial resolution estimations for local areas, it becomes problematic in detecting groundwater level fluctuations with limited observations over arid and mountainous regions (Yin et al. 2018). So, it requires better awareness about groundwater table and geospatial distribution of pumping wells. For instance, in the United States, a network of 850,000 operating monitoring wells providing fundamental measurements of groundwater quantity and quality (Taylor and Alley 2001) are not quite sufficient for regional and local level studies (Faunt 2009; Faunt et al. 2016; Miro and Famiglietti 2018). The networks of groundwater observation wells in the world often lack adequacy to provide spatial and temporal coverage of Groundwater Storage (GWS) change (Shah et al. 2000; Mogheir et al. 2005). Global Positioning System (GPS) and Gravity Recovery and Climate Experiment (GRACE) mission are perfect tools to overcome the drawbacks of the traditional groundwater monitoring like spatial limitations as sparsing, uneven and time-consuming (Chen et al. 2019), as the technologies mentioned above consist of all-weather monitoring capability, high precision and continuous space-time monitoring (Rodell et al. 2007; Castellazzi et al. 2018).

GRACE is a collaborative mission of NASA's Jet Propulsion Laboratory (JPL), the German Aerospace Center (DLR), the Center for Space Research at the University of Texas at Austin (CSR) and Germany's National Research Center for Geosciences, Potsdam (GFZ), launched on 17 March 2002 (Sarkar et al. 2020). It measures the change in ice sheets and glaciers, near-surface and subsurface GWS changes, as well as sea-level changes by GRACE 1 mission and GRACE, Follow on (FO) mission (Chen et al. 2019). Most of the researches based on GRACE satellite data to monitor GWS changes over a large-scale area as continental or regional achieved successful consequences (Swenson et al. 2003; Su et al. 2020). Also, the studies about the accuracy of GRACE derived GWS variations proved that the error of GRACE derived solutions over the study realm is larger than 400,000 km^2 but less than 1 cm (Swenson et al. 2003). Although the past decade GRACE studies exhibited that GRACE solution is capable of developing accurate quantitative estimations for GWS scenarios with the high temporal resolution. Still, it restricts only to continental or regional scale studies. Therefore, most of the recent studies took the step for effective downscaling of GRACE data (Gautam et al. 2017; Kannaujiya et al. 2020).

14.2 Study Realm

The Mehsana district lies between 23° 15' to 23° 53' North and 72° 07' to 72° 26' East in the Northern part of the Gujarat alluvium plains. The semi-arid Mehsana district significantly depends on the subsurface water resource and the rate of groundwater development 151.17%. It segregates into the alluvial plain, dissected hilly terrain and piedmont plain with inselbergs. A narrow belt of 20–30 km width in North-Eastern portion of the district specializes with moderate relief alluvium, gravel beds and occasional inliers of older. The aquifer system of the district comprises multi-layers and formed beneath the Precambrian hard rocks, semi-consolidated Mesozoic and tertiary formations and unconsolidated quaternary alluvial deposits (Gupte 2011) as confined, semi-confined and phreatic. The groundwater occurs as fissured formation (hard rock) as well as porous formation (sedimentary) in the district. The soft rock formation of groundwater occurs consists of two major upper layer aquifer units in unconfined condition, denoted as A. But some regions of the aquifer system are semi-confined to confine. The lower layer of the soft rock groundwater formation is located a few hundred metres below as alternate clay and sand layers. The lower formation subdivides into B, C, D and E units composed of coarse to fine-grained sand and occasional post-Miocene gravel beds. The F and G units of the lower aquifer system comprises fine to medium-grained sand, sandstone with interbedded clay and Miocene siltstone sediments. The groundwater in the alluvium plain is extensively developed through dug wells, tube wells and dug cum bore wells (Table 14.1).

14.3 Methodology

14.3.1 Effective Downscaling of Grace Data

The previous studies, based on effective downscaling of GRACE TWS pixels, suggest two main methods for spatial reductions that are statistical downscaling and dynamic downscaling method (Chen et al. 2014). The dynamic model has used complex data which have been obtained from multiple resources to generate high spatial resolution regional numerical models. The long term linear or non-linear relationship between two data sets develops a statistical downscaling model (Yin et al. 2018). Also, some additional data are required for multivariate statistical regression records (Piles et al. 2011). Therefore, with respect to dynamic models, statistical models are widely used due to simple and less time-consuming. The researches based on statistical downscaling employs linear, non-linear, multivariate and machine learning techniques to support vector machine model, Artificial Neural Network (ANN) and Random Forest (RF) model. Support vector machine model which had been proposed initially by Vapnik based on classification machine learning algorithms are based theoretically on Vapnik Chervonenks dimension (VC) and structural risk minimized inductive (Vapnik 1999). ANN is a processing system to identify non-linear information which stands on the simulation of the human brain, simplification and the abstraction (Ghorbani et al. 2013). The effective downscaling of GRACE-derived TWS employs a statistical regression model (Ning et al. 2014).

Table 14.1 Observed water table decline over Mehsana and surrounding districts

District	Phreatic aquifer (m)	Semi-confined aquifer (m)	Confined aquifer (m)
Mehsana	5–7	12–23	18–50
Patan	2.5–9	25–36	13–45
Banaskantha	9–18	21–28	15–111

Source Central Groundwater Board, Ahmedabad, Sinha (2014)

14.3.2 Detailed Process of Downscaling

The present study emphasizes the statistical downscaling model that has been developed based on the regression relationship between GRACE SH-derived TWS, and water mass balanced equation manipulated TWS. The time lag effect of GRACE data is reduced by using two months later, GRACE SH data for considering the month of the hydrological fluxes. The steps of developing downscaling equation are as follows:

- The spatial resolution of GLDAS hydrological parameters are 0.25° (P 0.25°), 0.25° (ET 0.25°) and 0.25° (SR 0.25°) for precipitation, evapotranspiration and storm surface runoff, respectively. These parameters have been aggregated to the low spatial resolution of GRACE SH as P 1°, ET 1° and SR 1° by using average pixel method.
- Calculate monthly TWS change (ΔS1°) with the water mass balance equation.

$$\Delta S_t^{i^\circ} = P_t^{i^\circ} - ET._t^{i^\circ} - SR._t^{i^\circ} \quad (14.1)$$

where i° denotes the spatial resolution in degree and t represents the month.

- Level 3 JPL, CSR and GFZ model data of RL06 data product have been average to derive GRACE SH TWS 1°. The level 3 data of RL 06 data product pre-processed as TWS anomaly based on time-mean of 2004–2009 and the data has been multiplied by the corresponding scaling factor to minimize the leakage error. The missing month's data are calculated with linear interpolation of the relevant contiguous months.
- Estimate a linear empirical regression in between hydrological cycle derived ΔS1° and GRACE SH-derived TWS 1°. Liner regressions have been obtained from the several trials and select the equation which has given dominant highest R^2 value in all trails. Therefore, the estimated equation is only applicable to the post-monsoon season of the period from 2003 to 2019.

$$\hat{y} = \beta_1 \cdot x + \beta_\circ \quad (14.2)$$

$$GRACE_{t+2}^{i^\circ} = \Delta S_t^{i^\circ} . \beta_\circ + \beta_1 \quad (14.3)$$

$$GRACE_{t+2}^{i^\circ} = 0.656 \cdot \Delta S_t^{i^\circ} + 0.034 \quad (14.4)$$

where i° denotes the spatial resolution in degree, t denotes the month, x or ΔS_t^i is regressor (predictor), y or $GRACE_{t+2}^{i^\circ}$ is dependent (predictor), β_1 is slope and $\beta_\circ$ is the intercept of the linear regression equation.

- Manipulating ΔS0.25° with Eq. (14.1) by using P 0.25°, ET 0.25°, and SR 0.25° and then calculate GRACE SH-related TWS 0.25° from Eq. (14.4).
- In the downscaling process, removing system residual is an essential step. The results generated from Eq. (14.4) are subtracted from the system residual derived from the difference between GRACE TWSA 1° obtained from ΔS 1° and GRACE SH directly derived TWS 1°.

14.4 Implications from Grace Data Downscaling in Mehsana District, Gujarat

The results of downscaled high spatial resolution GRACE SH pixels (0.25° × 0.25°) gets validated with the results procured from JPL mascon solution pixels (0.5° × 0.5°) of GRACE and groundwell data from the Mehsana district (Fig. 14.1). This downscaled GRACE-derived GWSA (Fig. 14.2a) varies from −7.00 ± 5.00 to 45.00 ± 5.00 mm/year, and JPL mascon derived GWSA (Fig. 14.2c) varies from −80.00 ± 5.00 to −0.00 ± 5.00 mm/year. According to the distribution of both solutions, Mehsana district experiences groundwater depletion throughout the considered period. But downscaled GRACE SH-derived GWS distribution has exhibited detailed picture with respect to JPL mascon-derived GWS distribution. GRACE -derived TWS solutions consist of uncertainties due to inherent data errors and data leakage errors. Therefore, GRACE-derived GWS solutions include GRACE errors and error accumulated within the global land surface (GLDAS) model. Figure 14.2b and d represents two-dimensional maps of GWSA-related uncertainties of downscaled GRACE SH and JPL mascon, respectively. The maximum uncertainty accumulation of downscale GRACE SH derived-GWSA is ± 15.04 mm/year, and JPL Mascon-derived GWSA is ±8.50 mm/year.

The research work carried out recently over Ahmedabad and Gandhinagar using integrated GRACE SH (Level 3) and GPS study exhibits the GWS depletion rate of −0.6 mm/year and deformation rate of −5.20 mm/year from 2009 to 2017 (Chopra et al. 2013). The current research values obtained from JPL mascon for GWS change are 5.71 ± 5.50 mm/year, −8.14 ± 5.50 mm/year and −1.43 ± 5.50 mm/year for 2003–2007, 2008–2013 and 2014–2019, respectively, which correlates very well with the GPS measurements of the previous work. Whereas for the same periods the downscaled GRACE SH pixels-derived GWS changing rates are −39.30 ± 7.04 mm/year, −80.2 ± 7.04 mm/year and −104.91 ± 7.04 mm/year showing more enhanced measurements with respect to GPS, GRACE SH and JPL mascon (Table 14.2).

Based on early facts and history, GRACE solution doesn't share direct analogue with the ground-based values. Still, in recent research, it has been justified that GRACE solution does correlate with in situ well data, mainly over water-stressed regions in India (Sarkar et al. 2020). So, downscaled GRACE SH and JPL mascon-derived GWSA correlates with in situ well data-derived groundwater level change anomaly, across Mehsana district from 2005 to 2015. With respect to JPL mascon-derived GWSA, the downscaled GRACE SH-derived GWSA has shown great accord with groundwater level change anomaly along with the same upliftment and depletions (Fig. 14.3). But due to high temporal resolution of GRACE solution, it has not coincided with the real-time, and the upliftment or depletion observed by ground well data got detected one year later by GRACE. The same phenomena repeat the observations also for the low altitude areas (Ning et al. 2014). A possible cause is that in tropical regions precipitation is the most significant component for TWS change, and the combined effect of precipitation, evapotranspiration and surface runoff does not complete in a short period like GRACE temporal resolution (Fig. 14.3).

14.5 Conclusion

Though JPL mascon has provided the regional scale solution for TWS change, it is not sufficient to derive GWSA for water management scale studies. Effective downscaling of GRACE pixels enhanced sensitivity of the data, and those data have more capability to measure GWS change accurately for water management. The present study develops a model for the post-monsoon season of the 15 years based on linear statistical regression implication between GRACE SH obtained TWS and hydrological parameters derived TWS. The comprehensive studies about

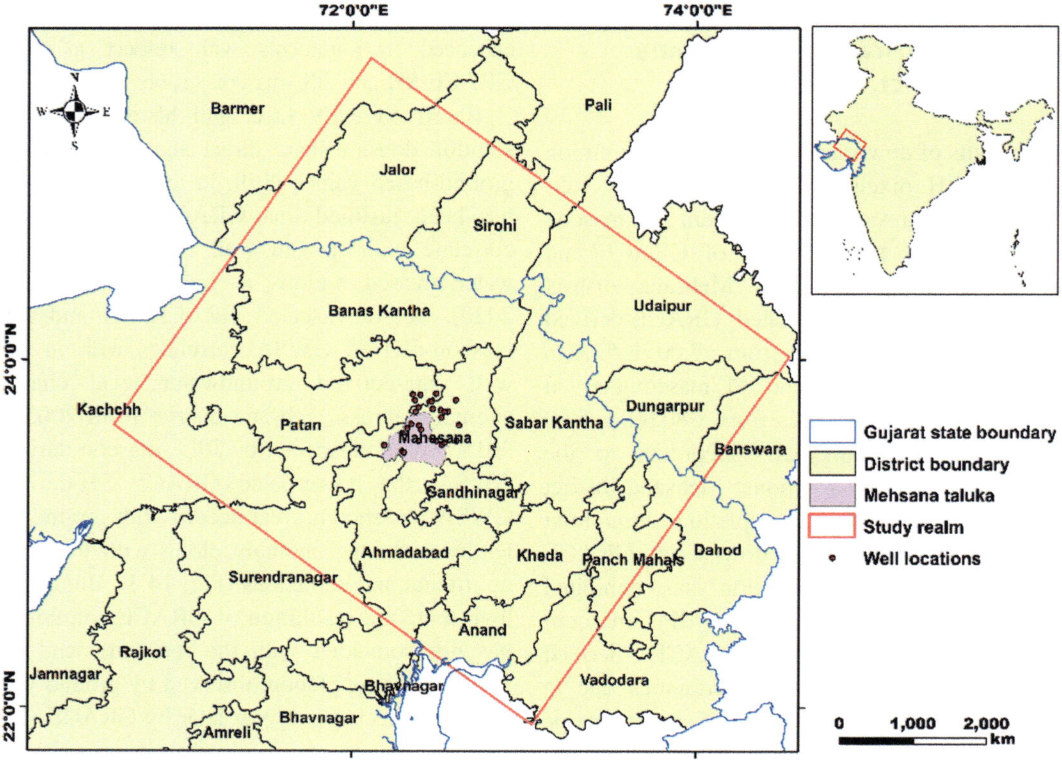

Fig. 14.1 The map represents the study realm (red rectangle) covering the Mehsana district, and a significant part of North Gujarat. (*Source* Authors Contribution)

Table 14.2 Downscaled GRACE SH and Mascon JPL manipulated groundwater storage change as flux and as quantity over Mehsana and surrounding districts

Time period	Trends as flux		Trends as storage	
	Downscale GRACE SH ±7.04 mm/year	Mascon JPL ±5.50 mm/year	Downscale GRACE SH ±0.54 km^3/year	Mascon JPL ±0.43 km^3/year
2003–2007	−39.38	-5.71	-3.05	−0.44
2008–2013	−80.22	-8.14	-6.22	−0.63
2014–2019	−104.91	-1.43	-8.13	−0.11

(*Source* GRACE data)

the capability of different GRACE solutions with the different basin sizes, climate, and intensity of usage of irrigation water had shown various merits with the mentioned variabilities. As an example, CSR mascon had shown higher uncertainty for large-scale basins with respect to JPL mass. Therefore, the current study has suggested innovating new downscaling models with different GRACE solutions based on different basin-scales, irrigation indexes and climate conditions.

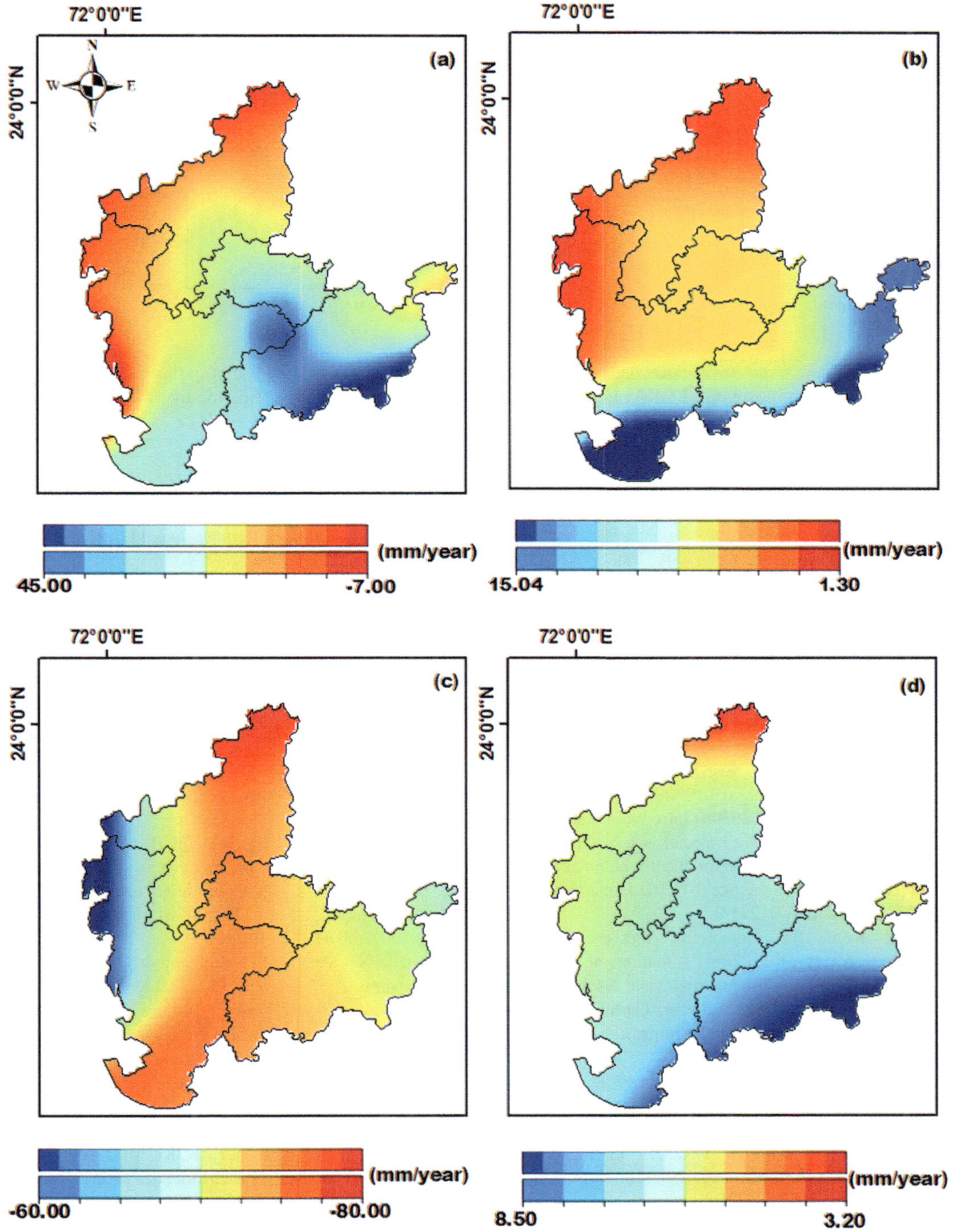

Fig. 14.2 **a** and **c** represents GWS change rate across Mehsana and surrounding districts in the post-monsoon season for a period of 2003–2019. **b** and **d** represents uncertainty distribution related to downscaling GRACE SH and JPL MASCON derived GWS change over the same area and the same period, respectively (*Source* GRACE data)

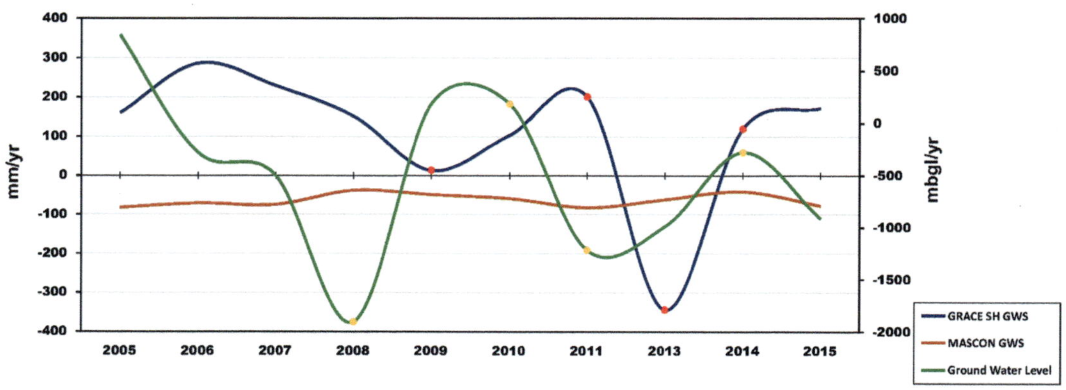

Fig. 14.3 It shows a correlation between different GWSA derived from downscaled GRACE SH solutions, JPL mascon solutions and the groundwater level change anomaly over Mehsana district during the post-monsoon period from 2005 to 2015. (*Source* GRACE data and CGWB India)

References

Adimalla N, Dhakate R, Kasarla A, Taloor AK (2020) Appraisal of groundwater quality for drinking and irrigation purposes in Central Telangana, India. Groundw Sust Dev 10:P100334. https://doi.org/10.1016/j.gsd.2020.100334

Adimalla N, Taloor AK (2020a) Hydrogeochemical investigation of groundwater quality in the hard rock terrain of South India using Geographic Information System (GIS) and groundwater quality index (GWQI) techniques. Groundw Sust Dev 10:P100288. https://doi.org/10.1016/j.gsd.2019.100288

Adimalla N, Taloor AK (2020b) Introductory editorial for 'Applied Water Science' special issue: "Groundwater contamination and risk assessment with an application of GIS". Appl Water Sci 10:216. https://doi.org/10.1007/s13201-020-01291-3

Bisht H, Kotlia BS, Kumar K, Arya PC, Sah SK, Kukreti M et al (2020) Estimation of suspended sediment concentration and meltwater discharge draining from the Chaturangi glacier, Garhwal Himalaya. Arab J Geosci 13(6):1–12. https://doi.org/10.1007/s12517-020-5204-w4

Castellazzi P, Longuevergne L, Martel R, Rivera A, Brouard C, Chaussard E (2018) Quantitative mapping of groundwater depletion at the water management scale using a combined GRACE/InSAR approach. Remote Sens Environ 205:408–418. https://doi.org/10.1016/j.rse.2017.11.025

Central Groundwater Board, Ahmedabad Sinha AK (2014) CENTRAL GROUNDWATER BOARD West Central Region Ahemdabad 2014

Chen J, Li J, Zhang Z, Ni S (2014) Long-term groundwater variations in Northwest India from satellite gravity measurements. Glob Planet Change 116:130–138. https://doi.org/10.1016/j.gloplacha.2014.02.007

Chen L, He Q, Liu K, Li J, Jing C (2019) Downscaling of GRACE-derived groundwater storage based on the random forest model. Remote Sens 11. https://doi.org/10.3390/rs11242979

Chopra S, Kumar D, Rastogi BK, Choudhury P, Yadav RBS (2013) Estimation of seismic hazard in Gujarat region, India. Nat Hazards 65:1157–1178. https://doi.org/10.1007/s11069-012-0117-5

Faunt CC (ed) (2009) Groundwater availability of the central valley aquifer. US Geological Survey Professional paper 1766, California

Faunt CC, Sneed M, Traum J, Brandt JT (2016) Water availability and land subsidence in the Central Valley, California, USA. Hydrogeol J 24:675–684. https://doi.org/10.1007/s10040-015-1339-x

Ghorbani MA, Khatibi R, Hosseini B, Bilgili M (2013) Relative importance of parameters affecting wind speed prediction using artificial neural networks. Theor Appl Climatol 114:107–114. https://doi.org/10.1007/s00704-012-0821-9

Gautam PK, Arora S, Kannaujiya S, Singh A, Goswami A, Champati PK (2017) A comparative appraisal of groundwater resources using GRACE-GPS data in highly urbanised regions of Uttar Pradesh. Springer International Publishing Switzerland, India. https://doi.org/10.1007/s40899-017-0109-4

Gupte P (2011) Groundwater scenario in major cities of India, 7–17

Haque S, Kannaujiya S, Taloor AK, Keshri D, Bhunia RK, Ray PKC, Chauhan P (2020) Identification of groundwater resource zone in the active tectonic region of Himalaya through earth observatory techniques. Groundw Sust Dev 10:P100337. https://doi.org/10.1016/j.gsd.2020.100337

Jasrotia AS, Taloor AK, Andotra U, Bhagat BD (2018) Geoinformatics based groundwater quality assessment for domestic and irrigation uses of the Western Doon valley, Uttarakhand, India. Groundw Sust Dev 6:200–212

Jasrotia AS, Kumar R, Taloor AK, Saraf AK (2019a) Artificial recharge to groundwater using geospatial and groundwater modelling techniques in North Western Himalaya, India. Arabian J Geosci 12 (24):774. https://doi.org/10.1007/s12517-019-4855-5

Jasrotia AS, Taloor AK, Andotra U, Kumar R (2019b) Monitoring and assessment of groundwater quality and its suitability for domestic and agricultural use in the Cenozoic rocks of Jammu Himalaya, India: a geospatial technology based approach. Groundw Sust Dev 8:554–566

Kannaujiya S, Gautam PKR, Chauhan P, Roy PNS, Pal SK, Taloor AK (2020) Contribution of seasonal hydrological loading in the variation of seismicity and geodetic deformation in Garhwal region of Northwest Himalaya. Quat Int. https://doi.org/10.1016/j.quaint.2020.04.049

Khan A, Govil H, Taloor AK, Kumar G (2020) Identification of artificial groundwater recharge sites in parts of Yamuna river basin India based on remote sensing and geographical information system. Groundw Sust Dev 11:P100415. https://doi.org/10.1016/j.gsd.2020.100415

Kim J, Lin S-Y, Singh S, Singh T, Tsai Y-L, Gupta S, Save H (2018) Surface deformations by ground water depletion over NW. India: local and global scale observations using InSAR and space geodesy and their geological implications. In: 20th EGU general assembly, proceedings from the conference held 4–13 Apr 2018 in Vienna, Austria, p 17305

Piles Maria, Camps Adriano, MerceVall-Ilosser IC (2011) Downscaling SMOS-Derived Soil Moisture Using MODIS Visible/Infrared Data. IEEE Trans Geosci Remote Sens 49(9):3156–3166

Miro ME, Famiglietti JS (2018) Downscaling GRACE remote sensing datasets to high-resolution groundwater storage change maps of California's Central Valley. Remote Sens 10. https://doi.org/10.3390/rs10010143

Mogheir Y, De Lima JLMP, Singh VP (2005) Assessment of informativeness of groundwater monitoring in developing regions (Gaza Strip case study). Water Resour Manag 19:737–757. https://doi.org/10.1007/s11269-005-6107-6

Ning S, IshidaiaA H, Wang J (2014) Statistical downscaling of Grace-derived terrestrial water storage using satellite and Gldas products. J Japan Soc Civ Eng Ser B1 (Hydraulic Eng 70:I_133–I_138). https://doi.org/10.2208/jscejhe.70.i_133

Rodell M, Chen J, Kato H, Famiglietti JS, Nigro J, Wilson CR (2007) Estimating groundwater storage changes in the Mississippi River basin (USA) using GRACE. Hydrogeol J 15:159–166. https://doi.org/10.1007/s10040-006-0103-7

Sarkar T, Kannaujiya S, Taloor AK, Champati Ray PK, Chauhan P (2020) Integrated study of GRACE data derived interannual groundwater storage variability over water stressed Indian regions. Groundw Sust Dev 10:100376. https://doi.org/10.1016/j.gsd.2020.100376

Shah T, Molden D, Sakthivadivel R, Seckler D (2000) The global groundwater situation: overview of opportunities and challenges. Glob Groundw Situ Overv Oppor chall. https://doi.org/10.5337/2011.0051

Singh AK, Jasrotia AS, Taloor AK, Kotlia BS, Kumar V, Roy S, Ray PKC, Singh KK, Singh AK, Sharma AK (2017) Estimation of quantitative measures of total water storage variation from GRACE and GLDAS-NOAH satellites using geospatial technology. Quat Int 444:191–200

Sneed M, Brandt JT (2015) Land subsidence in the San Joaquin Valley, California, USA, 2007–2014. In: Proceedings of the international association of hydrological sciences. Copernicus GmbH, pp 23–27

Su Y, Guo B, Zhou Z, Zhong Y, Min L (2020) Spatio-temporal variations in groundwater revealed by GRACE and its driving factors in the Huang-Huai-Hai Plain, China. Sensors (Switzerland) 20. https://doi.org/10.3390/s20030922

Swenson S, Wahr J, Milly PCD (2003) Estimated accuracies of regional water storage variations inferred from the Gravity Recovery and Climate Experiment (GRACE). Water Resour Res 39. https://doi.org/10.1029/2002WR001808

Taloor AK, Pir RA, Adimalla N, Ali S, Manhas DS, Roy S, Singh AK (2020) Spring water quality and discharge assessment in the Basantar watershed of Jammu Himalaya using geographic information system (GIS) and water quality Index(WQI). Groundw Sust Dev 10:P100364. https://doi.org/10.1016/j.gsd.2020.100364

Taylor CJ, Alley WM (2001) Ground-water-level monitoring and the importance of long-term water-level data. US Geol Surv Circ 1–68

Vapnik VN (1999) An overview of statistical learning theory. IEEE Trans Neural Netw 10(5):988–999. https://doi.org/10.1109/72.788640

Verma K, Katpatal YB (2020) Groundwater monitoring using GRACE and GLDAS data after downscaling within basaltic aquifer system. Groundwater 58 (1):143–151. https://doi.org/10.1111/gwat.12929

Wada Y, Van Beek LPH, Van Kempen CM, Reckman JWTM, Vasak S, Bierkens MFP (2010) Global depletion of groundwater resources. Geophys Res Lett 37:1–5. https://doi.org/10.1029/2010GL044571

Western AW, Blöschl G (1998) On the spatial scaling of soil moisture. J Hydrol 217:203–224

Yin W, Hu L, Zhang M, Wang J, Han SC (2018) Statistical downscaling of GRACE-derived groundwater storage using ET data in the North China Plain. J Geophys Res Atmos 123:5973–5987. https://doi.org/10.1029/2017JD027468

Ms. Anuradha A. Karunakalage is a graduate student in the Department of Geology in the University of Peradeniya, Sri Lanka. She is an alumnus of UN-affiliated Center for Space Science and Technology Education in Asia and Pacific (CSSTEAP) and Indian Institute of Remote Sensing (IIRS), ISRO, where she received her post-graduate diploma in Remote Sensing and GIS. She has been the recipient of CSSTEAP bronze medal in 2019/2020. Her research interests include hydrogeology, ground deformation, GRACE data applications, Synthetic Aperture Radar (SAR) applications and climate change

Mr. Suresh Kannaujiya has been working in Geosciences Department, Indian Institute of Remote Sensing (ISRO), Dehradun for the last 8 years. He was awarded M. Tech. in Applied Geophysics from IIT Roorkee. He is currently pursuing a Ph.D. in crustal deformation from IIT Dhanbad. He has been exposed to real-time challenges in the field of marine geophysics while being associated with Fugro India Pvt. Ltd. as a processing geophysicist for 5 months. He has authored many research articles published in national and international journals/periodicals in landslide modelling, groundwater depletion, TEC modelling, crustal deformation, hazard assessment, and remote sensing integration and geophysics for landslide demarcation. His current interest lies in modelling GNSS data for total electron content/strain/crustal deformation and active fault mapping through the geophysical survey

Dr. Rajat S. Chatterjee is working as a Senior faculty member of Geosciences and Geohazards Department and is Head of Disaster Management Science Department of IIRS (ISRO), Dehradun. The areas of specializations include microwave remote sensing applications in geosciences, remote sensing-based thermal anomaly detection and modelling, and structural geology and geodynamics. He is post-graduate in applied geology and has Ph.D. in geology and geophysics from IIT Kharagpur. He did the professional course DÉSS de Télédétection (master's in remote sensing) from the University of Pierre and Marie Curie (Paris VI University), France, and post-doctoral research for a brief duration in the University of Marne-la-Vallée, France. He was awarded with the Institute Silver Medal from IIT Kharagpur in 1992, P.R. Pisharoty Memorial Award (National Remote Sensing Award) in 2010 and ISRO Team Excellence Award in 2015. He has published more than 50 papers in journals and symposia proceedings, chapters in books and technical reports

Dr. Ajay Kumar Taloor has obtained his Doctorate in Remote Sensing and GIS applications in hydrogeology from University of Jammu, NAAC accredited A+ University of India. Thirteen years of research experience in the applicationsin geospatial technology for natural resources management of land and water resources. He has excelled twice with best paper presentation award in India. Being an expertof remote sensing applications in water science, cryosphere and climate change, tectonic and quaternary geomorphology, he is working on two major research projects on using space-based inputs for glacier mapping and climate change in Himalayas. He has also published many articles in tectonic and quaternary geomorphology in the recent years. He hashigh scientific temper and strong HR relations in science world, with high professional and managerial skills. He has edited many volumes in the top-rated journals in the Elsevier and Springer publishers, member of Editorial Board of the Quaternary Science Advances, and reviewer of the many top-rated international journals in science world

Mr. Pranshu Pranjal is dedicated, detailed and capable senior research fellow with 5 years of experience in graduate-level remote sensing and GIS research. He has been working in the field of groundwater depletion-induced land subsidence using space technology, and currently pursuing Ph.D. from Indian Institute of Technology (Indian School of Mines) Dhanbad. Moreover, he has also been awarded his M. Tech. in Geomatics from Indian Institute of Technology (Indian School of Mines) Dhanbad. His current area of interest lies among spatio-temporal analysis of hydrological parameters and modelling from space bourn gravity anomalies, exploring land deformation with GPS/GNSS data processing using GAMIT/GLOBK, etc. Additionally, he also has understanding on various surveying techniques and instruments like, Optical and Vernior Theodolite, Auto Level, Digital Level and Total Station

Dr. Prakash Chauhan obtained his Ph.D. in Physics from Gujarat University, Ahmedabad and post-graduate degree in Applied Geophysics from Indian Institute of Technology (IIT), Roorkee. He was the former Group Director of Space Applications Centre, ISRO, Ahmedabad in 2014. He joined as a scientist in Indian Space Research Organisation (ISRO) in 1991. He initiated research activities for planetary remote sensing at Space Applications Centre to study solar system objects, mainly Earth's Moon and Mars, through Indian planetary missions. He has done lead work like lunar surface composition mapping in the Chandrayaan-1 mission and was appointed as the Principal Investigator for Infrared Imaging Spectrometer (IIRS) instrument on-board Chandrayaan-2 mission. He led a team of scientists for scientific analysis of data from Mars Orbiter Mission (MOM) instruments. He has published more than 100 research papers in both national and international journals. He is the recipient of prestigious Prof. P.R. Pisharoty memorial award in 2004 by Indian Society of Remote Sensing, Hari Om Ashram Prerit Dr. Vikarm Sarabhai Research Award by Physical Research Laboratory (PRL), Ahmedabad in 2009, ISRO merit award by Indian Space Research Organisation in 2010 and Satish Dhawan award by ISRS, Dehradun in 2016

Dr. Prashant Kumar Champati Ray with a post-graduate and Ph.D. degree from IIT Bombay, India, and MS and PDF from the University of Twente, Netherlands, and is actively involved in research, education and training in the field of applications of remote sensing, GNSS and GIS in geosciences, geohazards and planetary geology. Currently, he is Group Head of Geosciences and Disaster Management Studies (GDMS) Group and Head of Geosciences and Geohazards Department of IIRS. His research interests include monitoring and modelling of landslides, active fault mapping, seismic hazard assessment, geodynamics, crustal deformation, earthquake precursor studies, mineral exploration and planetary geology. His professional career spans over 30 years during which he has implemented 14 projects and guided more than 100 students, including 11 Ph.D. students; published around 200 papers, including 54 papers in peer-reviewed national and international journals, and most importantly delivered more than 100 invited presentations at national and international forum

Dr. Senthil Kumar obtained his Ph.D. from the Indian Institute of Science, Bangalore, in the field of image processing in 1990. He joined ISRO in 1991 and has been serving in Indian satellite programmes in various capacities. His research includes sensor characterization, radiometric data processing, image restoration, data fusion and soft computing. He is a former Director of the Indian Institute of Remote Sensing, Dehradun, and a former UN-affiliated Centre for Space Science and Technology Education in Asia and the Pacific. He has published about 80 technical papers in international journals and conferences, besides technical reports. He is a recipient of ISRO Team Awards for Chandrayaan-1 mission and Prof. Satish Dhawan Award conferred by the Indian Society of Remote Sensing

Water Quality of Himalayan Rivers in Uttarakhand

15

Deeksha Aithani, Jyoti Kushawaha, and S. R. Sreerama Naik

Abstract

Rivers are considered a significant source of fresh water and are vital for society's sustenance and well-being. Due to rapid urbanization and industrialization, these precious natural resources are being degraded. Melting of Himalayan glaciers and ice is the foremost source of water supply in the Indo Gangetic plains. Also, all major river systems of northern India originate from the Himalayas. The quality of river water in the Himalayas is worsening over the years because of changes in sediment balance, vegetation, land use land cover (LULC) accompanied by anthropogenic activities, inadequately structured sewerage, and drainage system, dumping of treated or untreated effluents, etc. The Himalayan glacial system serves an essential role in creating the headwater streams of the important Himalayan Rivers. Some of which flows across the national boundaries and some through the states of India. This chapter discusses the physical and chemical water quality parameters of some of the Himalayan Rivers in Uttarakhand. The total 11 parameters are taken into account to get insight into the water quality of a few selected rivers in Uttarakhand because of the scarcity of data on the Himalayan River's water quality. Water quality parameters deliberate in this chapter specify that the water quality of Himalayan River is still good. However, developmental and other activities taking place in and along the rivers pose a threat to it.

Keywords

Himalayan river · LULC · Watershed · Water quality

15.1 Introduction

Rivers are a significant source of fresh water and are crucial for society's sustenance and well-being. Due to rapid urbanization, industrialization, and other developmental activities, these precious natural resources are being degraded (Bora and Goswami 2017; Singh et al. 2017; Khan et al. 2020; Sarkar et al. 2020; Haque et al. 2020; Adimalla and Taloor 2020a). Rivers as the source of fresh water are highly vulnerable to pollution due to anthropogenic activities like

D. Aithani · J. Kushawaha (✉)
School of Environmental Sciences, JNU, New Delhi 110067, India
e-mail: jyoti.vaishnavi@gmail.com

D. Aithani
e-mail: kumdiksha@yahoo.com

S. R. Sreerama Naik
Madurai Kamaraj University, Madurai, Tamilnadu 625021, India
e-mail: shree.rathod72@gmail.com

A. K. Taloor et al. (eds.), *Water, Cryosphere, and Climate Change in the Himalayas*, Geography of the Physical Environment, https://doi.org/10.1007/978-3-030-67932-3_15

disposal of wastewater, industrial effluents, sewage, pesticides, and fertilizers and also because of natural processes such as organic material, suspended sediments, and heavy metals contaminants from erosion and weathering of rocks in the catchment (Bisht et al. 2020a). Natural factors are similarly or more accountable for river water quality degradation (Jarvie et al. 1998). The observing trends show that industrial activities and wastewater disposal are a constant source of contamination factors for river water pollution due to high population growth. Simultaneously, natural processes such as erosion and weathering are external factors, i.e., air, water, precipitation, etc. Various agricultural and developmental activities taking place in the river basin's vicinity are also accountable for their degrading water quality. It is significant to gather scientifically reliable information about river water quality and conservation as they are the critical source of water required for various purposes such as industrial, domestic, and irrigation (Haritash et al. 2016; Khan et al.2017; Adimalla and Taloor 2020b). Over the years RS and GIS become an important tool to analyze the spatiotemporal water quality around the world (Jasrotia et al. 2018, 2019; Taloor et al. 2020a).

Some of the world's highest and most giant glaciers, having an average height of 6000 m, are found in great Himalayan ranges. Glaciers in the Indian Himalayas are spread over 3.8×104 km^2 and cover 17% of the mountain area (Pandey et al. 1999). The rivers in the Indo Gangetic plains acquire their water supply primarily from the Himalayas. All the major river systems of northern India originate from the melting of Himalayan glaciers and ice (Bisht et al. 2018). The average annual stream flows of the Indus, Ganga, and Brahmaputra are 206, 480, and 510 km^3 yr^{-1}, respectively, and probably more or less half of this water comes from snow and ice melt (Pandey et al. 1999). The upland catchment of the Himalayas, which is considered an unpolluted region, offers a basic unit to inspect the natural weathering and geochemical processes (Bisht et al. 2018). However, the quality of Himalayan Rivers is continuously degraded over the years because of anthropogenic activities, including disposal of untreated effluents, poorly designed sewerage and drainage systems, tourism, etc. (Haritash et al. 2016; Bisht et al. 2020a). Himalayan Rivers majorly contribute to the headwater streams of the important rivers which flow through states of India, Pakistan, Bangladesh, etc. Despite the extreme importance of the Himalayan region as a water reservoir, the representative hydrological and geochemical data on the headwater streams are generally not available.

Uttarakhand is mostly a hilly state, established on 9th November 2000 as the 27th state of India, after separating from the northern Uttar Pradesh. It is geographically positioned at the Himalaya's foothill zone, having a geographical area of 53,566 km^2. Its geo-coordinates lie from 30° 19′ 48″N to 78° 3′ 36″E. Uttarakhand shares its international boundaries with China (Tibet) in the north and Nepal in the east. It is also a spiritual place for Hindu religions and rich in natural resources, mostly water, glaciers, mountains, forest. It is mainly dominated by hilly, glaciers, and forest areas; only 15% area is loaded with most commercial activities (State profile).

15.2 Watershed and LULC Map

The SRTM DEM (30 m) data has been used to create the river stream and watershed map of the selected Himalayan Rivers in Uttarakhand. Further, the Land Use Land Cover (LULC) map has been generated from the LANDSAT 8 data. The source of the DEM and LANDSAT data is USGS. There are several steps to generate the watershed, first fill the raw DEM to remove small gaps from the data, followed by flow direction using fill DEM, flow accumulation, stream order. Then raster to vector conversion of stream order, followed by pour point generation, create a snap pour point to generate watershed. Landsat 8 data (30 m) is classified by supervised classification using the signature method (Taloor et al. 2020b).

The map of LULC of Uttarakhand and watershed of selected rivers are presented in Fig. 15.1. The Uttarakhand has been categorized into 8 significant land uses such as snow, water,

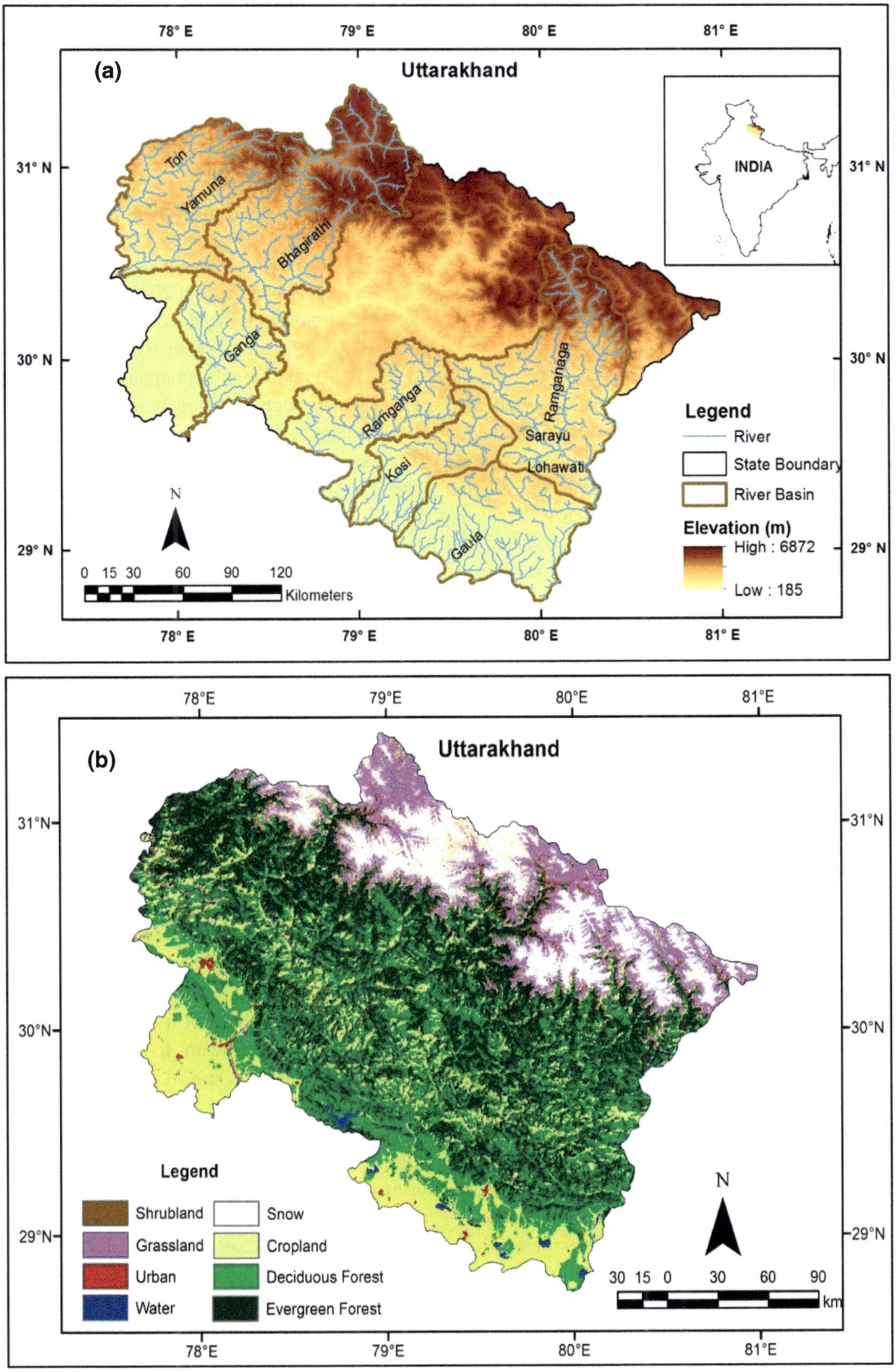

Fig. 15.1 **a** Watershed of selected rivers and **b** LULC map of Uttarakhand (*Source* SRTM DEM)

cropland, shrubland, grassland, deciduous forest, evergreen forest, urban, based on the LULC map. The evergreen and deciduous forest cover the significant part followed by snow and grassland. The cropland also covers Uttarakhand parts followed by urban and water while the least part is covered with shrubland. The watershed map has shown the catchment area and tributaries of the selected rivers of Uttarakhand.

15.3 Water Quality of Himalayan Rivers

This research work discusses the physical and chemical quality of some of the Himalayan Rivers, exclusively Ton, Bhagirathi, Ramganga, Ganga River (Rishikesh), and five crucial Kumaun Rivers of Uttarakhand. This chapter is exasperated to verify the water quality of Himalayan Rivers and its degradation due to urbanization and other developmental activities. Himalayan Rivers which are considered for the discussion are as follows.

15.3.1 Ton River

It is Yamuna's principal tributary and merges with it near Assan Barrage in Dehradun adjacent to the boundary of Uttarakhand and Himachal State. Ton is one of the chief perennial Indian Himalayan Rivers. It flows in the Garhwal region of Uttarakhand, touching Himachal Pradesh. It originates from Bandarpunch mountain at 6315 m. The water source for this River is mainly glacial melts; besides, it also receives water from its tributaries named as Bhitrall River, Kiarkuli River, Nalhota River, Noon River, and Nimi River. A massive stretch of this River passes through the western part of Dehradun, the Uttrakhand state's capital. Along its stretch in Dehradun, there are situated Indian Military Academy Campus, Forest Research Institute Campus, cantonment area, a famous picnic spot named Robber's Cave, and the famous Tapkeshwer Mahadev temple. The river water is an ultimate source for irrigation, drinking, and domestic deeds for nearby towns and villages. This river is vital for villagers and tourists (Khanna et al. 2010).

15.3.2 Bhagirathi River

It originates from Gangotri glacier (30 km long and 20 km wide) at Gomukh on the western slope of Chaukhamba group of peaks in Uttarkashi district (Bisht et al. 2020b). It is one of the largest and most significant source streams of the Ganga fluvial system (Kumar et al. 2020). It forms the River Ganga's mountainous catchment and flows through the deep gorges of the Garhwal Himalayas along with Alaknanda (originate from Satopanth and Bhagirath Kharak glacier). The Bhagirathi River joins with the Alaknanda at Devprayag and forms the River Ganga. Before its confluence with Alaknanda, it crosses around 225 km across the Himalayas. The catchment of the River lies in between 30°100′ to 30°300′ N latitudes and 78°100′ to 79°150′ E longitudes. The entire catchment area of the basin is around 7811 km^2, out of which 2328 km^2 is snowbound. The upper part of the Bhagirathi River basin lies between Gomukh (3812 m) to Harsil (2620 m) has a very steep gradient of 1192 m in a zone of 42 km (Pandey et al. 1999).

15.3.3 Ramganga River

It is one of the significant tributaries of the mighty Ganga River and initiates in the Chamoli district of Uttarakhand from Dudhotali mountain. The total channel length of it is 642 km from its starting to its union with River Ganges. The basin of Ramganga is spread in between 30°06′02.22′′N to 27°10′42.11′′N and 79°16′59.22′′E to 79°50′16′′E, having a catchment area of about 22685 km^2 with a mean elevation of 1530 m from the mean sea level (Khan et al. 2017). It traverses about 158 km in Kumaon Himalayas including the Jim Corbett National Park. Ramganga River is the only river that flows through the Jim Corbett National Park and fulfills the demand of the peoples and animals living in that region (Khan et al. 2017). Then, at Kalagarh it arrives to the Ganga flood plains, where the Ramganga Dam was built. After the Kalagarh, it flows through various parts of Uttar Pradesh through Bijnor, Moradabad, Rampur, Bareilly, Shahjahanpur, Hardoi, and Farrukhabad including industries and agricultural land (CWC 2012). The water of Ramganga River is also utilized for irrigation purposes besides fulfilling the demand for drinking water in these areas. There are also industrial setups in the towns mostly in Moradabad which disposed their effluents into the river (Khan et al. 2017).

15.3.4 Ganga River, Rishikesh

Ganga River is one of the important rivers of India, originates from Gangotri glacier which lies in between 30°43′22′′–30°50′49′′ North latitude and 79°4′4′′–79°16′34′′ East longitude at about 4,100 m above mean sea level. This glacier is of valley-type having a length of 30.20 km, and width changing from 0.5 to 2.5 km (NRCD 2009) located at the Garhwal Himalaya (Uttarkashi district). The total channel length of the Ganga River is 2525 km having a catchment area of 8,61,404 km^2. It is the fourth largest in the world and has the largest basin among river basins in India. The channel length of the Ganga River in Rishikesh includes Mini Goa Beach to Bhardwaj Sthal which is upstream and downstream of Rishikesh, respectively. Ganga Action Plan (GAP) Phase-I has been framed to clean the polluted stretch of River Ganga and Rishikesh is the first town on River Ganga which was taken up under this scheme. Under the GAP, the sewage outfall is tapped through an appropriate pumping station and diverted to a pond type Sewage Treatment Plants (STP) at Lakkarghat between Haridwar and Rishikesh. There are still lots of temples, ashrams, residences, hotels, and other commercial establishments along both banks of River Ganga. Amillion liters of sewage generated per day from this massive human activity within a narrow band along the river are discharged directly into River Ganga. There are no major industrial or tourist or other anthropogenic activities upstream to the Rishikesh. It is the first pilgrimage-cum-tourist destination after the river enters the plains (Haritash et al. 2016). Hence the water quality of River Ganga in the Rishikesh may reflect the effect of anthropogenic or developmental activities on the quality of Himalayan Rivers.

15.3.5 Kumaun Rivers

There are five important rivers of Kumaun named Gaula, Kosi, Saryu, Ramganga, and Lohawati, which have drained from the famous cities of the Uttarakhand and play a significant role in drinking and irrigation purposes (Seth et al. 2016). The water quality of these rivers has been discussed here.

Gaula River is a spring-fed river and originated from the Lesser Himalayas. Haldwani and Kathgodam town of Kumaun region have fulfilled their drinking and irrigation needs from this river water. This river is drained at Kathgodam which is also famous for illegal mining. The continuous erosion and illegal mining of the Gaula River create an uneven situation for the forest corridor and ultimately the wildlife (i.e., tigers, elephants, etc.) in the Terai region of Kumaun. The Ramganga River starts its journey from the Namik Glacier situated in Pithoragarh district. The river is east flowing through the numerous dense forest regions of Kumaun. At last it joins the Saryu River at Rameshwar near Ghat of Pithoragarh after fed by some small and big rivers in meanwhile. The Holy Saryu River originates from the Himalaya. It meets the Yamuna River in Ayodhya and has an ancient significance in Ramayana and Vedas (Seth et al. 2016). The Bageshwar district of Uttarakhand is situated on the banks of the Saryu River. Kosi River is a foremost arm of the River Ganga. It originates from spring source at Rudradhari (district Almora, Kumaun division, Uttarakhand) having a total catchment area of 3,420 km^2 and length 240 km. This river is used by locals for several purposes including domestic and agricultural purposes. Lohawati River originates nearby regal Vanasur Ka Kila at 7 km from

Table 15.1 Physico-chemical properties of the Himalayan Rivers in Uttarakhand (Mean ± standard deviation (Range))

Rivers	TDS (mg/l)	Turbidity (NTU)	pH	Conductivity (µS/cm)	BOD (mg/l)	Alkalinity (mg/l)/	Hardness (mg/l)	Calcium (mg/l)	Magnesium (mg/l)	Chloride (mg/l)	Sulfate (mg/)	References
Tons (Dehradun, Uttarakhand)	267.21 ± 59.64	–	7.94 ± 0.09	–	3.06 ± 0.56	237.49 ± 11.19	396.66 ± 11.27	126.24 ± 4.98	47.32 ± 1.48	36.80 ± 5.15	–	Khanna et al. (2011)
Bhagirathi River	83 (54–122)	–	7.8 (7.1–8.4)	104 (68–134)	–	–	–	281 (323–1001)	281 (147–488)	24 (4–49)	409 (167–871)	Pandey et al. (1999)
Ramganga River, Summer	208.24 ± 106.97	43.01 ± 37.38	7.31 ± 0.45	347.06 ± 178.28	15.01 ± 6.86	–	–	26.85 ± 8.73	9.91 ± 5.07	8.55 ± 9.51	14.39 ± 15.38	Khan et al. (2017)
Ramganga River, Monsoon	0.15 ± 0.06	617.19 ± 1028.77	7.81 ± 0.34	274.94 ± 112.14	2.77 ± 2.55	–	–	36.16 ± 14.74	8.85 ± 2.84	5.36 ± 3.04	10.3 ± 6.01	Khan et al. (2017)
Ramganga River, Winter	0.18 ± 0.08	4.77 ± 4.96	7.29 ± 0.24	353.75 ± 166.29	4.48 ± 2.41	–	–	37.54 ± 15.10	11.58 ± 6.19	10.6 ± 10.57	17.34 ± 12.63	Khan et al. (2017)
Ganga, Rishikesh	38.6 ± 17.94 (18-85)	–	9.4 ± 0.36 (9 – 10.5)	85.2 ± 35.54 (38 – 170)	9.8 ± 5.45 (4.3 – 19.5)	70.4 ± 29.55 (32 – 144)	101.2 ± 44.86 (64 – 212)	25.2 ± 8.84 (17.6 – 49.6)	6.6 ± 8.0 (0 – 37)	22.1 ± 6.51 (10 – 32.5)		Haritash et al. (2016)
Gaula, Nainital, Pre-monsoon	884 ± 124.88	5.9 ± 2.62	8.48 ± 0.14	–	–	461 ± 12.57	570 ± 86.16	119 ± 75.72	36 ± 7.81	25 ± 8.06	39 ± 2.87	Haritash et al. (2016)
Gaula, Nainital, Post-monsoon	224 ± 17.69	28.4 ± 3.93	7.68 ± 0.14	–	–	115 ± 11.12	135 ± 12.66	37 ± 3.90	12 ± 0.01	10.2 ± 3.76	16 ± 5	Seth et al. (2016)
Kosi River, Almora Pre-monsoon	469 ± 39.02	5.4 ± 1.08	8.09 ± 0.14	–	–	222 ± 14.41	342 ± 19.69	62 ± 43.83	12 ± 5.19	40.3 ± 29.12	21 ± 15.42	Seth et al. (2016)
Kosi River, Almora Post-monsoon	127 ± 19.91	20.7 ± 5.86	7.61 ± 0.22	–	–	57 ± 10.13	70 ± 23.18	19 ± 8.14	6 ± 0.73	12.3 ± 7.08	5 ± 2.60	Seth et al. (2016)
Ramganga, Pithoragarh, Pre-monsoon	555 ± 92.66	5.9 ± 1.60	8.46 ± 0.13	–	–	325 ± 52.60	372 ± 41.78	86 ± 7.89	25 ± 10.28	17.3 ± 1.92	15 ± 6.61	Seth et al. (2016)
Ramganga, Pithoragarh, Post-monsoon	344 ± 43.05	7.4 ± 1.61	8.07 ± 0.30	–	–	186 ± 34.79	206 ± 35.77	34 ± 11.05	20 ± 8.94	14 ± 1.41	8 ± 2.49	Seth et al. (2016)
Saryu River, Bageshwar, Pre-monsoon	621 ± 85.93	14.3 ± 3.10	8.42 ± 0.05	–	–	290 ± 34.24	454 ± 31.72	109 ± 14.94	28 ± 12.02	20.5 ± 4.92	61 ± 54.95	Seth et al. (2016)
Saryu River, Bageshwar, Post-monsoon	254 ± 70.07	47.3 ± 8.12	8.12 ± 0.34	–	–	138 ± 48.94	154 ± 42.81	36 ± 10.5	16 ± 4.07	14.7 ± 4.37	4 ± 2.55	Seth et al. (2016)
Lohawati River, Champawat, Pre-monsoon	427 ± 18.87	5.6 ± 0.25	7.66 ± 0.27	–	–	218 ± 7.60	354 ± 18.89	64 ± 8.26	9 ± 2.38	14.8 ± 1.92	6 ± 4.92	Seth et al. (2016)
Lohawati River, Champawat, Post-monsoon	318 ± 14.40	27 ± 6.60	7.75 ± 0.11	–	–	113 ± 6.63	154 ± 26.05	15 ± 3.17	6 ± 6.36	13 ± 1.87	7 ± 2.17	Seth et al. (2016)
BIS limit 2012, Acceptable	500	1	6.5-8.5	–	2 or less	200	200	75	30	250	200	BIS (2012)
BIS limit 2012, Permissible	2000	5	No relaxation	–	–	600	600	200	100	1000	400	BIS (2012)

(*Source* Authors collection)

Lohaghat and 20 km from Champawat district of Kumaun division. Lohaghat town which has historical and mythological importance lies on the bank of this river. This river is the lifeline for the numerous villagers of diverse parts of Uttarakhand. The water from this river is used for hydroelectric purposes of the state besides fulfilling the need for drinking and irrigation especially for seasonal vegetables (Seth et al. 2016).

15.4 Water Quality Parameters and Their Status in Himalayan Rivers in Uttarakhand

To determine the water quality of any region, various physical, chemical, and biological parameters are taken into considerations. These parameters help to identify the status and quality of water for the domestic, irrigation, and industrial uses (Adimalla et al. 2020; Adimalla and Taloor 2020a, b; Jasrotia et al. 2013). Only a few of the physico-chemical parameters to determine the water quality of rivers are taken into account due to the scarcity of data. These parameters are also presented in Table 15.1. These parameters include the following.

15.4.1 pH

It is the physical characteristics of all water/solutions. It has no health implications except that extreme values will show unnecessary acidity/alkalinity, with organoleptic consequences. The pH influences other water parameters such as ammonia toxicity, chlorine disinfection efficiency, and metal solubility. The extremely high values of pH have a corrosive effect on the water system and the palatability of water. It also disturbs the biological life of an aquatic system such as its effect on fish. Though, the most frequently encountered range is 6.5–8.0 (EPA 2001). The pH range in the Himalayan Rivers of Uttarakhand is within the BIS limit for drinking water except for Ganga, Rishikesh where it is slightly exceeding the range toward more alkaline pH.

15.4.2 Conductivity

It reflects the mineral salt content of water with indirect health significance. It is also referred to as electrical conductivity and is associated with ionic content of sample which is a function of the dissolved (ionizable) solids concentration. It is a precious pointer of the range into which hardness and alkalinity values are likely to occur. It also indicates the dissolved solids content of the water (EPA 2001). The highest and lowest conductivity is found in the Ramganga River during winter seasons and Bhagirathi River, respectively.

15.4.3 Total Dissolved Solids (TDS)

TDS is demarcated as dissolved solids in water which may comprise ionized and nonionized matter. It is occurring in water as a natural or added solutes having mostly organoleptic implications. High TDS values in water indicate that water may be saline (EPA 2001). In the Himalayan Rivers in Uttarakhand, the highest and lowest TDS values are for the Gaula River during pre-monsoon and Ramganga River during monsoon, respectively. TDS value exceeds the acceptable limit of BIS drinking water standard only in the Gaula River, Nainital (Pre-monsoon), Ramganga, Pithoragarh (pre-monsoon), and Saryu River, Bageshwar (pre-monsoon) whereas it is within the limit in other rivers.

15.4.4 Turbidity

Turbidity arises in water due to the presence of very finely divided solids such as clay particles, sewage solids, silt and sand washings, organic and biological sludges, etc. Its direct effects on health depend on the specific composition of materials causing the turbidity and also have other health consequences. The turbidity causing particles may also hinder the treatment process of water and in the case of the disinfection, results could be severe. If the turbidity is caused by sewage then, there is a great risk due to pathogenic organisms because they can be safeguarded by the turbidity causing particles and hence escape the action of the disinfectant (EPA 2001).

There is very high turbidity in all the selected rivers exceeding both acceptable and permissible limits of BIS standard for drinking water quality. Turbidity is high during monsoon or post-monsoon season whereas it is low during winter and summer season or pre-monsoon. It may be because of the high erosional rate of a Himalayan River in the mountains.

15.4.5 Alkalinity

Alkalinity measures the capacity of the water to neutralize acids hence reflecting the so-called buffer capacity which is the intrinsic resistance to pH change. Water having low or very low alkalinity has a poor buffer capacity hence is vulnerable to pH decrease, for example, "acid rain". It occurred in natural water because of the occurrence of bicarbonates formed by reactions in the soils through which the water permeates.

The Alkalinity values in rivers may be found up to 400 mg/l $CaCO_3$ and are without implication in the milieu of the water quality. Alkalinity is associated with the significant effects of eutrophication [over-enrichment] of waters. Eutrophication of water brings about a high degree of photosynthesis and higher consumption of carbon dioxide by algae. Highly alkaline water may consequence the unpalatability of water (EPA 2001). The values of alkalinity exceed the BIS limit for drinking water in all rivers during the pre-monsoon season except the Ganga River in Rishikesh. The alkalinity in rivers is below the limit during the post-monsoon season.

15.4.6 Hardness

It is a natural characteristic of water and helps to enhance the palatability and consumer acceptability for drinking use. It occurred in water due to rock formations—limestone, etc. Total hardness includes the calcium and magnesium concentrations and measured as mg/l $CaCO_3$. The concentration of barium, strontium, and iron can also contribute to hardness but, they are usually overlooked because of their low concentrations in the context of hardness. These metals have extensive abundance in rock formations which may result in a very high level of hardness in river and groundwater. Based on the hardness which is measured as concentration of $CaCO_3$, there is an arbitrary classifications of water such as Soft water up to 50 mg/l $CaCO_3$, Moderately hard for 151–250 mg/l $CaCO_3$, Moderately soft for 51–100 mg/l $CaCO_3$, Hard for 251–350 mg/l $CaCO_3$, Slightly hard for 101–150 mg/l $CaCO_3$, Excessively hard over 350 mg/l $CaCO_3$ (EPA 2001).

The hardness value exceeds the BIS acceptable standard for drinking water in Himalayan Rivers of Uttarakhand but it is below the permissible limit in all rivers during pre-monsoon season except for Ganga in Rishikesh. The post-monsoon values for hardness are below the BIS limit maybe because of dilution effects during the monsoon season.

15.4.7 Calcium (Ca^{2+})

It is very abundant in nature and occurs in rocks, bones, shells, etc. It may be beneficial at high concentrations because it aids in the palatability of water. However, a high level of Ca causes hardness and hard water is not good for laundry and industrial appliances (EPA 2001). The value for Ca is below the acceptable BIS limit except for Ton, Bhagirathi, Gaula (post-monsoon), Ramganga in Pithoragarh (Pre-monsoon), Saryu (pre-monsoon), and Lohawati (pre-monsoon) River.

15.4.8 Magnesium (Mg^{2+})

Like calcium, magnesium is also abundant and one of the major constituents of geological formations. It is the second key component of hardness comprising the 15–20% of the total hardness measured as $CaCO_3$. Its concentration is very substantial when considered in concurrence with that of sulfate (EPA 2001). The Mg values are well below the acceptable limit of the BIS drinking water standard except for Ton, Bhagirathi, and Gaula (pre-monsoon) River.

15.4.9 Chloride (Cl^-)

It occurs in all-natural water. The concentrations of chloride vary very extensively and attained a maximum level in seawater (up to 35,000 mg/l Cl). Soil and rock formations, sea spray, and waste discharges are the major sources for chloride concentration in fresh water. Sewage and some industrial effluents encompass huge amounts of chloride. It does not pose a health hazard to humans and the primary consideration is with palatability.

The water will start to taste salty at a concentration above 250 mg/l Cl and become progressively obnoxious as the concentration upsurges further. A higher concentration of chloride in freshwater makes water unsuitable for agricultural irrigation. High chloride concentration is an indicator of sewage contamination because sewage is such a rich source of chloride. Natural levels in rivers and other fresh waters typically vary between 15 and 35 mg/l Cl which is much below drinking water standards (EPA 2001). The chloride concentration in all discussed rivers is very low than the BIS acceptable limit for drinking water standard indicating fresh water and negligible contamination from sewage effluent.

15.4.10 Sulfate (SO_4^{2-})

It occurs in nearly all-natural waters. The concentrations of sulfate in rivers vary according to the nature of the terrain through which they flow. They are often consequent from the sulfides of heavy metals such as iron, nickel, copper, and lead. It occurred in water due to rocks, geological formations, discharges, and so on. The high concentration of sulfate has a health implication such as a laxative effect, especially in combination with magnesium and/or sodium. The Mg and sodium in water combine with sulfate and their combination enhanced the laxative effect to greater or lesser magnitude depending on their concentration. High sulfate concentration harshly limits the use of water for domestic purposes. Additional complications because of high sulfate include its readily reduction to sulfide causing noxious odors in polluted water in which dissolved oxygen is zero. Waters containing sulfates in surplus will also attack the fabric of concrete sewer pipes (EPA 2001). The values for sulfate are well below the BIS acceptable limit in all the rivers except the Bhagirathi River.

15.4.11 Biochemical Oxygen Demand (BOD)

It is an important indicator of overall water quality and it determines the amount of oxygen which will be required by microorganism for the breakdown of organic matter. In the presence of a huge quantity of organic matter, the dissolved oxygen in the water will be consumed at a faster rate than its replenishment from the atmosphere and photosynthetic activity, consequently the water body will become anaerobic. In case of low or negligible dissolved oxygen, bacterial degradation of the waste will occur and offensive products, for example, hydrogen sulfide will be generated. Anaerobic conditions may consequence the mortality of fish and other aquatic organisms (EPA 2001). The BOD of Himalayan Rivers is quite high than the acceptable BIS limit. It may be because of high organic material in river water as mountainous soil is rich in organic matter.

15.5 Conclusion

There is a data scarcity on the water quality of the Himalayan River. Water quality parameters discussed here indicate that the water quality of the Himalayan River is still good but developmental and other activities taking place in the vicinity of rivers are posing threat to it. For example, the Ganga River stretch in Rishikesh is good at the upper segment but the mid and lower parts had high concentration of contaminants because of tourism and pilgrims' activities. Likewise, Rivers which are close to major towns and tourist place such as Ton River, Gaula River has shown deteriorating water quality. The suitability of the water of Kumaun Rivers (Gaula, Saryu, Kosi, Lohawati, and Ramganga) for drinking is assessed by water quality parameters. These parameters indicate that rivers water in both the seasons (pre-monsoon and post-monsoon)needs pretreatment. However, the water quality of Kumaun Rivers is suitable for irrigation. The key factors accounting for the worsening of water quality of Himalayan Rivers might be eutrophication, tourism, anthropogenic, and geogenic processes. Consequently, to reestablish the liveliness and water quality of these rivers, appropriate water resource development program needs to be established.

References

Adimalla N, Dhakate R, Kasarla A, Taloor AK (2020) Appraisal of groundwater quality for drinking and irrigation purposes in Central Telangana, India. Groundw Sust Dev 10:P100334. https://doi.org/10.1016/j.gsd.2020.100334

Adimalla N, Taloor AK (2020a) Hydrogeochemical investigation of groundwater quality in the hard rock terrain of South India using Geographic Information System (GIS) and groundwater quality index (GWQI) techniques. Groundw Sust Dev 10:P100288. https://doi.org/10.1016/j.gsd.2019.100288

Adimalla N, Taloor AK (2020b) Introductory editorial for 'Applied Water Science'special issue: "Groundwater contamination and risk assessment with an application of GIS". Appl Water Sci 10:216. https://doi.org/10.1007/s13201-020-01291-3

Bisht H, Arya PC, Kumar K (2018) Hydro-chemical analysis and ionic flux of meltwater runoff from Khangri Glacier, West Kameng, Arunachal Himalaya, India. Environ Earth Sci 77:1–16. https://doi.org/10.1007/s12665-018-7779-6

Bisht H, Kotlia BS, Kumar K, Arya PC, Sah SK, Kukreti M et al (2020a) Estimation of suspended sediment concentration and meltwater discharge draining from the Chaturangi glacier, Garhwal Himalaya. Arab J Geosci 13(6):1–12. https://doi.org/10.1007/s12517-020-5204-w4

Bisht H, Kotlia BS, Kumar K, Joshi LM, Sah SK, Kukreti M (2020b) Estimation of the recession rate of Gangotri glacier, GarhwalHimlaya (India) through kinematic GPS survey and satellite data. Environ Earth Sci 13(6):1–12. https://doi.org/10.1007/s12665-020-090780

Bora M, Goswami DC (2017) Water quality assessment in terms of water quality index (WQI): case study of the Kolong River, Assam, India. Appl Water Sci 7 (6):3125–3135

Bureau of Indian Standards (BIS) (2012) Indian standard drinking water specification (second revision) BIS 10500:2012, New Delhi

CWC (2012) Environmental evaluation study of Ramganga major irrigation project. Central Water Commission 1:16

Environmental Protection Agency (EPA) (2001) Parameters of water quality interpretation and standards. Published by the Environmental Protection Agency, Ireland. ISBN 1-84096-015-3

Haque S, Kannaujiya S, Taloor AK, Keshri D, Bhunia RK, Ray PKC, Chauhan P (2020) Identification of groundwater resource zone in the active tectonic region of Himalaya through earth observatory techniques. Groundw Sust Dev 10:P100337. https://doi.org/10.1016/j.gsd.2020.100337

Haritash AK, Gaur S, Garg S (2016) Assessment of water quality and suitability analysis of River Ganga in Rishikesh, India. Appl Water Sci 6(4):383–392

Jarvie HP, Whitton BA, Neal C (1998) Nitrogen and phosphorus in east coast British rivers: speciation, sources and biological significance. Sci Total Environ 210:79–109

Jasrotia AS, Bhagat BD, Kumar A, Kumar R (2013) Remote sensing and GIS approach for delineation of groundwater potential and groundwater quality zones of Western Doon Valley, Uttarakhand, India. J Indian Soc Remote Sens 41(2):365–777

Jasrotia AS, Taloor AK, Andotra U, Bhagat BD (2018) Geoinformatics based groundwater quality assessment for domestic and irrigation uses of the Western Doon valley, Uttarakhand, India. Groundw Sust Dev 6:200–212

Jasrotia AS, Taloor AK, Andotra U, Kumar R (2019) Monitoring and assessment of groundwater quality and its suitability for domestic and agricultural use in the Cenozoic rocks of Jammu Himalaya, India: a geospatial technology based approach. Groundw Sust Dev 8:554–566

Khan MYA, Gani KM, Chakrapani GJ (2017) Spatial and temporal variations of physicochemical and heavy metal pollution in Ramganga River—a tributary of River Ganges, India. Environ Earth Sci 76(5):231

Khan A, Govil H, Taloor AK, Kumar G (2020) Identification of artificial groundwater recharge sites in parts of Yamuna river basin India based on remote sensing and geographical information system. Groundw Sust Dev 11: P100415. https://doi.org/10.1016/j.gsd.2020.100415

Khanna DR, Bhutiani R, Matta G, Singh V, Tyagi P, Tyagi B, Ishaq F (2010) Water quality characteristics of River Tons at District-Dehradun, Uttarakhand (India). Environ Conser J 11(1–2):119–123

Khanna DR, Singh S, Saxena N, Bhutiani R, Matta G, Kulkarni DB (2011) Physico-chemical and microbiological characterization of the ground water across the city Bareilly (UP), India. J Appl Nat Sci 3(2):315–318

Kumar D, Singh AK, Taloor AK, Singh DS (2020) Recessional pattern of Thelu and Swetvarn glaciers between 1968 and 2019, Bhagirathi basin, Garhwal Himalaya, India. Quat Int. https://doi.org/10.1016/jquaint.2020.05.017

National River Conservation Directorate (NRCD) (2009) Status paper on River Ganga: state of environment and water quality. Ministry of Environment and Forests, New Delhi

Pandey SK, Singh AK, Hasnain SI (1999) Weathering and geochemical processes controlling solute acquisition in Ganga headwater–Bhagirathi river, Garhwal Himalaya, India. Aqua Geochem 5(4):357–379

Seth R, Mohan M, Singh P, Singh R, Dobhal R, Singh KP, Gupta S (2016) Water quality evaluation of Himalayan rivers of Kumaun region, Uttarakhand, India. Appl Water Sci 6(2):137–147

Sarkar T, Kannaujiya S, Taloor AK, Ray PKC, Chauhan P (2020) Integrated study of GRACE data derived interannual groundwater storage variability over water stressed Indian regions. Groundw Sustain Dev 10: P100376. https://doi.org/10.1016/j.gsd.2020.100376

Singh AK, Jasrotia AS, Taloor AK, Kotlia BS, Kumar V, Roy S, Ray PKC, Singh KK, Singh AK, Sharma AK (2017) Estimation of quantitative measures of total water storage variation from GRACE and GLDAS-NOAH satellites using geospatial technology. Quat Int 444:191–200

State profile: Uttarakhand (2019). https://uk.gov.in/pages/view/93-state-profile Accessed 1 Nov 2019

Taloor AK, Kumar V, Singh VK, Singh AK, Kale RV, Sharma R, Khajuria V, Raina G, Kouser B, Chowdhary NH (2020a). Land use land cover dynamics using remote sensing and GIS techniques in Western Doon Valley, Uttarakhand, India. In: Geoecology of landscape dynamics. Springer, Singapore, pp 37–51. https://doi.org/10.1007/978-981-15-2097-6_4

Taloor AK, Pir RA, Adimalla N, Ali S, Manhas DS, Roy S, Singh AK (2020b) Spring water quality and discharge assessment in the Basantar watershed of Jammu Himalaya using geographic information system (GIS) and water quality Index(WQI). Groundw Sust Dev 10:P100364. https://doi.org/10.1016/j.gsd.2020.100364

Ms. Deeksha Aithani, Ph.D. research scholar at School of Environmental Sciences, Jawaharlal Nehru University, India. She did her Bachelor in Botany (Hon.) from Gargi College, Delhi University and Master's in Environment Management from Guru Gobind Singh Indraprastha University, Delhi. Currently, she is working on her Ph.D. research work related to trace metal pollution in Yamuna River floodplain and assessment of associated human health risk. She has authored publications in international peer-reviewed journal and conference. Her research interest includes environmental pollution, health risk assessment, ecology and solid waste management

Ms. Jyoti Kushawaha is a Ph.D. research scholar at School of Environmental Sciences, Jawaharlal Nehru University, India. She did her Bachelors in Geology (Hons.) from Banaras Hindu University, Varanasi. She has completed her Master's and M.Phil. in Environmental Sciences from Jawaharlal Nehru University, New Delhi. Her research area includes hydrogeology, geochemistry, groundwater modelling, remote sensing and GIS applications for natural resource management. She has authored publication in international peer-reviewed journal and conferences

Mr. S. R. Sreerama Naik, M.Sc., M. Tech., is a Ph.D. Research Scholar at Madurai Kamaraj University, Madurai, Tamil Nadu and is currently working as a Senior Research Fellow at Jawaharlal Nehru University, New Delhi. He is involved strongly in one of the Groundwater Analyst and Earth observation activities (satellite image analysis) using GIS and remote sensing application. He has completed M.Sc. Geography from Bangalore University and M. Tech. Geo-Informatics (remote sensing and GIS) from Bharathidasan University, Tiruchirappalli, Tamil Nadu, India. He is pursuing his Ph.D. on groundwater quality and coastal vulnerability. He has been working as a remote sensing and GIS professional for 5 years. He has authored publication in international peer-reviewed journal and conferences

Sources of Solute and Hydrochemical Analysis of Gangotri Glacier Meltwater

16

Harish Bisht, Bahadur Singh Kotlia, Kireet Kumar, Ajay Kumar Taloor, Pooja Chand, Jeewan Singh Bisht, Yogesh Maithani, Manmohan Kukreti, and Mohit Tewari

Abstract

In this study, we have examined the ionic and physical properties of Gangotri glacier meltwater as well as the prominent weathering process to determine the sources of solute. The meltwater samples were collected throughout the ablation periods of 2015 and 2016 near the snout of the glacier. The results, obtained by chemical analysis of the meltwater indicate that it is somewhat acidic in nature with $CaSO_4$-type water. In the meltwater, Ca^{2+} is the foremost cation followed by Mg^{2+}, Na^{+}, and K^{+} as well as SO_4^{2-} is the leading anion followed by HCO_3^{-}, Cl^{-}, and F^{-} during both the ablation periods. Based on the calculated denudation rate of the ions, we conclude that denudation rates of cation were 20.24 and 18.66 ton/km^2/ablation in 2015 and 2016 correspondingly, while the anion denudation rates were 89.01 and 92.13 ton/km^2/ablation during years 2015 and 2016 correspondingly.

Keyword

Gangotri glacier · Chemical · Weathering · Hydrogeochemistry · Ionic flux · Meltwater

16.1 Introduction

Himalaya contains the highest amount of snow and glacier outside the polar caps (Bolch et al. 2012; Sood et al. 2020). The Himalaya encompasses around 9575 glaciers which cover an area of about 40,000 km^2 (Raina and Srivastava 2008). Himalayan snowcaps and glaciers are

H. Bisht (✉) · B. S. Kotlia · P. Chand · J. S. Bisht · M. Kukreti
Centre of Advanced Study in Geology, Kumaun University, Nainital 263002, Uttarakhand, India
e-mail: harishbisht890@gamil.com

B. S. Kotlia
e-mail: bahadur.kotlia@gamil.com

P. Chand
e-mail: poojachand10june@gmail.com

J. S. Bisht
e-mail: jsinghbisht66@gamil.com

M. Kukreti
e-mail: manmohankukreti12@gmail.com

K. Kumar · M. Tewari
G. B. Pant National Institute of Himalayan Environment and Sustainable Development Kosi-Katarmal, Almora 263643, Uttarakhand, India
e-mail: kireet@gbpihed.nic.in

M. Tewari
e-mail: tewari.mohit258@gmail.com

A. K. Taloor
Department of Remote Sensing and GIS, University of Jammu, Jammu 180006, India
e-mail: ajaytaloor@gmail.com

Y. Maithani
Department of Geology, L. M. S. PG College Rishikesh, Dehradun 249201, Uttarakhand, India
e-mail: yogeshmaithani@yahoo.com

A. K. Taloor et al. (eds.), *Water, Cryosphere, and Climate Change in the Himalayas*, Geography of the Physical Environment, https://doi.org/10.1007/978-3-030-67932-3_16

known for frozen resources of water (Gupta et al. 2005; Singh et al. 2020). The frozen ice caps and the cover of snow in the Himalayas are constantly melting and playing a crucial role in the refilling of most of the streams and rivers. Numerous rivers of Asia, e.g., Ganges, Brahmaputra, Indus, Yellow River, Yangtze, Salween, and Mekong are fed through meltwater coming out of the Himalayan glaciers (Barandum et al. 2018). The glacier meltwater is also an important reserve, which continuously exploited for irrigation, drinking purpose, hydroelectricity (Bogen 1989; Bisht et al. 2018). Valley glaciers of Himalaya are known as the most vital features as they appreciably supply the continental solute budget (Taloor et al. 2020a).

The meltwater runoff from glaciers is the major freshwater source for downstream peoples. Therefore, hydrochemical study of glacier meltwater is significant because of the increasing requirement of freshwater in downstream sectors (Bisht et al. 2018). The hydrochemical characterization of glacier meltwater varies from one to other glaciers. Variability in the ionic concentration of meltwater is generally affected by the movement of water from different lithologies (Collins 1979). Different ions in meltwater are acquired by the rock weathering of diverse composition with a small contribution from the atmosphere (Kumar et al. 2009; Tiwari et al. 2018). Rock weathering is controlled by a variety of agents, e.g., temperature, rainfall, lithology, runoff, relief, and type of soil cover (Khan et al. 2020; Haque et al. 2020). In recent times, weathering processes has been increased because of higher glacier melting and precipitation possibly due to global warming (Kumar et al. 2019).

The weathering processes are stronger in glacial regions compared to that in the tropics (Souchez and Lemmens 1987). The higher amount of weathering in such areas is caused by the long time period of glacier meltwater to the lithology (Singh et al. 2012). The glaciated regions are the perfect locations to recognize the interaction between water and rock as the man-made effect is often very less (Brown 2002; Jasrotia and Kumar 2014; Sah et al. 2017; Taloor et al. 2020c). According to the authors (Clow and Mast 2010), the chemical weathering of rock-forming mineral gives ions and generally manage the hydrochemistry of meltwater stream although a significant variability in hydrochemical characteristics of streams emerging from the same lithological setup (Oliva et al. 2003), it is perhaps because of input from the atmosphere. The dissolved ionic and elemental composition carried through streams/rivers were regulated through processes of chemical weathering, while particulate transportation is assumed by physical weathering processes of rocks (Krishnaswami and Singh 2005). Some researches (Ahmad and Hasnain 2001; Kumar et al. 2009; Singh et al. 2014a) have been carried out in Gangotri glacier to assess the hydrochemical characteristics of glacier meltwater. However, none of these discussed in detail because of the limited dataset and be short of seasonal meltwater sampling. Therefore, for the fulfillment of this gap, we examined the chemistry of major ions, ionic flux, denudation rate of the ions, processes of chemical weathering together with solute sources in the meltwater originating from the Gangotri glacier based on the entire ablation (June to September) data of years 2015 and 2016.

16.2 Study Area

The Gangotri glacier (30° 43′ 10″–30° 55′ 50″ N and 79° 4′ 55″–79° 17′ 18″ E) is located in Garhwal Himalaya (Fig. 16.1). It is a 30.2 km long northwest-flowing valley-type glacier (Kaul 1999; Bisht et al. 2020b), occupying a catchment area of approximately 772.7 km^2 (Orr et al. 2019). The snout, called Gamukh, lies at 4000 m above mean sea level from where the Bhagirathi River originates. The shape of the snout keeps varying due to the splitting of ice blocks from the terminus and subsidence in the glacier portal (Bisht et al. 2020b). The regional climate of this region is exaggerated by Indian Winter Monsoon (IWM) and Indian Summer Monsoon (ISM) (Dimri et al. 2016; Kumar et al. 2018) whereas the micro-climate is influenced by the valley aspect as well as altitude (Naithani et al. 2001). Presently, the Gangotri glacier contains four active (connected with trunk) tributaries (Swachhand glacier, Ghanohlm glacier, Kirti glacier, and Maiandi glacier) and two inactive tributary glaciers (Raktavarna glacier and Chaturangi glacier) (Bisht et al. 2020b). These entire active and inactive tributary glaciers unite together and are called the Gangotri glacier system. From all the tributary glaciers, Chaturangi is the longest one, which is situated on the right flank of the valley (Bisht et al. 2019, 2020a). Various evidences of neotectonic activity were observed in the Gangotri glacier region, which is mostly responsible for the formation of present-day landform (Taloor et al. 2020a; Bisht et al. 2020b, c).

16.2.1 Regional Geology of Study Site

The study site is situated over the Main Central Thrust (MCT) (Fig. 16.2) and comprises quartzite, quartz-biotite schist, biotite, granite, and leucogranite (Metcalfe 1993). The upper margin of MCT corresponds to the Vaikrita thrust (MCT II) and the lower boundary to the Munsiari thrust (MCT I) (Valdiya 1980; Arita 1983; Jasrotia et al. 2018; Taloor et al. 2017, 2020b). It separates the Higher Himalayan Munsiari group of rocks from the Lesser Himalayan rocks of the Berinag Formation. The Vaikrita thrust separates the Pandukeshwar Formation from the rocks of the Pindari Formation. A smaller part of the Pindari thrust also separates the Munsiari Formation rocks from Joshimath Formation rocks (see Fig. 16.2). A variety of faults, thrusts, and joints are common features in the Himalayan terrain due to which the nearby rocks are fractured, sheared, and crushed and cause highly unstable hill slopes (Bisht et al. 2020c). In the Gangotri glacier area, rocks are also extremely crushed and sheared all along the strike of the faults. Toward northeast from Gangotri, the mica-schist intruded by massive granite called Gangotri granite (~21 Ma) (Jain et al. 2002). Further southeast, the area comprises tourmaline granite invasive into the composite rock of migmatite schist. Two distinct granite plutons occur in higher reaches of the River Bhagirathi and nearby at snout. Structurally underlying pluton is biotite granite and overlying pluton is aluminous S-type tourmaline granite (Stern et al. 1989).

16.2.2 Sample Collection

To understand the hydro-geochemistry of Gangotri glacier meltwater the samples were collected ~ 1 km downward from the current position of snout (Fig. 16.1). A total of 244 meltwater samples (122 in each year in 2015 and 2016) were collected daily in sample bottles during the ablation period June–September. The Ostrem (1975) method was used for the collection of meltwater samples. For comprehensive hydrogeochemical analysis, the collected sample bottles were carried to the Water Processing and Management (WPM) lab of G.B. Pant National Institute Kosi-Katarmal Almora, Uttarakhand. Along with these meltwater samples some rock samples were also collected near the snout to identify the contribution of ions from rocks into the glacier meltwater.

16.2.3 Onsite Measurements

Onsite measurement of pH, electrical conductivity (EC), and dissolved solids (TDS) of glacier meltwater were carried by APHA (2005) method. pH and EC were analyzed by portable pH and EC meter (New Professional Trimeter), while TDS of the meltwater was measured by TDS meter (PCS tester 35).

16.2.4 Laboratory Analysis

16.2.4.1 Hydrochemical Analysis

The hydrochemical analysis of Gangotri glacier meltwater samples was carried out using different standard methods and instruments. The potassium (K^+) and sodium (Na^+) ions concentration were resolved by the Flame Photometer (Systronics flame photometer 128). The calcium (Ca^{2+}) and magnesium (Mg^{2+}) ions concentration were carried out by EDTA titration, while the

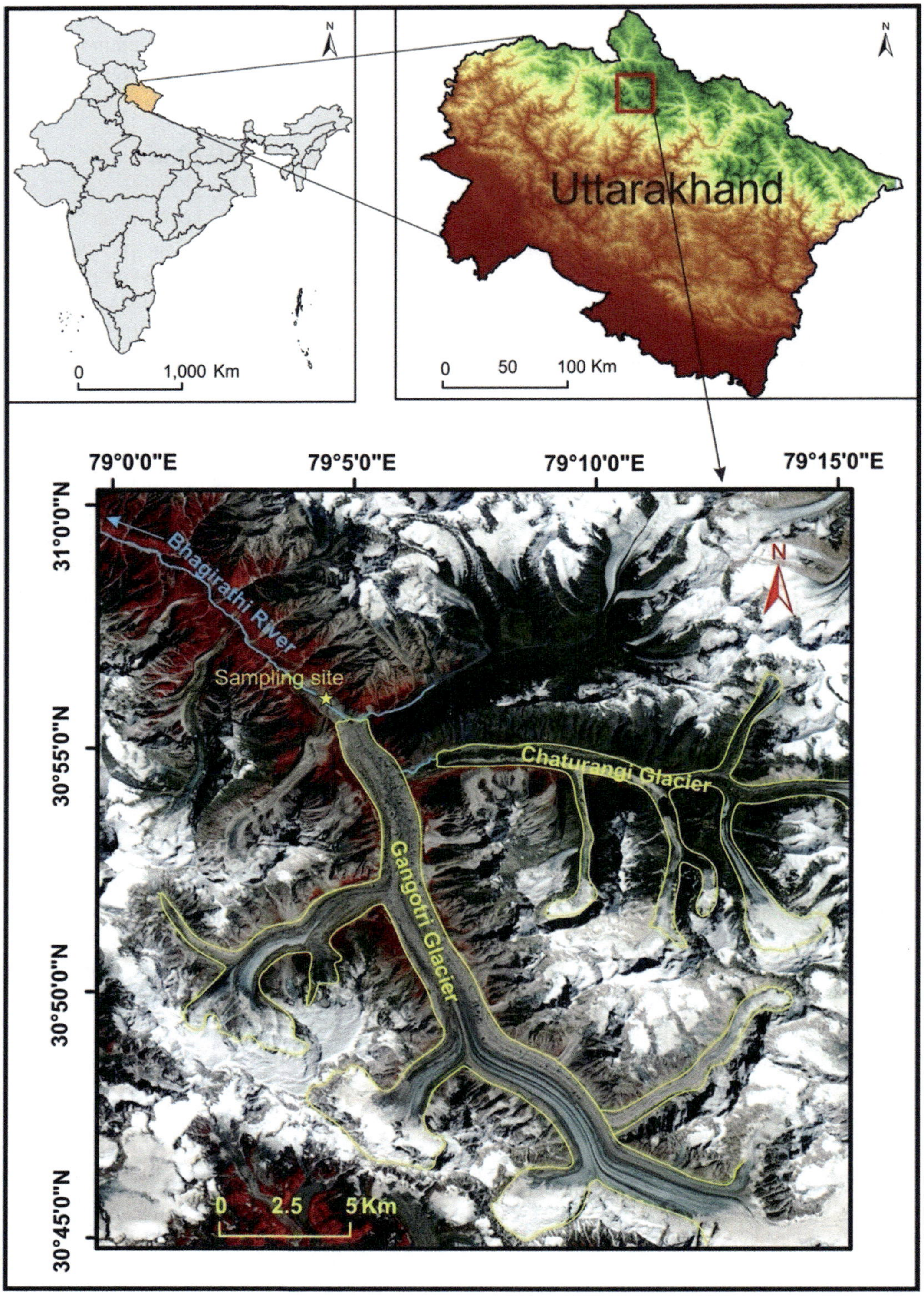

Fig. 16.1 Study area: the location map (*Source* Authors)

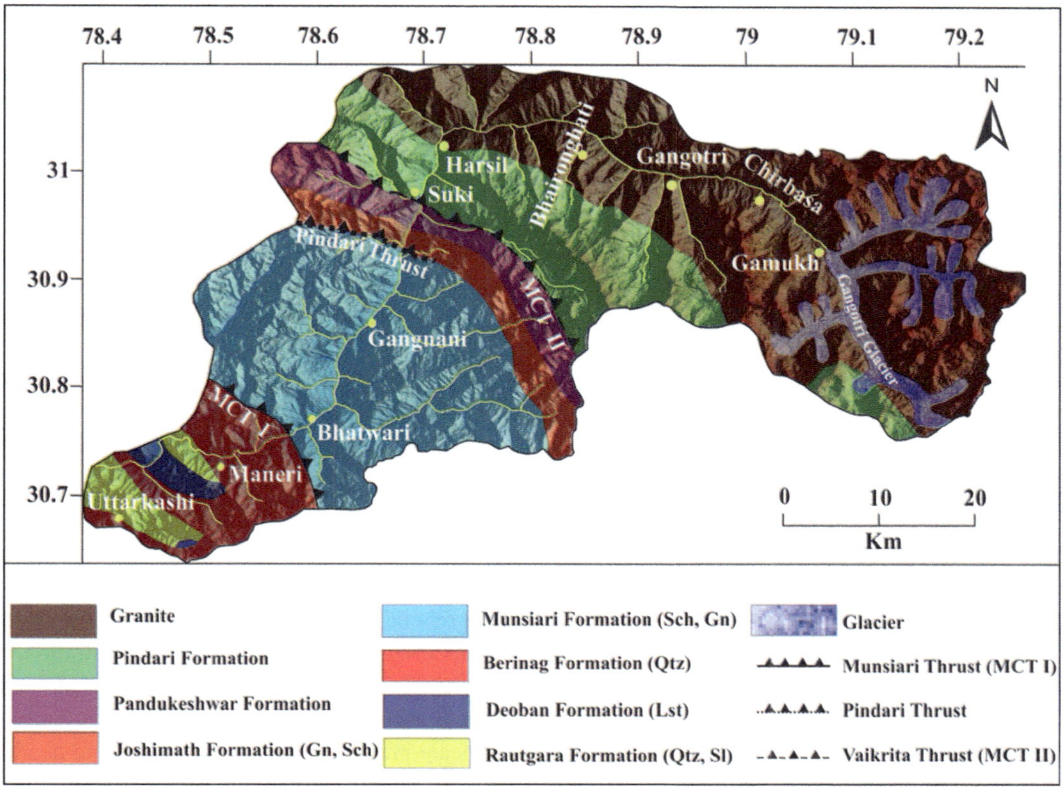

Fig. 16.2 A simplified geology of the Gangotri glacier system (*Source* Valdiya 1980)

bicarbonate (HCO_3^-) and chloride (Cl^-) ions were calculated through acid titration technique. Sulfate (SO_4^{2-}) and fluoride (F^-) ions concentration were analyzed using a photometer (Paqualb photometer 5000) and spectrophotometer (Eppendorf AG 22331), respectively. Rock Ware, software version 1.5 was applied for making Piper plot to identify the water type or hydrochemical facies, and SPSS, software version 10.5 was employed for statistical analysis.

16.2.5 Petrographic Analysis

Thin sections of the rock samples present near the glacier snout were prepared for petrographic analysis. The petrographic study of these thin sections was then undertaken in the petrographic microscope in transmitted light. Furthermore, the petrographic work was conducted and the mineral assemblages were determined by the method given by Gazzi (1966) and Dickinson (1970).

16.3 Results and Discussion

16.3.1 Hydrochemistry of Glacier Meltwater

Hydrochemical examination of Gangotri glacier meltwater was determined to identify the role of weathering processes in solutes and major ion chemistry. The analyzed physicochemical parameters of glacier meltwater with maximum, minimum, average, and standard deviation (STDEV) are shown in Table 16.1. The measured pH values of the glacier meltwater range between 4.97 and 6.08 with an average value of 5.64 ± 0.24 in the year 2015 and 5.28 and 6.57 with an average value of 6.07 ± 0.37 in the year

2016 (Table 16.1). The results obtained from both the years show that the nature of meltwater is somewhat acidic. EC ranges between 50.32 and 79.05 µS/cm with an average of 66.66 ± 9.43 µS/cm for the year 2015 and 45.62 and 76.21 µS/cm with an average of 62.75 ± 9.82 µS/cm for the year 2016 (Table 16.1). TDS varied from 25.45 to 40.2 mg/L with a mean of 33.83 ± 4.84 mg/L in 2015 and 23.05–38.73 mg/L with a mean of 31.82 ± 5.04 mg/L in 2016. Therefore, the variability in EC and TDS indicates that the hydrochemistry of catchment is mainly controlled due to rock water interaction followed by meteorological parameters (Bisht et al. 2017; Jasrotia et al. 2019; Adimalla et al. 2020). Result also indicates that the average EC and TDS values are higher in the year 2015 than in 2016. This might be due to lower values of discharge volume during the entire ablation period in 2015 (3.36×10^{11} L) than in 2016 (3.83×10^{11} L) (Bisht et al. 2017, Tables 16.2 and 16.3). During the low release, the glacier meltwater has more concentration of dissolved ion due to high residence time and extended contact of meltwater with local rock types (Singh et al. 2014a).

Average cations concentration Ca^{2+}, Mg^{2+}, $Na^{+,}$ and K^{+} was found as 23.21 ± 11.88, 11.41 ± 2.50, 6.81 ± 3.12, and 5.03 ± 1.51 mg/L, respectively, in the year 2015 and 18.62 ± 11.01, 10.05 ± 2.99, 4.74 ± 2.74, and 4.25 ± 1.96 mg/L, respectively, in the year 2016. Furthermore, the average concentration of anions SO_4^{2-}, HCO_3^{-}, $Cl^{-,}$ and F^{-} was 115.23 ± 15.99, 86.56 ± 18.90, 2.19 ± 1.14, and 0.41 ± 0.10 mg/L, respectively, in the year 2015 and 107.51 ± 15.74, 76.34 ± 17.82, 1.78 ± 0.94, and 0.31 ± 0.10 mg/L, respectively, in the year 2016. The results obtained from the analysis of these ionic concentrations indicate Ca^{2+} is a prominent cation and subsequently followed by Mg^{2+}, $Na^{+,}$ and K^{+} which contributing 49.96%, 24.56%, 14.66%, and 10.83% of the total cations, respectively, in the year 2015 and 49.44%, 26.69%, 12.58%, and 11.29% of the total cations, respectively, in year 2016. Among the anions, SO_4^{2-} is the foremost anion followed by HCO_3^{-}, $Cl^{-,}$ and F^{-} which contributing 56.38%, 42.35%, 1.07%, and 0.20% of the total anions, respectively, in 2015 and 57.82%, 41.05%, 0.96%, and 0.17% of the total anions, respectively, in 2016.

16.3.2 Sources of Solutes and Hydrochemical Characteristics of Meltwater

Hydrochemical characteristics of glacier meltwater play a significant role to determine the sources of solutes (Sharma et al. 2019). Rock weathering is a vital source behind the fluctuation of the ionic composition of natural water (White 2002). Therefore, the mineralogical attribute of catchment rocks is important to comprehend the interaction of rock and water and its involvement in the ionic concentration of meltwater (Bisht et al. 2018). The geological investigation shows that mainly the granitic rocks are present in the upper reaches and nearby area of glacier snout (Fig. 16.2). Mainly two types of granitic rocks are their one is biotite granite and another is tourmaline granite. The petrographic analysis shows that the biotite [$K(Mg,Fe)_3AlSi_3O_{10}(OH)_2$], alkali feldspar [$KAlSi_3O_8$], plagioclase feldspar [$NaAlSi_3O_8$ or $CaAl_2Si_2O_8$], and quartz [SiO2] are the most abundant minerals in the biotite granite (Fig. 16.3a), while the tourmaline [$Na(Mg,Fe)_3Al_6(BO_3)_3Si_6O_{18}(OH)_4$], alkali feldspar [$KAlSi_3O_8$], plagioclase feldspar [$NaAlSi_3O_8$ or $CaAl_2Si_2O_8$], and quartz [SiO_2] are the most abundant mineral in tourmaline granite (Fig. 16.3b).

The results suggest that K^{+} has definitely entered into the Gangotri glacier meltwater is due to chemical weathering of biotite and alkali feldspar. The Na^{+} is entered into the meltwater due to alkali feldspar, plagioclase feldspar, and tourmaline, while Ca^{2+} has contributed to the meltwater due to plagioclase feldspar. The Mg^{2+} is added to the meltwater is due to the weathering of biotite and tourmaline. In addition, the replacement of ions of crystal lattice of micas and other minerals probably is the source of F^{-} in

Table 16.1 Physicochemical parameters and elemental ratios of Gangotri glacier meltwater for ablation the period 2015 and 2016

Parameters	2015 (n = 122)				2016 (n = 122)			
	Maximum	Minimum	Average	STDEV	Maximum	Minimum	Average	STDEV
EC	79.05	50.32	66.66	9.43	76.21	45.62	62.75	9.82
pH	6.08	4.97	5.64	0.24	6.57	5.28	6.07	0.37
TDS	40.2	25.45	33.83	4.84	38.73	23.05	31.82	5.04
Ca^{2+}	42	2.4	23.21	11.88	34.81	1.96	18.62	11.01
Mg^{2+}	13.56	3.72	11.41	2.50	13.14	2.85	10.05	2.99
Na^{+}	10.58	0.69	6.81	3.12	9.06	0.48	4.74	2.74
K^{+}	6.24	0.39	5.03	1.51	6.94	0.4	4.25	1.96
F^{-}	0.57	0.15	0.41	0.10	0.49	0.13	0.31	0.10
Cl^{-}	4.15	0.35	2.19	1.14	3.39	0.43	1.78	0.94
SO_4^{2-}	144.37	94.08	115.23	15.99	133.54	86.17	107.51	15.74
HCO_3^{-}	124.44	64.05	86.56	18.90	112.31	57.25	76.34	17.82
(Ca+Mg)/(Na+K)	6.45	1.67	2.96	0.49	6.85	2.66	3.39	0.66
(Na+K)/TZ^{+}	0.37	0.13	0.26	0.03	0.27	0.13	0.23	0.03
(Ca+Mg)/TZ^{+}	0.86	0.62	0.74	0.03	0.87	0.73	0.77	0.03
K/Cl	6.01	0.75	2.69	1.15	3.97	0.78	2.46	0.57
Na/Cl	6.46	1.60	3.24	0.84	3.74	1.00	2.54	0.44
C-Ratio	0.46	0.39	0.42	0.02	0.46	0.37	0.41	0.02
S-Ratio	0.60	0.53	0.57	0.02	0.63	0.54	0.59	0.02
Ca/Na	5.06	1.55	3.31	0.65	10.16	1.66	3.91	1.14
Mg/Na	8.43	1.28	2.14	1.23	7.25	1.43	2.81	1.39
HCO_3/Na	94.58	8.36	17.55	13.98	120.42	11.11	24.63	19.97

Unit Analytes are in mg/L excluding pH and EC, EC unit is (μS/cm) (*Source* Authors)

water (Sadat 2012). Factors controlling the hydrochemistry of glacier meltwater and major sources of dissolved ions are also explained by the relationship between the ions and their ratios (Sharma et al. 2013). To understand the weathering process and sources of solutes, the relationship between the ions were examined. The scatter graph of TZ^{+} and Na+K (Fig. 16.4) indicate most points present above the 1:1 equiline and low elemental ratios were 0.26 ± 0.03 in 2015 and 0.23 ± 0.03 in 2016 (Table 16.1). Furthermore, the scatter diagram of TZ^{+} and (Ca+Mg) (Fig. 16.5) indicate most points present over the 1:1 equiline with average ratios (0.74 ± 0.03) in 2015 and (0.77 ± 0.03) in 2016 (Table 16.1). The (Ca+Mg)/(Na+K) ratios were 2.96 ± 0.49 in 2015 and 3.39 ± 0.66 in 2016. The low Na+K/TZ ratios and high ratios of Ca+Mg/TZ and (Ca+Mg)/(Na+K) in both the years suggest that weathering of carbonate is the foremost process affecting meltwater hydrochemistry of the Gangotri glacier.

The scatter graph of (Ca+Mg) and SO_4^{2-} (Fig. 16.6) indicates most of the points lie over 1:1 equiline, revealed that the $MgSO_4$ and $CaSO_4$ rich rocks are the main sulfate sources in the meltwater of Gangotri. Singh et al. (2014b) suggested that oxidation of sulfide minerals could be another probable cause of sulfate in the glacier meltwater. The chemical weathering of surrounding rocks is the main source of solutes in meltwater with a small contribution from atmosphere (Singh et al. 2014b). The higher amount of chloride concentration has been observed in

Table 16.2 Estimation of daily and entire ablation periods ionic flux and ion denudation rate through glacier meltwater in Gangotri catchment in the year 2015

	Analytes (ton)								
	Ca^{2+}	Mg^{2+}	Na^{+}	K^{+}	F^{-}	Cl^{-}	SO_4^{-2}	HCO_3^{-}	Total
IIF (ton day^{-1})	64.03	31.47	18.79	13.88	1.13	6.04	317.82	238.76	**691.92**
IIF (ablation period)	7811.35	3839.66	2292.24	1693.22	137.99	736.74	38774.11	29128.49	**84413.80**
Fx^{+} (ton day^{-1})	128.17								
Fx^{+} (ablation period)	15636.47								
Fx^{-} (ton day^{-1})					563.75				
Fx^{-} (ablation period)					68777.33				
R^{+} (ton km^{-2} day^{-1})	0.16								
R^{+} (ablation period)	20.24								
R^{-} (ton km^{-2} day^{-1})					0.73				
R^{-} (ablation period)					89.01				
Discharge of water	**[b]2.76 × 10^9 L/d; 3.36 × 10^{11} L during the entire ablation period**								

[a]Catchment Area (772.70 km^2)

Note Ablaiton period, 1 June–30 September (122 days); Data taken from [a]Orr et al. (2019), [b]Bisht et al. (2017). (IIF)-Individual Ion Flux, (F^-)-anion flux, (F^+)-cation flux, (R^-)-denudation rate of anion, and (R^+)-denudation rate of cation (*Source* Authors)

sea and ocean water when compared to rocks, therefore, the atmospheric contribution of ions to the glacier meltwater is best explained by measurement of ratios (i.e., element-to-chloride) (Kumar et al. 2009). To determine the contribution of atmospheric input into the Gangotri glacier meltwater, the element-to-chloride ratio was studied in the years 2015 and 2016. The mean ratio Na/Cl ratio in Gangotri glacier meltwater were found as 3.24 ± 0.84in 2015 and 2.54 ± 0.44 in 2016, while the average K/Cl ratio were 2.69 ± 1.15 in 2015 and 2.46 ± 0.57 in 2016. These ratios appear much higher than marine aerosols (Na/Cl = 0.850 and K/Cl = 0.018) in both the years (Table 16.1). It clearly indicates a comparatively smaller input of ions in the meltwater from atmosphere during the study period. In addition, the scatter diagram of Cl^- and Na+K (Fig. 16.7) indicates Na+K are high to chloride in both the ablation periods, which also suggests that low atmospheric contribution of ion in the glacier meltwater. Similar results of low atmospheric contribution of ions in other Himalayan glaciers meltwater are reported by several authors (Singh et al. 2015; Singh and Ramanathan 2017; Singh et al. 2019). It clearly shows that the concentration of dissolved ions is chiefly affected by rock weathering processes. Meltwater drained through the silicate rocks shows less ionic ratios Mg/Na = 0.24, Ca/Na = 0.35, and HCO_3/Na = 2 (Gaillardet et al. 1999), whereas, water drained from carbonate rocks shows higher Mg/Na (10), Ca/Na (50), and HCO_3/Na (120) ratios as well as high concentration of Mg^{2+} and Ca^{2+} ions (Negrel et al. 1993). We observed the ratios as HCO_3/Na (17.55 ± 13.98 in 2015 and 24.63 ± 19.97 in 2016), Mg/Na (2.14 ± 1.23 in 2015 and

Table 16.3 Estimation of daily and entire ablation periods ionic flux and denudation rate of ions through meltwater in Gangotri catchment in the year 2016

	Analytes (ton)								
	Ca^{2+}	Mg^{2+}	Na^{+}	K^{+}	F^{-}	Cl^{-}	SO_4^{-2}	HCO_3^{-}	Total
IIF (ton day^{-1})	58.44	31.55	14.86	13.34	0.99	5.59	337.38	239.56	**701.71**
IIF (ablation period)	7129.71	3849.08	1813.40	1627.81	120.32	682.58	41160.79	29226.53	**85610.22**
Fx^{+} (ton day^{-1})	118.19								
Fx^{+} (ablation period)	14420.00								
Fx^{-} (ton day^{-1})					583.52				
Fx^{-} (ablation period)					71190.22				
R^{+} (ton km^{-2} day^{-1})	0.15								
R^{+} (ablation period)	18.66								
R^{-} (ton km^{-2} day^{-1})					0.76				
R^{-} (ablation period)					92.13				
Discharge of water	**[b]3.14 × 10⁹ L/d; 3.83 × 10¹¹ L during entire ablation period**								

[a]Catchment Area (772.70 km^2)

Note Ablation period, 1 June–30 September (122 days); Data taken from [a]Orr et al. (2019), [b]Bisht et al. (2017). (*Source* Authors)

2.81 ± 1.39 in 2016), and Ca/Na (3.31 ± 0.65 in 2015 and 3.91 ± 1.14 in 2016). The result indicates that the ratios Mg/Na, Ca/Na, and HCO_3/Na are in accordance that both the lithologies are mainly controlling the hydrochemistry of Gangotri glacier meltwater.

Acid hydrolysis is a very significant technique, which is one of the responsible factors for the chemical weathering of rocks (Raiswell 1984). In the meltwater stream, the concentration of SO_4^{2-} and HCO_3^- reflects the two prominent sources of hydrous protons, mainly driving the sub-glacial reactions for weathering (Brown et al. 1996; Hasnain and Thayyen 1999).

$$4FeS_2(s) + 16CaCO_3(s) + 15O_2(aq) + 14H_2O\ (l) \leftrightarrow 16Ca^{2+}(aq) + 16HCO_3^-(aq) + 8SO_4^{2-}(aq) + 4Fe(OH)_3(s) \quad (16.1)$$

$$CO_2 + H_2O \leftrightarrow H_2CO_3 \quad (16.2)$$

$$H_2CO_3 \leftrightarrow H^+ + HCO_3 \quad (16.3)$$

$$CaCO_3(s) + H_2CO_3(aq) \leftrightarrow Ca^{2+}(aq) + 2HCO_3^-(aq) \quad (16.4)$$

The significance of two reactions of proton formation (i.e., sulfide oxidation and carbonation) being an important factor for the carbonate rock weathering are best explained by C-ratio [$(HCO_3/(HCO_3+SO_4)$] (Huang et al. 2008). If it is closer to 0.5 signifies the coupled reaction involving sulfide oxidation and carbonate dissolution (Eq. 16.1) and if it is closer to 1 signifies chemical weathering due to carbonation reactions (Eqs. 16.2, 16.3, and 16.4) (Brown et al. 1996). Similarly, if S-ratio [$SO_4/(SO_4+HCO_3)$] is closer to 0.5, it shows weathering by two reactions involving sulfide oxidation and carbonate dissolution, while if ratio is close to 0, it indicates

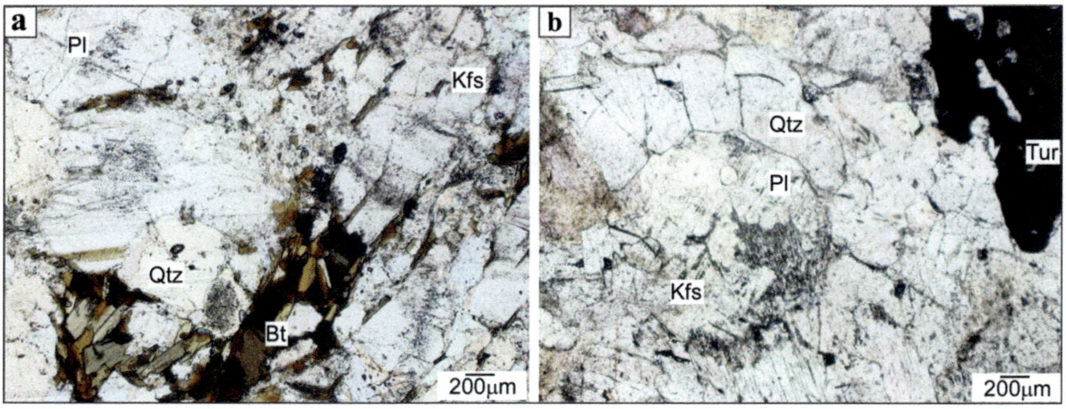

Fig. 16.3 Microphotographs of the rocks nearby area of the glacier snout. **a** Biotite granite comprising Bt, Kfs, Qtz, and Pl. **b** Tourmaline granite comprising Tur, Kfs, Qtz, and Pl. (The Bt—biotite; Kfs—alkali feldspar; Pl—plagioclase feldspar; Qtz—quartz; Tur—tourmaline) (*Source* Authors)

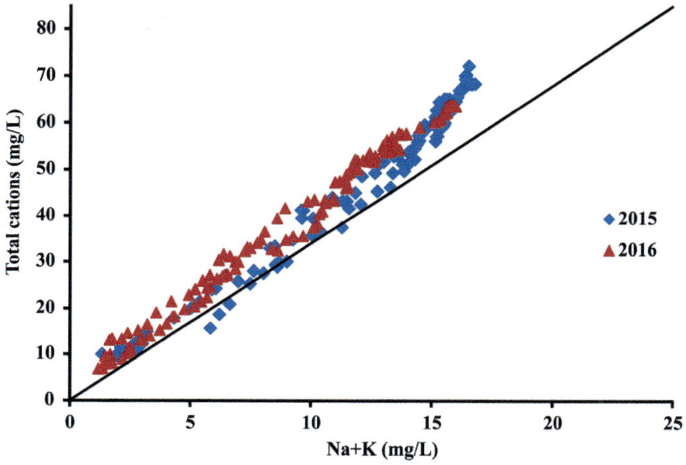

Fig. 16.4 Scatter graph of total cations and (Na+K) for Gangotri glacier meltwater (*Source* Authors)

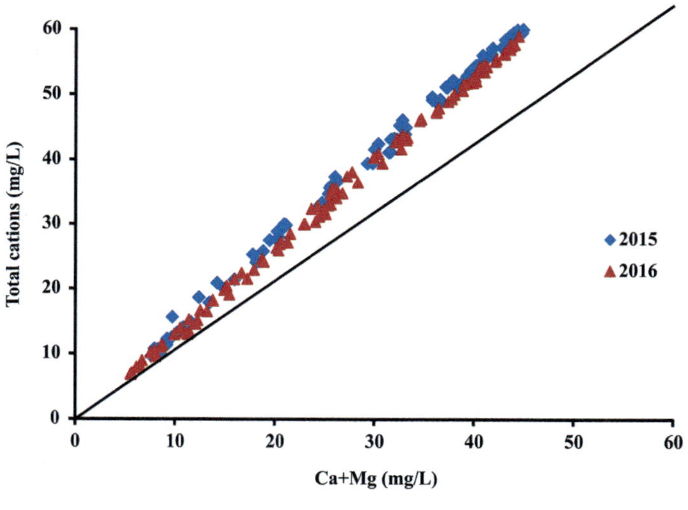

Fig. 16.5 Scatter graph of total cations and (Ca+Mg) for Gangotri glacier meltwater (*Source* Authors)

chemical weathering by reaction of carbonation (Tranter et al. 1997). Mean C-ratio of meltwater of Gangotri glacier was 0.42 ± 0.02 in 2015 and 0.41 ± 0.02 in 2016. It indicates that the weathering of rocks around the glacier is generally controlled by the coupled reactions. In addition, the average S-ratio for the meltwater was determined as 0.57 ± 0.02 in 2015 and 0.59 ± 0.02 in 2016. It also confirms weathering is controlled by both the reactions (i.e., carbonate dissolution and sulfide oxidation).

16.3.3 Ionic Flux Computation and Denudation Rate of the Ions

In the present study, the individual ionic flux and denudation rate of the ions through glacier meltwater were calculated in 2015 and 2016. Cationic flux (Fx^+) and denudation rate of cations (R^+) is obtained with the help of the formula given by Fang et al. (2012)

$$Fx^+ = \sum_{i=1}^{n} C_i^+(t) \times Q(t) \tag{16.5}$$

$$R^+ = {Fx^+}/{m} \tag{16.6}$$

where C_i^+ is the concentration of cations, Q is the discharge of meltwater, t is a time period, m represents the catchment area. Likewise, the anionic flux (Fx^-) and the rate of denudation anions (R^-) were estimated.

Ionic flux, ions denudation rate, total area of catchment, and meltwater discharge of Gangotri glacier during 2015 and 2016 are given in Tables 16.2 and 16.3, respectively. The catchment of Gangotri glacier covers an area of about 772.70 km^2 (Orr et al. 2019) with an average discharge of 3.36 × 10^{11} L and 3.83 × 10^{11} L during the entire ablation season in 2015 and 2016, respectively (Bisht et al. 2017). The average anionic flux of 373.94 ton per day and 44127.12 ton per ablation period were observed in 2015, while 299.72 ton per day and 36566.63 ton per ablation season were observed in 2016. In addition, the average cationic flux, 128.17 ton per day and 15636.47 ton per ablation period were observed in 2015, while 118.19 ton per day and 14420.00 ton per ablation period were observed in 2016. The result also shows that, in both the years, the calcium contributes the highest cationic flux and the sulfate contributes the highest anionic flux. Furthermore, the anion denudation rate 0.73 ton/km^2/day and 89.01 ton/km^2/ablation were observed in 2015, whereas 0.76 ton/km^2/day and 92.13 ton/km^2/ablation in 2016. Similarly, the cation denudation rates of 0.16 ton/km^2/day and 20.24 ton/km^2/ablation were observed in 2015, whereas 0.15 ton/km^2/day and 18.66 ton/km^2/ablation in 2016.

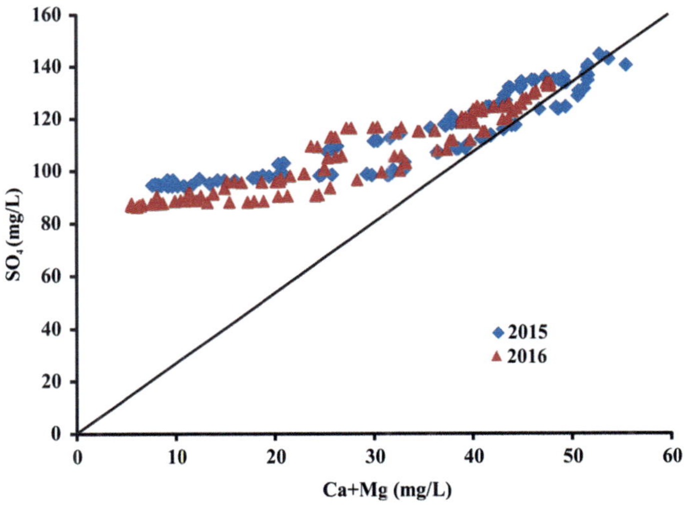

Fig. 16.6 Scatter plot between SO_4 and (Ca+Mg) to determine the supply of sulfate in the Gangotri glacier meltwater (*Source* Authors)

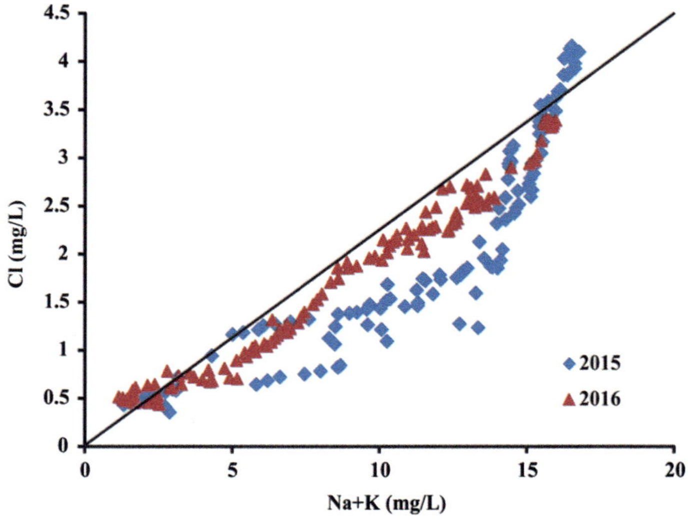

Fig. 16.7 Scatter diagram between Cl and (Na+K) to determine atmospheric input of ions in the meltwater (*Source* Authors)

The results derived from the present study reveals that cation denudation rate in glacier meltwater is higher compared to different glaciers of Himalaya and other parts of the globe (Table 16.4).

16.3.4 Hydrochemical Facies of Meltwater and Partial Pressure of CO_2

The Piper plot (Piper 1944) is an appropriate technique to categorize the hydrogeochemical facies based on various ionic compositions (Hem 1985). It is also important to establish a relationship between rock type and composition of water by plotting major ions values in this diagram (Singh et al. 2014a). The piper diagram was plotted to identify the hydro-geochemical facies of Gangotri glacier meltwater (Fig. 16.8). The piper plot contains one rhombus-shaped diagram at the top and two triangular shaped diagrams on its right and left at the bottom (Fig. 16.8). By using the piper diagram, our results show that the $CaSO_4$ dominated hydro-chemical facies were present. Figure 16.8 also shows that the strong acid (SO_4+Cl) exceeds over weaker acid (HCO_3) and alkaline earth metals (Ca+Mg) exceed over

Table 16.4 The cation denudation rate of meltwater and compared to different glaciers in Himalaya and other parts of the globe

Glaciers	Time period	Area of Catchment (Km^2)	Rates of cation denudation (ton/km^2/ablation)	References
Gangotri	2015/2016	772.70*	20.24/18.66	Present study
Glacier No. 1	2006/2007	3.34	11.46/13.90	Fang et al. (2012)
Kuannersuit	2001	258	15.9	Yde et al. (2005)
Dokriani	1992	23	9.7	Hasnain and Thayyen (1999)
South Cascade	1992	6.1	14.1	Anderson et al. (1997)
Haut	1990	11.7	13.7	Sharp et al. (1995)
ChhotaShigri	1987	40	17.4	Hasnain et al. (1989)

Source Singh et al. (2014a)

alkalis (Na+K) during both the years. Such a combination allows us to conclude that weathering of carbonates is the main solute source in meltwater. Effective CO_2 pressure or partial pressure of carbon dioxide (pCO_2) of glacier meltwater has been used to explain the various environments of weathering (Sharp et al. 1995; Wadham et al. 1998). pCO_2 can be calculated with the help of pH and HCO_3 values of the water (Stumm and Morgan 1981). The high pCO_2 values in the solution were appeared when the supply of proton (H^+) is higher than they consumed, whereas the low pCO_2 values were appeared when the need for protons is higher than the CO_2 diffusion rate (Tranter et al. 1993). In the intense cold glacial environment, the CO_2 has a high rate of solubility in water than discharge in atmosphere (Stumm and Morgan 1981). The average pCO_2values for Gangotri glacier meltwater were $10^{-1.37}$ in 2015 and $10^{-1.69}$ in 2016 which is more than pCO_2 values of atmosphere ($10^{-3.5}$). The higher pCO_2 values of meltwater than atmosphere in both the years are might be due to low temperature of the area, high turbulence in the stream, and carbonate weathering processes (Adimalla and Taloor 2020). Finally, the results suggest that the meltwater of Gangotri glacier is in a disequilibrium state compared to the surrounding.

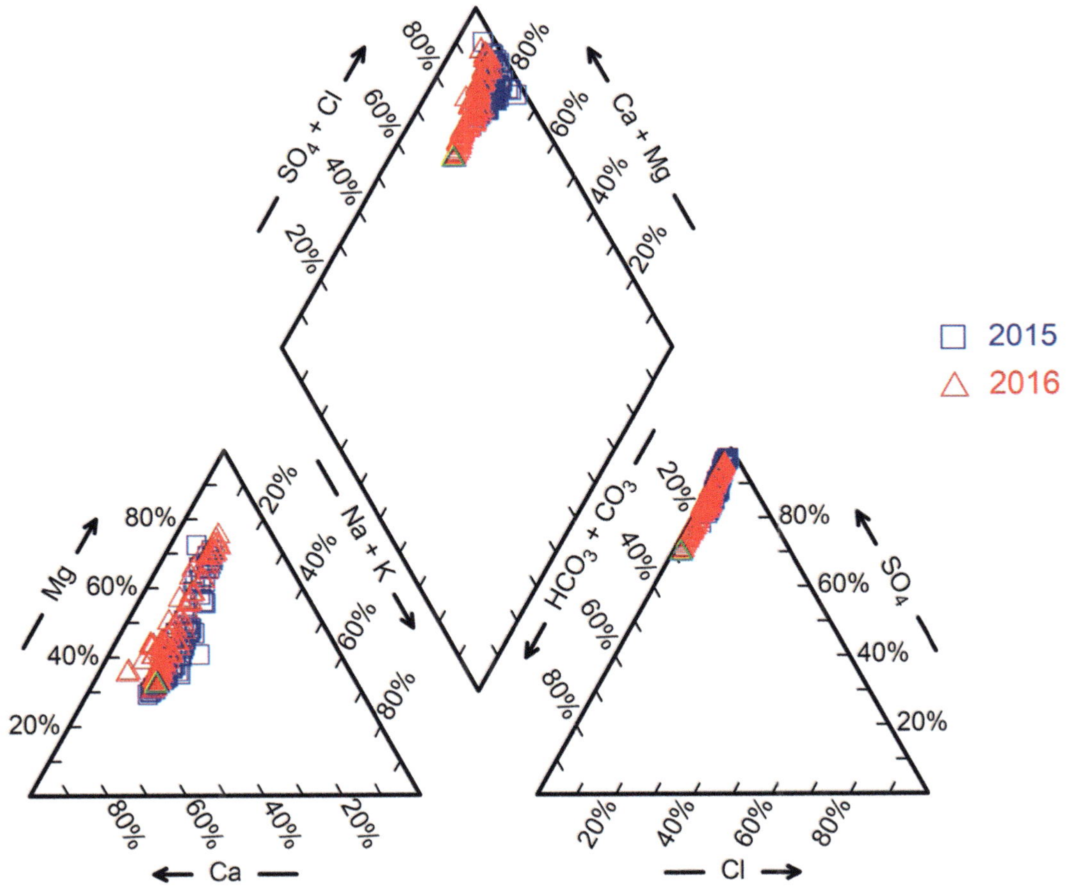

Fig. 16.8 Piper diagram for dominant ionic concentrations to determine the hydrochemical facies and chemical characterization of Gangotri glacier meltwater (*Source* Authors)

16.3.5 Statistical Analysis

The correlation matrix of different hydrochemical parameters of the meltwater for the year 2015 (a) and 2016 (b) is presented in Table 16.5. The correlation matrix shows that the EC is positively correlated with TDS in both the years (Table 16.5), indicating a direct relation of the EC with the ions dissolved in the meltwater through weathering of surrounding rocks and atmospheric input. A high positive correlation was observed between the different cations and anions (Table 16.5), which indicates the ions derived from the same source, i.e., weathering of carbonate.

Table 16.5 Correlation matrix between different physicochemical parameters during entire ablation (a) 2015 and (b) 2016

	EC	pH	TDS	Ca	Mg	Na	K	F	Cl	SO_4	HCO_3
(a)											
EC	1										
pH	0.81	1									
TDS	0.99	0.81	1								
Ca	0.96	0.86	0.96	1							
Mg	0.80	0.92	0.80	0.81	1						
Na	0.97	0.86	0.97	0.97	0.84	1					
K	0.73	0.85	0.73	0.83	0.88	0.83	1				
F	0.91	0.89	0.91	0.97	0.87	0.95	0.91	1			
Cl	0.95	0.82	0.95	0.95	0.78	0.92	0.74	0.91	1		
SO_4	0.95	0.70	0.95	0.92	0.72	0.91	0.69	0.87	0.93	1	
HCO_3	0.90	0.64	0.90	0.86	0.66	0.83	0.59	0.80	0.92	0.97	1
(b)											
EC	1										
pH	0.86	1									
TDS	0.99	0.86	1								
Ca	0.99	0.88	0.99	1							
Mg	0.81	0.94	0.81	0.82	1						
Na	0.97	0.91	0.97	0.98	0.85	1					
K	0.93	0.97	0.93	0.94	0.95	0.96	1				
F	0.97	0.94	0.97	0.97	0.88	0.97	0.97	1			
Cl	0.96	0.90	0.96	0.97	0.86	0.99	0.96	0.98	1		
SO_4	0.92	0.93	0.92	0.94	0.82	0.97	0.94	0.96	0.97	1	
HCO_3	0.91	0.83	0.91	0.94	0.74	0.96	0.88	0.92	0.95	0.95	1

(*Source* Authors)

16.4 Conclusions

The hydrochemical analysis of Gangotri glacier meltwater helped in understanding the processes of weathering that regulate the chemistry of main ions during ablation seasons 2015 and 2016. Furthermore, the study also insight into the chemical weathering process and petrography of surrounding rocks to estimate the sources of ions and the agents controlling the ionic composition. The hydrochemical study of Gangotri glacier meltwater indicates that the meltwater has a somewhat acidic nature with $CaSO_4$ hydrochemical facies. The anions concentration shows that SO_4^{2-} is a prominent anion and subsequently followed by HCO_3^-,Cl^- and F^- in both the years. Among the cations, Ca^{2+} is the leading cation, associated with Mg^{2+}, Na^+ and K^+ in both years. The higher proportion of (Ca+Mg) in total cations, higher (Ca+Mg)/(Na+K) ratio, and low (Na+K)/TZ^- ratio suggest weathering of carbonate is a major geochemical process to control the meltwater chemistry of Gangotri glacier. The correlation matrix also indicates that the ions are derived from a similar source, i.e., carbonate weathering. The mean values of C-ratio and S-ratio of Gangotri glacier meltwater show that sulfide oxidation and carbonation are affecting the weathering of rocks. In addition, mean equivalent ratio of K/Cl and Na/Cl in meltwater is significantly greater than marine aerosols, suggesting a little input of ions from atmosphere. Based on the petrographic analysis, we assume that the rocks near the glacier snout are rich in alkali feldspar, plagioclase feldspar, quartz, biotite, and tourmaline, which contribute to the major cations in Gangotri glacier meltwater. The high pCO_2 values of Gangotri glacier meltwater than atmospheric values suggested that the open system of weathering and higher solubility of CO_2 in the meltwater. Further, we believe that the meltwater of Gangotri glacier is in a disequilibrium state, compared to the atmosphere of the surrounding. An estimation of solute flux suggests that the denudation rate of cations of Gangotri glacier meltwater was found to be 20.24 and 18.66 ton/km^2/ablation in 2015 and 2016 correspondingly. Likewise, the denudation rate of anion was observed as 89.01 and 92.13 ton/km^2/ablation during 2015 and 2016 correspondingly.

Acknowledgements The authors would like to thank DST, New Delhi, for economic support to carry out the current research. We are also grateful to the Director, G.B. Pant National Institute, Kosi-Katarmal, Almora for extending us the required working facilities in the field and laboratory.

References

Adimalla N, Dhakate R, Kasarla A, Taloor AK (2020) Appraisal of groundwater quality for drinking and irrigation purposes in Central Telangana, India. Groundw Sust Dev 10:P100334. https://doi.org/10.1016/j.gsd.2020.100334

Adimalla N, Taloor AK (2020) Introductory editorial for 'Applied Water Science'special issue: "Groundwater contamination and risk assessment with an application of GIS". Appl Water Sci 10:216. https://doi.org/10.1007/s13201-020-01291-3

Ahmad S, Hasnain SI (2001) Chemical characteristics of stream draining from Dudu glacier: an Alpine meltwater stream in Ganga Headwater, Garhwal Himalaya. J China Univ Geosci 12(1):75–83

Anderson SP, Drever JI, Humphrey NF(1997) Chemical weathering in glacial environments. Geology 25 (5):399–402. https://doi.org/10.1130/0091-7613(1997)025%3c0399:CWIGE%3e2.3.CO;2

APHA (2005) Standard methods for examination of water and wastewater, 21st edn. American Public Health Association, Washington, DC

Arita K (1983) Origin of the inverted metamorphism of the lower Himalayas, Central Nepal. Tectonophysics 95(398):43–60. https://doi.org/10.1016/0040-1951(83)90258-5

Barandum M, Huss M, Usubaliev R, Azisov E, Berthier E, Kääb A, et al (2018) Multi-decadal mass balance series of three Kyrgyz glaciers inferred from modelling constrained with repeated snow line observations. The Cryosphere 12:1899–1919. https://doi.org/10.5194/tc-12-1899-2018

Bisht H, Sah S, Kumar K, Arya PC, Tewari M (2017) Quantification of variability in discharge and suspended sediment concentration of meltwater of Gangotri Glacier, GarhwalHimalaya. ENVIS Bull Himal Ecol 25:10–16

Bisht H, Arya PC, Kumar K (2018) Hydro-chemical analysis and ionic flux of meltwater runoff from Khangri Glacier, West Kameng, Arunachal Himalaya, India. Environ Earth Sci 77:1–16. https://doi.org/10.1007/s12665-018-7779-6

Bisht H, Rani M, Kumar K, Sah S, Arya PC (2019) Retreating rate of Chaturangi glacier, Garhwal Himalaya, India derived from kinematic GPS survey and satellite data. CurrSci 116:304–311. https://doi.org/10.18520/cs/v116/i2/304-311

Bisht H, Kotlia BS, Kumar K, Arya PC, Sah SK, Kukreti M, Chand P (2020a) Estimation of suspended sediment concentration and meltwater discharge draining from the Chaturangi glacier, Garhwal Himalaya. Arab J Geosci 13(6):1–12. https://doi.org/10.1007/s12517-020-5204-w4

Bisht H, Kotlia BS, Kumar K, Joshi LM, Sah SK, Kukreti M (2020b) Estimation of the recession rate of Gangotri glacier, Garhwal Himalaya (India) through kinematic GPS survey and satellite data. Environ Earth Sci 13(6):1–12. https://doi.org/10.1007/s12665-020-090780

Bisht H, Kotlia BS, Kumar K, Dumka RK, Taloor AK, Upadhyay R (2020c) GPS derived crustal velocity, tectonic deformation and strain in the Indian Himalayan arc. Quat Int. https://doi.org/10.1016/j.quaint.2020.04.028

Bogen J (1989) Glacial sediment production and development of hydroelectric power in glaciered areas. Ann Glaciol 13:6–11

Bolch T, Kulkarni A, Kääb A, Huggel C, Paul F, Cogley JG, et al (2012) The state and fate of Himalayan glaciers. Science 336(6079):310–314. https://doi.org/10.1126/science.1215828

Brown GH (2002) Glacier meltwater hydrochemistry. Appl Geochem 17:855–883

Brown GH, Tranter M, Sharp M (1996) Subglacial chemical erosion- seasonal variations in solute provenance, Haut glacier d'Arolla, Switzerland. Ann Glaciol 22:25–31. https://doi.org/10.3189/1996AoG22-1-25-31

Clow DW, Mast MA (2010) Mechanisms for chemostatic behavior in catchments: implications for CO2 consumption by mineral weathering. Chem Geol 269:40–51. https://doi.org/10.1016/j.chemgeo.2009.09.014

Collins DN (1979) Hydrochemistry of meltwater draining from an Alpine glacier. Arctic, Antarctic and Alpine Res 11:307–324. https://doi.org/10.3189/S0022143000029877

Dickinson WR (1970) Interpreting detrital modes of greywake and arkose. J Sediment Pet 40:695–707

Dimri AP, Yasunari T, Kotlia BS, Mohanty UC, Sikka DR (2016) Indian winter monsoon; present and past. Earth Sci Rev 163:297–322. https://doi.org/10.1016/j.earscirev.2016.10.008

Fang F, Zhongqin L, Shuang J, Jhiwen D, Feiteng W (2012) Hydrochemical characteristics and solute dynamics of meltwater runoff of urumqi glacier no. 1, Eastern Tianshan, Northwest China. J Mt Sci 9:472–482. https://doi.org/10.1007/s11629-012-2316-7

Gaillardet J, Dupre B, Louvat PI, Allegre CJ (1999) Global silicate weathering and CO2 consumption rates deduced from the chemistry of large rivers. Chem Geol 159:3–30. https://doi.org/10.1016/S0009-2541(99)00031-5

Gazzi P (1966) Le Arenarie del Flysch Sopracretaceodell' Appennino Modenese: correlazioni con ilFlysch di Monghidoro. Mineralogicae Petrographica Acta 12:69–97

Gupta RP, Haritashya UK, Singh P (2005) Mapping dry/wet snowcover in the Indian Himalayas using IRS multispectral imagery. Remote Sens Environ 97:458–469. https://doi.org/10.1016/j.rse.2005.05.010

Hasnain SI, Subramanian V, Dhanpal K (1989) Chemical characteristics and suspended sediment load of meltwaters from a Himalayan Glacier in India. J Hydrol 106:99–108. https://doi.org/10.1016/0022-1694(89)90168-6

Hasnain SI, Thayyen RJ (1999) Discharge and suspended-sediment concentration of meltwaters, draining from the Dokriani glacier, Garhwal Himalaya, India. J Hydrol 218:191–198. https://doi.org/10.1016/s00221694(99)00033-5

Haque S, Kannaujiya S, Taloor AK, Keshri D, Bhunia RK, Ray PKC, Chauhan P (2020) Identification of groundwater resource zone in the active tectonic region of Himalaya through earth observatory techniques. Groundw Sust Dev 10:P100337. https://doi.org/10.1016/j.gsd.2020.100337

Hem JD (1985) Study and interpretation of the chemical characteristics of natural water. 3rd ed. In: U.S. Geol Survey Water-Supply Paper, vol 2254, pp 65–100

Huang X, Sillanpää M, Duo B, Gjessing ET (2008) Water quality in the Tibetan Plateau: metal contents of four selected rivers. Environ Pollut 156(2):270–277

Jain AK, Singh S, Manickavavasagam RM (2002) Himalayan collision tectonics. Gondwana Res 7: 1–14

Jasrotia AS, Kumar A (2014) Groundwater Quality Mapping Based on the Geographical Information System (GIS) of Jammu District, Jammu and Kashmir India. J Spatial Hydrol 12(1):1–21

Jasrotia AS, Taloor AK, Andotra U, Bhagat BD (2018) Geoinformatics based groundwater quality assessment for domestic and irrigation uses of the Western Doon valley, Uttarakhand, India. Groundw Sust Dev 6:200–212

Jasrotia AS, Taloor AK, Andotra U, Kumar R (2019) Monitoring and assessment of groundwater quality and its suitability for domestic and agricultural use in the Cenozoic rocks of Jammu Himalaya, India: a geospatial technology based approach. Groundw Sust Dev 8:554–566

Kaul MK (1999) Inventory of Himalayan glaciers. Geol Surv India 34:136–137

Khan A, Govil H, Taloor AK, Kumar G (2020) Identification of artificial groundwater recharge sites in parts of Yamuna river basin India based on remote sensing and geographical information system. Groundw Sust

Dev 11:P100415. https://doi.org/10.1016/j.gsd.2020.100415
Krishnaswami S, Singh SK (2005) Chemical weathering in the river basins of the Himalaya, India. Curr Sci 89 (5):841–849
Kumar K, Miral MS, Joshi S, Pant N, Joshi V, Joshi LM (2009) Solute dynamics of meltwater of Gangotri glacier, Garhwal Himalaya, India. Environ Geol 58 (6):1151–1159. https://doi.org/10.1007/s00254-008-15926
Kumar R, Kumar R, Singh A, Singh S, Bhardwaj A, Kumari A, Sinha RK, Gupta A (2019) Hydrogeochemical analysis of meltwater draining from BilareBanga glacier, Western Himalaya. Acta Geophys 67 (2):651–660. https://doi.org/10.1007/s11600-019-00262-w
Kumar R, Kumar R, Singh S, Singh A, Bhardwaj A, Kumari A, Randhawa SS, Saha A (2018) Dynamics of suspended sediment load with respect to summer discharge and temperatures in Shaune Garang glacierized catchment, Western Himalaya. Acta Geophys 66:1109–1120. https://doi.org/10.1007/s11600-018-0184-4
Metcalfe RP (1993) Pressure, temperature and time constraints on metamorphism across the MCT zone of High Himalaya slab in the Garhwal Himalaya. In: Treloar RJ, Searle MP (eds) Himalaya tectonics. Geological Society of London, pp 495–509
Naithani AK, Nainwal HC, Sati KK, Prasad C (2001) Geomorphologicalevidence of retreat of the Gangotri glacier and itscharacteristics. Curr Sci 80(1):87–88
Negrel P, Allegre CJ, Dupre B, Lewin E (1993) Erosion sources determined by inversion of major and trace element ratios in river water: the Congo basin case. Earth Planet Sci Lett 120:59–76. https://doi.org/10.1016/0012-821X(93)90023-3
Oliva P, Viers J, Dupré B (2003) Chemical weathering in granitic environments. Chem Geol 202:225–256
Orr EN, Owen LA, Saha S, Caffee MW (2019) Rates of rockwall slope erosion in the upper Bhagirathi catchment, Garhwal Himalaya. Earth Surf Process Landforms 44:3108–3127. https://doi.org/10.1002/eps.4720
Ostrem G (1975) Sediment transport in glacial meltwater stream. In: Jopling AV, McDonald BC (eds) Glaciofluvial and Glacio-lacustrine Sedimentation. Society of Economic palaeontologists and Mineralogists, Oklahoma, USA. Spec Pub No 23, pp 101–122
Piper AM (1944) A graphical procedure in the geochemical interpretation of water analysis. Trans Am Geophys Union 25:914–923
Raina VK, Srivastava D (2008) Glacieratlas of India. Geol Soc India, Bangalore
Raiswell R (1984) Chemical models of solute acquisition in glacial meltwater. J Glaciol 30(104):49–57. https://doi.org/10.1017/s0022143000008480
Sadat N (2012) Study of fluoride concentration in the river (Godavari) and groundwater of Nanded city. Int J Eng Invent 1(1):11–15
Sah S, Bisht H, Kumar K, Tiwari A, Tewari M, Joshi H (2017) Assessment of hydrochemical properties and annual variation in meltwater of Gangotri glacier system. ENVIS Bull Himal Ecol 25:17–23
Sharma MK, Thayyen RJ, Jain CK, Arora M, Lal S (2019) Assessment of system characteristics of Gangotri glacier headwater stream. Sci Total Environ 662:842–851. https://doi.org/10.1016/j.scitotenv.2019.01.229
Sharma P, Ramanathan AL, Pottakkal J (2013) Study of solute sources and evolution of hydrogeochemical processes of the ChotaShigri Glacier meltwaters, Himachal Himalaya, India. Hydrol Sci J 58(5):1128–1143. https://doi.org/10.1080/02626667.2013.802092
Sharp M, Tranter M, Brown GH, Skidmore M (1995) Rates of chemical denudation and CO2 drawdown in a glacier-covered alpine catchment. Geology 23(1):61–64. https://doi.org/10.1130/00917613(1995)023%3c0061:ROCDAC%3e2.3.CO;2
Singh S, Sood V, Taloor AK, Prashar S, Kaur R (2020) Qualitative and quantitative analysis of topographically derived CVA algorithms using MODIS and Landsat-8 data over Western Himalayas. Quat Int, India. https://doi.org/10.1016/j.quaint.2020.04.048
Singh VB, Keshari AK, Ramanathan AL (2019) Major ion chemistry and atmospheric CO2 consumption deduced from the Batal glacier, Lahaul-Spiti valley, Western Himalaya, India. Environ Dev Sustain 19:1–19. https://doi.org/10.1007/s10668-019-00501-6
Singh VB, Ramanathan AL (2017) Hydrogeochemistry of the Chhota Shigri glacier meltwater, Chandra basin, Himachal Pradesh, India: Solute acquisition processes, dissolved load and chemical weathering rates. Environ Earth Sci 76(5):223. https://doi.org/10.1007/s12665-017-6465-4
Singh VB, Ramanathan AL, Pottakkal JG, Kumar M (2014a) Seasonal variation of the solute and suspended sediment load in Gangotri Glacier meltwater, central Himalaya, India. J Asian Earth Sci 79:224–234. https://doi.org/10.1016/j.jseaes.2013.09.010
Singh VB, Ramanathan AL, Pottakkal JG, Kumar M (2014b) Hydrogeochemisty of meltwater of the Chaturangi glacier, Garhwal Himalaya, India. Proc Natl Acad Sci India 85(1):187–195. https://doi.org/10.1007/s40010-014-0181-1
Singh VB, Ramanathan AL, Pottakkal JG, Sharma P, Linda A, Azam MF, Chatterjee C (2012) Chemical characterisation of meltwater draining from Gangotri glacier, Garhwal Himalaya, India. J Earth Syst Sci 121 (3):625–636. https://doi.org/10.1007/s12040-012-01777
Singh VB, Ramanathan AL, Sharma P, Pottakkal JG (2015) Dissolved ion chemistry and suspended sediment characteristics of meltwater draining from Chhota Shigri Glacier, Western Himalaya, India. Arab J Geosci 8:281–293. https://doi.org/10.1007/s12517-013-1176-y
Sood V, Gusain HS, Gupta S, Taloor AK, Singh S (2020) Detection of snow/ice cover changes using subpixel-

based change detection approach over Chhota-Shigri glacier, Western Himalaya, India. Quat Int. https://doi.org/10.1016/j.quaint.2020.05.016

Souchez RA, Lemmens MM (1987) Solutes. In: Gurnell AM, Clark MJ (eds) Glacio-fluvial sediment transfer. Wiley, UK, pp 285–303

Stern CR, Kligfield R, Schelling D, Virdi NS, Futa K, Peterman ZE, Amini H (1989) The Bhagirathi Leucogranite of the High Himalaya (Garhwal India); Age, Petrogenesis, and tectonic implications. Geol Soc Am Spec Pap 232:33–45

Stumm W, Morgan JJ (1981) An introduction emphasizing chemical equilibria in natural waters. In: Stumm W, Morgan JJ (eds) Aquatic chemistry, 2nd edn. Wiley, New York, pp 230–258

Taloor AK, Joshi LM, Kotlia BS, Alam A, Kothyari GC, Kandregula RS, Singh AK, Dumka RK (2020a) Tectonic imprints of landscape evolution in the Bhilangana and Mandakini basin, Garhwal Himalaya, India: a geospatial approach. Quat Int. https://doi.org/10.1016/j.quaint.2020.07.021

Taloor AK, Kumar V, Singh VK, Singh AK, Kale RV, Sharma R, Khajuria V, Raina G, Kouser B, Chowdhary NH (2020b). Land use land cover dynamics using remote sensing and GIS techniques in Western Doon Valley, Uttarakhand, India. In: Geoecology of landscape dynamics 2020. Springer, Singapore, pp 37–51. https://doi.org/10.1007/978-981-15-2097-6_4

Taloor AK, Pir, RA, Adimalla N, Ali S, Manhas DS, Roy S, Singh AK (2020c) Spring water quality and discharge assessment in the Basantar watershed of Jammu Himalaya using geographic information system (GIS) and water quality Index(WQI). Groundw Sust Dev 10:P100364. https://doi.org/10.1016/j.gsd.2020.100364

Taloor AK, Ray PK, Jasrotia AS, Kotlia BS, Alam A, Kumar SG, Kumar R, Kumar V, Roy S (2017) Active tectonic deformation along reactivated faults in Binta basin in Kumaun Himalaya of north India: inferences from tectono-geomorphic evaluation. Z Geomorphol 61(2):159–180

Tiwari SK, Kumar A, Gupta AK, Verma A, Bhambri R, Sundriyal S, Yadav J (2018) Hydrochemistry of meltwater draining from Dokriani Glacier during early and late ablation season, West Central Himalaya. Himal Geol 39(1):121–132

Tranter M, Sharp MJ, Brown GH, Willis IC, Hubbard BP, Nielsen MK, Smart CC, Gordon S, Tulley M, Lamb HR (1997) Variability in the chemical composition of in-situ subglacial meltwaters. Hydrol Proc 11:59–77. https://doi.org/10.1002/(SICI)1099-1085

Tranter M, Brown GH, Raiswell R, Sharp MJ, Gurnell AM (1993) A conceptual model of solute acquisition by Alpine glacier meltwaters. J Glaciol 39(133): 573–581. https://doi.org/10.3189/S0022143000016464

Valdiya KS (1980) Geology of the Kumaun lesser Himalaya Wadia Institute of Himalayan Geology, Dehradun, pp 291

Wadham JL, Hodson AJ, Tranter M, Dowdeswell JA (1998) The hydrochemistry of meltwater draining a polythermal-based, high Arctic glacier, South Svalbard. Hydrol Proc 12:1825–1849. https://doi.org/10.1002/(SICI)10991085(19981015)12:12%3c1825::AIDHYP669%3e3.0.CO;2-R

White AF (2002) Determining mineral weathering rates based on solid and solute weathering gradients and velocities: application to biotite weathering in saprolites. Chem Geol 190:69–89. https://doi.org/10.1016/S0009-2541(02)00111-0

Yde JC, Knudsen NT, Ole BN (2005) Glacier hydrochemistry, solute provenance, and chemical denudation at a surge-type glacier in Kuanner suit Kuussuat, Disko Island, West Greenland. J Hydrol 300:172–187. https://doi.org/10.1016/j.hydrol.2004.06.008

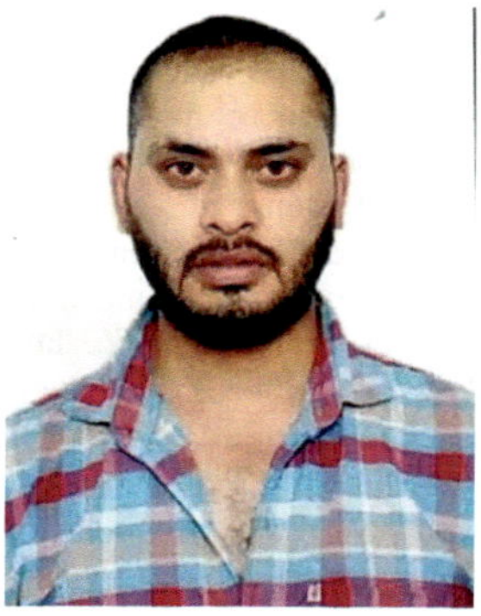

Mr. Harish Bisht is pursuing Ph.D. in Geology from Centre of Advanced Study, Department of Geology, Kumaun University, Nainital. He has worked in various DST sponsored projects in G.B. Pant National Institute of Himalayan Environment, Kosi-Katarmal, Almora. He is a recipient of CSIR-NET JRF and GATE award. He has published a total of three research papers in national journal and four research papers in international journals. He has actively participated in three international conferences and four national conferences

Dr. Bahadur Singh Kotlia is Research Scientist (Professor) at Centre of Advanced Study, Department of Geology, Kumaun University, Nainital. He received his Ph.D. degree in Geology from Panjab University, Chandigarh. He is a recipient of Alexander von Humboldt Fellowship. He received his Post-Doctoral degree from University of Bonn, West Germany, Institute of Palaeontology. He has executed various projects in the field of Earth Sciences. He has published over 80 research papers in international journals and 50 research papers in national journals

Er. Kireet Kumar is Scientist-G at G.B. Pant National Institute of Himalayan Environment, Kosi-Katarmal, Almora. He received his M.Tech. degree in Environmental Engineering from I.I.T. Kanpur, Uttar Pradesh. His areas of specialization are environmental engineering, hydrology and water resource management. He is a Fellow of Leadership for Environment and Development (LEAD) International, UK. He has executed various projects in the field of glaciology and hydrology. He has published several research papers in peer-reviewed journals

Dr. Ajay Kumar Taloor has obtained his Doctorate in Remote Sensing and GIS applications in hydrogeology from University of Jammu, NAAC accredited A^+ University of India. Thirteen years of research experience in the applications in geospatial technology for natural resources management of land and water resources. He has excelled twice with best paper presentation award in India. Being an expert of remote sensing applications in water science, cryosphere and climate change, tectonic and quaternary geomorphology, he is working on two major research projects on using space-based inputs for glacier mapping and climate change in Himalayas. He has also published many articles in tectonic and quaternary geomorphology in the recent years. He has high scientific temper and strong HR relations in science world, with high professional and managerial skills. He has edited many volumes in the top-rated journals in the Elsevier and Springer publishers, member of Editorial Board of the Quaternary Science Advances, and reviewer of the many top-rated international journals in science world

Smt. Pooja Chand is pursuing Ph.D. in Geology from Centre of Advanced Study, Department of Geology, Kumaun University, Nainital. She is a recipient of Department of Science and Technology Inspire Fellowhip. She has published one research paper in International Journal. She has participated in two international conferences and three national conferences

Mr. Jeewan Singh Bisht received his M. Sc. degree in Geology from Centre of Advanced Study, Department of Geology, Kumaun University, Nainital. He did his M.Sc. dissertation work in hydrogeochemical studies of the lakes of Kumaun lesser Himalaya. He has a good knowledge of geological field work and mapping. He has participated in one national conference. He is actively involved in various projects related to environmental issues

Mr. Manmohan Kukreti is pursuing Ph.D. in Geology from Centre of Advanced Study, Department of Geology, Kumaun University, Nainital. He received his M.Sc. degree in Geology from Department of Geology, L.M.S. P.G. College, Rishikesh, Uttarakhand. He has a 2-year teaching experience in subject of Geology. He has published two research papers in international journals. He has participated in two international conferences and two national conferences

Mr. Yogesh Maithani received his M.Sc. degree in Geology from Department of Geology, L.M.S. P.G. College, Rishikesh, Uttarakhand. He has a 5-year teaching experience in subject of Geology. He has a good knowledge of geological field work. He has attended various training programmes in field of geology

Mr. Mohit Tewari is presently working as a Lab Assistant in G.B. Pant National Institute of Himalayan Environment, Kosi-Katarmal Almora. He received his B.Tech. degree in Biochemical from Kumaun Engineering College, Dwarahat, Uttarakhand. He has a 5-year experience in field of soil and water chemistry

Geochemical Characterization and Evolution of Groundwater in Parts of Kashmir Valley, Western Himalaya

17

Khurshid Ahmad Lone, Riyaz Ahmad Mir, and Nadeem Ahmad Bhat

Abstract

The present study describes the hydrogeochemistry and chemical evolution of groundwater in parts of Srinagar and Ganderbal districts of Kashmir valley, Western Himalaya. A systematic and seasonal groundwater sampling from bore wells, dug wells and springs was carried out during December 2004 (premelting season) and June-July 2005 (post melting season). A total of 140 samples were collected to assess the physico-chemical characteristics, sources of major ions (Ca^{2+}, Mg^{2+}, Na^{+}, K^{+}, HCO_3^{-}, Cl^{-}, SO_4^{2-} and NO_3^{-}), chemical evolution and quality of the groundwater. On the basis of chemical characterization, three groups of groundwater were identified and designated as G-I, G-II and G-III respectively. The G-I indicated the abundance of Ca^{2+}, Mg^{2+} and HCO_3^{-} ions, thereby, indicating the dissolution of mainly carbonates whereas the G-II revealed also the abundance of Na^{+} ions that indicated the influence of weathering and dissolution of primary/or secondary silicates in the area. The G-III, although similar to G-II indicated a feeble enrichment of Cl^{-} content in addition to Na^{+} that is attributed to anthropogenic sources. Thus, the G-I and G-II indicated dominantly a lithological control on the source of major ions whereas, the enrichment of Cl^{-} ions in G-III documented the control of anthropogenic factors (a non-lithological control) as well. Further, the chemical imprint taken by groundwater through a complex interplay of hydrogeological processes, flow patterns and higher time of residence of ions in aquifer matrix characterized the groundwater dominantly as of primary facies type (i.e., Ca-HCO_3 and Ca-Mg-HCO_3 water types). The secondary facies/hybrid types (i.e., Ca-Mg-Na-HCO_3 and Ca-Mg-Na-HCO_3-Cl water types) were also present at places in the area. Overall, the chemical characteristics of groundwater indicated that the groundwater has largely retained its meteoric character that is suitable for domestic uses and the system is in its primary stage of evolution with a limited migratory history. Further, the predominance of bicarbonate waters documented the existence of open nature of groundwater system in the area.

K. A. Lone
Department of Geology, Government Boys Higher Secondary School, Handwara-Kupwara, Jammu and Kashmir 193221, India
e-mail: khurshidgeo37@gmail.com

R. A. Mir (✉)
Department of Earth Sciences, Indian Institute of Technology Roorkee, Roorkee 247667, India
e-mail: riyazgsi@gmail.com

N. A. Bhat
Department of Earth Sciences, University of Kashmir, Srinagar 190006, India
e-mail: mnadeem83@gmail.com

A. K. Taloor et al. (eds.), *Water, Cryosphere, and Climate Change in the Himalayas*, Geography of the Physical Environment, https://doi.org/10.1007/978-3-030-67932-3_17

Keywords

Groundwater · Premelting and postmelting · Hydrochemistry · Evolution · Kashmir valley

17.1 Introduction

The uneven distribution of freshwater around the globe has stimulated man right from ancient times to search for the alternative and reliable sources of water during the times of water scarcity. Groundwater development dates back to ancient times (Biswas 1970, Sarkar et al. 2020; Haque et al. 2020; Adimalla and Taloor 2020a). Utilization of groundwater for various purposes greatly preceded its understanding of its origin, occurrence and movement. Initially, the Greek and the Roman philosophers put forward groundwater theories that ranged from fantasy to nearly correct accounts (Adams 1928; Barker and Horton 1936). Later contributions during the nineteenth century greatly emphasized over the hydraulics of groundwater development. In the present era, the advent of digital technology has produced a competence for development and management of groundwater resources that was non-existent here to before (Singh et al. 2017; Adimalla and Taloor 2020b; Adimalla et al. 2020; Khan et al. 2020).

In the valley of Kashmir, no systematic and detailed scientific study has been carried out to give a detailed account or as a base lines study on the hydrogeochemical characterization of the groundwater vis-à-vis its hydrogeological conditions and evolution. The valley of Kashmir though bestowed with abundant water resources in the form of snow and glaciers, freshwater lakes, rivers and springs, yet a great concern is voiced for supply and rationing of drinking water particularly in upland and urban areas (Mir and Mir 2019; Mir and Jeelani 2015a; Mir 2018). During recent years, the Central Groundwater Board (CGWB) has drilled bore wells in all districts of the Kashmir valley for the assessment of the groundwater resource. The CGWB has reported a groundwater potential of 2400 million m^3/year in the valley which is being currently exploited at a higher rate (Mir and Lone 2020). During recent past, almost all the districts of Kashmir valley particularly the Srinagar district and its surrounding areas/or districts have seen a great deal of growth, development and rapid urbanization with the establishment of new housing colonies and/or human settlements. As a consequence of these developments an increase in the consumption of surface water and deterioration of its quality has been witnessed (Mir and Jeelani 2015a; Mir et al. 2016; Mir and Gani 2019). The surface water resources are facing a huge demand and stress due to the rapid population growth and development in the area. In addition, the water supply has become more erratic and inadequate particularly during the summer and dry periods leading to local water crisis (Mir et al. 2016; Mir and Gani 2019). As a result, the people of valley have been stimulated to search for alternative and reliable sources of water in the area. Thus, the tapping of groundwater has become very popular and is preferred over surface water because of impure quality of the later. Regarding groundwater, a general perception prevails also among the masses in the area that the surface-soil-strata act as a natural filter thereby providing safe and pure water for consumption. Keeping these considerations in view, the people of the area have installed a number of bore wells throughout the valley.

Considering the widespread use of groundwater in all parts of the Srinagar and Ganderbal districts, it becomes therefore very imperative to study its chemical nature and evolution as well as its quality controlling factors. No systematic and planned work on scientific grounds has been carried out so far in this area to unravel the natural phenomena that governs and controls the hydrochemical nature of groundwater or the anthropogenic factors that presently affect it. Thus, the present study attempts to characterize the chemical nature of groundwater, identify the sources of chemical ions and its evolution, to unravel the possible influence of natural and anthropogenic sources on it etc. Nevertheless, the dataset produced in this study may also serve as the baseline for any future hydrogeochemical and groundwater monitoring studies in the area.

17.2 Study Area

The study area covers parts of Srinagar and Ganderbal districts of the Kashmir valley, northwest Himalaya, India. The study area is located within the core of the Kashmir valley comprising both the urban and rural areas. It lies between 34° 3′–34° 25′N latitude and 74°70′–75°0′E longitude and covers an area of 2,228 km^2 (Fig. 17.1). However, it is important to mention that the data was collected before the bifurcation of old Srinagar district into Srinagar and Ganderbal districts in 2006. The study area is connected by all-weather motorable roads and by National Highway (NH 1A) and the air route to the rest of India. In the study area major portion of cultivable land has a network of canals which provide desirable irrigation facility during the cropping season. The total population of the Srinagar district (major part of area) is ~12, 02,447 which comprises of a rural population of ~17,313 281 and urban population of ~121, 9516 (Census 2001). The population density in Srinagar is 401/Km^2, which is the highest in the state. Hand-pumps are widely used to pump out groundwater for domestic and other related purposes. Most of the wells are shallow, hence, liable to contamination from the sources like sewage drains, septic tanks, polluted surface water bodies and agricultural farmlands and/or floating gardens of the Dal lake where huge pesticides are being used.

The area witnesses a continental climate system, characterized with marked seasonality linked with the mechanism of weather in the Indian sub-continent (Mir and Jeelani 2015a; Mir et al. 2016). In general, four seasons as winter season (November- February), spring season (March-mid-May), summer season (mid-May-mid-September) and autumn season (mid-September- mid-November) have been described for the area. January is the coldest month whereas; the month of July is generally the warmest. The temperature varies between −5 and >30 °C whereas the precipitation is higher during winter season when snowfall occurs. The total annual precipitation of ~1100 mm is reported in the area (Mir et al. 2016). The March month receives maximum rainfall and October the least and September–November is usually a dry season (Mir and Jeelani 2015a). The weather is highly variable largely owing to the variation in altitude and aspect (Neve 1933; Bhat et al 2019).

17.2.1 Geology and Geomorphology

The main lithological units exposed in the area include, the Agglomeratic Slate, Nidhatbagh beds, Panjal Volcanics, Gangamopteris beds, Zewan Formation, Triassic Formation, Karewa Formation and recent Alluvial deposits (Fig. 17.2). The Agglomeratic Slate consists of pyroclastic slates, conglomerates and forms the lower part of the Panjal Volcanic Series. The Agglomeratic Slate is overlain by Panjal Volcanics that are generally basaltic to andesitic in composition. Near Nishat garden, the Nishatbagh beds are also present over the Agglomeratic Slate. Thin Gangamopteris beds of light colored shale overlie the Panjal Volcanics which are overlain itself by fossiliferous limestones and shales of Zewan Formation. The Zewan Formation is followed by Triassic formations that are mainly composed of homogeneous compact, light grey colored massive limestone with thin shale and sandstone. The Karewas flanking the surrounding mountain precipices overlie the folded Triassic and pre-Triassic rocks. The Karewas are fluviatile and lacustrine deposits that are classified into Nagum and Dilpur Formations. The Nagum Formation comprises blue, grey and buff silts and sands and has been intermixed with conglomerates. The Dilpur Formation comprises alternating laminated yellow marls and silts and sands. The recent Alluvium is present in the low-lying areas deposited by the Jhelum River and its tributaries. The alluvium consists of finely compacted detrital sediments such as the loam, clay, silt and sand (Mir and Jeelani 2015b; Mir et al. 2016; Alam et al. 2017, 2018).

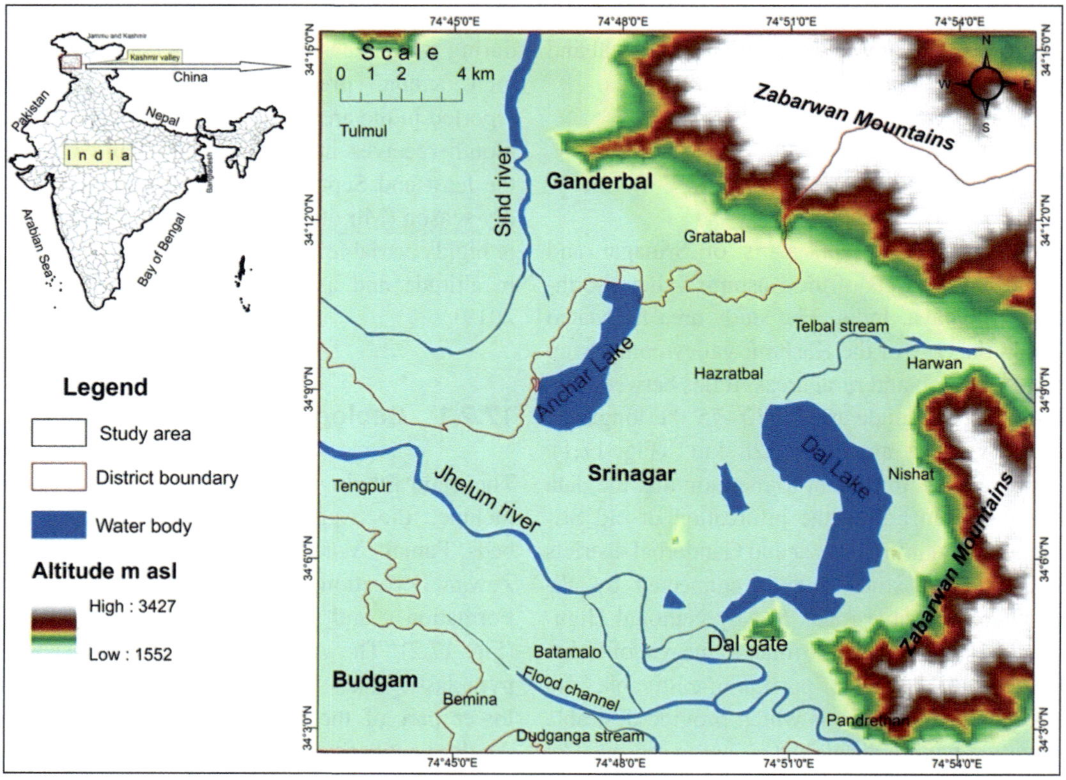

Fig. 17.1 Location map of the study area (*Source* USGS data)

The study area has conspicuous physiographic variations comprising moderately high hills, mountain ranges and alluvial tract. It occupies the central sector of the valley sandwiched between the Pir-Panjal Range towards the southwestern side and the Greater Himalayan Range towards the northeastern side. In the study area, the parts of the Greater Himalayan Mountain Range are called as Zabarwan Mountains. Nearly 50% of the area is covered by high hills characterized by hilly rugged and undulating topography. Flat alluvial terrain occurs in the flood plains of Jhelum River and its major streams. Alluvial fans are predominant feature occurring at foothills. In the peripheral areas typical Karewa table lands are present. The general topographic slope in the northern and southern parts is towards the south and north while in the central part the master slope is towards west.

17.3 Materials and Methods

In this study, a total of 140 groundwater samples were collected during premelting and postmelting seasons in one-liter polyethylene bottles. The collection of samples included 70 samples each in premelting/or winter season (December 2004) and postmelting/or summer season (June/July, 2005) between 10.00 and 15.00 h from different sampling stations (Fig. 17.3). The samples were collected from 11 springs, 5 dug wells and 54 bore wells representing the various sources of groundwater used in rural as well as urban areas (Table 17.1). To avoid the effect of rust of pipes and stagnant water in bore-wells, pumping was done for five to ten minutes continuously prior to sample collection. In case of springs and dug wells the samples were collected at depths greater than 30 cm below the water surface to avoid the

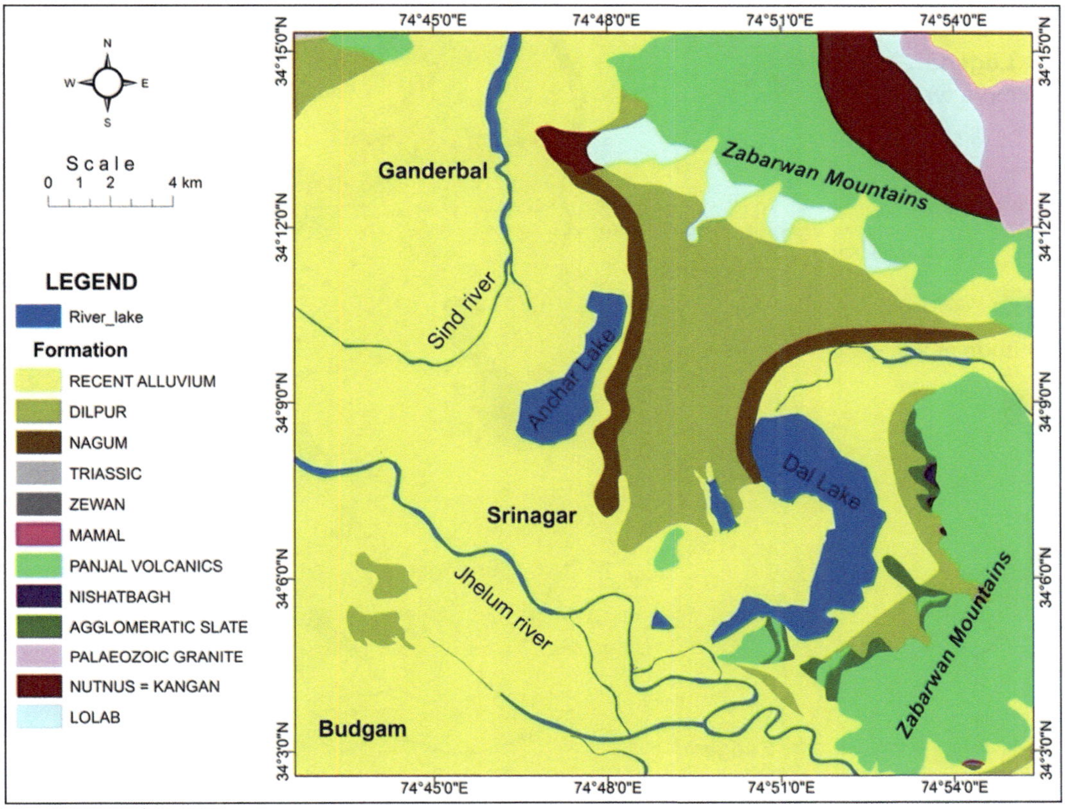

Fig. 17.2 Geological map of the area (after Mir et al. 2016)

collection of floating debris. The sample containers were also washed with conc. HNO_3 followed by a complete removal with distilled water followed by washing with the water that was to be sampled. The water temperature, pH and conductivity were measured in situ whereas the other parameters were determined in Laboratories at Center of Research for Development (CORD) and Department of Geology and Geophysics, University of Kashmir, Srinagar.

The chemical analysis was carried out after standard methods (Trivedy and Goel 1984; APHA 2001; Jasrotia et al. 2013; Sah et al. 2017; Bisht et al. 2018; Taloor et al. 2020). Water temperature was determined by the potable laboratory thermometer and pH with the help of digital pH meter that was standardized before use with a buffer solution having pH of 7. EC was determined with the help of a Conductivity meter that was standardized with KCl (0.01 M) solution. The estimation of Ca^{2+} and $Mg^{2+}$was done by EDTA titration using Erichrome black T and murexide as indicators, where for the Cl^- estimation the $AgNO_3$ (0.02 N) titration using potassium chromate (5%) as an indicator was used. The HCO_3^- was determined by HCl (0.01 N) titration where methyl orange was used as an indicator. Na^+ and K^+ions were determined by Flame Emission Photometry. Spectrophotometric method was used for the estimation of SO_4^{2-} and NO_3^-. The results expressed in mg/L were converted into meq/Lunits for further processing, plotting and interpretation.

17.4 Results and Discussion

17.4.1 Geochemical Characteristics of Groundwater

The various physicochemical parameters analyzed include the pH, electrical conductivity

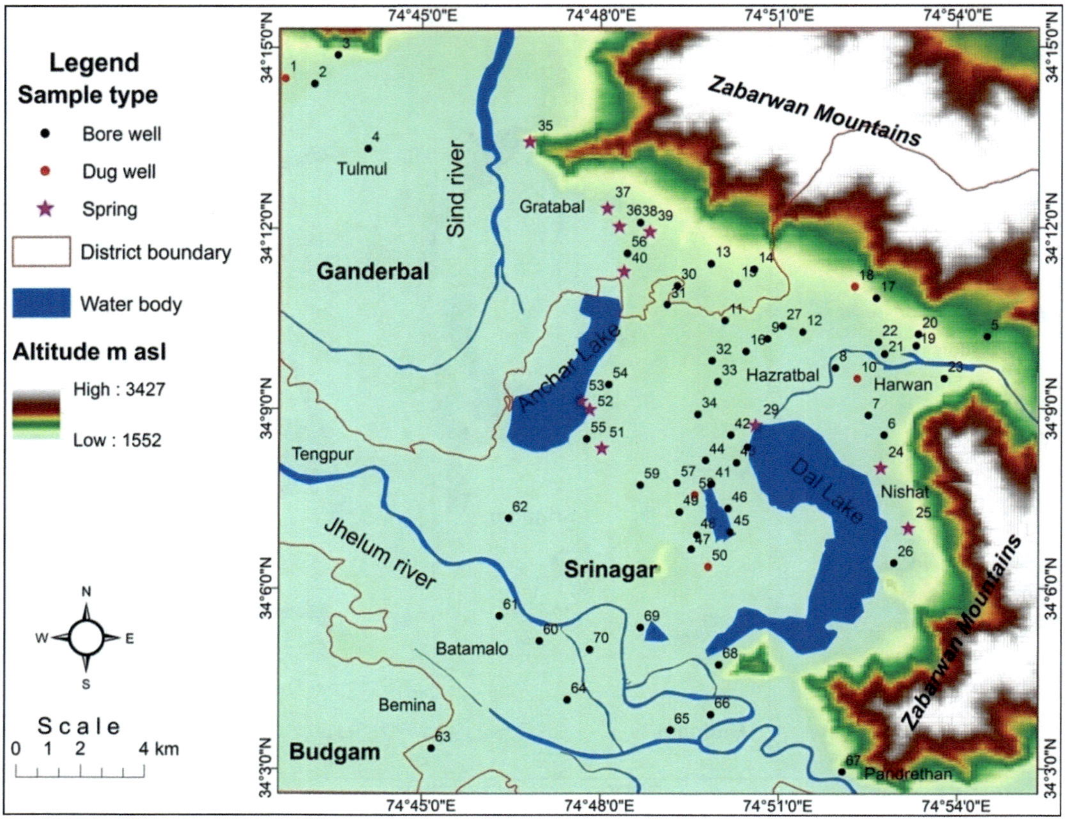

Fig. 17.3 Sampling site map of the area (*Source* Authors data)

(EC), total dissolved solids (TDS), total hardness (TH), Ca^{2+}, Mg^{2+}, Na^{+}, K^{+}, Cl^{-}, SO_4^{2-}, HCO_3^{-} and NO_3^{-}. The statistical overview of the results is given in (Table 17.2a and b). The groundwater temperature ranged from 11 to 15 °C with a mean of 13.06 °C. Temperature showed least variation in the study area. In springs, the highest temperature was recorded at location 51 (15 °C) and 53 (14 °C) and lowest at location 29 and 35 (11 °C). In dug wells and bore wells highest temperature was recorded at location 10 (13 °C) and 48 and 49 (15 °C) and lowest at location 1 and 18 (10 °C) and at location 4, 13, 14, 15, and 38 (12 °C) (Fig. 17.4a). However, there was a slight increase in temperature during postmelting season being relatively higher in wells at location 50 (21° C); 45, 46, 47, 48 (20 °C) that are either shallow or located around urban limits.

The TDS ranged from 107 to 710 mg/L with a mean concentration of 382.21 mg/L. TDS was lower in springs (107–429 mg/L), moderate in dug wells (141–704 mg/L) and high in tube wells (136–710 mg/L). Among springs, the high TDS was found at locations 29, 37, 39, 40 and 51 varying between 383 and 429 mg/L respectively whereas, the low TDS was found at locations 24 and 25 (115 and 107 mg/L). Among dug wells, the high TDS was found at locations 50 and 58 (704 and 592 mg/L) while lowest concentration was found at location 10 (141 mg/L) (Fig. 17.4 b). Among tube wells high TDS was present at a number of locations whereas the lowest values were found at locations 4, 6, 8, 14 and 15 varying between 151 mg/L to166 mg/L. As per the Carrols TDS classifications (1962), the groundwater falls in fresh water category.

The EC varied between 168 and 1110 μS/cm with a mean of 587.65 μS/cm. EC was lower in springs (168–670 μS/cm) and higher in dug wells (220–1100 μS/cm) and tube wells (212–1110

Table 17.1 Details of the sampling locations of the area

S. no	Location no	Location name	Sample type	Area	S. no	Location no	Location name	Sample type	Area Type
1	35	Gandarbal	Spring	Rural	36	38	Garatbal(1)	Bore well	Rural
2	36	Nagabal	Spring	Rural	37	54	Umarhair	Bore well	Rural
3	37	Darind	Spring	Rural	38	56	Pandach	Bore well	Rural
4	39	Garatbal(2)	Spring	Rural	39	6	Shalimar(1)	Bore well	Urban
5	40	Pandach	Spring	Rural	40	7	Shalimar(2)	Bore well	Urban
6	24	Ishber	Spring	Urban	41	16	Mirakshah colony	Bore well	Urban
7	25	Laam	Spring	Urban	42	23	Harwan	Bore well	Urban
8	29	Habbak(2)	Spring	Urban	43	26	Bren	Bore well	Urban
9	51	Naushera	Spring	Urban	44	28	Habbak(1)	Bore well	Urban
10	52	Buchpora(1)	Spring	Urban	45	32	Zakura(1)	Bore well	Urban
11	53	Buchpora(2)	Spring	Urban	46	33	Zakura(2)	Bore well	Urban
12	1	Badampora	Dug well	Rural	47	34	Malabag	Bore well	Urban
13	10	Telbal(2)	Dug well	Rural	48	41	Professor colony	Bore well	Urban
14	18	Chatraham(2)	Dug well	Rural	49	42	Nasemabad	Bore well	Urban
15	50	Khajarbal(2)	Dug well	Urban	50	43	Nasembag	Bore well	Urban
16	58	Sadarbal(2)	Dug well	Urban	51	44	Kanitar	Bore well	Urban
17	2	Barsu	Bore well	Rural	52	45	Nageen(1)	Bore well	Urban
18	3	Wandhama	Bore well	Rural	53	46	Nageen(2)	Bore well	Urban
19	4	Tulmula	Bore well	Rural	54	47	Khajarbal(1)	Bore well	Urban
20	5	New Theed	Bore well	Rural	55	48	Shahabad	Bore well	Urban
21	8	Telbal(1)	Bore well	Rural	56	49	Bagwanpora	Bore well	Urban
22	9	Batpora	Bore well	Rural	57	55	Anchar	Bore well	Urban
23	11	Sadpora	Bore well	Rural	58	57	Sadarbal(1)	Bore well	Urban
24	12	Gassu	Bore well	Rural	59	59	Lalbazar	Bore well	Urban
25	13	Shehama	Bore well	Rural	60	60	Chatabal	Bore well	Urban
26	14	Bakura(1)	Bore well	Rural	61	61	Qamawari	Bore well	Urban
27	15	Bakura(2)	Bore well	Rural	62	62	Noorbagh	Bore well	Urban
28	17	Chatraham (1)	Bore well	Rural	63	63	Bemina	Bore well	Urban
29	19	Danihama(1)	Bore well	Rural	64	64	Batamaloo	Bore well	Urban
30	20	Danihama(2)	Bore well	Rural	65	65	Jawharnagar	Bore well	Urban
31	21	Mulphak	Bore well	Rural	66	66	Rajbagh	Bore well	Urban
32	22	Burzhama	Bore well	Rural	67	67	Pandhrathan	Bore well	Urban
33	27	Wanihama	Bore well	Rural	68	68	Dalgate	Bore well	Urban
34	30	Gulabag(1)	Bore well	Rural	69	69	Babdemb	Bore well	Urban
35	31	Gulabag(2)	Bore well	Rural	70	70	Nawabazar	Bore well	Urban

(*Source* Authors data)

Table 17.2 Statistical overview of hydrogeochemical characteristics during **a** premelting season (December 2004) and **b** during postmelting season (June–July, 2005) of the area

(a)

Parameter/constituent	Springs		Dug wells		Bore wells		Overall range	Overall mean
	Range	Mean	Range	Mean	Range	Mean		
pH	7.10–8.07	7.85	7.16–8.03	7.62	7.10–8.30	7.58	7.10–8.30	7.62
Temp. (°C)	11–15	12.64	10–13	12.4	12–15	13.20	11–15	13.06
EC (μS/cm)	168–670	515.18	220–1100	635	212–1110	598.03	168–1110	587.65
TDS (mg/L)	107.429	329.54	141–704	406.4	136–710	387	107–710	382.21
Ca^{2+} (mg/L)	36–105	66.18	27–100	66.80	20–127	73.77	20–127	72.09
Mg^{2+} (mg/L)	5–25	14.27	8–28	19.40	6–35	17.77	5–35	17.34
Na^{+} (mg/L)	3–42	17.63	7–70	37.0	6–80	30.93	3–80	28.28
K^{+} (mg/L)	0.5–9.0	3.07	2.2–8.0	5.94	0–10	3.58	0–9	3.67
Cl^{-} (mg/L)	13–60	30.09	9–93	40.20	7–120	14.61	7–120	39.7
HCO_3^{-}(mg/L)	120–350	266.81	185–455	326	95–450	304.17	95–455	299.87
SO_4^{2-} (mg/L)	1.5–4.0	5.59	2.1–4.2	3.24	1.5–5.0	2.65	1.5–5.0	3.15
NO_3^{-}(mg/L)	1.4–4.0	2.59	0.9–8.5	4.54	0.5–6.8	2.55	0.5–8.5	2.7
TH(mg/L)	115–340	231.18	105–365	250.20	86–410	250.73	86–410	247.64

(b)

Parameter/constituent	Springs		Dug wells		Bore wells		Overall range	Overall mean
	Range	Mean	Range	Mean	Range	Mean		
pH	6.95–7.91	7.61	7.01–8.27	7.52	6.98–8.36	7.47	6.95–8.36	7.50
Temp. (°C)	15–19	16.90	15–21	17.20	14–20	16.27	14–21	16.20
EC (μS/cm)	140–698	431.18	160–1290	598.0	143–1305	543.13	140–1305	529.46
TDS (mg/L)	90–447	276.09	102–826	383.0	92–835	345.75	90–835	345.75
Ca^{2+} (mg/L)	25–97	42.82	21–75	52.0	23–128	58.39	21–128	55.50
Mg^{2+} (mg/L)	3–22	11.82	9-55	22.20	5–56	16.77	3–56	16.38
Na^{+} (mg/L)	2.1–30	14.69	4–74	30.0	3.5–106	28.39	2.1–106	26.34
K^{+} (mg/L)	0.25–6.5	2.17	0.9–8.9	4.90	0–11.8	2.86	0–11.8	2.74
Cl^{-} (mg/L)	7.2–43	21.68	7–95	36.80	3–105	31.10	3–105	30.3
HCO_3^{-}(mg/L)	85–315	187.64	145–515	288	105–525	237.15	85–505	265
SO_4^{2-} (mg/L)	1.1–3.3	2.11	1.15–3.1	2.27	0.3–4.15	2.38	0.3–4.15	2.33
NO_3^{-}(mg/L)	0.9–4.3	2.03	0.3–8.8	4.05	0–6.3	2.02	0.0–8.5	2.17
TH(mg/L)	79–300	214.54	94–405	224.8	80–410	215.51	79–410	211.79

(*Source* Authors data)

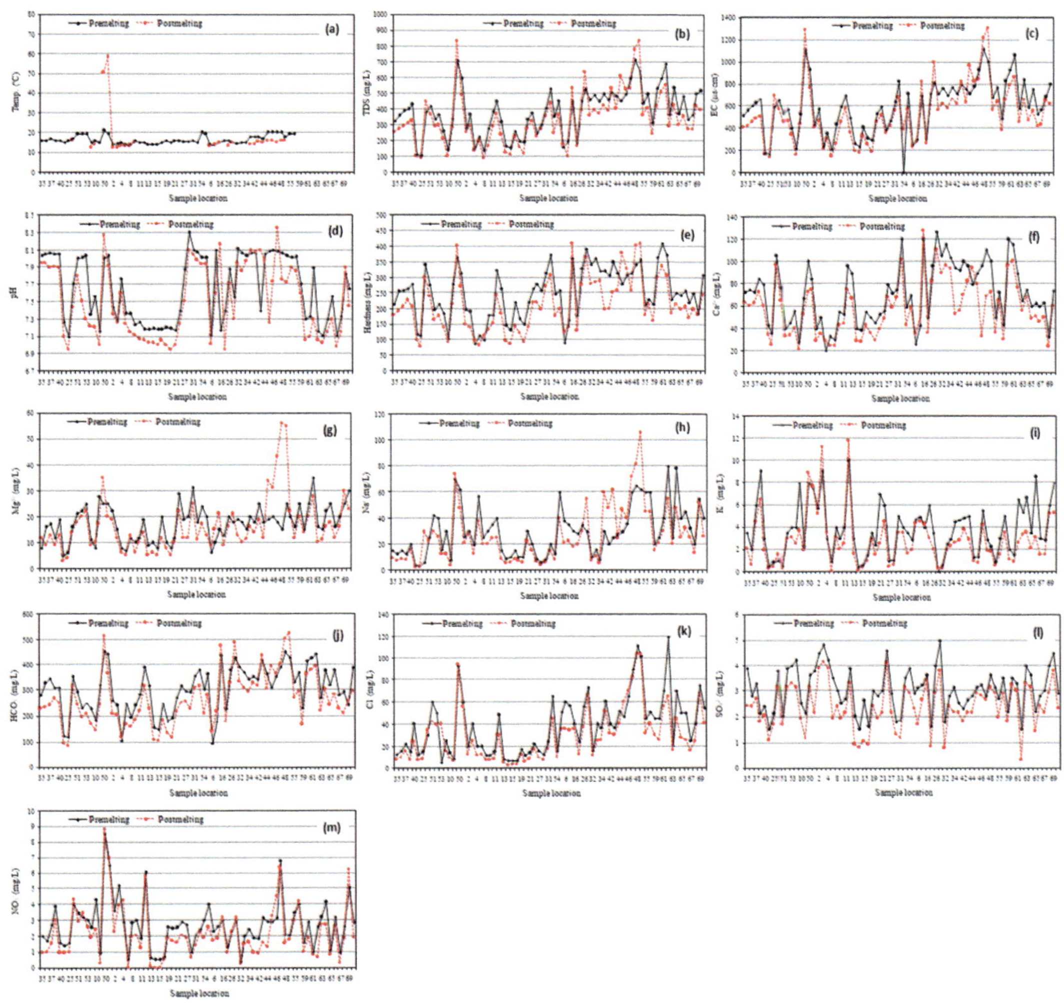

Fig. 17.4 Spatio–temporal variation of physicochemical parameters of groundwater of the area (*Source* Authors data)

µS/cm). In springs, high EC was found at locations 29, 39, 40, 51 varying between 599 and 670 µS/cm whereas low EC was found at locations 24 (168 µS/cm) and 25 (180 µS/cm). In dug wells the location 50 (1100 µS/cm) and 58 (925 µS/cm) showed higher EC while low EC was found at location 10 (220 µS/cm). In tube wells, a large number showed high EC while the low EC was found at locations 8, 4, 6, 14 and 15 varying from 212 to 260 µS/cm (Fig. 17.4c). The pH ranged between 7.10 and 8.30 with an average of 7.62. The pH of springs, dug wells and tube wells ranged from 7.1 to 8.07, 7.16 to 8.03 and 7.10 to 8.30. The pH of the area does not mark any significant variation (Fig. 17.4d). The TH ranged between 86 mg/L and 410 mg/L with an average of 218 mg/L. The lower values were found in springs except the location no. 29 (340 mg/L) and highest in tube wells with the exception of location 4 (94 mg/L) (Fig. 17.4e). The groundwater is found to be moderately hard to very hard as per the Sawyer and MaCarthy (1967).

Among the cations, the Ca^{2+} concentration varied from 20 to 127 mg/L with an average of 73.77 mg/L. The Ca^{2+} was high in tube wells (20–127 mg/L) with the exception of location 4 (20 mg/L), moderate in springs (36–105 mg/L) with the exception of location 29 (105 mg/L) and relatively low in dug wells (27–100 mg/L) with the exception of locations 50 and

58 (100 mg/L) (Fig. 17.4f). Mg^{2+} concentration ranged from 5 to 35 mg/L with an average of 17.34 mg/L. Mg^{2+} did not fluctuated much and was found low in springs (5 mg/l–25 mg/L) with the exception of location 53 (25 mg/L), intermediate in dug wells (8–28 mg/L) and relatively high in tube wells (6–35 mg/L) with the exception of locations 5 (7 mg/L), 6 and 7 (6 mg/L) (Fig. 17.4g). Na^{+} concentration varied from 3 mg/L to 80 mg/L with an average value of 28.28 mg/L. It was high in tube wells (6–80 mg/L) with the exception of few locations (Fig. 17.4h), moderate in dug wells (7–70 mg/L) with the exception of location 50 (70 mg/L) and low in springs (3–42 mg/L) with the exception of location 52 (42 mg/L). K^{+} concentration varied from 0–10 mg/L with a mean of 3.67 mg/L. K^{+} showed uniformity in spatial variation, ranging in springs from 0.5–9 mg/L with relatively high at location 39 (9 mg/L). It varied from 2.2 to 8.0 mg/L in dug wells being higher at location 10 and 50(8 mg/L) and ranged from 0-10 mg/L in tube wells being higher at location 12 (Fig. 17.4i).The mean concentration was relatively high in dug wells (5.94 mg/L), moderate in tube wells (3.58 mg/L) and low in springs (3.07 mg/L).

Among the anions, HCO_3^- was the dominant anion and represented the total alkalinity of the groundwater. HCO_3^- concentration varied between 95 and 455 mg/L with an average value of 2.99.87 mg/L. It was high in dug wells (185–455 mg/L) with the exception of location 10 (185 mg/L), moderate in tube wells (95-450 mg/L) with the exception of locations 6 and 7 (95, 105 mg/L) respectively and low in springs (120-350 mg/L) with the exception of 29 (350 mg/L) and 37 (345 mg/L) (Fig. 17.4j). The Cl^- content varied from 7 mg/l to 120 mg/L with a mean of 39.7 mg/L. The higher values were found in tube wells with the exception of locations 14, 15, 17 (7 mg/L) and 13 (9 mg/L), moderate in dug wells with anomalous values at locations 50 (93 mg/l) and 58 (60 mg/L) and low in springs with the exception of locations 51 (60 mg/L), 52 (50 mg/L) and 53 (55 mg/L) (Fig. 17.4k). $SO_4^{2-}$was found to be at lower levels ranging from 1.5 to 6.6 mg/L with the mean value of 3.15 mg/L (Fig. 17.4l). NO_3^- levels varied from 0.4 to 8.5 mg/L with a mean concentration of 2.70 mg/L. The lower values were found in springs (0.9–8.5 mg/L) with relatively high values at locations 29 (4 mg/L) and 39 (3.9 mg/L), high in dug wells (0.9–8.5 mg/L) with the exception of location 18 (0.9 mg/L), moderate in tube wells (0.4–6.8 mg/L) with relatively high values at certain locations (Fig. 17.4m). Overall the results indicated that most of the physicochemical parameters are well below the standards of WHO (2006) and ISI (2012) as given in (Table 17.3).

17.4.2 Spatio-Temporal Variation of Geochemical Characteristics of Groundwater

The spatio-temporal variation of physicochemical parameters of springs, dug wells and bore wells are shown in (Fig. 17.4a–m). The springs in general showed a decrease in the concentrations of various ions during postmelting season thereby indicating the effect of dilution due to the infiltration and recharging of abundant water into the ground resulting from the melting of seasonal snow/ice in the area. However, the spring 29 located around the Habbak near Hazratbal urban area showed an increase in concentration of Na^{+}, Cl^- and NO_3^-. The increase in these ions reflected the significant influence of anthropogenic activities. This is an open spring and collects the sewage from a nearby municipal drain close to it. It is also inferred that the input of sewage from various municipal drains in this area has probably increased the levels of TDS and EC with a consequent reduction in the pH of groundwater.

The dug wells showed almost similar trend of major ions during both the seasons. However, a general decrease in the concentration levels of certain ions during postmelting season is observed. For example, a feeble increase in Mg^{2+} ion is noticed at location 10 that is an open nature and shallow well possibly due to some local/point source. The higher Mg^{2+} may also be

Table 17.3 Comparison of hydrogeochemical characteristics of the area with the National and International permissible standards

Parameter	Range		WHO (2006)		ISI (2012)	
	Premelting	Postmelting	Acceptable level	Max. Permissible level	Acceptable level	Max. permissible level
Temp	11.0–15.0	14–21	–	–	–	–
TDS	107–710	90–835	500	1500	500	3000
EC	168–1110	140–1305	–	1600	800	4800
pH	7.10–8.36	6.95–8.36	7–8.5	6.5–9.2	6.5–9.2	9.2
Ca^{2+}	20–127	21–128	75	200	75	200
Mg^{2+}	5–35	3–56	<30 (if SO_4. is 250 mg/L	150 (if SO_4. is 250 mg/L	30	100
Na^+	3–80	2.1–106	–	200	–	–
K^+	0–9	0–11.8	–	12	–	–
Cl^-	7–120	3–105	200	250	250	1000
HCO_3^-	30–455	85–505	–	–	–	–
SO_4^{2-}	1.5–6.61	0.2–6.5	200	400	150	400
NO_3^-	0.5–8.5	0–8.8	–	50	45	–
TH	86–410	79–410	100	500	300	600

(*Source* Authors data)

due to the precipitation of Ca^{2+} as a result of a decrease in pCO_2 of groundwater in the well. Similarly, the well at location 50 located SW bank of Nagin Lake showed a pronounced increase in Mg^{2+}, HCO_3^- and TH coupled with a slight increase in EC, TDS, pH, Na^+, K^+, Cl^-, NO_3^- levels during the postmelting season. The increase in the concentration of these ions is attributed to upwelling of water table to near surficial levels where the influx and leaching down of nutrients from the domestic wastes and vegetable gardens into the shallow water levels is occurring. The process is also causing an increase in the concentration of Na^+, K^+, Cl^- and NO_3^-ions in the groundwater of the area. The increase in Mg^{2+} and HCO_3^- may also be attributed to the decomposition of aquatic plants in the nearby Nagin Lake where groundwater apparently showed higher hardness. However, Ca^{2+} level showed conspicuous decrease which was attributed to the lower concentration of Ca^{2+} in the nearby recharging source area. The increased biological activity in this source area during postmelting season probably consumes a considerable amount of Ca^{2+}. Location 58 indicated a decrease in concentration level of all ions except a slight increase in K^+ and NO_3^-. This increase in K^+ and NO_3^- levels indicated an obvious influence of leaching of nutrients from nearby vegetable gardens during the postmelting season. During this season a lot of fertilizers are also added to vegetable gardens for soil amendments that may also contribute to higher concentration of these ions in the groundwater.

All the bore wells in rural areas also showed the effect of dilution except the locations 3, 4, 8, 12. In urban areas also the dilution effects were observed in 50% of the sampling locations. But, a differential increase in certain constituents reflected a significant impact of human activities on groundwater in addition to the sediment–water interaction. The increase in K^+ and NO_3^- levels in the above mentioned wells manifest the influence of anthropogenic activities through influx of local agricultural discharges. However, the increase in Ca^{2+}, HCO_3^- and TH at well 4 was obviously due to the influx of fresh recharging water from a stream that may probably be responsible for high pCO_2 of the groundwater. The well 8 and 10 located in the same locality also showed the same trend and indicated the influence of sub-surface environment and hydrogeochemical processes in this locality as well. Among bore wells nested in rural areas various physico-chemical parameters indicated the normal dilution effect in the postmelting season with certain exceptions. For examples, well 3 showed increase in K^+, NO_3^-, well 4 showed increase in Ca^{2+}, HCO_3^-, TH, K^+ and NO_3^-, well 8 showed a slight increase in Mg^{2+}, well 12 showed increase K^+ concentration and more or less constant concentration NO_3^-. Overall, the fluctuations with respect to alkaline earths may be attributed to the increased partial pressure of CO_2 that favored the dissolution of carbonate minerals in the area (Jasrotia et al. 2018, 2019).

In the urban localities of Srinagar district, the concentration levels of the physico-chemical parameters generally showed a decline. But, in about 50% of the wells an increasing trend was noticed in various constituents. Location 6 and 7 also indicated an increase in the concentration levels of Ca^{2+}, Mg^{2+}, HCO_3^- and TH in the postmelting season (Fig. 17.4d, e, i). whereas, the rest of the parameters showed a decreasing trend. The well at location 16 indicated an increase in the concentration levels of various physico-chemical parameters with the exception of Na^+, Cl^- and SO_4^{2-}. At location 28 an anomalous behavior was observed with respect to Mg^{2+}, Na^+, K^+, HCO_3^- and NO_3^- with an increase in EC and TDS as well. The rest of the parameters indicated the dilution effect. At location 34, there was an increase in the concentration of Ca^{2+}, Mg^{2+}, HCO_3^- and TH in the postmelting season whereas the rest of the parameters decreased in concentration levels. The well numbers 41 and 42 indicated an increase in Na^+ only (Fig. 17.4f). The pH of the bore wells showed least variation while other parameters show dilution influence. This indicated the possible effect of cation exchange reaction in the area.

At location 43, an increase in the concentration levels of Na^+, HCO_3^-, pH, EC and TDS was observed in postmelting season while other parameters showed decreasing concentrations. Leaching and dissolution of clay minerals may be the dominant mode of mineral release around these locations. At location 45 the concentrations of the constituents showed an increase in the postmelting season except pH, K^+ and SO_4^{2-}. The well at location 46 nested in the same locality indicated the same trend except Ca^{2+} which showed a decrease in concentration. The well no. 47 indicated an increase in pH, Mg^{2+}, Na^+ and HCO_3^- only. This trend was similar to well 50. The possible mechanism for the higher concentration of all these ions in these localities seems to be the excessive withdrawal of groundwater that results in the formation of cone of depression at the well sites and creates favorable conditions for induced flow from nearby polluted surface water bodies. The influx of polluted water results in the alteration of the chemistry of groundwater and/or mixing of waters in the area. The wells at locations 48 and 49 reflected the same trend with enrichment of Mg^{2+}, Na^+, TH, HCO_3^-, EC and TDS in the postmelting season. Besides, an increase in the concentration of Cl^- was also observed at well 49. The rest of the parameters showed a decrease in concentration. At location 55, only Mg^{2+} concentration showed an increase whereas the well no. 57 indicated an increase in NO_3^- levels only. Similarly, at location 69 an increasing trend was observed in Mg^{2+}, HCO_3^- and NO_3^-ions (Fig. 17.4e, i, l). The increase in Mg^{2+} content at well 55 was attributed possibly to the longer flow path followed by water. The Mg^{2+}also remains in solution for relatively longer time because of its solubility.

17.4.3 Sources of Major Ions in Groundwater

The infiltrating water pickup a bulk of the ionic species from the rock/sediment matrix and follows a trend corresponding to its environment (Jeelani 2004). In the present area, the general order of cations in the groundwater during premelting season was (i) $Ca^{2+} > Mg^2$

$^{+} > Na^{+} > K^{+}$ (61%) (ii) $Ca^{2+} > Na^{+} > Mg^{2+} > K^{+}$ (37%) (iii) $Na^{+} > Ca^{2+} > Mg^{2+} > K^{+}$ (2%) whereas the order of anions was $HCO_3^{-} > Cl^{-} > SO_4^{2-}$. During the postmelting season a similar cationic and anionic trends prevailed with 63% and 37% of the total samples falling in the first two dominant orders respectively. However, at certain locations the Mg^{2+} and Na^{+} were found to replace each other in order of dominance. At location 6 the dominance of cations ($Na^{+} > Ca^{2+} > Mg^{2+} > K^{+}$) and anions ($Cl^{-} > HCO_3^{-} > SO_4^{2-}$) during premelting season indicated a shift in dominance of cations ($Ca^{2+} > Na^{+} > Mg^{2+} > K^{+}$) as well as anions ($HCO_3^{-} > Cl^{-} > SO_4^{2-}$) during the postmelting season. Furthermore, it is observed that the TDS ranged from 107 to 710 mg/L during premelting season and from 90 to 835 mg/L during postmelting season while the weight ratio of Na/Na + Ca was 0.04: 0.55 and 0.06:0.56 during the pre- and postmelting season respectively. The range of TDS and the ratio of Na/Na + Ca suggested the confinement of the sampling points in the rock dominance category (Gibbs 1970). These observations indicated that various litho-units/rock types and their weathered products have mostly contributed ionic composition to the groundwater.

Moreover, the scatter plots of Ca + Mg versus TZ^{+}, Ca + Mg versus HCO_3, Ca + Mg vs. $HCO_3 + SO_4$, Ca + Mg versus Na + K, Na + K versus Cl (Fig. 17.5a-f). suggested that Ca + Mg are the dominant cations and HCO_3 is the dominant anion. The Ca + Mg versus HCO_3 plot (Fig. 17.5a). and Ca + Mg versus Na + K (Fig. 17.5d), suggested the contribution from carbonate lithology to the major ions significantly. However, the trend line of Ca + Mg versus $HCO_3 + SO_4$ plot (Fig. 17.5b) suggested some contribution of ions from the silicates or sulphides also. This is also evident from the plot of Ca + Mg versus TZ^{+} where points fall just below the 1:1 equiline and spread nearly in somewhat linear fashion thereby suggesting contribution of certain ions possibly from silicates and/or sulphides. In order to assess the bonding affinity of Na + K and Cl, their equivalent concentrations were also plotted assuming that alkalis occur mainly associated with chlorides. This scatter showed excess in Na + K ions and therefore implied that the alkali ions might be possibly combining with HCO_3^{-}as well as

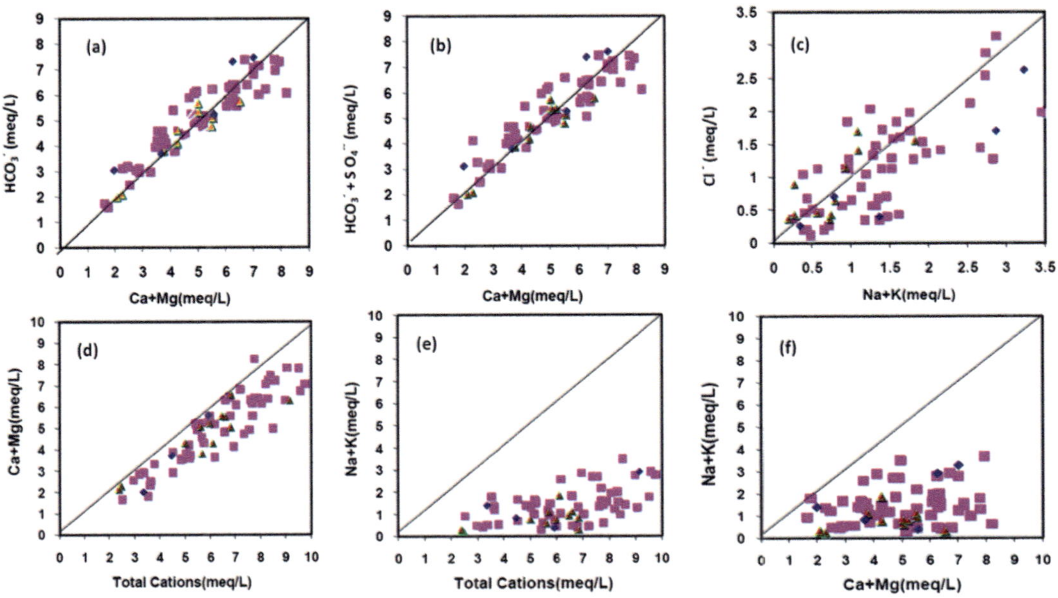

Fig. 17.5 Scatter diagrams of major ions for premelting season of the area. (*Note* purple square symbol = bore wells, blue symbol = dug wells and red triangle symbol = springs)(*Source* Authors data)

SO_4^- ions and indicated the influence of silicate weathering and dissolution on the chemical composition of groundwater. Further, the points falling close to the 1:1 trend line indicated the input of certain ions from water-logged areas, through evaporation processes and/or anthropogenic sources as well. Overall, the higher concentration of Ca and HCO_3 indicated the intense chemical weathering and dissolution of mainly carbonate rocks. The interaction of carbonate rocks with carbonic acid liberates abundance of HCO_3^- ions and that phenomenon may release Ca^{2+} ions resulting in the high concentration of Ca^{2+} in the water. The dissolved $Ca^{2+}$is derived by the dissolution of carbonate minerals particularly calcite whereas the origin of Mg^{2+} may be released through the dissolution of alumino-silicates, pyroxenes and amphiboles of the volcanic rocks and their weathered products as well in the area. Further, the low Ca + Mg/HCO_3 ratios could be the result of either Ca/Mg depletion by cation-exchange, chlorite dissolution, or bicarbonate enrichment, due to dissolution of secondary silicates.

During the postmelting season, the scatter plots (Fig. 17.6a-f), indicated almost similar trend of chemical composition of groundwater. The clustering of points toward relatively lower equivalent levels indicated the effect of general dilution on ionic constituents of groundwater. The effect of dilution was also indicated by the plot of Na + K/Cl where the points clustered toward lower concentrations around the equiline. But, the higher equivalent concentrations of Na + K showed a pronounced departure from 1:1 trend. Cl^- also exhibited an increasing tendency at higher concentrations with increasing Na^+, K as TDS increased. The Na^+ ions originate from the interaction of meteoric water with primary silicates and/or secondary alumino-silicates (clay minerals) with some modifying effect by the cation-exchange reactions. However, some enrichment from water-logged bodies and anthropogenic sources particularly in the low-lying areas at locations 6, 7, 47, 48, 49, 50, 62, 69 is also indicated from Na + K/Cl plot. Overall, these observations indicated that the weathering of silicate rocks as well as the anthropogenic activities as the sources of ions to groundwater. K^+ generally remained at low concentration levels during both pre- and post-melting seasons in the area. However, the higher concentration of K^+ at

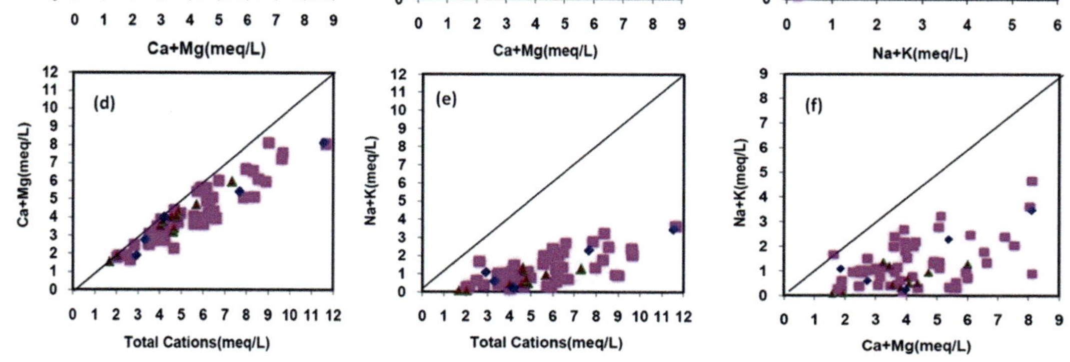

Fig. 17.6. Scatter diagrams of major ions for postmelting season of the area (*Note* purple square symbol = bore wells, blue symbol = dug wells and red triangle symbol = springs) (*Source* Authors data)

some locations (2, 3, 10, 12, 21, 23, 37, 39, 50, 58, 62, 64, 66, 69, 70) indicated an alarming anthropogenic influence. Soil amendments applied on paddy fields and floating gardens are also the possible secondary sources of increased K^+ concentration. Further, the effect of plants and fertilizers being added to the soil may be a significant source of K^+ in some intensively cultivated areas (Hem 1985; Jasrotia et al. 2018). Generally, the K^+ is liberated with greater difficulty from silicate minerals and exhibits a strong tendency to be reincorporated into solid weathering products especially certain clay minerals.

Relatively low concentration of Cl^- manifest low background levels in the lithological source of the area. No defined linear relationship between concentration of Cl^- and Na^+ indicated the negligible control of dissolution of this ion from lithology/or halite. However, the anomalous concentrations in urban localities and near polluted surface water bodies and/or water-logged bodies indicated the influence of increasing anthropogenic activities such as the effect of domestic wastes via improper sewage disposal, faulty drainage systems. For identifying the source of the NO_3^- levels in the water, the criteria after (Madison and Brunett 1984) was used. The NO_3^- levels were generally low and fall in the concentration ranges of <0.2 mg/L that represent the possible low natural/background levels. However, high concentrations of NO_3^- (0.21–3.0 mg/L) occurring in isolated wells and/or localized areas at locations 12, 29, 39, 47, 50, 56, 5758, 64, 69 indicated the human influences also. The higher concentration of K^+ and NO_3^- during both the seasons coincides at certain locations generally close to areas of agricultural activities. The domestic wastes and floating gardens in the close proximity to surface water bodies may also be responsible for its higher concentration. Overall, it is inferred that the weathering and dissolution of carbonate as well as silicates rocks are mainly influencing the major ion chemistry of groundwater of the area. However, the contribution of silicate weathering is less significant than the carbonate dissolution whereas, the anthropogenic activities have a modifying influence on the chemical composition of groundwater.

17.4.4 Hydrogeochemical Facies

In this study, the concept of hydrogeochemical facies was used to denote the diagnostic chemical character of water solutions in hydrogeochemical systems (Morgan and Winner 1962; Ophori and Toth 1989; Back 1966; Mir et al. 2016). The facies reflect the effect of chemical process in the lithological environment and groundwater flow patterns (Back 1966). The trilinear piper diagrams (Piper 1944) were used to show the chemical character of groundwater. From this plot the analysis and inference were drawn based

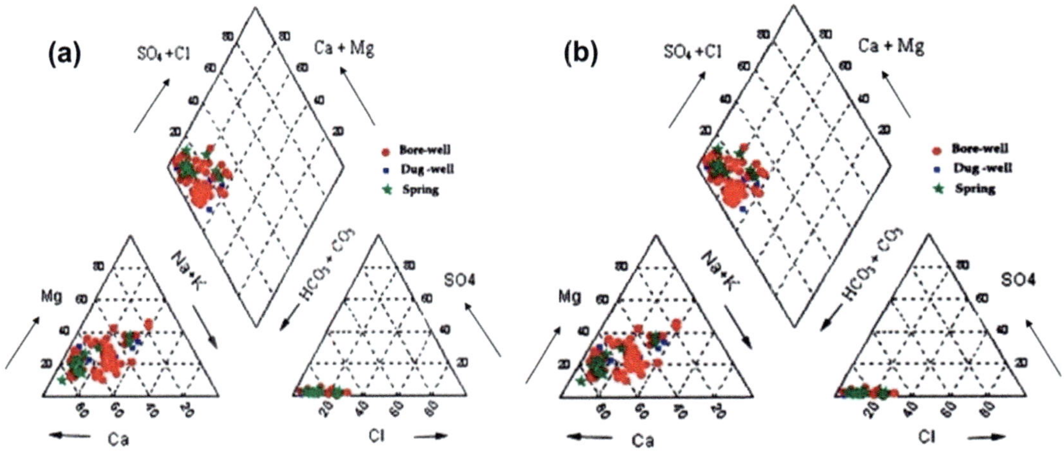

Fig. 17.7 Piper-Trilinear diagram for **a** premelting season and **b** postmelting season of the area (*Source* Authors data)

Table 17.4 Hydrochemical facies at different locations during the **a** premelting season and **b** postmelting season of the area

(a) Premelting season			(b) Postmelting season		
$Ca\text{-}HCO_3$	Ca–Mg–HCO_3	Hybrid waters	$Ca\text{-}HCO_3$	Ca–Mg–HCO_3	Hybrid waters
1, 9, 12, 13, 14, 16, 19, 20, 24, 25, 26, 27, 28, 29, 30, 32, 33, 34, 35, 36, 37, 38, 39, 40, 41, 42, 43, 44, 45, 46, 51, 60, 61, 63, 67	2, 17, 18, 21, 22, 23, 31, 5459, 65, 6, 68, 70	3, 4, 5, 8, 10, 11, 47, 48, 49, 50, 52, 535, 56, 57, 58, 64, 6, 7,62, 69	1, 4, 9, 11, 12, 13, 14, 15, 19, 20, 24, 25, 26, 27, 28, 29, 30, 32, 33,34, 35, 36, 37, 38, 39,40, 4, 51, 60, 61, 63, 67	2, 7, 17, 18, 21, 22, 23, 31, 45, 46, 54, 59, 65,66, 68, 70	3, 5, 8, 10, 41, 41, 43, 47, 48, 49, 50, 52, 53, 55, 56, 57, 58,62, 64, 6, 69

(*Source* Authors data)

on hydrogeochemical facies concept (Back and Hanshaw 1965). The Piper-trilinear diagrams for premelting and postmelting seasons (Fig. 17.7a, b) revealed the analogies, dissimilarities and different types of waters (Table 17.4a, b) as the data falls within the fields of 1, 3, and 5. The observation suggested that the alkaline earths exceeded alkalis and weak acids exceeded strong acids. Three hydrochemical facies/water types were broadly identified during the pre- and post-melting seasons as, (1) $Ca\text{-}HCO_3$ type (Ca > Mg), (2) Ca-(Mg)-HCO_3 type/hybrid waters, and (3) Hybrid waters/mixed-cation-bicarbonate waters. However, it is important to note that the above mentioned water types were obliterated (i.e., water type changed seasonally) at certain locations from premelting to postmelting seasons respectively. For instance, the shift was observed from (1) $Ca\text{-}HCO_3$ type to Ca-Mg-Na-HCO_3 type/hybrid waters at locations 41, 42 and 43, (2) $Ca\text{-}HCO_3$ to Ca-(Mg)-HCO_3 at locations 45 and 46 (3) Ca-(Mg)-HCO_3 to $Ca\text{-}HCO_3$ at locations 4 and11 and (4) from hybrid waters or mixed cation–anion water type of Ca-(Mg)-Na-HCO_3-Cl to Ca-(Mg)-HCO_3 at location 7.

17.4.5 Chemical Classification and Relationships of Groundwater Types

The trilinear diagrams are considered useful in bringing out chemical relationships among the groundwater samples than with other possible plotting methods as these diagrams are more definite in plotting (Walton 1970). The identified water types of the area are discussed as below.

17.4.5.1 Ca–HCO_3 and Ca-(Mg)-HCO_3

The characterization of the groundwater with respect to the hydrochemical facies of $Ca\text{-}HCO_3$ and Ca-(Mg)-HCO_3 indicated that the overall chemical character of groundwater was falling within the normal alkaline-earth water type group. This water type was found predominantly $\geq$ 70% in the area. The dominance of these two facies also indicated the influence of carbonate lithology on the chemistry of subsurface water. The occurrence of Ca- and Ca-(Mg)-HCO_3 type of water with high concentration of bicarbonates ($HCO^-{}_3$) and high Ca/Mg and Ca/Na ratio probably corresponds to its source from upper zone that is from shallow aquifer system (Freeze and Cherry 1979). This facies also reflected the movement of recharging meteoric water relatively shallow at depths with a limited mixing trend. These observations possibly reflected a primary stage in the evolution of groundwater with a limited migratory history in shallow aquifers. Ophori and Toth (1989) have described such type of waters as young recharging water that is occurring in recharge areas with HCO_3^- and Ca^{2+} ions as the dominant constituents.

17.4.5.2 Hybrid/Mixed Cation-Bicarbonate Water

This water type showed the enrichment of Na^+ an index of dissolution of silicates and/or soil-salts from the silt and clay strata in the area. The fine-grained clay and silt layers/lenses induce low permeability and retard the groundwater movement that enhances subsurface residence time of water thereby increasing the water–sediment interaction processes. This phenomenon favors release of Na into solution by exchange reactions. The Na^+ may exchange with other cations such as Ca or Mg within the clay minerals and other related minerals that form parts of the aquifer matrix. The exchange reactions are observed at most of the sites in the area. In this way, the enriching water type with Na-HCO_3 occurs. The exchange process can be represented as.

$$\mathrm{Ca/Mg(CaCO_3)_2} + \mathrm{Na_2}E \rightarrow \mathrm{2NaHCO_3} + \mathrm{Ca/MgE}.$$

(E = Exchanger).

An additional enrichment with respect to Cl^- has also been observed. A possible input of these ions is presumed to be from water-logged areas and/or anthropogenic sources such as from domestic wastes and agricultural wastes where nutrients usually leach down into the shallow horizons of the aquifer system.

17.4.6 Chemical Evolution of Groundwater

The chemical quality and evolution of groundwater along its underground flow paths and movement is dependent on the chemical and physical properties of surrounding host rocks, the quantitative, qualitative properties of infiltrating water in the recharge areas, the surface water bodies and the waste products of human activities (Matthess 1982). In the present area, the changes and evolution of the water chemistry along the flow paths in the shallow subsurface horizons was investigated. For this, eight flow-paths were chosen in the northwest, northeast and north-central portions of the study area. The flow-paths were selected approximately in the direction of flow that was determined on the basis of water level gradient observed at the sampling sites (Lone and Mir 2021). The flow-paths assumed continuity of flow from the topographic high points that is from recharge areas to the low-lying areas and were chosen to coincide with a maximum number of wells (Fig. 17.8).

Along the flow path 1 lying NW of the area, the groundwater chemistry indicated the tendency to evolve toward bicarbonate nature that is Ca-HCO_3 and Ca-Mg-HCO_3 water types. However, near location 3 (Wandhama) and 4 (Tulmul) the groundwater evolved toward a mixed cation–anion/hybrid water type. A slight shift from a mixed cation–anion/hybrid water type (Ca-Mg-Na-HCO_3) to Ca-Mg-HCO_3 (as indicated by the individual well no. 4) was observed in the postmelting season. This influence was probably due to the influx of fresh recharging water from a nearby local stream into the aquifer system. The fresh water builds up high partial pressure and enhances the dissolution of carbonate minerals in the aquifer matrix. It also obliterates the mixed-cation-bicarbonate character of groundwater in the area. Along the flow-path 2, the groundwater acquires a hybrid chemical character in the sub-surface environment. For example, the mixed water type of Ca-Mg-Na-$HCO_3$was observed in the fringe area at location 5. At this location the groundwater flow showed enrichment of Na^+. It is possibly due to the mixing of Na-rich water with meteoric water equilibrating at depth with sodium rich lithology via some deep fractures or joints or lineaments. Besides, at location 6 (Shalimar) there was slight enrichment of Cl^- indicating the possible localized anthropogenic influence.

Along the path 3 originating in the north near foot-hills of the Zabarwan Mountains, the groundwater evolved to Ca-HCO_3, Ca-Mg-HCO_3 and then to hybrid water of Ca-Mg-Na-HCO_3 near location 53. This facies indicated the addition of Na^+ possibly from clay and silt horizons. Along the path 4 and 5, the groundwater indicated an evolutionary trend toward alkaline water type of Ca- and Ca-Mg-HCO_3 facies and finally toward mixed cation–anion-bicarbonate water type with some enrichment of Cl^- as well. Along the path 6, the groundwater chemistry changed from Ca-Mg-HCO_3 to Ca-HCO_3 from well 18 to 20 and then again to Ca-

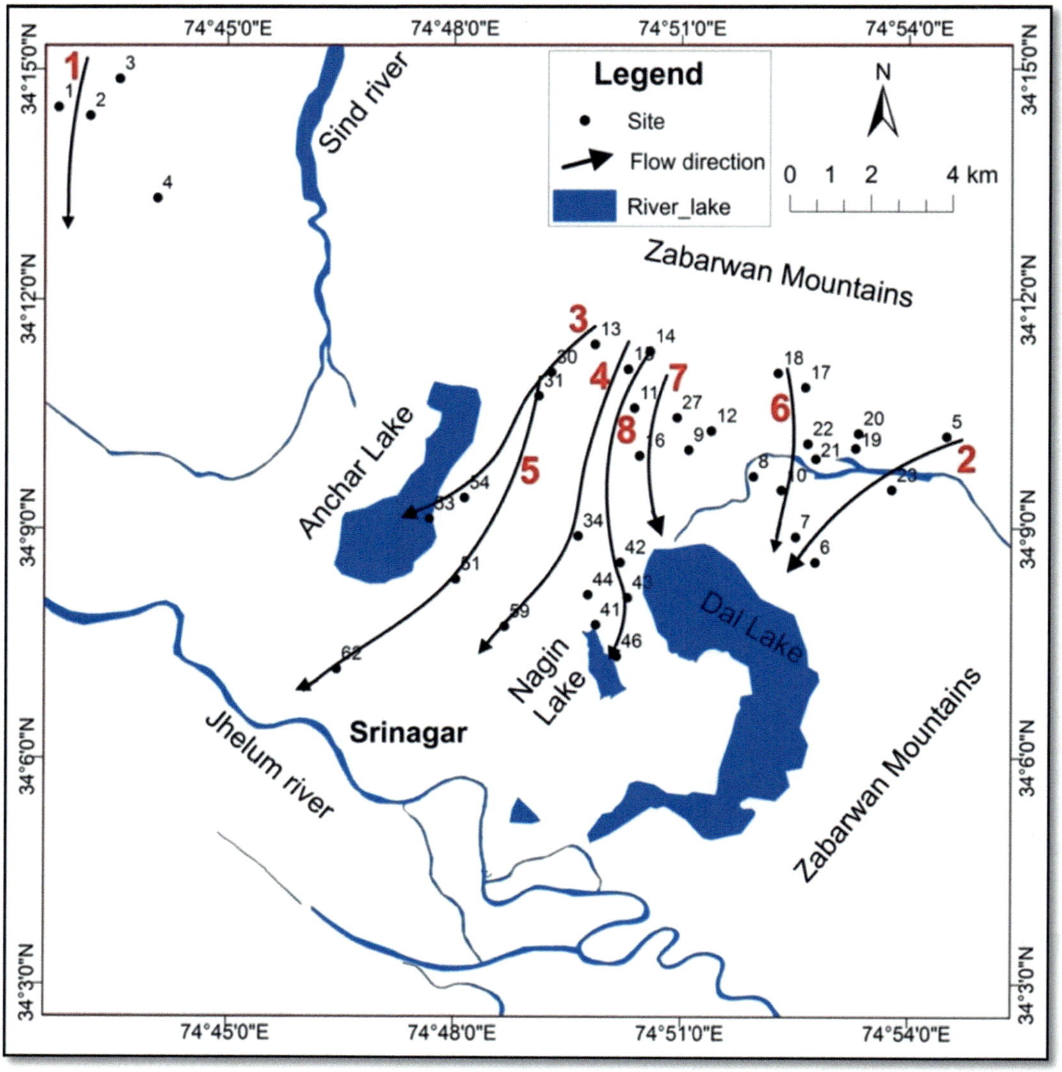

Fig. 17.8 Groundwater flow paths showing chemical evolution trends in the area (*Source* Authors data)

Mg-HCO_3 near well 21 where from it later evolved to mixed-cation-bicarbonate water type. A slight variation in the water chemistry was observed in the postmelting season with a slight addition of Mg^{2+} to Ca-HCO_3 water type. The flow path 7 indicated a tendency to evolve toward alkaline water type of Ca-HCO_3. The water retained this chemical character almost all along the down flow direction. The only change along the flow-path was impacted by the addition of Na^+ near the well 11 resulting in the formation of a mixed-cation-bicarbonate water type facies of Ca-Mg-Na-HCO_3. However, this hybrid chemical nature shifted to the dominant primary facies type of Ca-HCO_3, later along the flow path. Along the flow path 8 the groundwater evolved towards alkaline water type of Ca-HCO_3 that shifted to secondary facies type of mixed cation–anion-bicarbonate water type (Ca-Mg-Na-HCO_3) during the postmeltingseason at 43, 42, 41 locations.

Overall, the groundwater chemistry patterns along their flow-paths indicated the evolution of water toward alkaline waters with Ca-HCO_3 and

Ca-Mg-HCO_3 water types as the dominant primary hydrochemical facies. The predominance of bicarbonate waters in the area documented the open nature of groundwater system and the retention of largely meteoric character by the groundwater. The water types Ca-HCO_3 and Ca-Mg-HCO_3 with high concentration of bicarbonates and high Ca/Mg and Ca/Na ratio reflected the shallow groundwater flow system. Furthermore, dolomite dissolution is responsible for the higher concentration of HCO_3^- and Mg^{2+} in the groundwater and for the observed shift in the HCO_3/Ca ratio towards values 2:1. An analysis of Ca/Mg molar ratio also confirmed the dissolution of calcite and dolomite in the subsurface areas. However, there are some pockets along groundwater flow paths where groundwater exhibits a tendency to evolve toward mixed-cation bicarbonate water types of Ca-Mg-Na-HCO_3^- and Ca-Mg-Na-HCO_3-Cl (hybrid waters). These locations lie towards western areas around Nagin Lake. At places, the subsurface environment also furnished Na^+ ions by cation exchange reactions thereby enriching the water with Na-HCO_3 character. In addition, the water type Ca-Mg-Na-HCO_3-Cl almost similar to Ca-Mg-Na-HCO_3 revealed some additional enrichment of Cl^- at certain sites. Since, the Cl^- and SO_4^{2-} are not significant constituents in the lithology of the area; therefore, there is no tendency of water towards development of SO_4^{2-} or Cl^- facies. These findings indicated that the Chebotarev hydrochemical evolution sequence (Chebotarev 1955) is not relevant in this area.

17.4.7 Factors Controlling the Behavior of Major Ions

The overall chemical characteristics of groundwater and the relative abundance of the major ions have been evaluated on Langlier-Ludwig diagram (Fig. 17.9a, b).The trends of chemical alteration of meteoric water were classified into three chemical groups designated as G-I, G-II and G-III controlled by natural as well as the anthropogenic factors. G-I involved the dominance of Ca^{2+}, Mg^{2+} and HCO_3^- ions thereby indicating the dissolution and/or precipitation of carbonates in the area. Most of the points/locations falling in G-I are scattered all over in the area particularly along the peripheral Karewas and low-lying areas. About 80% of the locations showed the dominance of G-I classified typically as Ca-Mg-HCO_3water type. This local meteoric water (LMW) Ca-Mg-HCO_3 type indicated the percolation of water to shallow depths with a very little chemical alteration. This type of chemical character is possibly due to the local groundwater flow paths and a consequent lesser contact time of the water with the sub-surface aquifer matrix/or lithology. Furthermore, the G-I indicated the dissolution of carbonate lithology thereby contributing the Ca^{2+}, Mg^{2+}, and HCO_3^-ions to the groundwater.

The G-II waters were identified mainly by the abundance of HCO_3^-, Na^+ and/or Ca^{2+}, Mg^{2+} ions in the water. G-II showing the enrichment of Na^+ indicated a significant contribution from weathering and dissolution of primary and/or secondary silicates (soil-salts) in the area. This mixed chemical character was related to increased residence time of groundwater in the silt or clay strata within the sub-surface horizons. Generally, the intercepting silt and clay layers/lenses retard the frequent movement of groundwater and enhance the water–sediment interaction and hence, the residence time of ions in soil matrix. The favorable time for interaction releases excess of Na^+ions into the water. Similarly, the G-III almost similar to G-II showed a feeble increase of Cl^- ion in addition to Na^+ ion in the water. Overall, this group reflected the influence of anthropogenic activities. The Cl^- and Na^+ ions may have been discharged into the groundwater probably through the influx of contaminated water from domestic and/or polluted surface water bodies or water-logged bodies. The relatively high Cl^- concentration may be relate to input from sewage pollution and leachate, percolation and/or evaporation of water-logged bodies. The hydrogeochemical characteristics indicated that the meteoric character of groundwater is not completely obliterated. Further, the sources of the ions are

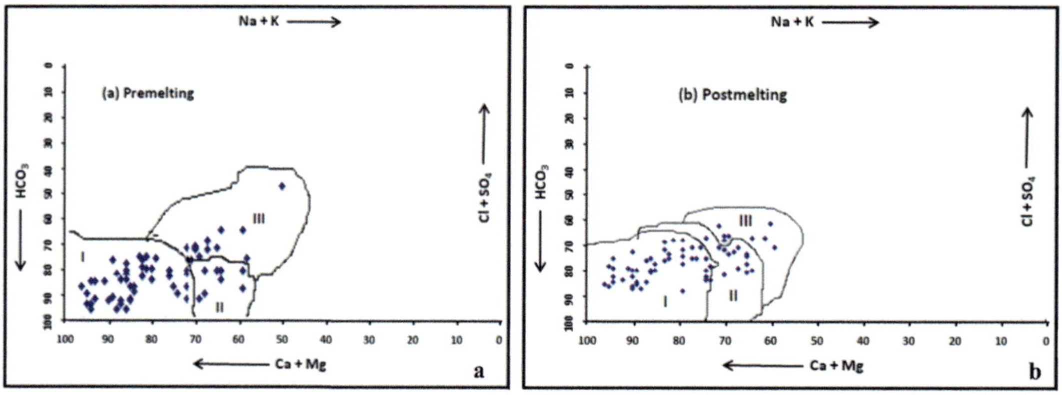

Fig. 17.9. Langlier—Ludwig diagram for **a** premelting season and **b** postmelting season of the area (*Source* Authors data)

dominantly controlled by the natural sources (lithological control). To a large extent the anthropogenic factors also influence the chemical nature of water (non lithological control). Overall, it is inferred that the areal distribution of different chemical groups in the area is governed by the disposition of sub-surface litho-units, surface water bodies and/or local topography.

17.5 Conclusions

In this study, the chemical typology combined with total mineralization of groundwater from different sources indicated that the carbonate and silicate (primary and secondary) weathering plays a dominant role in the characterization and chemical evolution of the groundwater. The dissolution of calcite, dolomite, intermediate basalts and/or clay minerals etc., largely determined the major element composition of the groundwater. The major element composition didn't reveal any significant spatio-temporal variation from the upland source or aquifer recharge areas to discharge points. However, enrichment with respect to certain ions at certain locations such as NO_3^-, K^+, Cl^-, Na^+, SO_4^{2-} due to human perturbation have a modifying effect on the ionic composition of groundwater. These chemical ions are mainly released from domestic wastes, improper sewage disposal and chemical fertilizers and animal waste on farm-lands and vegetable gardens, polluted surface water bodies and/or water-logged areas. The chemical characterization classified the groundwater into three groups designated as G-I, G-II dominantly controlled by lithological factors and G-III influenced by non-lithological factors that is anthropogenic activities. Although the groundwater is suitable for domestic purposes at current stage, but, the increased discharge and leaching of the anthropogenic nutrients may in future result in the contamination of these shallow groundwater resources in the area. The groundwater belongs to primary facies of Ca-HCO_3, Ca-Mg-HCO_3 water types with hybrid water types present locally at certain locations. The dominance of primary facies indicated that the groundwater is in its primary stage of evolution with a limited migratory history. This signature also indicated a limited flow of water in this area. The topographic high points enclosing this area favor circulation of water at relatively shallow depths. Overall, the observed chemical character revealed that the local flow system might be dominating in recharging the aquifers than the regional flow. However, to unravel the depth and lateral extent of regional groundwater circulation further exhaustive studies are required and recommended in this area.

Acknowledgements The authors would like to thank HoD, Department of Geology and Geophysics, and Director, CORD (Centre for Research and Development) University of Kashmir, Srinagar for their support during the work for M.Phil program.

References

Adams FD (1928) Origin of springs and rivers-a historical review. Fenia 50(1):16

Adimalla N, Dhakate R, Kasarla A, Taloor AK (2020) Appraisal of groundwater quality for drinking and irrigation purposes in Central Telangana, India. Groundw Sust Dev 10:100334. https://doi.org/10.1016/j.gsd.2020.100334

Adimalla N, Taloor AK (2020a) Hydrogeochemical investigation of groundwater quality in the hard rock terrain of South India using geographic information system (GIS) and groundwater quality index (GWQI) techniques. Groundw Sust Dev https://doi.org/10.1016/j.gsd.2019.100288

Adimalla N, Taloor AK (2020b) Introductory editorial for 'Applied Water Science' special issue: "Groundwater contamination and risk assessment with an application of GIS". Appl Water Sci 10:216. https://doi.org/10.1007/s13201-020-01291-3

Alam A, Bhat MS, Kotlia BS, Ahmad B, Ahmad S, Taloor AK, Ahmad HF (2017) Coexistent pre-existing extensional and subsequent compressional tectonic deformation in the Kashmir basin, NW Himalaya. Quat Int 444:201–208

Alam A, Bhat MS, Kotlia BS, Ahmad B, Ahmad S, Taloor AK, Ahmad HF (2018) Hybrid tectonic character of the Kashmir basin: Response to comment on "Coexistent pre-existing extensional and subsequent compressional tectonic deformation in the Kashmir basin, NW Himalaya (Alam et al (2018) 2017)" by Shah (2017). Quat Int 468:284–289

APHA (2001) Standard methods for the examination of water and waste. American Public Health Association, Washington DC

Back W (1966) Hydrochemical facies and groundwater flow patterns in northern part of Atlantic Coastal Plain. US Geological Survey Professional Paper 498-A:42

Back W, Hanshaw BB (1965) Chemical Geohydrology. Adv Hydro Sci 2:49–109

Baker MN, Horton RE (1936) Historical development of ideas regarding the origin of springs and groundwater. Eos Trans Am Geophy Union 17(2):395–400

Bhat MS, Alam A, Ahmad B, Kotlia BS, Farooq H, Taloor AK, Ahmad S (2019) Flood frequency analysis of river Jhelum in Kashmir basin. Quat Int 507:288–294

Bisht H, Arya PC, Kumar K (2018) Hydro-chemical analysis and ionic flux of meltwater runoff from Khangri Glacier, West Kameng, Arunachal Himalaya, India. Environ Earth Sci 77:1–16. https://doi.org/10.1007/s12665-018-7779-6

Biswas AK (1970) History of hydrology, 336 p, North-Holland Publishing Co, Amsterdam

Carrol D (1962) Rainwater as a chemical agent of geologic processes-a review, US Geol. Sur. Water-supply Paper, 1535-G. 18

Census Report (2001) District census book, Srinagar, p 26

Chebotarev II (1955) Metamorphism of natural water in the crust of weathering. Casmochemica Acta 8:22–48, 137–170, 190–212

Freeze RA, Cherry JA (1979) Groundwater. Prentice Hal, Englewood Cliffs, NJ

Gibbs RJ (1970) Mechanism controlling world water chemistry. Science 170:1088–1090.

Haque S, Kannaujiya S, Taloor AK, Keshri D, Bhunia RK, Ray PKC, Chauhan P (2020) Identification of groundwater resource zone in the active tectonic region of Himalaya through earth observatory techniques. Groundw Sust Dev 10:100337. https://doi.org/10.1016/j.gsd.2020.100337

Hem JD (1985) Study and interpretation of the chemical characteristics of natural Department of the Interior. US Geological Survey, Water, p 2254

ISI (2012) Indian Standards Institute Indian Standard Specification for drinking water IS: 10500

Jasrotia AS, Bhagat BD, Kumar A, Kumar R (2013) Remote sensing and GIS approach for delineation of groundwater potential and groundwater quality zones of Western Doon Valley, Uttarakhand, India. J Indian Soc Remote Sens 41(2):365–377

Jasrotia AS, Taloor AK, Andotra U, Bhagat BD (2018) Geoinformatics based groundwater quality assessment for domestic and irrigation uses of the Western Doon valley, Uttarakhand, India. Groundw Sust Dev 6:200–212

Jasrotia AS, Taloor AK, Andotra U, Kumar R (2019) Monitoring and assessment of groundwater quality and its suitability for domestic and agricultural use in the Cenozoic rocks of Jammu Himalaya, India: A geospatial technology based approach. Groundw Sust Dev 8:554–566

Jeelani G (2004) Effect of subsurface lithology on hydrochemistry of springs of a part of Kashmir Himalaya. Himal Geol 25(2):145–151

Khan A, Govil H, Taloor AK, Kumar G (2020) Identification of artificial groundwater recharge sites in parts of Yamuna river basin India based on remote sensing and geographical information system. Groundw Sust Dev 11:P100415. https://doi.org/10.1016/j.gsd.2020.100415

Lone KA, Mir RA (2021) A preliminary study on aquifers and its possible geometry in parts of Kashmir valley, India. J Geol Soc India (In press)

Madison RJ and Brunett JO (1984) Overview of the occurrence of nitrate in groundwater of the U.S., in National Water Summary 1984: USGS, Water Supply Paper 2275

Matthess G (1982) The properties of groundwater. Wiley, New York, p 498
Mir RA, Lone KA (2020) A recent scenario of groundwater quality in Kashmir, northwest Himalaya, India. Springer Nature Switzerland AG 2020. R. A. Bhat, K. R. Hakeem (eds.), Bioremediation and Biotechnology, Vol 4, https://doi.org/10.1007/978-3-030-48690-7_3
Mir RA, Jeelani G (2015a) Hydrogeochemical assessment of river Jhelum and its tributaries for domestic and irrigation purposes, Kashmir valley, India. Curr Sci 109(2):311–322
Mir RA, Jeelani G (2015b) Textural characteristics of sediments and weathering in river Jhelum basin located in Kashmir valley western Himalaya. J Geol Soc India 86(4):445–458
Mir RA, Jeelani G, Dar FA (2016) Spatio-temporal variations and factors controlling the hydrogeochemistry of river Jhelum located in Kashmir valley, western Himalayas. Environ Mont Assess 188:438
Mir RA (2018) Recent changes of two parts of Kolahoi Glacier and its controlling factors in Kashmir basin, western Himalaya. Remote Sen Appl Soc Env 11:265–281
Mir IA, Mir RA (2019) Geochemistry of surface sediments in parts of Bandipora-Ganderbal areas, Kashmir valley, western Himalaya: Implications for provenance and weathering. J Earth SystSci 128:223
Mir RA, Gani MK (2019) Evaluation of water quality of upper part of river Jhelum basin using multivariate statistical techniques. Arab J Geosci 12:445
Morgan CO and Winner MD (1962) Hydrochemical facies in the "400-feet" and "600-feet" sands of the Baton Rouge area, Louisiana. In: Short papers in geology, hydrology, and topography. US GeolSurv Profess Paper 450-B:120–121
Neve A (1933) The tourist's guide to Kashmir, Ladakh. Skardu etc, Lahore
Ophori OU, Toth SZ (1989) Pattern of groundwater chemistry in Ross Greek Basin, Aberta Canada. Groundwater 27(1):20–26
Piper AM (1944) A graphic procedure in the geochemical interpretation of water analyses. Trans Am Geophy Union 25:914–923
Sah S, Bisht H, Kumar K, Tiwari A, Tewari M, Joshi H (2017) Assessment of hydrochemical properties and annual variation in meltwater of Gangotri glacier system. ENVIS Bull Himalayan Ecol 25:17–23
Sarkar T, Kannaujiya S, Taloor AK, Ray PKC, Chauhan P (2020) Integrated study of GRACE data derived interannual groundwater storage variability over water stressed Indian regions. Groundw Sust Dev 10:100376. https://doi.org/10.1016/j.gsd.2020.100376
Sawyer GN, McCarthy DL (1967) Chemistry of sanitary engineers, 2nd edn. McGraw Hill, New York
Singh AK, Jasrotia AS, Taloor AK, Kotlia BS, Kumar V, Roy S, Ray PKC, Singh KK, Singh AK, Sharma AK (2017) Estimation of quantitative measures of total water storage variation from GRACE and GLDAS-NOAH satellites using geospatial technology. Quat Int 444:191–200
Taloor AK, Pir RA, AdimallaN AS, Manhas DS, RoyS SAK (2020) Spring water quality and discharge assessment in the Basantar watershed of Jammu Himalaya using geographic information system (GIS) and water quality Index(WQI). Groundw Sust Dev 10:100364. https://doi.org/10.1016/j.gsd.2020.100364
Trivedy RK, Goel PK (1984) Chemical and biological methods for water pollution studies Environment Publications. Karad, India p, p 215
Walton WC (1970) Groundwater resource evaluation. McGraw-Hill, New York
WHO (2006) International standards for drinking water. World Health Organization, Geneva

Mr. Khurshid Ahmad Lone is working as a Lecturer at Government Boys Higher Secondary Institute, Handwara, J & K. He has completed his Master's Degree in Applied Geology from University of Kashmir, Srinagar. He has qualified the CSIR-NET examination with an All India rank of 02 in Earth Sciences. After obtaining Master's degree, he has also completed his degree of Mater in Philosophy (M. Phil) with a specialization in hydrogeology from the University of Kashmir. He has worked as a Lecturer at the same University of Kashmir for about 3 years. Later, he has qualified the J&K, Public Service Commission Recruitment examination and was appointed as a Lecture in Geology for Intermediate Level where he has been actively engaged in teaching the subject of Geology for last 8 years. He has several publications related to Hydrochemistry and Water quality of National/International repute to his credit. He has also presented his work at J&K Science Congress and other conferences of National repute. He has also co-authored a book chapter (a Springer Publication)

Dr. Riyaz Ahmad Mir is working as a Sr. Geoscientist in Geological Survey of India for the last 8 years. He has been honored with high ranks in joint CSIR-UGC (NET+JRF) examination and UPSC Geologist/Geoscientists Examination. After receiving his Master's degree in Geology from Bundelkhand University, Jhansi, he completed his M.Phil. in Hydrogeochemistry from University of Kashmir, Srinagar and his Ph.D. in Geoinformatics from Indian Institute of Technology (IITR), Roorkee. He has also worked with the National Institute of Hydrology (NIH), Roorkee for his Doctorate degree. He has been awarded with best Paper Presentation (SiD-2018) at Cluster University, Srinagar. He has a good working experience in the areas of hydro-and-sediment geochemistry, remote sensing and GIS and its applications in glacier, Glacier Lake Outburst Flood (GLOF), landslide studies and its modelling, climate change studies and its effect of water resources. He has also a vast field expertise in Geochemical Mapping (GCM), Specialized Geological Thematic Mapping (STM) and landslide zonation and susceptibility mapping, etc. He has published about 20 research articles in refereed international/national journals, about 10 extended abstracts in national and international conferences/symposiums and 1 book chapter (Springer publication). He has also completed about 7 Geological Reports of different research projects at GSI. He has published two articles in Hindi Magazine 'Tawi' published by GSI, J&K and is also writing to daily locals regarding geoscientific issues. He has delivered several guest lectures at Degree College and Intermediate level pertaining to the role of Earth Sciences in society

Dr. Nadeem Ahmad Bhat is working as a Senior Geologist in Geological Survey of India. He has completed his M. Phil from University of Kashmir, Srinagar on Isotope Hydrology and Ph.D. from Banaras Hindu University, Varanasi, India on Sediment Geochemistry. He is honoured with high ranks in UPSC Geologist Examination. He has more than 15 years of professional experience in the fields of hydrology, geochemistry, field geology, geological mapping, mineral investigations, Himalayan geology and engineering geology. He has published 15 papers in reputed national and international journals, 7 technical reports and 13 abstracts in national and international conferences. He has served as a core faculty at GSI Field Training Centre, Aishmuqam, imparting training on Himalayan and Engineering Geology to newly recruited Geoscientists of Geological Survey of India

18 Delineation of Groundwater Potential Recharge Zone Using Remote Sensing and GIS Techniques—A Case Study of Rampur Tehsil, Shimla District, Himachal Pradesh India

C. Prakasam, R. Aravinth, and R. Saravanan

Abstract

The study is carried out to assess the potential sites for groundwater recharge in Rampur Tehsil, Shimla district, Himachal Pradesh. The Weighted Overlay Model (WOM) in Geographic Information System (GIS) environment was used to delineate the groundwater potential recharge sites in the study area. Various layers such as soil, geology, geomorphology, land use land cover (LULC), slope, and lineaments have been used as delineating factors. These factors were then assigned individual values based on their degree of association for groundwater occurrence. The output derived is classified into five types, namely, very poor, poor, moderate, good, and very good. The results indicate that most of the areas fall under good conditions for an artificial mode of groundwater recharge (53.29%), and moderate category covers about (39.51%). Nearly 7% of the area comes under very poor conditions. Most of the sites suitable for groundwater recharge present along the forest and agricultural lands. The study opens up new vistas for rapid assessment of potential groundwater recharge sites and serves as the database for decision-making and planning for sustainable water management practices.

Keywords

Groundwater potential recharge zones · GIS · Remote sensing · Weighted overlay model

18.1 Introduction

In recent years, there has been an increased development in various water-based sectors such as agriculture, industry, and urban sectors, especially in countries like India. This has led to increased water demand in water supply which ultimately leads to the overexploitation of the water resources (Prabhu and Venkateswaran 2015; Singh et al. 2017; Adimalla and Taloor 2020; Sarkar et al. 2020; Sood et al. 2020). An increase in the developmental plan for water resources has increased irrigation capacity, steady economy, and quality life. In India, there are over 17 million wells providing irrigation water to 50% of the agriculture sector. It provides 80% of drinking water to rural and 50% to urban areas. About 34% of groundwater consumption is reported annually

C. Prakasam (✉) · R. Aravinth · R. Saravanan
Department of Civil Engineering, Chitkara University, Kallujhanda, Himachal Pradesh, India
e-mail: cprakasam@gmail.com

R. Aravinth
e-mail: aravinthraja662891@gmail.com

R. Saravanan
e-mail: saravana8689@gmail.com

A. K. Taloor et al. (eds.), *Water, Cryosphere, and Climate Change in the Himalayas*, Geography of the Physical Environment, https://doi.org/10.1007/978-3-030-67932-3_18

by (Magesh et al. 2012; Taloor et al. 2020a). Unmonitored groundwater consumption for various household and industrial purposes leads to acute shortage of groundwater and availability of natural resources in many parts of the country (indiawaterportal.org). The remote sensing datasets with their high temporal and spatial resolution play an important role in analyzing terrain conditions and are used for various purposes such as feasibility of recharge sites, evaluation of groundwater and surface water resources (Bisht et al. 2019). Satellite imageries provide results on various parameters such as lineaments, drainage, LULC, etc. which are essential in determining groundwater recharge sites (Thilagavathi et al. 2015). Many researchers and scientists have attempted to map groundwater potential zones using various methods such as Weighted Overlay Model (WOM), Multi-Criteria Decision Analysis (MCDA), Analytic Hierarchy Process (AHP), etc. (Singh et al. 2020; Taloor et al. 2020; Golla et al. 2018; Ibrahim-Bathis and Ahmad 2016; Nanda et al. 2017; Prabhu and Venkateswaran 2015; Srinivasa Rao and Jugran 2003; Chaudhary and Kumar 2018; Singh et al. 2018; Haque et al. 2020).

The present research was carried out to assess the potential sites for groundwater recharge in Rampur Tehsil, Shimla district, Himachal Pradesh. Various layers such as soil, geology, geomorphology, LULC, slope, and lineaments have been used as delineating factors. Results were derived using WOM in GIS environment. The objectives of the research are to prepare various thematic layers such as soil, geology, geomorphology, slope, LULC, drainage, and lineaments consequently to determine the potential groundwater recharge zones using WOM.

18.2 Study Area

Rampur Tehsil has a total geographical area of 987 km^2. It extends between 31°15’3”–31° 44’10” N and 77°30’19”–77°59’21” E. The soil is mostly coarse loamy and the area is mostly covered by forest and barren lands. The temperature varies between 0° and 40° in the summer. Horticulture and agriculture are the important economic activities in this region. The location map study area is shown in Fig. 18.1.

18.3 Materials and Methods

The delineation of potential groundwater recharges zone requires an assessment of various physical parameters such as lithology, geomorphology, LULC, etc. In order to prepare a base map of the study area the important features such as settlements, roadways, rivers, etc., were digitized using the Survey of India Toposheets as well as the Soil and Land Use Survey of India (SLUSI). The physical parameters are derived from maps retrieved from the Soil and Land Use Survey of India Department. The Advanced Spaceborne Thermal Emission and Reflection Radiometer (ASTER) and Global Digital Elevation Model (GDEM) data were used to extract the slopes and lineaments of the study area. Different data sets used in the present study are given in Table 18.1.

Moreover, an assessment of both physical parameters and anthropogenic activities are also required for the delineation of these recharge sites (Jasrotia et al. 2019; Khan et al. 2020). Values ranging from 1 to 5 were assigned to individual causative factors in order to implement the WOM model (Al-ruzouq et al. 2019; Nithya et al. 2019; Murmu et al. 2019). A maximum likelihood supervised classifier was applied to prepare the LULC map of the study area using Landsat-8, images. Layers such as geology, soil, and geomorphology were digitized in the GIS environment. The datasets were processed in the GIS environment and results were retrieved to locate suitable areas for groundwater potential recharge zone (Fig. 18.2).

18.4 Results and Discussion

18.4.1 Landuse and Landover

Landuse plays a vital role in the selection of suitable sites for groundwater recharge, features

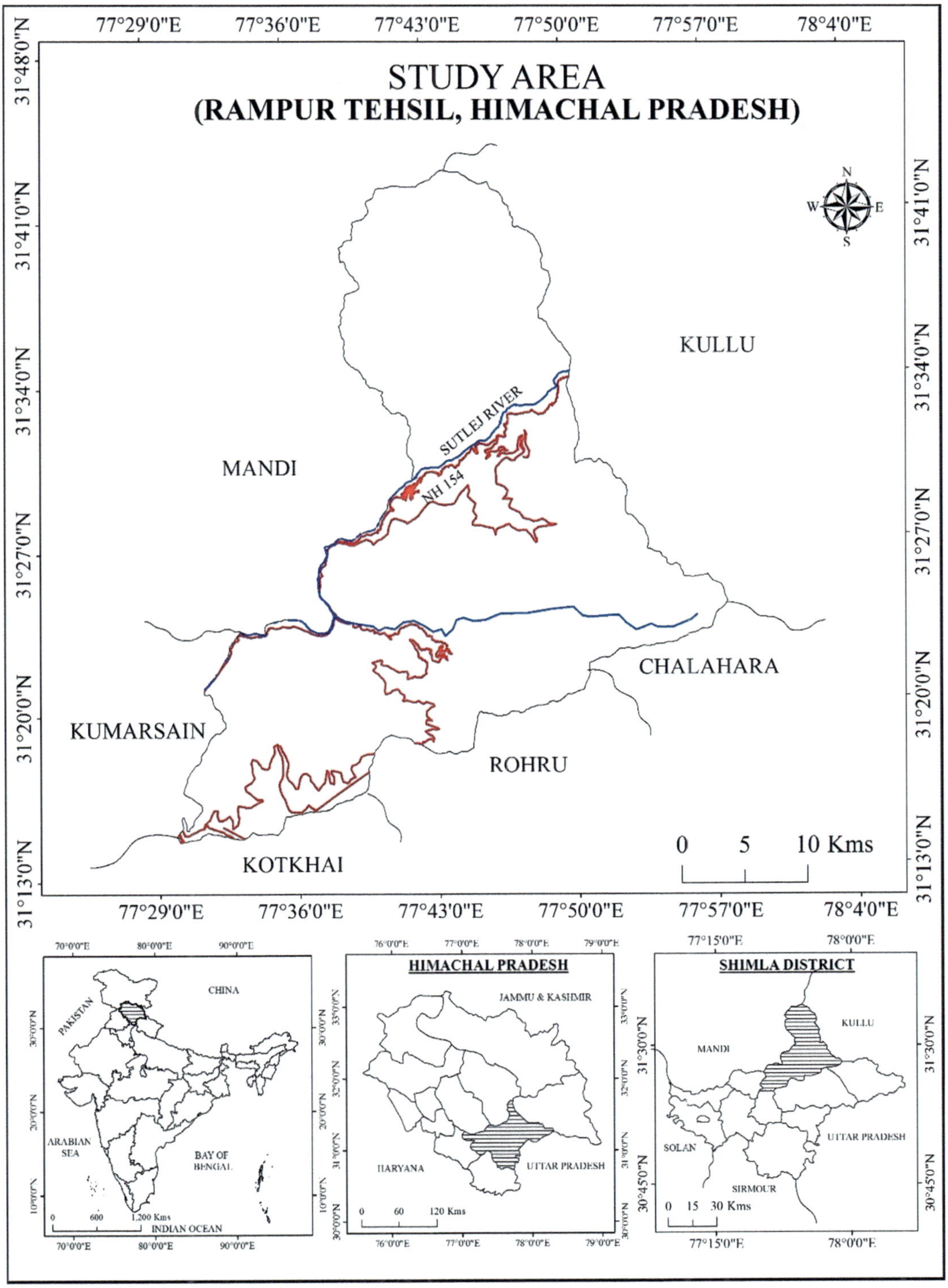

Fig. 18.1 Location map of the study area (*Source* Survey of India, Toposheets)

Table 18.1 Data used in the present study

S. no.	Data	Source	Year	Resolution
1	LULC	Landsat -8 OLI	2018	30 m
2	Geology	SLUSI	2018	1:50000
3	Geomorphology	SLUSI	2018	1:50000
4	Soil	SLUSI	2018	30 m
5	Slope	ASTER GDEM	2009	30 m
6	Lineaments	ASTER GDEM	2009	30 m

(*Source* Landsat-8 OLI, SLUSI, NASA)

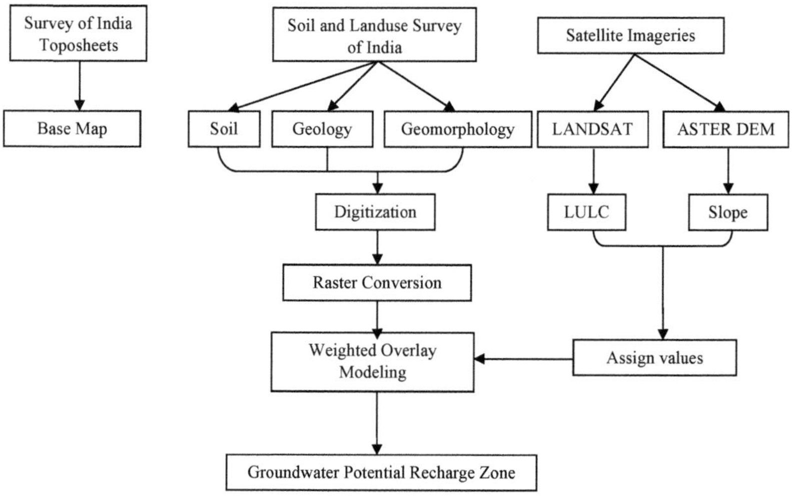

Fig. 18.2 Research methodology used in the present study (*Source* Authors)

such as agricultural land, forest, and barren land allows the water to percolate through the soil making them effective causative factors for increasing the groundwater table. The resultant land use map was differentiated into five classes, namely, forest, agricultural land, barren land, settlements, and glacier (Fig. 18.3). Forest covers most of the study area with 66.99%. The agriculture and barren land cover about 25.5% together of the total study area. These areas are high potential zones for water recharge. During monsoon season,these areas can retrieve more rainfall and allow the rainwater to percolate through the soil making them highly effective for increasing the groundwater table. The built-up land glaciers cover about 1.27% and 6.13% (Table 18.2), respectively, of the study area. Built-up lands prevent water from recharging the groundwater table due to the construction of buildings and concrete structures.

18.4.2 Geology

Geology plays a key role in groundwater recharge as the rock types present in an area could hugely affect the amount of water entering the groundwater table. The study area is covered with five major rock types namely schist, slate, alluvium, glacier, and habitation (Fig. 18.4). The area is mainly dominated by two types of rock formation, namely, slate and schist that comprise 89% (Table 18.3) of the study area. These rocks vary from moderate to strong in nature in the GSI index. Waters can percolate through the cracks within the rocks or even between them. Fractures and joints formed along with the rock surface act as perfect carriers for rainwater into the groundwater table. Alluvium material accounts for only 0.5% but since they are fine-graded rocks that form along the river banks found to be the most effective carrier for groundwater recharge.

Fig. 18.3 Land use and land cover classification map (*Source* Authors calculation)

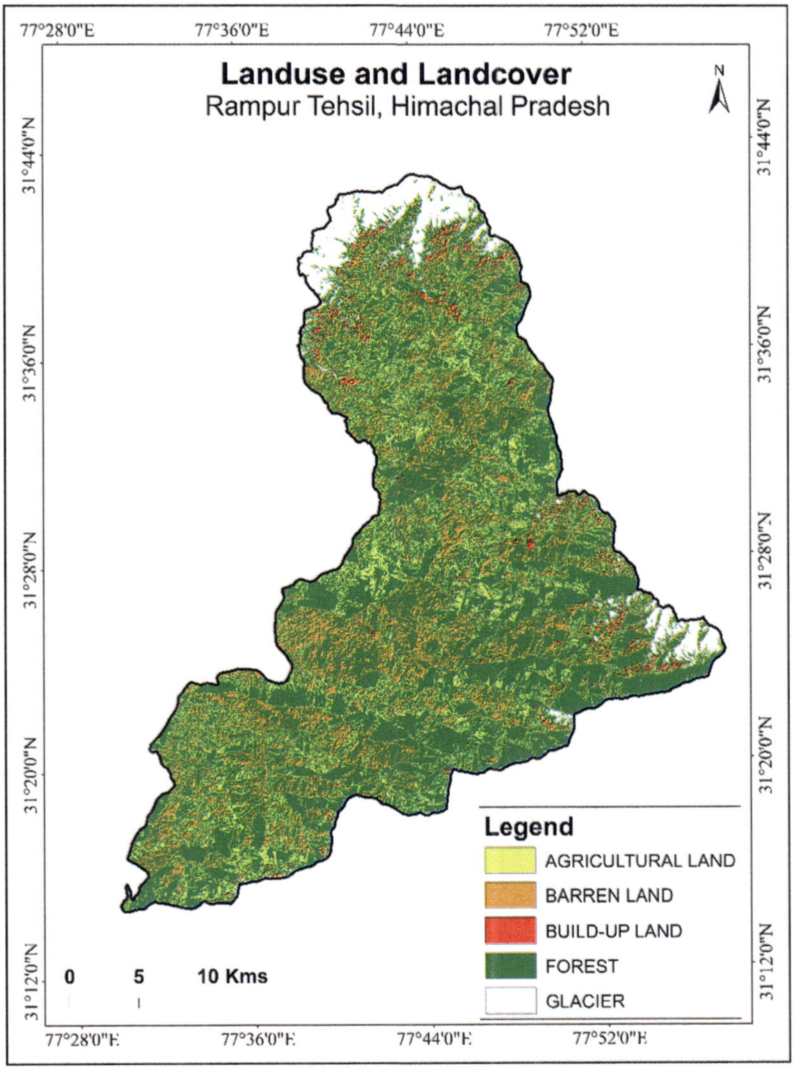

Table 18.2 Areal distribution of land use and land cover classes and weightage for WOM

S. no.	LULC	Area (km^2)	Percent (%)	Weightage
1	Agricultural land	114.14	11.55	4
2	Barren land	138.89	14.06	5
3	Build-up land	12.58	1.27	2
4	Forest	661.77	66.99	4
5	Glacier	60.52	6.13	1
	Total	**987**	**100.00**	

(*Source* Landsat-8 OLI, SLUSI, NASA)

18.4.3 Soil

Soil refers to the grain size of the individual rock particles. This variation in size decides the amount of groundwater entering through the soil. The fine loamy, coarse loamy, glacier, and habitation are the four major classes found here (Fig. 18.5). For example, rocks with high granular size will allow the easy infiltration of water compared to that of rocks that are small in size. The entire study area is more prominent to fine loamy soil particles in which 40–40–20% composition of sand, silt, and clay is reported and accounts for about 84.5% of the total (Table 18.4). These types of soils are moderately allowing the water to infiltrate the soil. The coarse loamy accounts for only 5.0% of the geographical area and allows a greater amount of water infiltration capacity. The glaciers and habitation are poor in nature because of their water holding capacity.

18.4.4 Geomorphology

The geomorphological features of the study area are classified into glacier, river terraces, undifferentiated hillside, and mountainside slopes (Fig. 18.6). The hillside and mountainside slopes

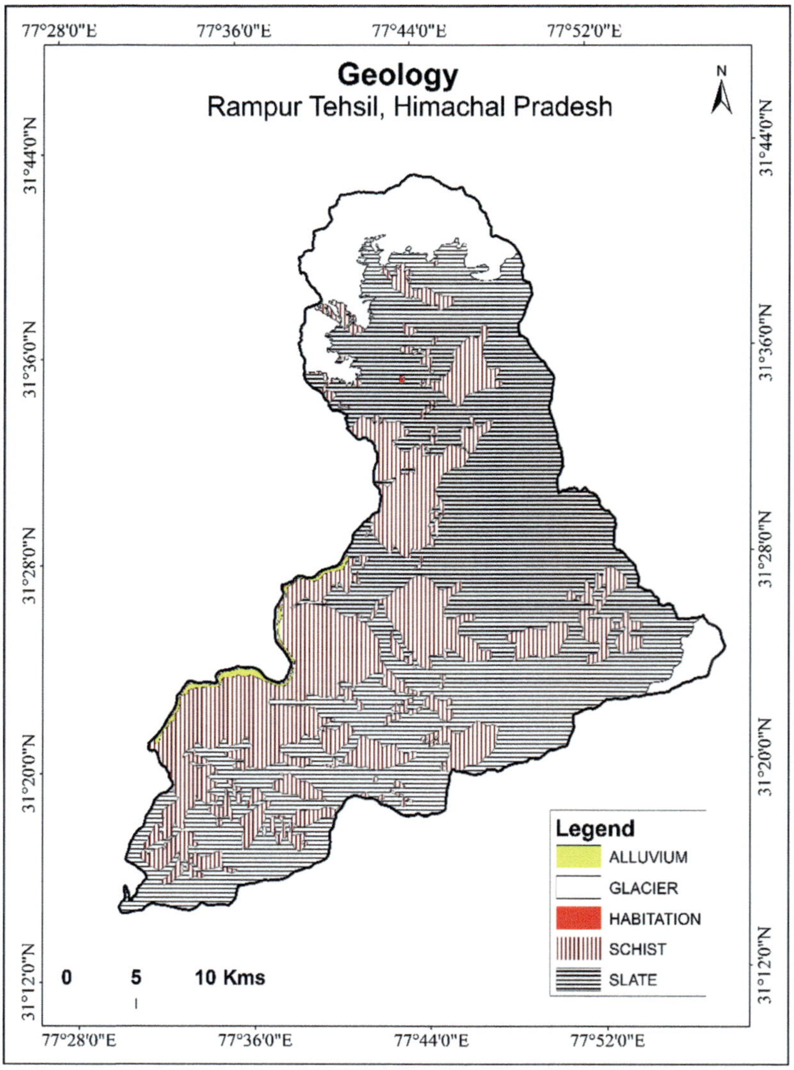

Fig. 18.4 Geology map of the study area (*Source* Authors calculation)

Table 18.3 Areal distribution of geological material classes and weightage for WOM

S. no.	Geology	Area (km^2)	Percent	Weightage
1	Alluvium	5.08	0.51	4
2	Glacier	102.65	10.39	1
3	Habitation	0.14	0.01	2
4	Schist	291.23	29.48	3
5	Slate	588.80	59.60	3
	Total	**987**	**100**	

(*Source* Authors calculation)

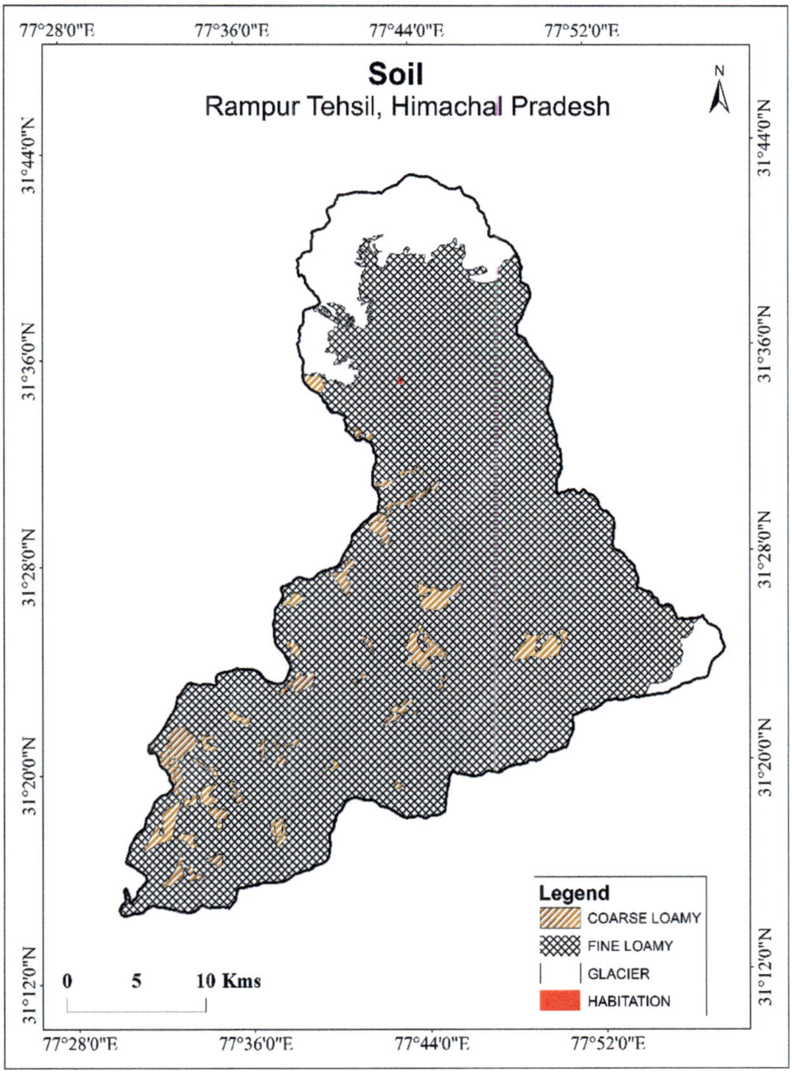

Fig. 18.5 Soil map of the study area (*Source* Authors calculation)

account for 89% of the study area and provide moderate infiltration capacity. The river terrace accounts for only 0.53% but facilitates greater water percolation capacity (Table 18.5). Glacier is composed of snow, and water present over it gets freezes and therefore, reflects poor infiltration.

Table 18.4 Areal distribution of soil texture classification and weightage for WOM

S. no.	Soil	Area (km^2)	Percent	Weightage
1	Coarse loamy	49.69	5.03	4
2	Fine loamy	835.22	84.55	3
3	Glacier	102.85	10.41	1
4	Habitation	0.14	0.01	2
	Total	**987**	**100.00**	

(*Source* Authors calculation)

18.4.5 Slope

Slope determines the running capacity of the water; if a slope is too steep the water gets discharged into river quickly and if a slope is gentle facilitates high infiltration capacity. The sloping conditions of the study area are classified into five classes (Fig. 18.7). The results indicate that moderate to steep slopes cover 91% of the study area and appear to have lesser water infiltration capacity. The next higher class is covered by a slope with more than 41° (243 Km^2) and 21–30° (78 Km^2) as given in Table 18.6.

18.4.6 Lineaments

Lineaments are linear features that are present along the earth's surface. They constitute features such as fractures, faults, and joints. The lineaments are hotspots for groundwater recharge. Water can infiltrate through cracks and reach the water table. The more the number of lineaments, the higher will be the water infiltration capacity of the area. Lineaments were digitized from a hill shade map derived from ASTER DEM data. The lineament density analysis was carried out and the results are shown in (Fig. 18.8). The lineament density was classified between 0 and 0.93 with 0 being no lineament presence and 0.93 indicating the highest presence of lineaments (Table 18.7).

18.5 Groundwater Potential Recharge Zones

Once the contributing factors were calculated, individual weights have been assigned to each causative factor. These factors were then weighted and overlaid in GIS environment. The output derived is classified into four types ranging from very poor, poor, good, and extremely good (Fig. 18.9). The results indicate that most of the area falls under good conditions for an artificial mode of groundwater recharge (53.29%), and moderate category covers about (39.51%). Nearly 7% of the area comes under very poor and poor conditions. These areas are present along with the glaciers where recharge of groundwater is highly difficult. Most of the sites found suitable for groundwater recharge are present along the forest and agricultural lands (Table 18.8).

18.6 Conclusions

The research was carried out to facilitate the sites that are highly active to groundwater recharge. The WMO was used to delineate the potential sites for groundwater recharge sites. The results indicate that most of the area falls under good conditions for the artificial mode of groundwater recharge (53.29%), and moderate category covers about (39.51%). Nearly 7% of the area comes under very poor and poor conditions. These areas are present along with the glaciers where recharge of groundwater is highly difficult. Most of the sites suitable for groundwater recharge present along the forest and agricultural lands. The methodology used in the study serves as a good example of rapid assessment of potential groundwater recharge sites in the mountainous region. The results derived from the research will be helpful for various governmental and NGO organizations in decision-making during the planning of sustainable water management.

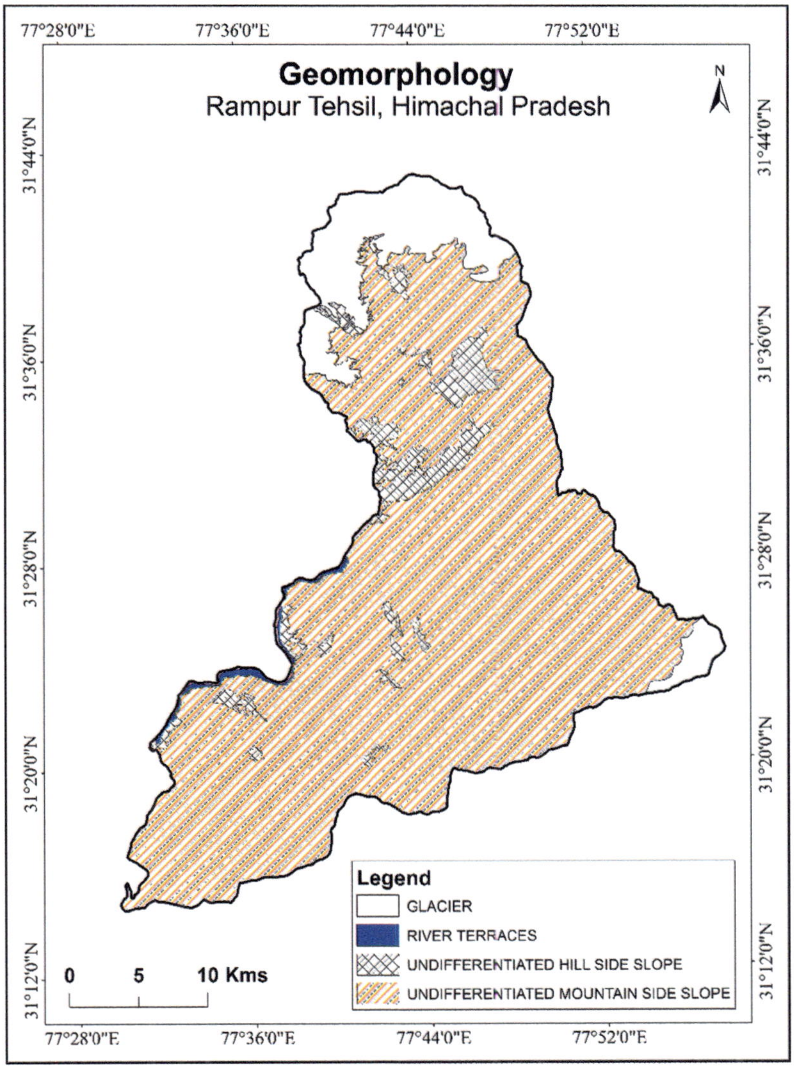

Fig. 18.6 Geomorphology map of the study area (*Source* Authors calculation)

Table 18.5 Areal distribution of geomorphology weightage for WOM

S. no	Geomorphology	Area (km^2)	Percent	Weightage
1	Glacier	102.85	10.41	1
2	River Terraces	5.19	0.53	4
3	Undifferentiated hillside slope	54.77	5.54	3
5	Undifferentiated mountainside slope	825.08	83.52	3
	Total	987	100	

(*Source* Authors calculation)

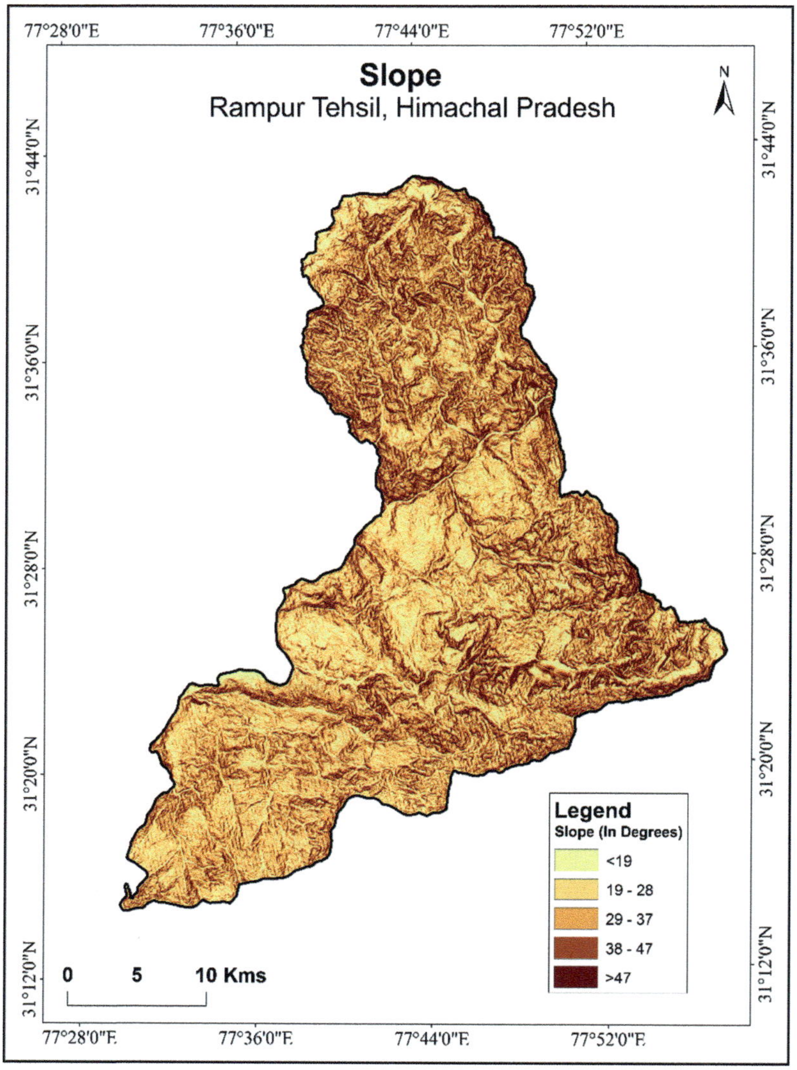

Fig. 18.7 Slope map of the study area (*Source* Authors calculation)

Table 18.6 Areal distribution of slope and weightage for WOM

S. no.	Slope (Degree)	Area (km^2)	Percent	Weightage
1	0–10	31	3.14	5
2	11–20	49	4.96	4
3	21–30	78	7.89	3
4	31–40	587	59.41	2
5	41–63	243	24.60	1
	Total	987	100	

(*Source* Authors calculation)

Fig. 18.8 Lineament density map of the study area (*Source* Authors calculation)

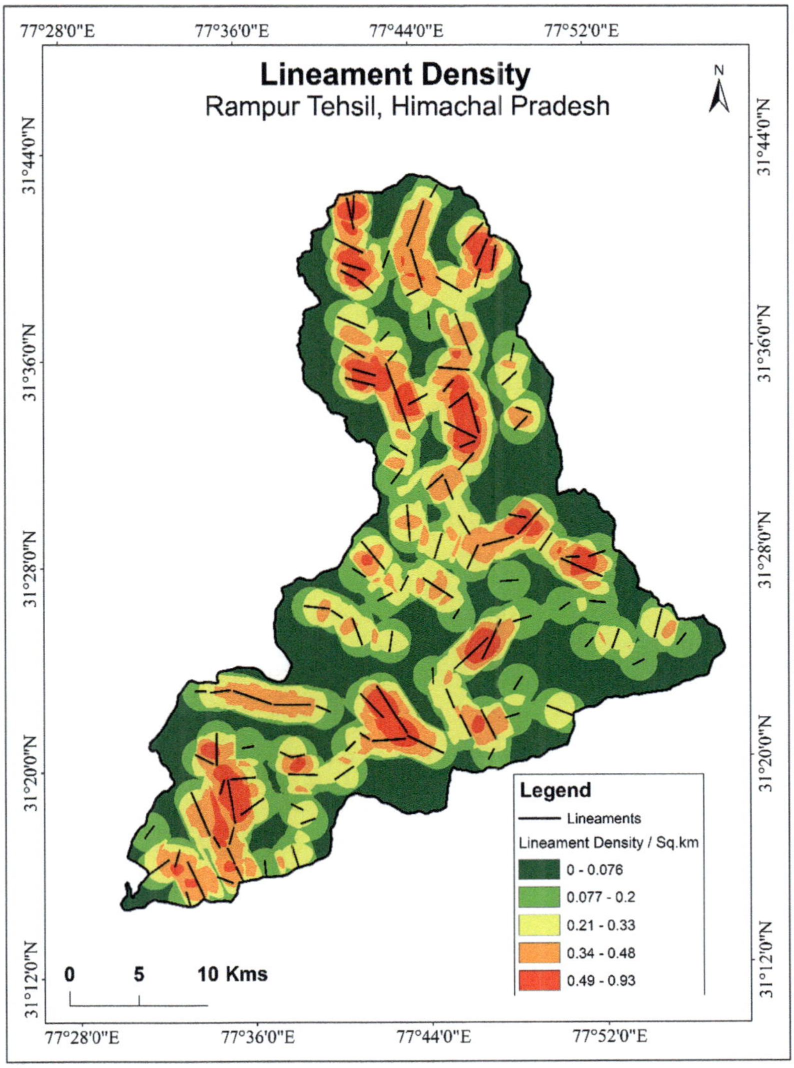

Table 18.7 Areal distribution of lineament density classification and weightage for WOM

S. no	Density/(km^2)	Weightage
1	0–0.076	5
2	0.077–0.2	4
3	0.21–0.33	3
4	0.34–0.48	2
5	0.49–0.93	1

(*Source* Authors calculation)

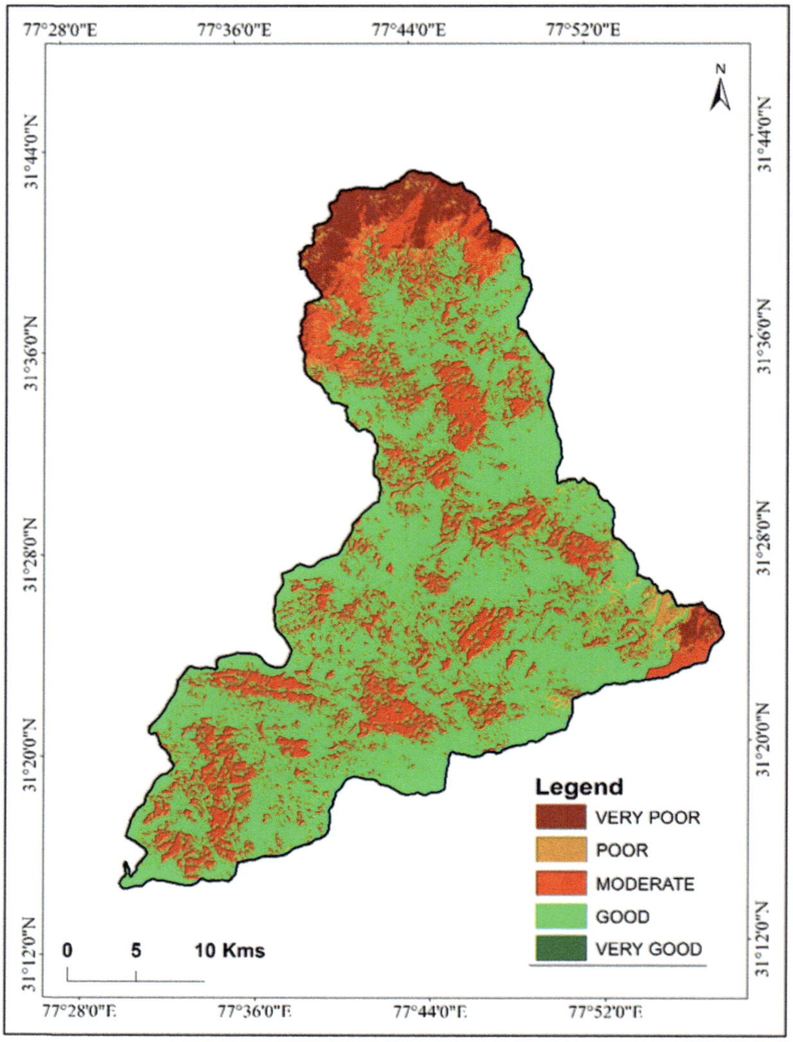

Fig. 18.9 Potential recharge groundwater zones along the study area (*Source* Authors calculation)

Table 18.8 Aerial distribution of groundwater potential recharge zone

S. no.	Category	Weightage	Area (km^2)	Percent
1	Very poor	1	44	4.46
2	Poor	2	27	2.74
3	Moderate	3	390	39.51
4	Good	4	526	53.29
5	Very good	5	0	0.00
Total			**987**	**100.00**

(*Source* Authors calculation)

Acknowledgements We would like to thank NRDMS—DST, New Delhi for funding the research. We would like to thank CSIR, New Delhi, GOI for sponsoring the SRF—Direct, scholarship for pursuing Research Work.

References

Adimalla N, Taloor AK (2020) Hydrogeochemical investigation of groundwater quality in the hard rock terrain of South India using Geographic Information System (GIS) and groundwater quality index (GWQI) techniques. Groundw Sust Dev 10:100288. https://doi.org/10.1016/j.gsd.2019.100288

Al-Ruzouq R, Shanableh A, Merabtene T, Siddique M, Khalil MA, Idris A, Almulla E (2019) Potential groundwater zone mapping based on geo-hydrological considerations and multi-criteria spatial analysis: North UAE. CATENA 173:511–552. https://doi.org/10.1016/j.catena.2018.10.037

Bisht H, Rani M, Kumar K, Sah S, Arya PC (2019) Retreating rate of Chaturangi glacier, Garhwal Himalaya, India derived from kinematic GPS survey and satellite data. CurrSci 116:304–311. https://doi.org/10.18520/cs/v116/i2/304-311

Chaudhary BS, Kumar S (2018) Identification of groundwater potential zones using remote sensing and GIS of KJ Watershed India. J Geol Soc India 91(6):717–721. https://doi.org/10.1007/s12594-018-0929-3

Golla V, Etikala B, Veeranjaneyulu A, Subbarao M, Surekha A, Narasimhlu K (2018) Data sets on delineation of groundwater potential zones identified by geospatial tool in Gudur area, Nellore district, Andhra Pradesh, India. Data Brief 20:1984–1991. https://doi.org/10.1016/j.dib.2018.09.054

Haque S, Kannaujiya S, Taloor AK, Keshri D, Bhunia RK, Ray PK, Chauhan P (2020) Identification of groundwater resource zone in the active tectonic region of Himalaya through earth observatory techniques. Groundw Sust Dev 100337. https://doi.org/10.1016/j.gsd.2020.100337

Ibrahim-Bathis K, Ahmed SA (2016) Geospatial technology for delineating groundwater potential zones in Doddahalla watershed of Chitradurga district, India. Egypt J Remote Sens Space Sci 19(2):223–234

Jasrotia AS, Kumar R, Taloor AK, Saraf AK (2019) Artificial recharge to groundwater using geospatial and groundwater modelling techniques in North Western Himalaya, India. Arabian J Geosci 12:774. https://doi.org/10.1007/s12517-019-4855-5

Khan A, Govil H, Taloor AK, Kumar G (2020) Identification of artificial groundwater recharge sites in parts of Yamuna river basin India based on remote sensing and geographical information system. Groundw Sustain Dev 100415. https://doi.org/10.1016/j.gsd.2020.100415

Magesh NS, Chandrasekar N, Soundranayagam JP (2012) Delineation of groundwater potential zones in Theni district, Tamil Nadu, using remote sensing, GIS and MIF techniques. Geosc Front 3(2):189–196. https://doi.org/10.1016/j.gsf.2011.10.007

Murmu P, Kumar M, Lal D, Sonker I, Singh SK (2019) Delineation of groundwater potential zones using geospatial techniques and analytical hierarchy process in Dumka district, Jharkhand, India. Groundw Sustain Dev 9:100239. https://doi.org/10.1016/j.gsd.2019.100239

Nanda S, Annadurai R, Barik KK (2017) Geospatial decipherment of groundwater potential of Kattankolathur block of Tamil Nadu using MCDM techniques. Remote Sens Appl: Soc Environ 8:240–250

Nithya CN, Srinivas Y, Magesh NS, Kaliraj S (2019) Assessment of groundwater potential zones in Chittar basin, Southern India using GIS based AHP technique. Remote Sens Appl: Soc Environ 15:P100248. https://doi.org/10.1016/j.rsase.2019.100248

Prabhu MV, Venkateswaran S (2015) Delineation of artificial recharge zones using geospatial techniques in Sarabanga Sub Basin Cauvery River, Tamil Nadu. Aquatic Procedia 4:1265–1274. https://doi.org/10.1016/j.aqpro.2015.02.165

Sarkar T, Kannaujiya S, Taloor AK, Ray PK, Chauhan P (2020) Integrated study of GRACE data derived interannual groundwater storage variability over water stressed Indian regions. Groundwater Sustain Develop. P 100376. https://doi.org/10.1016/j.gsd.2020.100376

Singh AK, Jasrotia AS, Taloor AK, Kotlia BS, Kumar V, Roy S, Ray PK, Singh KK, Singh AK, Sharma AK (2017) Estimation of quantitative measures of total water storage variation from GRACE and GLDAS-NOAH satellites using geospatial technology. Quat Int 444:191–200

Singh LK, Jha MK, Chowdary VM (2018) Assessing the accuracy of GIS-based multi-criteria decision analysis approaches for mapping groundwater potential. Ecol Indic 91:24–37. https://doi.org/10.1016/j.ecolind.2018.03.070

Singh S, Sood V, Taloor AK, Prashar S, Kaur R (2020) Qualitative and quantitative analysis of topographically derived CVA algorithms using MODIS and Landsat-8 data over Western Himalayas, India. Quat Int. https://doi.org/10.1016/j.quaint.2020.04.048

Sood V, Gusain HS, Gupta S, Taloor AK, Singh S (2020) Detection of snow/ice cover changes using subpixel-based change detection approach over Chhota-Shigri glacier, Western Himalaya, India. Quat Int. https://doi.org/10.1016/j.quaint.2020.05.016

Srinivasa Rao Y, Jugran DK (2003) Delineation of groundwater potential zones and zones of groundwater quality suitable for domestic purposes using remote sensing and GIS. Hydrol Sci J 5:821–833. https://doi.org/10.1623/hysj.48.5.821.51452

Taloor AK, Kumar V, Singh VK, Singh AK, Kale RV, Sharma R, Khajuria V, Raina G, Kouser B, Chowdhary NH (2020a). Land use land cover dynamics using remote sensing and GIS techniques in Western Doon

Valley, Uttarakhand, India. In: Geoecology of landscape dynamics 2020 (pp 37–51). Springer, Singapore
Taloor AK, Pir RA, Adimalla N, Ali S, Manhas DS, Roy S, Singh AK (2020b) Spring water quality and discharge assessment in the Basantar watershed of Jammu Himalaya using geographic information system (GIS) and water quality Index(WQI). Groundw Sust Develop 10:100364. https://doi.org/10.1016/j.gsd.2020.100364
Thilagavathi N, Subramani T, Suresh M, Karunanidhi D (2015) Mapping of groundwater potential zones in Salem Chalk Hills, Tamil Nadu, India, using remote sensing and GIS techniques. Environ Monit Asses 187 (4):164

Dr. C. Prakasam obtained Ph.D. and M. Tech. degree in Remote Sensing and GIS field. He is an alumnus of The University of Burdwan, West Bengal. He carries more than 13 years of research, teaching and administrative experience. He has published three books, a significant number of research papers in reputed national and international journals and conferences and filed four patents. He has completed two research projects funded by Govt. Of India and Govt. Of H. P. His area of interest is landslide engineering, EIA and engineering geosciences and geospatial sciences. He is an active member of various societies like Indian Society of Remote Sensing (ISRS), Indian Society of Geomatics, Indian Society for Technical Education, Indian Institute of Geomorphologists, The Institution of Engineers, ICI, etc.

Mr. R. Aravinth currently working as Doctoral fellow in the Department of Civil Engineering at Chitkara University, Himachal Pradesh. He obtained his M. Tech. in Geoinformatics from the University of Madras, Chennai, India. He worked as Junior Research Fellow in a DST, Government of India sponsored project for 2 years. He is also recipient of Senior Research Fellowship (SRF) from the distinguished CSIR, New Delhi, GOI. He has published more than eight papers as a part of edited books and a research article in a reputed national and international journal. His areas of interest are remote sensing and GIS, landslide hazard modelling, slope stability, environmental modelling, etc.

Mr. R. Saravanan is currently working as Doctoral Fellow in the Department of Civil Engineering at Chitkara University, Himachal Pradesh. He completed his B.E. Civil Engineering and M.E. Hydrology and Water Resources Engineering from Anna University. He worked as Junior Research Fellow in a DOEST, Government of Himachal Pradesh, India sponsored project for 2 years. He has published seven papers in reputed national and international journals and conferences and filed two patents. His areas of interest are hydrology modelling, remote sensing, environmental flow monitoring, etc.

19 Geospatial Approach for Water Quality Index Mapping for Drinking Purpose in Guna District, Madhya Pradesh, India

Ankita Bhardwaj and Suraj Kumar Singh

Abstract

The present study addresses the objective of mapping the Water Quality Index (WQI) using a geospatial method to ensure availability of healthy drinking water in parts of Madhya Pradesh Guna district. One thousand nine hundred seventy-two water samples were obtained in the presented study and tested in a laboratory to estimate the Physico-chemical and microbial contamination of drinking water. The GIS techniques were used for spatial analysis of WQI for all the blocks of Guna district to classify in very poor, poor, good and very good water quality categories. In the present study, eight parameters, i.e. pH, Turbidity, Total Dissolve Solid (TDS), Fluoride, Chloride, Iron, total Coliform, were taken into consideration in assigning weights. Higher weightage was assigned according to its water quality and vice versa. The overall assessment of the WQI shows that very good category covers 21.69% of the total area exhibited good category covers 5.57%, moderate category covers 7.3% poor category covers 12.53% and very poor WQI category covers 52.91%.

Keywords

Geospatial · Technology · GPS · Hydrochemistry · Water quality index mapping

19.1 Introduction

For the existence and survival of life on earth, water is necessary, we drink it, bathe it, relax it and irrigate plants, crops, etc. (Ponsadailakshmia et al. 2017; Singh et al. 2017a, b; Bhat et al. 2019; Viswanath et al. 2015; WHO 2012). Suitability of drinking water purposes, groundwater pollution and water quality are very important (Sawant et al. 2015; Singh et al. 2020; Thapa et al. 2017; Hasanet al. 2017; Bhat et al. 2019; Taloor et al. 2020a). Harmfull pollutants in water is a global threat now a days and affected the mankind widely over the years from one area to another in magnitude and form (Dhar and Sahoo 2015; Haque et al. 2020; Jhariyaa et al. 2017; Pandey et al. 2015; Prakasa et al. 2017; Jasrotia and Kumar 2014; Kumar et al. 2017, 2020; Bisht et al. 2018, 2020; Taloor et al. 2019; Khan et al. 2020). The study involves a valuable approach that includes field and laboratory testing of different drinking water sources in six blocks of the Guna district of Madya Pradesh.

A. Bhardwaj
Centre for Climate Change and Water Research, Suresh Gyan Vihar University, Jaipur, India
e-mail: ankita.49421@mygyanvihar.com

S. K. Singh (✉)
Centre for Sustainable Development, Suresh Gyan Vihar University, Jaipur, India
e-mail: suraj.kumar@mygyanvihar.com

A. K. Taloor et al. (eds.), *Water, Cryosphere, and Climate Change in the Himalayas*, Geography of the Physical Environment, https://doi.org/10.1007/978-3-030-67932-3_19

To estimate the concentration of chemical, physical and microbial contaminants in drinking water, 1972 samples were obtained and examined in laboratory. In the analysis, the safe and dangerous zone was delineated using the technique of spatial interpolation on the GIS platform. The field study included the collection of groundwater samples from municipal wells, hand pumps, health centre water supplies, and schools or other water sources used for drinking purposes by the majority of people. In the last two to three decades, industrialization and urbanization in Guna have undergone revolutionary, rapid industrial growth. As a result of this urbanization, groundwater supplies are increasingly threatened by rising demands and pollution in many countries (Prajapatiet al. 2017; Jasrotia et al. 2018, 2019; Taloor et al. 2020b).

According to Indian Standard BIS: IS 10500-1991, the impact of water pollution in the human body or society can have bad effects of using water with the concentration of Physico-chemical and microbial above the desirable/permissible limit according to drinking water specification. Physical pollution in drinking water: PH pollution in drinking water may cause vomiting, hand tremors, muscle twitching, tingling in the extremities, or confusion in the face. The effects of water contamination in the human body or culture can be as follows. The higher turbidity level can increase the risk of people developing gastrointestinal diseases due to turbidity contamination in drinking water. For immune-compromised persons, this is particularly troublesome since pathogens such as viruses or bacteria may bind to the suspended solids. The contamination of TDS in drinking water can give a bitter, salty, or brackish flavour to water. Water hardness, scale forming, and staining can be caused by calcium and magnesium, two minerals commonly found in TDS. There are many health issues with a high level of TDS in water (Jasrotia and Kumar 2014; Adimalla and Taloor 2020a; Adimalla et al. 2020).

Chemical pollution in drinking water: the pollution of iron in drinking water, the overload of iron in the liver and other organs, and the development of free radicals that damage cells and tissues can increase the risk of certain cancers. Taking high doses of iron can also cause stomach pain and vomiting. One of the most prevalent health issues is fluoride pollution in drinking water and tooth decay (Shekhar et al. 2012). The low fluoride concentration can decrease the prevalence of tooth decay in the local population. Dental fluorosis or skeletal fluorosis can result in too much fluoride, which can damage bones and joints. Chloride exposure in drinking water can lead to muscle weakness, breathing problems, constant vomiting, prolonged diarrhoea, excessive thirst, high blood pressure, and excessive fatigue. Contamination of Total Hardness (TH) in water, high magnesium and calcium may affect the organs and causing problems related to health. Gastrointestinal problems and skin irritations (Adimalla and Taloor 2020b) may be caused by complete alkalinity pollution in drinking water, potential side effects, and risks of alkaline water and an overall excess of alkalinity in the body.

Faecal bacteria are single-celled microorganisms, virtually always associated with faecal water contamination. When assessing (indicating) the microbial content of water, faecal coliforms, and indicator bacteria are used. In E-coli, the most common facultative pathogenic and disease-causing bacteria, E-coli occurs naturally in most warm-blooded animals, including humans, in the intestines and faeces, and is an indicator or direct consequence of faecal contamination when present in water.

19.2 Study Region

Guna district is situated in the northern portion of Madhya Pradesh, covering an area of approximately 6379.07 km^2 (Fig. 19.1) lies between 25°07'29.14" north latitude and 23°52'17.25" north and 76°47'12.70" east longitude and 77°45'15.27" east longitude covers total 1305 villages with a total population of 1240938. The study area within the drainage system of the Yamuna basin and drains along the rivers Parbati and Kuno, which are tributaries of the Chambal River. The river Sindh and Chambal drains the eastern and western portion in study region respectively.

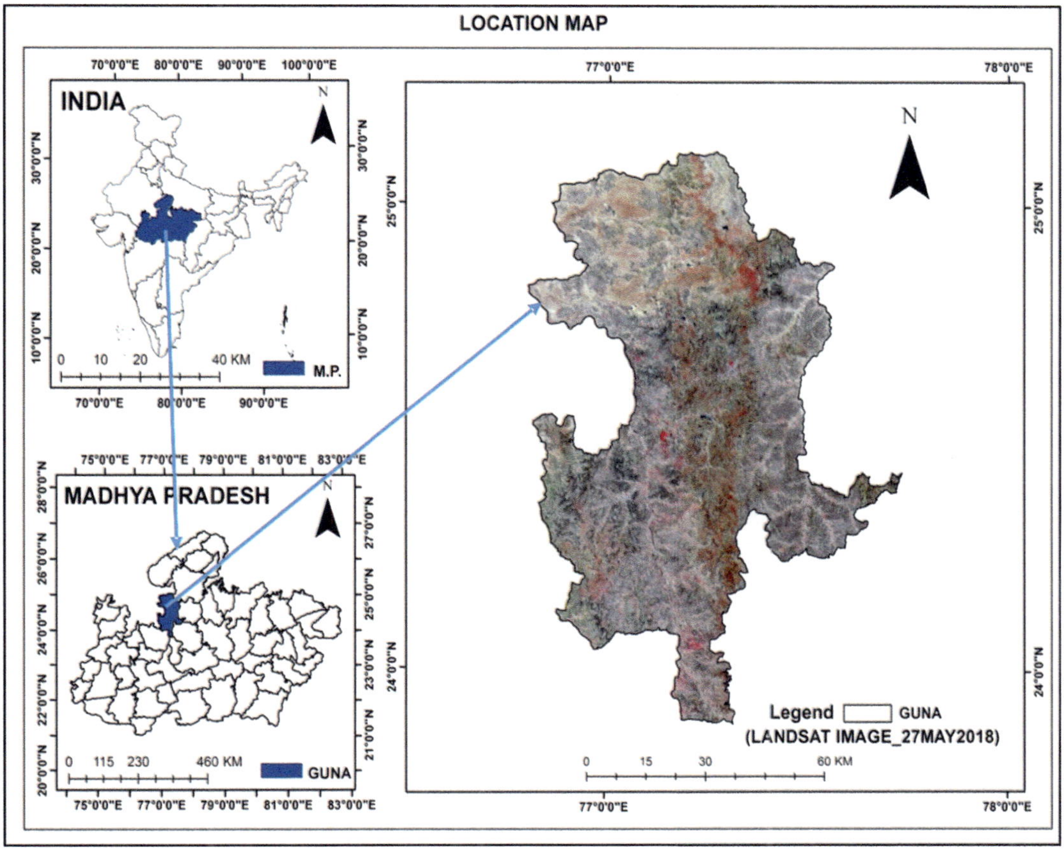

Fig. 19.1 Location map of study region (*Source* USGS)

19.3 Data Used

- The Landsat-8 OLI/TIRS (30m) satellite images of the study area acquired/download from USGS for thematic maps preparation.
- The SoI toposheet of 1:50000 acquired from SoI, Dehradun for georeferencing and base map preparation.
- The samples of groundwater were collected with the help of Madhya Pradesh Council of Science and Technology, Bhopal (MPCST Bhopal) India.

19.4 Methodology

The methodology (Fig. 19.2) was devised into three major categories, water sample collection from the field, water sample analysis in the laboratory, and water quality index mapping in the GIS environment.

19.4.1 Field Survey

The handheld GPS was used for the collection of water sources, such as hand pumps, open dug wells, overhead tanks/pools, tube wells with a hand pump, or a power pump and other drinking water sources (Fig. 19.3b). A weighted sample bottle/container or sampler was used for sampling water supply. The samples from the tube wells/hand pump, the source outlet, and the spirit-lights were sterilized under flames before the specimen was collected in the bottle for bacteriological analyses. The samples have been taken directly from open wells in the pre-sterilized glass bottles. The field survey

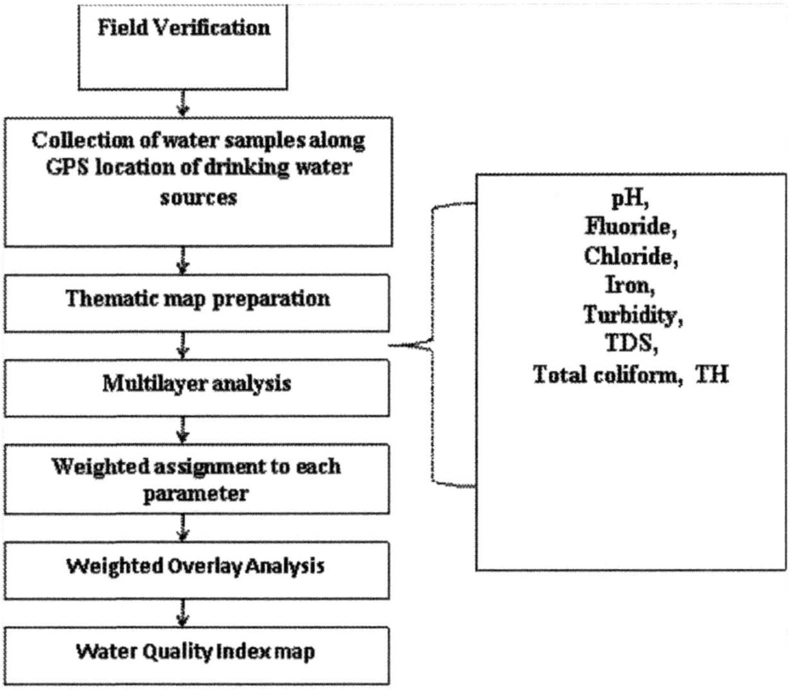

Fig. 19.2 Flowchart of the methodology used in the present study (*Source* Authors)

involves groundwater samples collections from field which are extensively used for drinking purposes includes households and public hand pumps, private wells and water sources in medical centres and colleges.

19.4.2 Water Quality Analysis

The water quality samples collected were transported to the laboratory for analysis and then the water samples immediately analyzed for Physico-Chemical (pH, TDS, turbidity, iron, fluoride, TH, and chloride). Merck's tool kit was used to evaluate microbial parameters (present/absent Coliform, Coliform/100 ml, and E. coli/100 ml) in which the water samples were allowed to stand for maximum reaction and colour production for 25–30 min. In a water sample, the existence of various colours suggests the presence of a higher concentration of microbes above the acceptable limit.

Stepwise water quality monitoring for bacteriological evaluation was done as per the standard operating procedures as under the first step was the detection of Coliform in water presence/absence test (Qualitative- by H2S vials field strip method, Transchem, Agritech Ltd.) at the source. Then the second step was the detection of total coliform/thermotolerant Coliform in water (Membrane Filtration Technique) at the Quality Assurance Laboratory of MPCSTBhopal. Then in the next step detection of E. coli from total coliform colonies by blue fluorescence of MUG, the hydrolysis method was done.

Water samples analysis is done according to drinking water measurement of Indian Standard BIS:IS 10500-2012 parameter within permissible/desirable limits given in the below Table 19.1.

19.4.3 GIS Processing

The GIS processing involves the transfer of the sample locations recorded by GPS (Fig. 19.3b) and was used for geospatial investigation of all parameters in different blocks. Then based on the spatial analysis of these parameters, delineation of safe and unsafe zones based on Physico-

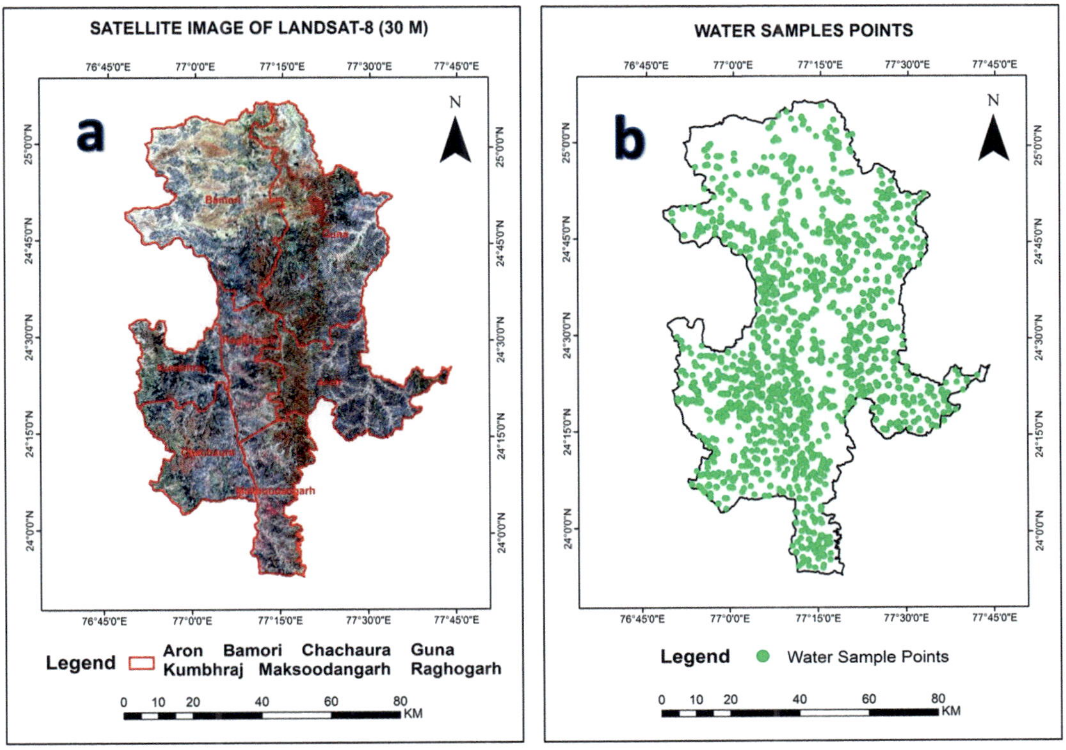

Fig. 19.3 **a** Map representing satellite image with the tehsil boundary of Guna district. **b** Water sample points locatedin the study area (*Source* Authors)

Table 19.1 Physicochemical and biological parameters within permissible/desirable limits

S. no.	Physicochemical and biological parameters	Desirable limits
1	Turbidity	<=5
2	pH	6.5–8.5
3	Total Hardness (as $CaCO_3$) ppm	<=300
4	Chlorides (as Cl) ppm	<=250
5	Total Dissolved Solids ppm	<=500
6	Iron(Fe) ppm	<=0.3
7	Fluoride (F) ppm	<=1.0
8	Total coliforms/100 ml	1–10/100
9	E. Coli	Absent/100 ml

(*Source* Indian Standard BIS: IS 10500-1991)

chemical and microbial contamination were done. The SoI toposheet of the 1:50000 scale was used for base map preparation and field survey planning of water sample collection on the ground. Water quality analysis result superimposed in the GIS platform and used to link all water quality results with their respective sources of water to further demarcate the water contamination or safe and unsafe water source/zone. The Inverse Distance Weighted (IDW) method used for demarcating and water quality index mapping.

Table 19.2 Relative weightage parameters

S. no	Parameter	Standards (BIS)	wi (Weight)	Wi (Relative weight)
1	pH	6.5-8.5	4	0.13
2	Turbidity	<=5	4	0.13
3	(TDS)	<=500	4	0.13
4	Fluoride	<=1.0	5	0.16
5	Chloride	<=250	3	0.09
6	Iron	<=0.3	5	0.16
7	Total Hardness as(CaCo3)	<=300	2	0.06
8	Total Coliform/100 mL	1-10/100	5	0.16
	Total		32	1

(*Source* Authors)

19.4.4 Water Quality Index (WQI)

Step-1: Via a knowledge-based methodology, the thematic layers were allocated acceptable weights and then incorporated into the GIS environment to prepare the study area's water quality chart. A total of eight sets of parameters were selected that were believed to influence the water quality in the region. Each factor was assigned an appropriate weight based on knowledge-based, and finally, a map of water quality was produced. All detrmined parameters has assigned a weight (Wi) as per their relative significance in overall water quality (Table 19.2). The max weight 5 was assigned to Total Coliform, Fluoride, and Iron because of significant and harmful effects of humans, while pH, TDS, turbidity, assigned weight 4, chlorides, assigned weight 3, TH were assigned weight 2, according to its value or harmful effects for drinking purposes (Saana et al. 2016).

Step-2: The (Wi) is estimated from Eq. 19.1:

$$\mathrm{Wi} = \mathrm{wi} / \sum_{i=1}^{n} \mathrm{wi} \quad (19.1)$$

where Wi is relative weight, wi is weight of each factor, and n is factor number.

Step-3: For each parameter, the quality rating scale is estimated by dividing concentration by its respective criteria (World Health Organization 2011) in each water sample and multiplying the results by 100 (Eq. 19.2).

$$\mathrm{Qi} = \frac{\mathrm{Ci}}{\mathrm{Si}} \mathrm{x} 100 \quad (19.2)$$

where Qi is rating quality, Ci is concentration of every chemical parameter in mg/ L; Si is standard for Indian drinking water in mg/ L in accordance with the BIS 10500:2012 guidelines. At first the SI calculated for WQI for every parameter by using the Eqs. 19.3 and 19.4.

$$\mathrm{SIi} = \mathrm{Wi} \ \mathrm{x} \ \mathrm{Qi} \quad (19.3)$$

$$\mathrm{WQi} = \sum_{i=1}^{n} \mathrm{SIi} \quad (19.4)$$

where SIi is sub-index of the ith parameter; Qi is ranking dependent on the parameter of the ith concentration, and n is parameter number.

Step-4: Using ARC Map 10.2.2 software, focused on spatial and statistical analysis in GIS. In the GIS environment, consistency maps of the different parameters, pH and TDS, turbidity, fluoride, iron, chloride, TH, and E.coliform were prepared using the IDW technique to interpolate their respective performance. The computed WQI values were categorized into five types (i.e. very poor, poor, moderate, good and very good).

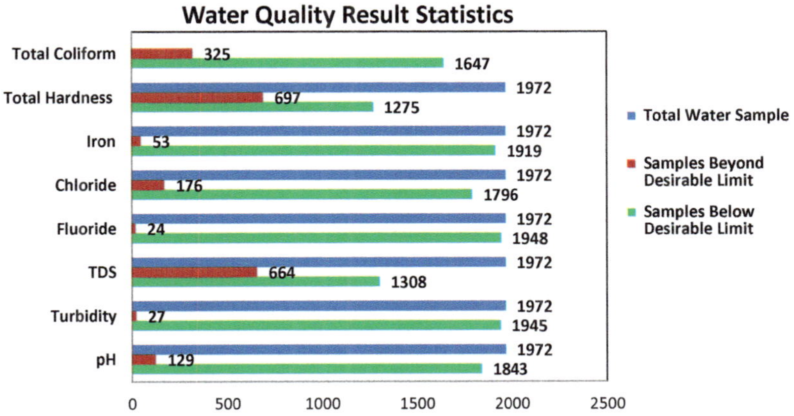

Fig. 19.4 Water sample result statistics of individual water quality parameters (*Source* Authors)

19.5 Results

In the current research work water quality of drinking water sources divided into three categories as per Indian Standard BIS:IS 10500-2012 that is water contamination due to physical parameter (pH, Turbidity, and TDS), water contamination due to chemical parameter (Chloride, Iron, Fluoride, and TH) and water contamination due to the microbiological parameter (Total coliform/100 ml and E.coli). The cumulative result of water contamination estimated due to physical parameter, chemical parameter, and microbial parameter, and a final cumulative water quality index map prepared according to all above three categories (physical, chemical and microbial) to find out those drinking water sources and demarcate safe drinking water zones which fulfil Indian Standard desirable limits criteria for supply of safe water for drinking purpose. Statistics of the individual parameter of physical, chemical, and microbial concentrations in laboratory tested drinking water samples of Guna district given below Table 19.3 (Figs. 19.4 and 19.5).

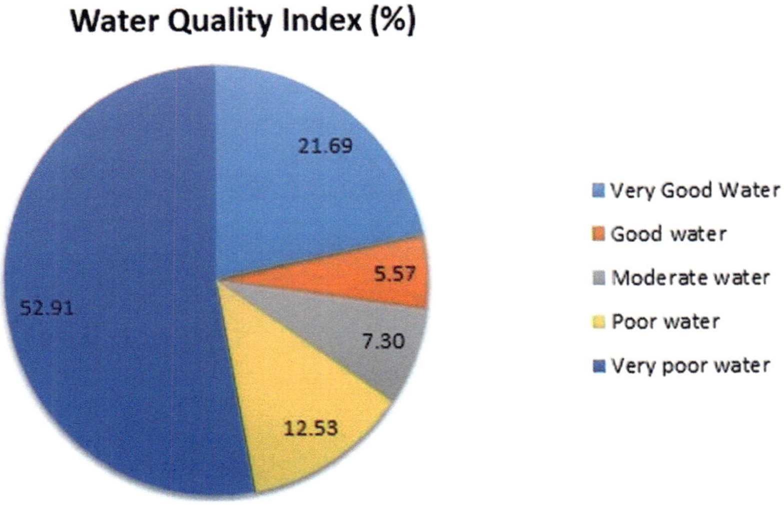

Fig. 19.5 Water quality index statistics of the cumulative result of all water quality analysis (*Source* Authors)

Table 19.3 Statistics of the individual parameter (physical, chemical, and microbial) concentrations in laboratory tested drinking water samples of Guna district

Water quality parameter	Indian Standard BIS:IS 10500-1991 desirable limit	Water sample result below the desirable limit	Water sample beyond the desirable limit	Total sample
pH	6.5–8.5	1843	129	1972
Turbidity	<=5	1945	27	
Total Dissolve Solid (TDS)	<=500	1308	664	
Fluoride (F)	<=1.0	1948	24	
Chloride (Cl)	<=250	1796	176	
Iron (Fe)	<=0.3	1919	53	
Total Hardness ($CaCo_3$)	<=300	1275	697	
Total Coliform	1–10/100 ml	1647	325	

(*Source* Authors)

19.5.1 Water Quality Index

Through applying the above technique, the total assessment of WQI was categorized as very good which covers 21.69%, good category covers 5.57%, moderate category covers 7.3%, poor category covers 12.53% whereas 52.91% of the total area was categorized under very poor category as per Indian Standard BIS:IS 10500-1991. These water sources are affected by physical, chemical, or microbial contamination and appear to need treatment. Statistics of the cumulative result of physical, chemical, and microbial concentrations in laboratory tested drinking water samples of study area are given below Table 19.4.

19.6 Discussion

19.6.1 pH

The pH is general water quality tests performed in the study presented, showing the acidity of the sample but calculating the possible activity of hydrogen ions (H^+). The pH distinguishes

Table 19.4 Arealstatistics water quality index of Guna district

WQI value (%)	Water quality	Area (%)
<35	Very good water	21.69
35–43	Good water	5.57
43–55	Moderate water	7.3
55–75	Poor water	12.53
>75	Very poor water	52.91

(*Source* Authors)

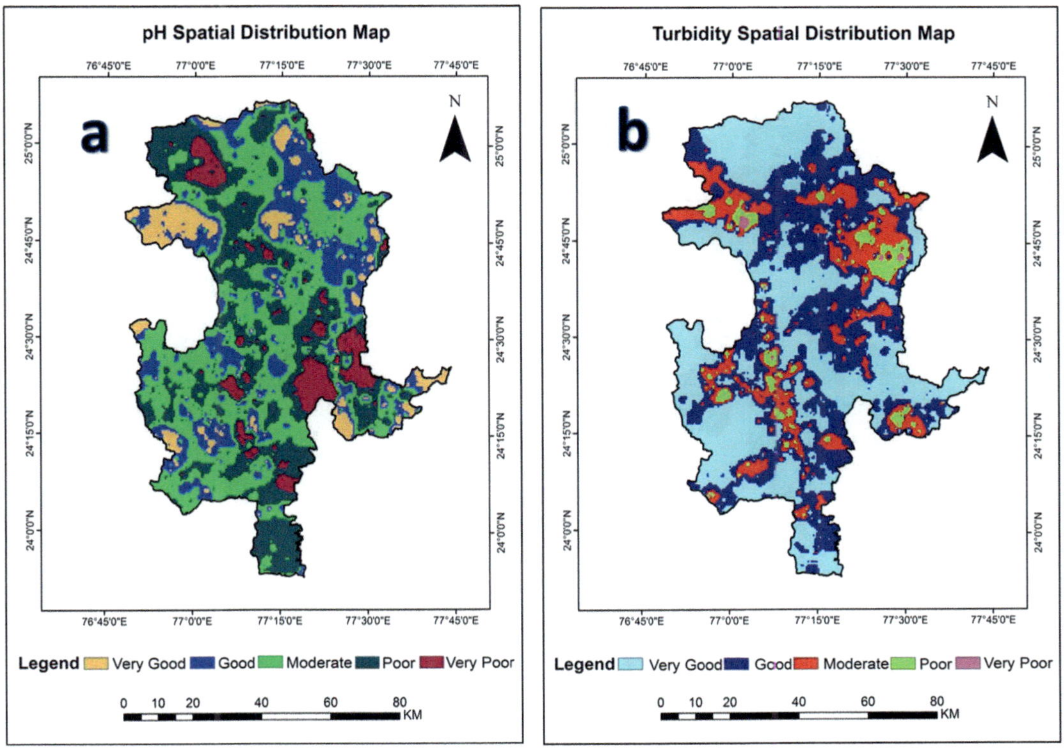

Fig. 19.6 **a** pH spatial distribution map. **b** Turbidityspatial distribution map (*Source* Authors)

groundwater acidity and alkalinity. Also, many dissolved organic and inorganic things are regulated by it. Within the desirable range (6.5–8.5), the pH value varies (Fig. 19.6a).

19.6.2 Turbidity

The measurement of turbidity is a vital key to water quality. The degree to which water loses its visibility, due to the presence of suspended particles, is turbidity. The higher the total solids suspended, the greater the turbidity. Turbidity is considered as an acceptable measure of water quality (Fig. 19.6b).

19.6.3 Fluoride

Fluoride is a chemical substance that can be added to water or toothpaste to help prevent bad teeth. Fluoride is a mineral that is found in many places naturally, including your teeth (Fig. 19.7a).

19.6.4 Chloride

Generally, chloride is not considered detrimental to human health. It regulates the water's salinity. In underground aquifers, natural formations containing groundwater, chloride comes into solution, but it has recently been found due to anthropogenic or human-caused factors such as road salt, waste pollution, and water softeners (Fig. 19.7b).

19.6.5 Total Hardness

It is usually expressed as calcium carbonate (mg) equivalent per litre. Calcium carbonate water is commonly considered soft at less than

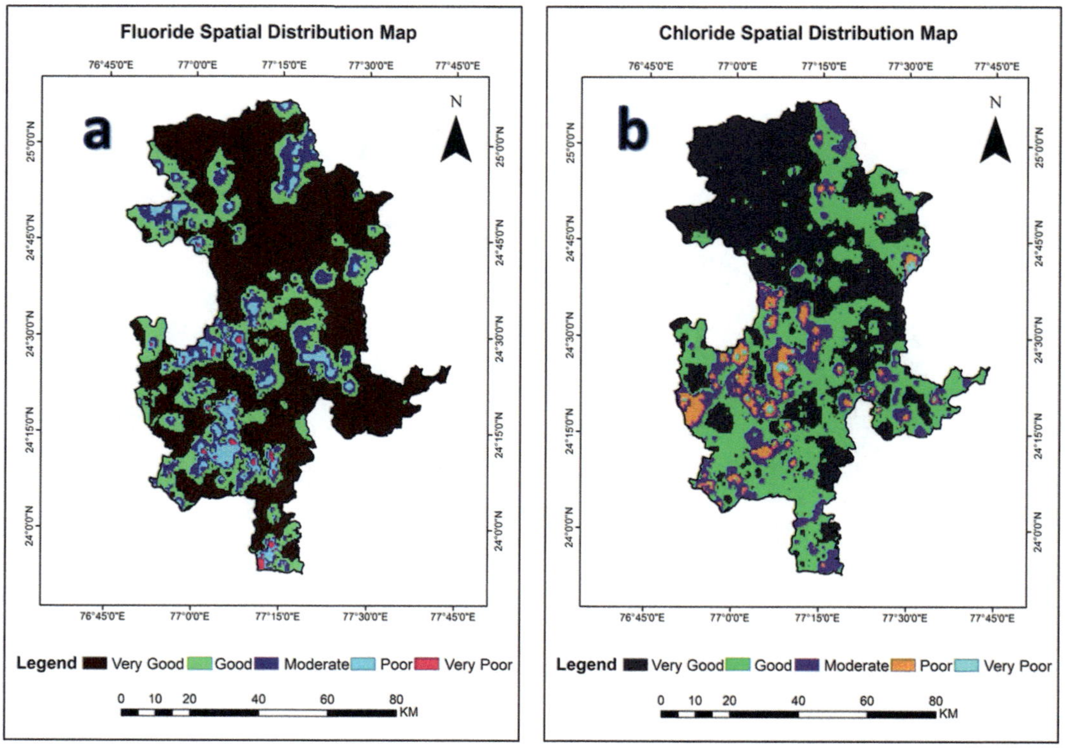

Fig. 19.7 **a** Fluoride spatial distribution map. **b** Chloridespatial distribution map (*Source* Authors)

60 mg/l; hard at 60–120 mg/l; and tough at more than 180 mg/l Fig. 19.8a

19.6.6 Iron

It is found in two forms (i.e. solube is ferrous and insoluble is ferric) in water. Ferrous water contain iron, since iron is dissolved into water, is transparent and colourless. The water becomes opaque when expose to atmosphere and air in the pressure tank. The reddish brown is oxidized or ferric iron which is not dissolved into water (Fig. 19.8b).

19.6.7 Total Coliform

In water bodies that are polluted with faeces from infected humans or animals, Complete Coliform can be found. Via various means, waste can reach the water, including drainage overflows, drainage systems that do not operate properly, contaminated storm water runoff, and agricultural runoff. There are three bacterial coliform classes, and each is a water quality measure, and each has a different degree of risk (Fig. 19.9a).

19.6.8 Total Dissolved Solid (TDS)

A calculation of total amount of inorganic and organic material is the total dissolved solids. Such solids are mostly salts, minerals and organic matter that can be water quality measure (Fig. 19.9b).

19.6.9 Water Quality Index (WQI) of Guna District

According to the results WQI region is categorized from very good to very poor water quality zones based on water samples obtained from the

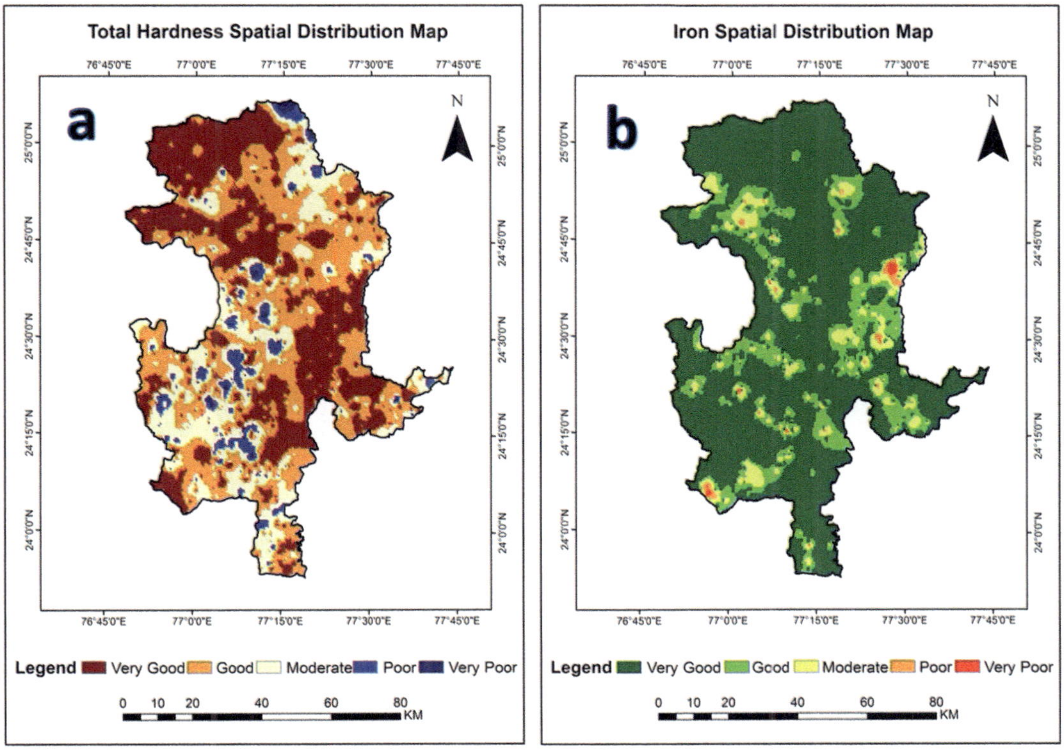

Fig. 19.8 **a** TH spatial distribution map. **b** Iron spatial distribution map (*Source* Authors)

entire district of Guna and meeting the drinking water safety standards criteria (Fig. 19.10).

19.7 Conclusions

With rapid economic, urban, and agricultural growth in India, water pollution is poised to increase; however, the phenomenon could vary spatially and temporally. Groundwater monitoring is expensive and time-consuming; however, dissemination of available data through such research studies is helpful corrective steps for various issues related to water quality and drinking water. The physical, chemical, and microbial water quality analysis offer valuable information on the quality and improvements in the quality of the water supply and the efficacy of the treatment process. A total of one thousand nine hundred and seventy water samples were analyzed in current research, and the results of the study show that only the thematic layers were given adequate weightage by knowledge-based techniques and then incorporated into the GIS environment to prepare the study area's water quality index chart.A total of eight sets of parameters were selected that were believed to influence the water quality in the region. An acceptable weight based on expertise was allocated to each factor, and a map of water quality was finally prepared. Use of geospatial techniques for WQI reveals that most part of study region 21.69% falls under the (very good), 5.57% (good), 7.3% (moderate), 12.53% (poor) whereas 52.91% of the area (very poor) category according to Indian Standard BIS:IS 10500-2012 desirable limit of water quality. These findings are based on data for drinking purposes available

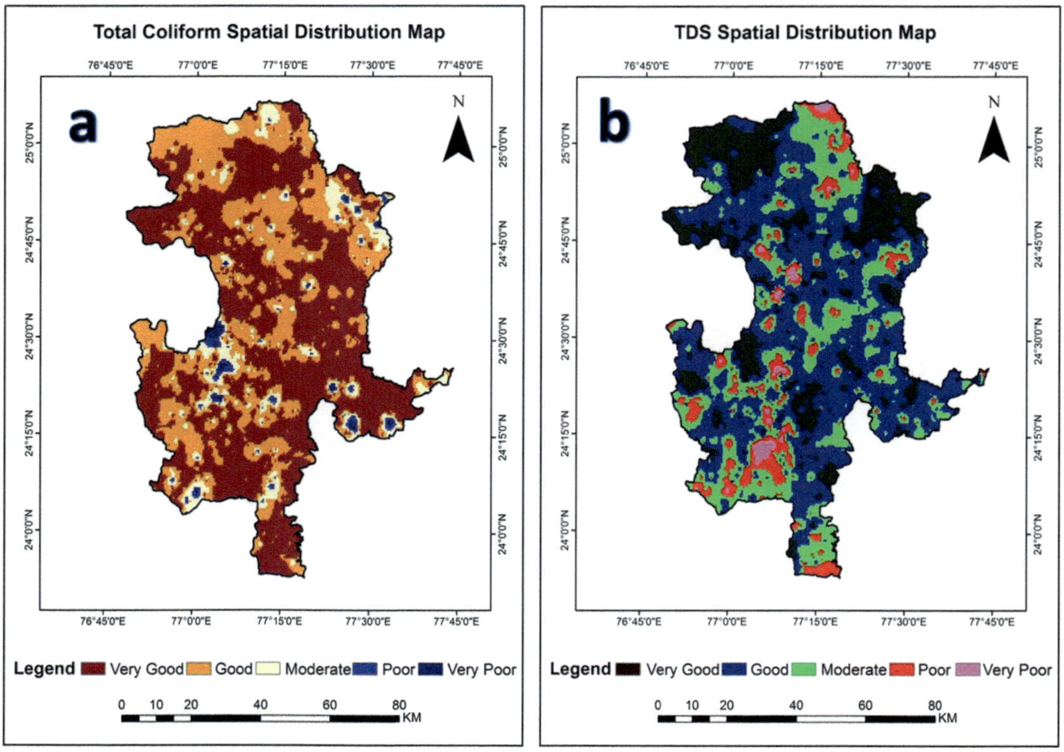

Fig. 19.9 **a** Total Coliform. **b** TDS spatial distribution map (*Source* Authors)

for water quality research. A further collection of water samples and ongoing monitoring of drinking water supplies is needed to assess the recent status of water quality for drinking purposes. Research currently underway is useful to prepare and maintain the potable water supply in the study area in the long term.

19.8 Recommendations

This chapter summarizes the main regulations for the control of water quality and presents some practical recommendations for keeping water quality healthy and safe.

- Clean water is a feature not only of natural processes but also of citizens' responsible social actions and government agencies' integrated and organized management.
- Monitoring should be directed to physical, chemical, and microbiological parameters for identifying possible contamination sources, using specific kits.
- Anthropogenic activities should be prohibited along with the water bodies (wells, tube wells).
- The dumping of industrial discharges into river water without treatment should be avoided.
- Animals should not be washed in the riverine environment either.

Acknowledgements We Sincerely thankful to Hon'ble Vice-Chancellorof the Suresh Gyan Vihar University, Jaipur, for the provision of institutional facilities and their support and guidance for the completion of this work. The authors are also thankful toMadya Pradesh Council of Science and Technology, Bhopal, India for help to analyse the water quality results.

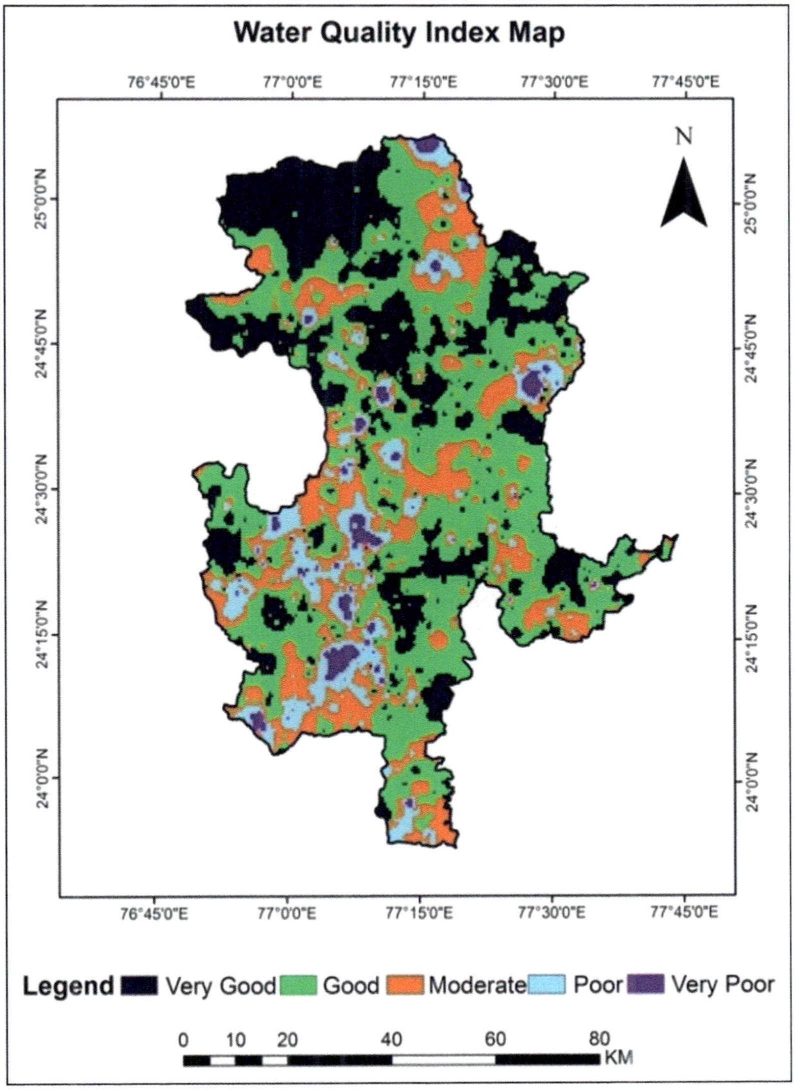

Fig. 19.10 Water Quality Index Map of Guna district (*Source* Authors)

References

Adimalla N, Dhakate R, Kasarla A, Taloor AK (2020) Appraisal of groundwater quality fordrinking and irrigation purposes in Central Telangana, India. Groundw Sust Dev 10:100334. DOI: doi.org/10.1016/j.gsd.2020.100334

Adimalla N, Taloor AK (2020a) Hydrogeochemical investigation of groundwater quality in the hard rock terrain of South India using Geographic Information System (GIS) and groundwater quality index (GWQI) techniques. Groundw Sust Dev 10:100288. DOI: doi.org/10.1016/j.gsd.2019.100288

Adimalla N, Taloor AK (2020b) Introductory editorial for 'Applied Water Science' special issue: "Groundwater contamination and risk assessment with an application of GIS". Appl Water Sci 10:216. https://doi.org/10.1007/s13201-020-01291-3

Bhat MS, Alam A, Ahmad B, Kotlia BS, Farooq H, Taloor AK, Ahmad S (2019) Flood frequency analysis of river Jhelum in Kashmir basin. Quat Int 507:288–294

BIS (2012) Indian Standard Specification for Drinking Water (IS 10500: 2012)

Bisht H, Arya PC, Kumar K (2018) Hydro-chemical analysis and ionic flux of meltwater runoff from Khangri Glacier, West Kameng, Arunachal Himalaya,

India. Environ Earth Sci 77:1–16. https://doi.org/10.1007/s12665-018-7779-6

Bisht H, Kotlia BS, Kumar K, Arya PC, Sah SK, Kukreti M, Chand P (2020) Estimation of suspended sediment concentration and meltwater discharge draining from the Chaturangi glacier Garhwal Himalaya. Arab J Geosci 13(6):1–12. https://doi.org/10.1007/s12517-020-5204-w4

Dhar A, Sahoo S (2015) Identification of groundwater potential zones considering water quality aspect. Environ Earth Sci 74:5663–5675

Hasan M, Shang Y, Akhter G, Jin W (2017) Evaluation of groundwater suitability for drinking and irrigation purposes in Toba Tek Singh District, Pakistan. Irrig Drain Sys Eng 6:185

Haque S, Kannaujiya S, Taloor AK, Keshri D, Bhunia RK, Ray PKC, Chauhan P (2020) Identification of groundwater resource zone in the active tectonic region of Himalaya throughearth observatory techniques. Groundw Sust Dev 10:100337. https://doi.org/10.1016/j.gsd.2020.100337

Jasrotia AS, Kumar A (2014) Groundwater quality mapping based on the geographical information system (GIS) of Jammu District, Jammu and Kashmir India. J Spat Hydrol 12(1):1–21

Jasrotia AS, Taloor AK, Andotra U, Bhagat BD (2018) Geoinformatics based groundwater quality assessment for domestic and irrigation uses of the Western Doon valley, Uttarakhand, India. Groundw Sust Dev 6:200–212

Jasrotia AS, Taloor AK, Andotra U, Kumar R (2019) Monitoring and assessment of groundwaterquality and its suitability for domestic and agricultural use in the Cenozoic rocks of Jammu Himalaya, India: a geospatial technology based approach. Groundw Sust Dev 8:554–566

Jhariyaa DC, Kumara T, Dewanganb R, Pal D (2017) Assessment of groundwater quality index for drinking purpose in the Durg District, Chhattisgarh using geographical information system (GIS) and multi-criteria decision analysis (MCDA) techniques. J Geol Soc India 89:453–459

Khan A, Govil H, Taloor AK, Kumar G (2020) Identification of artificial groundwater recharge sites in parts of Yamuna river basin India based on remote sensing and geographical information system. Groundw Sust Dev 11:100415. https://doi.org/10.1016/j.gsd.2020.100415

Kumar D, Singh AK, Taloor AK, Singh DS (2020) Recessional pattern of Thelu andSwetvarn glaciers between 1968 and 2019, Bhagirathi basin, Garhwal Himalaya, India. Quat Int. https://doi.org/10.1016/j.quaint.2020.05.017

Kumar R, Chauhan A, Rawat L (2017) Physico-chemical analysis of surface and ground water in selected sites of Dehradun, Uttarakhand, India. J Environ Anal Toxicol 6:420

Pandey AC, Singh SK, Dipankar Saha (2015) Geological and hydrogeomorphological control on iron-arsenic contamination in groundwater in part of Gangetic plain, India. Int J Adv Remote Sens GIS 4:55–63

Ponsadailakshmia S, Ganapathy S, Prasanna SM, Madhuramba G (2017) Evaluation of water quality suitability for drinking using drinking water quality index in Nagapattinam district, Tamil Nadu in Southern India. Groundw Sustain Dev 6:43–49

Prajapati M, Jariwala N, Agnihotri P (2017) Spatial distribution of groundwater quality with special emphasis on fluoride of Mandvi Taluka, Surat, Gujarat, India. Appl Water Sci 7:4735–4742

Prakasa R, Puttanna EVS, Sooryanarayana K, Biswas KR, Kumar A, Kumar JS (2017) Assessment of nitrate threat to water quality in India. In: Abrol YP, Adhya TK, Aneja VP, Raghuram N, Pathak H, Kulshrestha U, Sharma C, SinghB (eds), In book: The Indian Nitrogen Assessment. Elsevier, pp 323-333. https://doi.org/10.1016/B978-0-12-811836-8.00021-5

Saana SBBM, Fosu SA, Sebiawu GE, Napoleon J, Thomas K (2016) Assessment of the quality of groundwater for drinking purposes in the Upper West and Northern regions of Ghana. SpringerPlus 5:1–15. https://doi.org/10.1186/s40064-016-3676-1

Sawant RS, Jadhav SD, Godghate AG, Patil RS (2015) Water quality analysis of few villages from Gadhinglaj Tehsil, (MS) India. Int J Curr Res 7:17754–17757

Shekhar S, Pandey AC, Nathawat MS (2012) Evaluation of fluoride contamination in groundwater sources in hard rock terrain in Garhwa district, Jharkhand, India. Int J Environ Sci 3:1022–1030

Singh SK, Kumar V, Kanga S (2017a) Land use/land cover change dynamics and river water quality assessment using geospatial technique: a case study of Harmu River, Ranchi (India). Int J Sci Res Comput Sci Eng 5:17-24

Singh AK, Jasrotia AS, Taloor AK, Kotlia BS, Kumar V, Roy S, Ray PKC, Singh KK, Singh AK, Sharma AK (2017b) Estimation of quantitative measures of total water storage variation from GRACE and GLDAS-NOAH satellites using geospatial technology. Quat Int 444:191-200

Singh SK, Kanga S, Ghosh B (2020) Groundwater quality assessment using Geoinformatics. Publisher: LAP Lambert Academic Publishing. ISBN: 978-6200442079

Taloor AK, Kotlia BS, Jasrotia AS, Kumar A, Alam A, Ali S, Kouser B, Garg PK, Kumar R, Singh AK, Singh B (2019) Tectono-climatic influence on landscape changes in the glaciated Durung Drung basin, Zanskar Himalaya, India: A geospatial approach. Quat Int 507:262–273

Taloor AK, Kumar V, Singh VK, Singh AK, Kale RV, Sharma R, Khajuria V, Raina G, Kouser B, Chowdhary NH (2020a). Land use land cover dynamics using remote sensing and GIS techniques in Western Doon Valley, Uttarakhand, India. In: Geoecology of landscape dynamics 2020. Springer, Singapore, pp 37–51

Taloor AK, Pir RA, Adimalla N, Ali S, Manhas DS, Roy S, Singh AK (2020b) Spring water quality and discharge assessment in the Basantar watershed of Jammu Himalaya using geographic information system (GIS) and water quality Index(WQI). Groundw Sust Dev 10:100364. https://doi.org/10.1016/j.gsd.2020.100364

Thapa R, Gupta S, Reddy DV (2017) Application of geospatial modeling technique in the delineation of fluoride contamination zones within Dwarka Basin, Birbhum, India. Geosci Front 8:1105–1114

Viswanath N, Dileep CK, Ammad KK, Kumari U (2015) Ground water quality, and multivariate statistical methods. Environ Process J 2:347–360

WHO G (2012) Guidelines for drinking-water quality. World Health Org 216:303–304

Dr. Suraj Kumar Singh, Ph.D., is an Associate Professor and Coordinator at the Centre for Sustainable Development, Suresh Gyan Vihar University, Jaipur, India. He has published various research papers in national and international journals and participated and organized international conferences, workshops, symposiums and webinars. He is presently on the reviewer panel for several research journals, and he is supervising several Ph.D. students on their dissertations

Smt. Ankita Bhardwaj is a Ph.D. scholar in Geoinformatics, Centre for Climate Change and Water Research, Suresh Gyan Vihar University, Jaipur, India. She has done M.Sc. in Remote Sensing and GIS from Jiwaji University, Gwalior, India. She worked as Senior Research Fellow at Madhya Pradesh Council of Science and Technology, Bhopal, India. She has specialization in geospatial applications, i.e. water resources development and watershed management

20 Application of Environmental Isotopes and Hydrogeochemistry in Groundwater Management—A Case Study of Bringi Watershed, Kashmir Himalayas, India

Nadeem Ahmad Bhat, Ghulam Jeelani, and Riyaz Ahmad Mir

Abstract

Twenty-seven water samples including precipitation (3), streams (6) and springs (18) from Bringi watershed, southeast Kashmir were bimonthly collected for 1 year and analysed for ionic concentrations, stable isotopes and tritium. The objectives of the study were to recognize the site of recharge for Karst springs, components and mechanism of groundwater recharge. The local meteoric water line (LMWL) is $\delta D = 7.094 \times \delta^{18}O + 9.791$ ($r^2 = 0.82$) on the basis of monthly averages weighted amount. The winter precipitation isotopic composition (average = −10.4‰ for $\delta^{18}O$ and −58.2‰ for δD) is reflected in streams (average = −8.5‰ for $\delta^{18}O$ and −47.3‰ for δD) and spring water (average = −8.8‰ for $\delta^{18}O$ and −51.7‰ for δD) during summer and late spring, which is representative of winter snow melting. Mean elevation of recharge was estimated between 2500 and 2900 m above the mean sea level (amsl) using altitude effect ($\delta^{18}O$ = −0.27‰ per 100 m). Based on the isotopic mass balance equations, the average surface to groundwater contribution in peak flow time was 337.35 m^3/s, approximately 75% of total discharge from the stream and 7.5 m^3/s during lean period, which is approximately 18.6% of total runoff. In addition, average residence time of springs is very short (less than 1 year) and hence responds very quickly to the hydrological events. The quality of surface and groundwater is good for drinking, domestic and agricultural purposes.

Keywords

Bringi watershed kashmir · Environmental isotopes · Hydrogeochemistry · Karst springs · Recharge area

20.1 Introduction

Kashmir Valley is bestowed with adequate assets of water in variety of glaciers, snow, groundwater and surface water. Several springs of freshwater occur in southeast Kashmir, in Anantnag District ('Ananta' means infinite and 'Naga' means water springs) controlled by Karst terrain (Lawrence 1967; Bhat et al. 2014, 2019a; Alam et al. 2017). For decades, the springs are used for various purposes (i.e. drinking, agriculture, aquaculture, floriculture, tourism, etc.).

N. A. Bhat (✉) · G. Jeelani · R. Ahmad Mir
Department of Earth Sciences, University of Kashmir, Srinagar, India
e-mail: mnadeem83@gmail.com

G. Jeelani
e-mail: geojeelani@gmail.com

R. Ahmad Mir
e-mail: udfriyaz@gmail.com

A. K. Taloor et al. (eds.), *Water, Cryosphere, and Climate Change in the Himalayas*, Geography of the Physical Environment, https://doi.org/10.1007/978-3-030-67932-3_20

For the flourishing of any socioeconomic culture, water resources which includes glaciers, lakes, groundwater are of immense importance (Singh et al. 2017; Kumar et al. 2020; Taloor et al. 2020a, b)

Among the upper land catchments of River Jhelum, Bringi catchment is a karst terrain with replacement of water between Karst springs and streams (Coward et al. 1972; Jeelani et al. 2011, 2014). To reduce contamination of water resources of the area, it is important to demarcate the potential sites of springs recharge and their recharging mechanism (Jeelani et al. 2014; Gat 1971; Ford and Williams 1989). Environmental isotopes (δ^2H, $\delta^{18}O$ and ^3H) along with hydrogeochemistry and hydrogeology have been used by several workers (Ford and Williams 1989; Eyankware et al. 2018). The isotopic signature of meteoric water at a particular location serves as a basis for demarcating ground water recharge area (Gat 1971; Lee et al. 1999; Gonfiantini et al. 1976; Jeelani et al. 2010; McConville et al. 2001). On the other hand, as a result of interaction between rock and water, the chemistry of groundwater changes until a quasi-chemical equilibrium is reached especially HCO_3, Ca and Mg (Goldscheider and Drew 2007; Adimalla and Taloor 2020a; Freeze and Cherry 1979; Jasrotia and Kumar 2014; Fetter 1980; Jeelani et al. 2010; Sah et al. 2017; Bisht et al. 2018; Jasrotia et al. 2018, 2019; Adimalla and Taloor 2020b; Adimalla et al. 2020; Sarkar et al. 2020).

20.2 Area of Study

Bringi catchment, an upland watershed of River Jhelum in Kashmir Valley, lies between 33°20′ to 33°45′N latitudes and 75°10′ to 75°30′E longitudes (Fig. 20.1), with an area of approximately 595 km^2 (Bhat and Jeelani 2018). The elevation of watershed ranges 1650 m amsl at Achabal to >4000 m amsl at Sinthan top. Bringi Stream and its tributaries especially east Bringi and west Bringi which joins with the Jhelum River near Anantnag are drained in this watershed (Bhat et al. 2014).

The area is characterized by temperate climate with four well-developed seasons (Jeelani et al. 2014) and monthly variation in the average temperature and precipitation from 1990 to 2009 is shown in Fig. 20.2 (Bhat and Jeelani 2018).

20.3 Geology and Hydrogeology

Geologically, the area is covered by permo-Triassic rocks especially Triassic Limestone and Panjal Traps. Recent alluvium and Karewa deposits occurs towards low-lying area (Fig. 20.3) (Alam et al. 2017, 2018). Panjal Volcanics and rocks of Upper Palaeozoic are present along the marginal parts of study area. Triassic Limestone is >1000 m high deposit with different layers of sandstone and shale (Bhat et al. 2019c, d). The Karewa deposits are fluviatile and lacustrine sediments that lie on top of Triassic limestone, with unconsolidated coarse to fine-grained sand, dark grey clays, light grey sands, varved clays, brown loam, marl, gravel, silt, lignite, etc. (Bhat et al. 2019c, d). The fluvial deposits contain boulders, gravel, sand, silt, clay, which represent active flood plain sediments (Jeelani et al. 2010; Taloor et al. 2019).

Three main springs, namely, Achabalnag, Kokernag and Kongamnag hosted by the Triassic Limestone and Karewa deposits were studied (Fig. 20.4a–c) during the present work (Jeelani et al. 2014; Bhat and Jeelani 2018). Achabalnag is a Karst spring at the base of Sosanwar Hills, where the water gushes out from two sites about 150 m apart, with one major outlet carrying about 80–90% of the total spring discharge (Fig. 20.4a). The water is channelled through the Achabal Garden and Villages downstream. Kokernag, a set of seven springs, is another major Karst spring where the water gushes out at various places (Fig. 20.4b). The water is channelled through the Kokernag garden and the villages downstream. At Kongamnag, water forms a pool, at bottom of the hill of limestone (Fig. 20.4c) and the water is drained downstream through villages. Kokernag and Achabal springs are cold (8 °C–14 °C) as compared to Kongamnag with temperature ranging from 14 °C to 19 °C. Higher temperature of Kongamnag spring might be due to deep motion of infiltrating water. Kokernag discharge

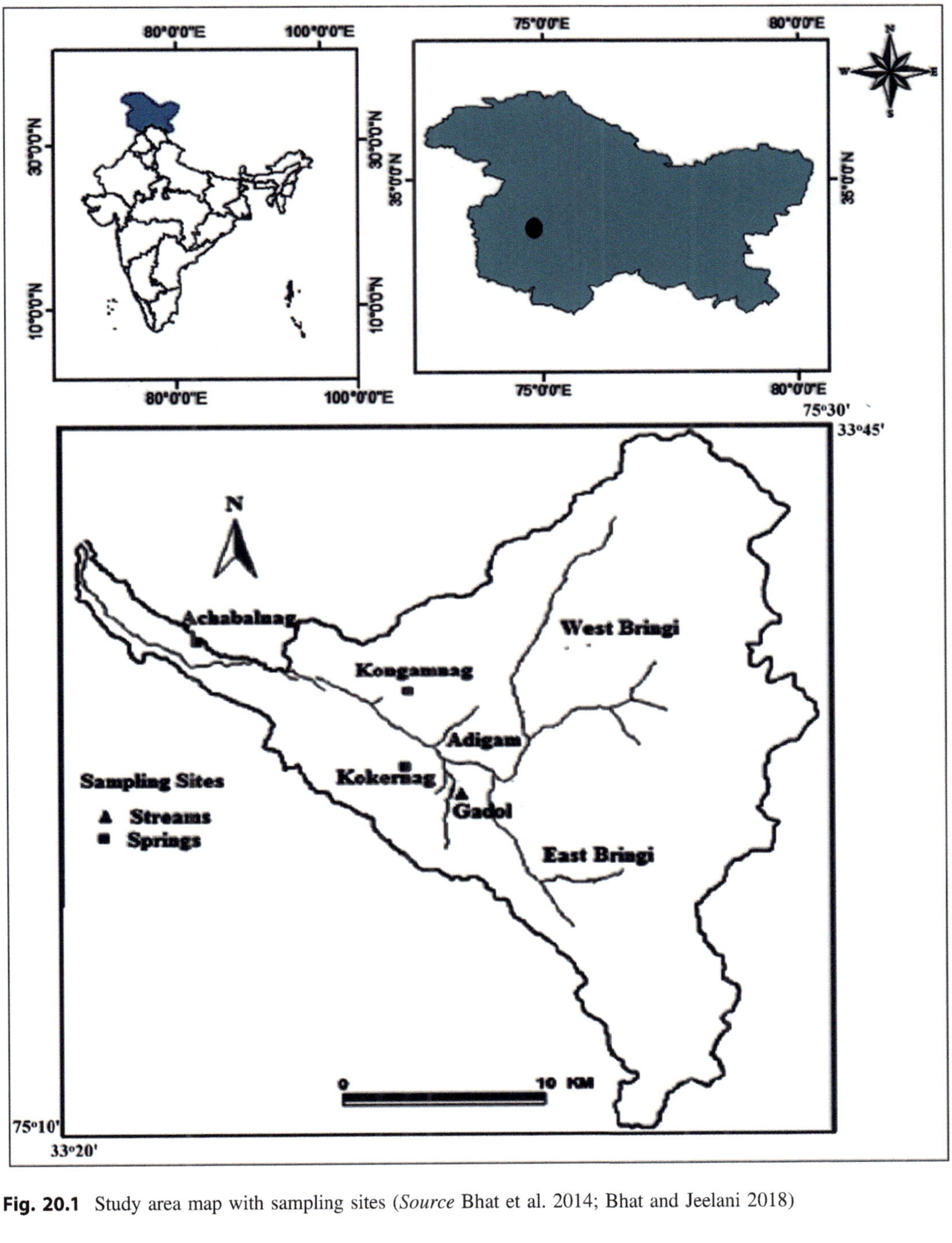

Fig. 20.1 Study area map with sampling sites (*Source* Bhat et al. 2014; Bhat and Jeelani 2018)

varies 600 L/s during winters to 8000 L/s during summers and Achabal discharge varies from 160 L/s in winter to 3000 L/s in summers. The variability in discharge is mainly due to fast response of spring to hydrological events. Discharge of Kongamnag varies 12L/s during winters to 20 L/s during spring and summer seasons.

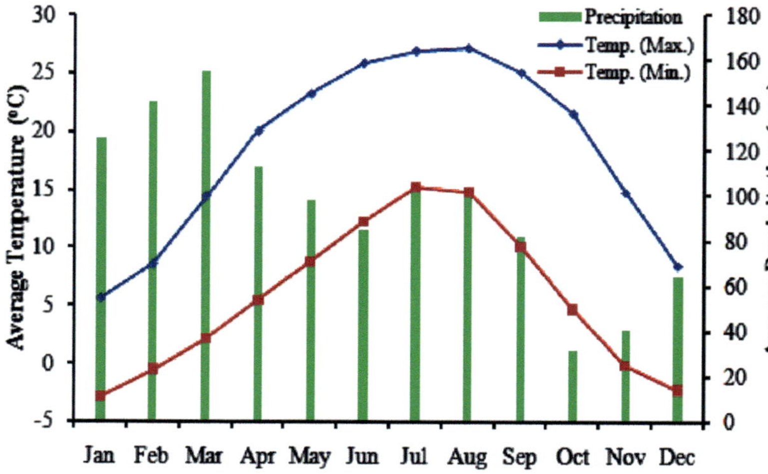

Fig. 20.2 Mean monthly precipitation and temperature from 1990 to 2009 (*Source* Bhat and Jeelani 2018)

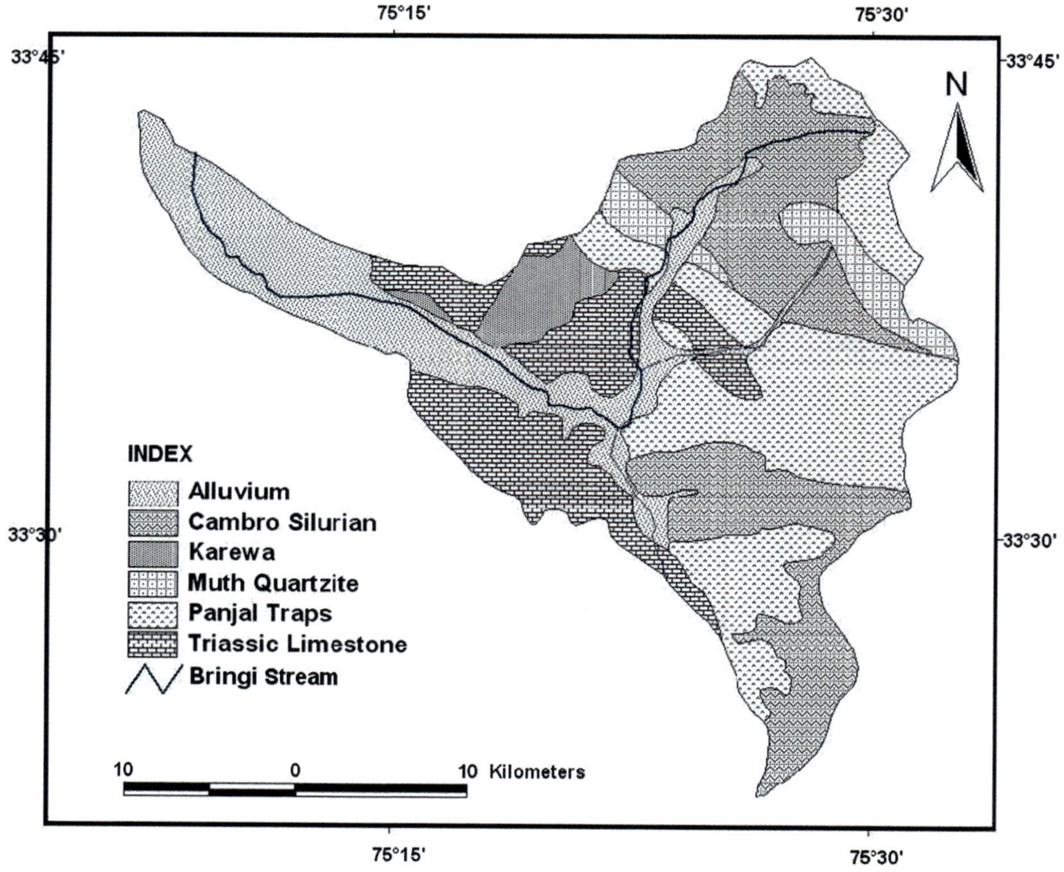

Fig. 20.3 Geological map of study site (*Source* Bhat et al. 2014)

Fig. 20.4 Photgraphs of Bringi springs catchment **a** Achabalnag, **b** Kokernag and **c** Kongamnag (*Source* Authors)

20.4 Methodology

Water samples from Bringi Stream (two sites), three sites for precipitation (Pindabal, Kokernag and Achabal) and springs (Kokernag, Achabalnag, Kongamnag) were collected bimonthly for one hydrological cycle March 2008 to January 2009 (Fig. 20.1). The samples were analysed for ions including Na^+, Mg^{2+}, Ca^{2+}, K^+, Cl^-, HCO_3^-, SO_4^{-2}, SiO_2, F^-, NO_3^- and isotopes ($\delta^{18}O$, δ^2H and ^3H) by using the techniques (APHA 2006; Epstein and Mayeda 1953). Precipitation collectors were installed for collection of precipitation samples for stable isotopes. About 2 ml glycerine was added into the containers, which form a thin film above water to avoid evaporation. For ^3H analysis, the samples were collected in 1 litre sampling bottles during January to September 2008. Master parameters including pH, electric conductivity (EC) and temperature were measured in the field. Major ion analysis was carried out in Hydrogeology Laboratory, Department of Earth Science, Kashmir University, Srinagar. To separate the suspended sediments, waters were filtered through <0.45 μm nucleopore filter paper. All the cations and anions were calculated using the titration methods, flame photometer and spectrophotometer. The tritium and oxygen isotope analysis was carried out in Bhaba Atomic Research Centre (BARC) Mumbai whereas hydrogen isotope in NIH Roorkee (Epstein and Mayeda 1953). In addition, precipitation and temperature data were collected from IMD, Srinagar, Kokernag Station (Table 20.1).

Table 20.1 Bimonthly precipitation, temperature and discharge data of watershed Bringi

Stations	Parameters	Mar-08	May-08	Jul-08	Sep-08	Nov-08	Jan-09
Kokernag	Temperature (°C)	18.3	23.6	27.3	25.4	16.1	5
Kokernag	Precipitation (mm)	25.8	134.7	62.4	107	9.9	195
Pindabal	Stream discharge (m^3/s)	321	633	719	470	220	124
Adigam		95	170	266	133	58	57

(*Source* Bhat and Jeelani 2018)

20.5 Results

The statistical results of physicochemical parameters are given in Table 20.2 and isotope ($\delta^{18}O$, δD and ^3H) details are presented in Table 20.3.

20.5.1 Precipitation

The precipitation samples are moderately alkaline with pH varying from 8.6 to 8.9 with mean of 8.75. EC ranges from 64 to 78 µS/cm with average of 70 µS/cm. Total dissolved solids show a narrow range of salinity (41–50 mg/l) with mean of 46 mg/l. Ca (66%) is higher cation followed by Mg (21%) > Na (12%) > K (1%) and anions as HCO_3(83%) > SO_4 (10%) > Cl (7%). The concentration of different ions may be due to dissolution of various gases and atmospheric pollutants (Bhat and Jeelani 2015).

Stable isotopes in precipitation show marked spatial and temporal variations (Table 20.3). δD and $\delta^{18}O$ ranged from −12.59 to −60.6‰ with a mean value of −31.1‰ and −2.1 to −10.6‰ with a mean of –5.8‰. Samples were enriched during summer and low altitudes and depleted during winter and at higher elevation. Higher values were observed in July ($\delta D = -12.6$‰; $\delta^{18}O = -2.1$‰) and lower in January (mean $\delta D = -58.2$‰, $\delta^{18}O = -10.4$‰). Stable isotopes in precipitation showed good correlation (Fig. 20.5) with precipitation amount and temperature ($r^2 = 0.97$).

20.5.2 Streams

Temperature of the stream water samples varied from 9.6°C to 17.4°C with an average of 12.48 °C. pH between 7.8 and 9.8 with mean of 8.2 indicate stream water is alkaline. Electrical conductivity varied from 140 to 235 µS/cm, with mean value of 175 µS/cm. Similarly, TDS varied from 89 to 193 mg/l with average 122 mg/l. About 50% of samples show cation order as Ca (62%) > Na (21%) > Mg (13%) > K (4%) and anion order as HCO_3 (88%) > SO_4 (8%) > Cl (4%). Remaining 50% showed cation as Ca (61%) > Mg (26%) > Na (12%) > K (1%) and anion as HCO_3 (89%) > Cl (7%) > SO_4 (4%).

Stable isotopes of water collected from stream showed small variation (Table 20.3), with δD from −36.4 to −47.3‰ with average −44.1‰ and $\delta^{18}O$ ranging from −6.8 to −8.5‰ with average −7.8‰. Water is enriched in early autumn (September) season ($\delta D = -36.43$‰ and $\delta^{18}O = -6.8$‰) and depleted in late spring (May) season ($\delta^2H = -47.33$‰ and $\delta^{18}O = -8.5$‰). The isotopic characteristic of summer precipitation (July) with enriched isotopes is not fully reflected in the streams (Fig. 20.5). However, the enriched signals of summer precipitation are clearly reflected in streams during autumn season. During winter, base flow is the main contributor to surface runoff due to low temperatures and negligible melting.

Table 20.2 Statistical overview of analytical results of different water samples of study area along with comparison with WHO and BIS standards

Sample type	Sampling date and number	Stat	pH	EC	Temp	TDS	SiO_2	F^-	NO_3^-	Ca^{2+}	Mg^{2+}	Na^+	K^+	HCO_3^-	SO_4^{2-}	Cl^-	Ca/Mg
				μS/cm	(°C)	mg/L				mmol/L							
Precipitation	Mar–Jul 2008, N = 3	Min.	NA	64.0	NA	41.0	NA	NA	1.38	0.10	0.04	0.01	0.0002	0.26	0.02	0.01	2.6
		Max.	NA	78.0	NA	50.0	NA	NA	1.73	0.14	0.05	0.03	0.0005	0.30	0.05	0.04	3.5
		Mean	NA	69.7	NA	44.5	NA	NA	1.55	0.12	0.04	0.02	0.0003	0.28	0.03	0.02	3.06
		St. Dev	NA	7.4	NA	4.2	NA	NA	0.17	0.02	0.00	0.01	0.0001	0.02	0.01	0.01	0.44
Streams	Mar-08 to Jan-09, N = 6	Min.	7.8	140	9	89	0.2	0.8	0.35	0.29	0.04	0.09	0.0002	0.69	0.03	0.01	2.01
		Max.	9.8	235	17.4	150.4	1.7	1.4	4.2	0.81	0.40	0.10	0.03	2.13	0.16	0.16	8.7
		Mean	8.3	174.7	12.4	110.7	1.2	1.0	1.61	0.42	0.15	0.10	0.02	1.12	0.07	0.06	3.96
		St. Dev	0.8	38.2	3.3	25.4	0.5	0.3	1.53	0.19	0.13	0.01	0.01	0.53	0.05	0.05	2.54
Springs	Mar-08, N = 3	Min.	7.1	200	10.9	128	1.04	0.83	0.65	0.51	0.18	0.09	0.0002	1.15	0.08	0.09	1.2
		Max.	7.6	530	13.1	339.2	1.6	0.9	3.18	0.96	0.80	0.57	0.0005	3.61	0.19	0.19	3.32
		Mean	7.3	314.0	11.7	201.0	1.3	0.85	2.0	0.69	0.41	0.30	0.0004	1.97	0.14	0.15	2.22
		St. Dev	0.3	187.2	1.2	119.8	0.3	0.04	1.27	0.24	0.34	0.25	0.0001	1.42	0.06	0.05	1.06
	May-08, N = 3	Min.	7.4	142	11.2	90.88	0.5	0.82	1.25	0.24	0.16	0.10	0.001	0.74	0.12	0.07	1.33
		Max.	8	442	13	282.8	0.7	0.86	4.2	0.92	0.48	0.52	0.003	2.79	0.13	0.17	2.00
		Mean	7.7	243.3	12.2	155.7	0.6	0.84	2.9	0.49	0.27	0.26	0.002	1.42	0.13	0.11	1.74
		St. Dev	0.3	172.1	0.9	110.1	0.1	0.02	1.5	0.37	0.18	0.23	0.005	1.18	0.01	0.05	0.36
	Jul-08, N = 3	Min.	7.2	236.0	10.4	151.0	0.5	0.82	1.2	0.55	0.16	0.09	0.001	1.56	0.04	0.01	0.98
		Max.	7.8	454.0	14.3	290.6	0.75	1.25	4.55	0.92	0.76	0.58	0.003	3.03	0.17	0.10	4.19
		Mean	7.4	338.7	12.8	216.7	0.6	1.09	3.06	0.74	0.38	0.27	0.002	2.24	0.11	0.04	2.88
		St. Dev	0.3	109.6	2.1	70.1	0.15	0.23	1.7	0.18	0.33	0.27	0.001	0.74	0.07	0.05	1.68
	Sep-08,	Min.	7.0	222.6	12.0	142.5	0.7	0.62	0.85	0.63	0.24	0.11	0.003	1.47	0.03	0.02	1.15

(continued)

Table 20.2 (continued)

Sample type	Sampling date and number	Stat	pH	EC	Temp	TDS	SiO_2	F^-	NO_3^-	Ca^{2+}	Mg^{2+}	Na^+	K^+	HCO_3^-	SO_4^{2-}	Cl^-	Ca/Mg
				μS/cm	(°C)	mg/L				mmol/L							
	N = 3	Max.	7.7	448.0	14.4		1.9			0.80	0.70		0.005	3.20	0.37		2.64
		Mean	7.3	319.7	13.4	204.6	1.2	0.65	3.4	0.71	0.41	0.32	0.004	2.13	0.18	0.10	2.07
		St. Dev	0.4	115.9	1.2	74.2	0.6	0.05	2.45	0.09	0.25	0.29	0.0009	0.93	0.17	0.12	0.81
	Nov-08, N = 3	Min.	7.3	212.2	11.9	135.8	0.6	0.74	0.72	0.34	0.36	0.13	0.002	1.23	0.10	0.04	0.94
		Max.	8.1	454.8	14.3	291.1	1.7	0.96	5.0	0.86	0.72	0.24	0.005	3.44	0.27	0.22	1.2
		Mean	7.6	305.7	13.3	195.7	1.0	0.85	3.34	0.57	0.50	0.19	0.0025	2.02	0.19	0.11	1.11
		St. Dev	0.4	130.5	1.2	83.5	0.6	0.11	2.29	0.27	0.19	0.06	0.001	1.23	0.09	0.10	0.14
	Jan-09, N = 3	Min.	7.6	280.2	8.2	179.3	0.17	0.86	0.68	0.81	0.34	0.11	0.004	1.97	0.03	0.04	1.38
		Max.	8.2	520.2	12.0	332.9	1.7	0.96	4.2	1.16	0.84	0.23	0.006	3.93	0.09	0.18	2.4
		Mean	7.9	367.5	9.7	235.2	0.7	0.9	2.89	0.93	0.53	0.17	0.005	2.68	0.06	0.09	1.93
		St. Dev	0.3	132.7	2.0	84.9	0.9	0.05	1.92	0.20	0.27	0.06	0.001	1.09	0.03	0.08	0.52
WHO 2006			7–8.5	–		–	–	1.5	45	1.875	1.233	8.7	0.307	–	2.08	5.64	–
BIS 2012			6.5–8.5	–		–	–	1.5	45	1.875	1.233	–	–	3.28	2.08	7.05	–

(*Source* Bhat et al. 2014)

Table 20.3 Statistical summary of isotopes ($\delta^{18}O$, δD and ^{3}H) in different water samples of Bringi Basin (AMSL—Above Mean Sea Level, Min—Minimum, Max—Maximum, Ave—Average, WM = Weighted Mean, Prec—Precipitation

Station	Sample type	Elevation (m AMSL)	$\delta^{18}O$ (‰)				$\delta^{2}H$ (‰)				Tritium (TU)
			Min	Max	Ave	WM	Min	Max	Ave	WM	
Achabal	Prec	1656	−10.0	−2.1	−4.8	−5.4	−60.6	−12.6	−26.2	−29.4	13.29
Kokernag	Prec	1890	−10.6	−2.4	−5.8	−6.2	−57.6	−23.2	−32.7	−35.0	14.65
Pindabal	Prec	2110	−10.4	−3.2	−6.6	−6.7	−56.5	−15.7	−34.4	−34.9	16.89
Bringi	Stream	2110	−8.7	−6.9	−7.9	−7.9	−47.3	−36.4	−44.1	−44.0	15.50
Achabal	Spring	1656	−8.6	−6.6	−7.8	−7.8	−47.3	−40.9	−44.0	−44.5	13.19
Kokernag	Spring	1890	−8.9	−7.2	−8.2	−8.2	−50.3	−40.7	−45.7	−45.8	15.48
Kongamnag	Spring	1922	−9.2	−8.1	−8.8	−8.9	−57.6	−54.6	−55.9	−55.8	16.65

(*Source* Bhat and Jeelani 2015)

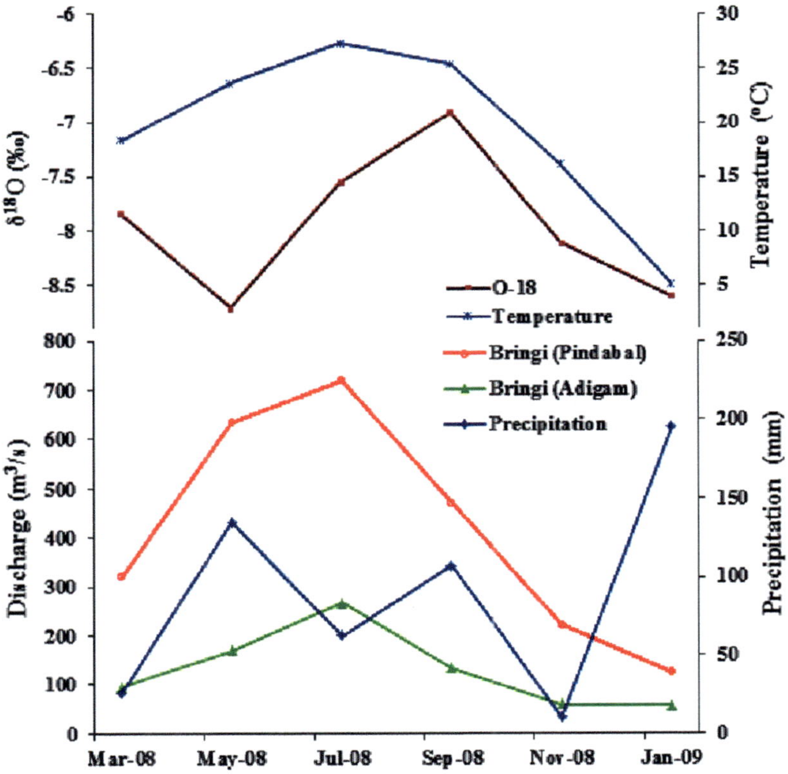

Fig. 20.5 Variation of $\delta^{18}O$, temperature, precipitation and stream discharge (*Source* Bhat and Jeelani 2018)

20.5.3 Springs

The water of springs is odourless and colourless. Water temperature varied from 8.2 °C to 14.4 °C with an average of 12.2 °C. The Karst springs including Kokernag and Achabalnag showed variability of annual temperature (~5.4 °C and ~3.8 °C) as compared to Kongamnag with annual inconsistency of 1.7 °C. The TDS varied between 90 and 339.2 mg/l, with average of 206 mg/l. Most of the samples show high Ca concentration than Mg while few show Na is higher than the Mg. While the anions showed high HCO_3 concentration.

The temporal and spatial variability of springs is given in Table 20.3, with δD ranging from −40.7 to −57.6‰ with average value of −48.5‰ and $\delta^{18}O$ ranging from −6.6 to −9.2‰ with a mean of −8.3‰. Achabalnag was enriched (average δD = −44.0‰; $\delta^{18}O$ = −7.8‰) and Kongamnag was depleted in heavy isotopes (average δD = −56.0‰; $\delta^{18}O$ = −8.8‰). The annual amplitude of $\delta^{18}O$ variations is 1.16 to 1.97, quite similar to the Karst springs of Meramec River Basin and Kapuz Karst springs (Kattan 1997; Fredrickson and Criss 1999) with amplitude varying from 1 to 2.

20.5.4 Tritium (^3H)

Tritium concentration of precipitation water varied from 13.3 to 16.9 TU and average 15.1 TU (Table 20.3). There is higher value of tritium in snow samples of winter season while less in summer season indicating different sources of precipitation (Jeelani et al. 2010).

20.6 Discussion

20.6.1 Geochemical Processes Controlling Chemistry of Water

To determine the evolution of water, ions are powerful tools (Eyankware et al. 2018; Bhat et al. 2016, 2019b, c, d). The hydrochemical data was plotted in Gibbs diagram (Fig. 20.6) to find out the source of sample which shows the rock dominance environment (Gibbs 1970).

Four hydrochemical water types were observed following the order of Ca–Mg–HCO3 > Ca–HCO_3 > Ca–Mg–Na– HCO_3 > Ca–Na–HCO_3 in Piper trilinear diagram (Piper 1994) (Fig. 20.7). The water types have resulted due to carbonate dissolution. The evolution of stream water is Ca–Mg–HCO_3 water type which is due to limited time for interaction between water and rock as well as easily carbonate mineral dissolution.

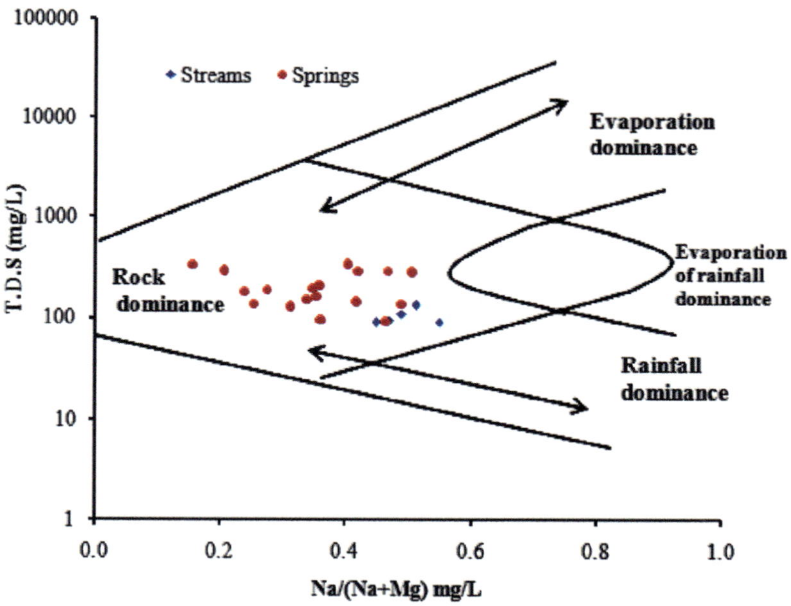

Fig. 20.6 Gibb's diagrams showing rock weathering as the main mechanism for groundwater evolution (*Source* Gibbs 1970; Bhat et al. 2014)

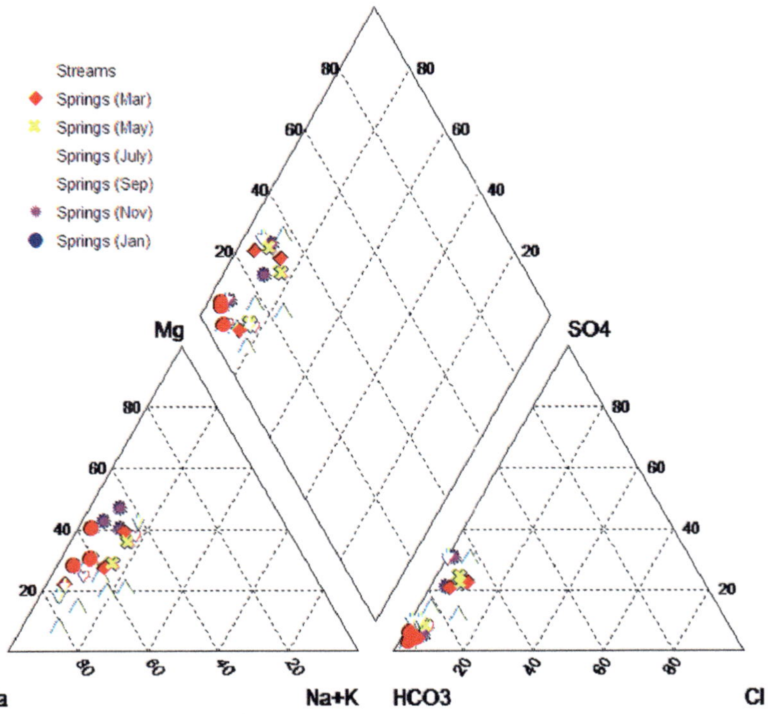

Fig. 20.7 Piper trilinear diagram showing carbonate weathering as the main mechanism for groundwater evolution (*Source* Piper1994; Bhat et al. 2014)

The spring water and stream water are characterized by Ca–Mg–HCO_3 trend in Langelier–Ludwig diagram (Langelier and Ludwig 1942) which shows the clear dominance by carbonate dissolution (Fig. 20.8).

Various binary diagrams were plotted for identifying lithological source in water. In (Ca + Mg) versus HCO_3 (Fig. 20.9a), most water samples of stream lie on to the trend line, which shows carbonate weathering as the main source of solute acquisition. However, some spring water samples fall away from trendline indicating other source for HCO_3 in addition to carbonate weathering. In scatter graph between (Na + K) and (Ca + Mg) (Fig. 20.9b), the stream samples fall close to the trendline which might be because of silicate weathering. However, the spring samples especially Achabalnag and Kokernag fall away from trendline suggesting less contribution of ions from silicate weathering. In some spring samples (Kongamnag), (Na + K) exceeds (Ca + Mg) due to the impact of silicate weathering. In Cl^- and (Na + K) plot (Fig. 20.9c), all samples present above the trendline show different source of Na (Eyankware et al. 2018). The plots given in Fig. 20.9b, c favour the higher contribution of silicate weathering for ionic concentration, mainly Na.

The binary plot of Mg and Ca/Mg (Fig. 20.9 d) represents decrease in molar ratio with increase in Mg concentration, indicating weathering of carbonate rocks especially dolomite is main source for Mg and Ca. In Ca/(Ca + Mg) versus $SO_4/(SO_4 + HCO_3)$ binary plot (Fig. 20.9e), high Ca/(Ca + Mg) molar ratio is due to interaction of water with calcite. The Ca/(Ca + Mg) molar ratios of 0.5 and 1 correspond to dissolution of pure calcite and stoichiometric dolomite (Frondini 2008).

20.6.2 Delineation Area of Recharge for Karst Springs Using Hydrogeological and Hydrogeochemical Approach

The analytical results suggest that Bringi Stream and the Karst springs have similar chemical

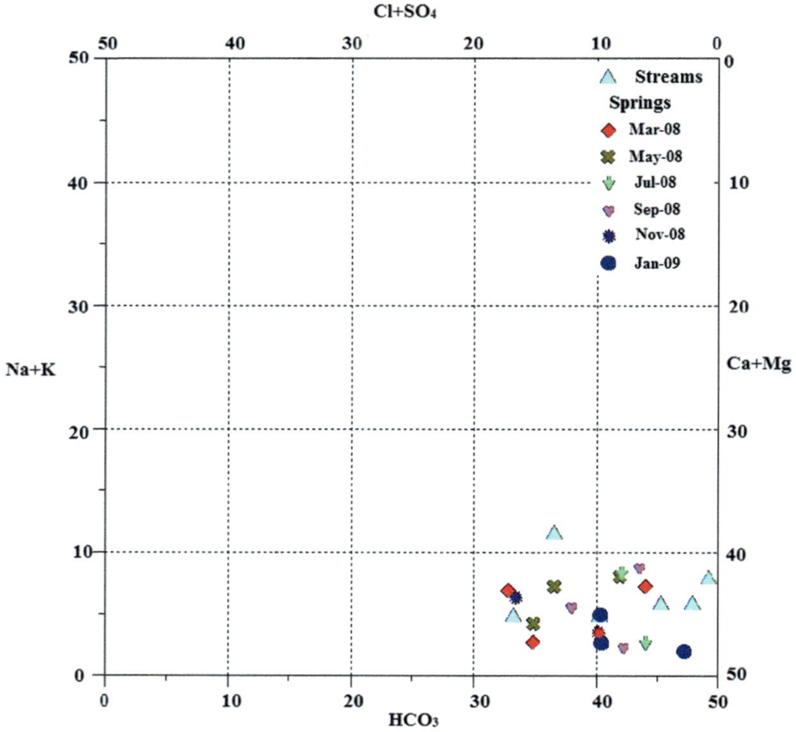

Fig. 20.8 Langelier–Ludwig diagram showing carbonate weathering as the main mechanism for groundwater evolution (*Source* Langelier–Ludwig 1942; Bhat et al. 2014)

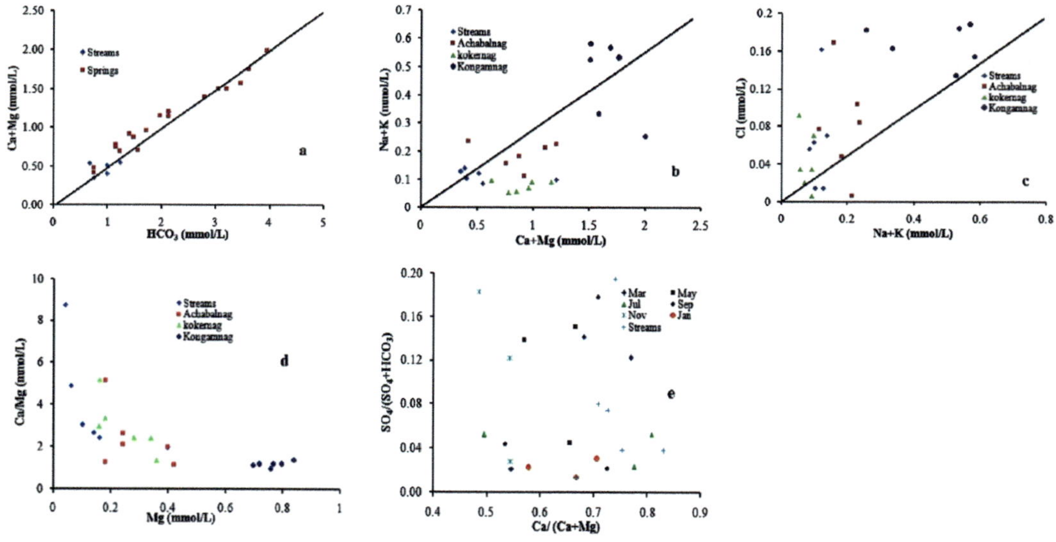

Fig. 20.9 Binary plots depicting ionic sources in streams and springs. **a** (Ca + Mg) versus HCO_3. **b** (Ca + Mg) versus Na + K. **c** (Na + K) versus Cl. **d** Ca/Mg versus Mg. **e** Ca/(Ca + Mg) versus $SO_4/(SO_4 + HCO_3)$ (*Source* Bhat et al. 2014)

composition. However, a subtle difference between the major ion chemistry of the springs is observed. Kokernag and Achabalnag springs have low concentration of ions as compared to Kongamnag. This is mainly due to greater high contact time of water with base lithology in

Kongamnag as compared to Kokernag and Achabalnag springs. Both springs and streams show a different summer maximum and winter minimum temperatures (Jeelani 2010; Jeelani et al. 2014). Similarly, discharge of springs and streams is low during winter and high during summer (Fig. 20.10).

The temporal chemographs of Ca^{2+}, TDS and HCO_3^- of streams showed high concentration in January and March while low in remaining months (Fig. 20.11). In summer season, there is significant stream discharge and less interaction between water and rock, which decreases dissolved ion concentration in spring (dilution effect).

However, the Karst springs especially Achabalnag and Kokernag confirm high concentration in July, which is mainly because of piston effect. The temporal plot of spring discharge and TDS (Fig. 20.12) represents both TDS and discharge raise concurrently in July.

20.6.3 Isotopic Approach

The relationship among $\delta^{18}O$ and δD in global precipitation is known as global meteoric water line (GMWL) (Craig 1961) and is given in Eq. 20.1.

$$\delta D = 8 \times \delta^{18}O + 10 \quad (20.1)$$

Rozanski et al. (1993) modified the GMWL, using more available data (Eq. 20.2)

$$\delta D = (8.20 \pm 0.07) \times \delta^{18}O + (11.27 \pm 0.65) \quad (20.2)$$

Based on the amount weighed mean monthly samples, the regression equation between δD and $\delta^{18}O$ (Fig. 12.3) known as local meteoric water line (LMWL) is given in Eq. 20.3.

$$\delta D = 7.094 \times \delta^{18}O + 9.791\,(r^2 = 0.82) \quad (20.3)$$

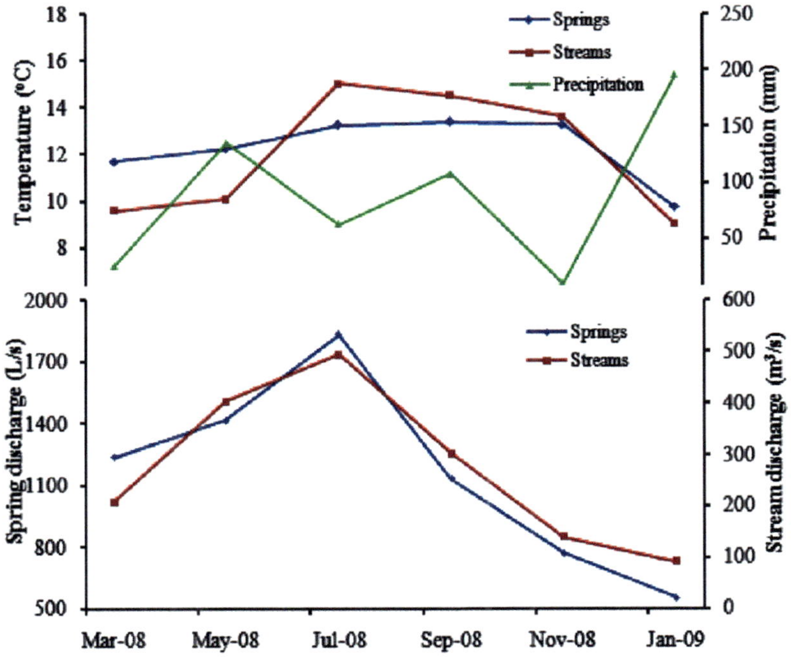

Fig. 20.10 Bimonthly variation of discharge, precipitation amount, stream and spring temperatures (*Source* Bhat et al. 2014)

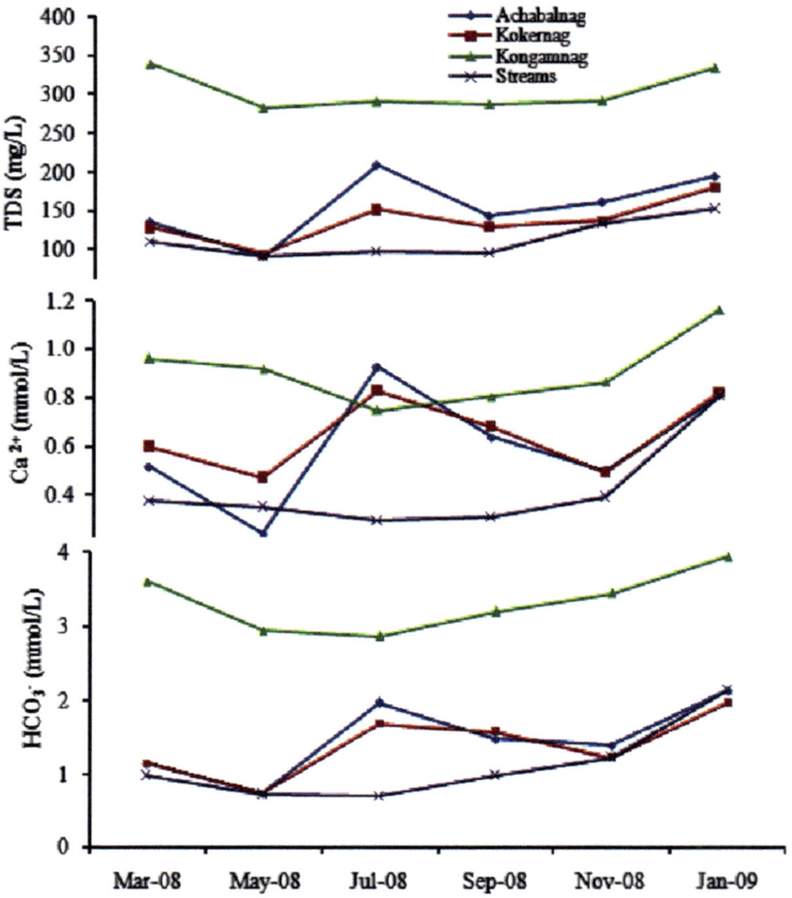

Fig. 20.11 Correlation of Ca, TDS, HCO_3 and chloride between Karst springs and Bringi Stream (*Scource* Bhat et al. 2014)

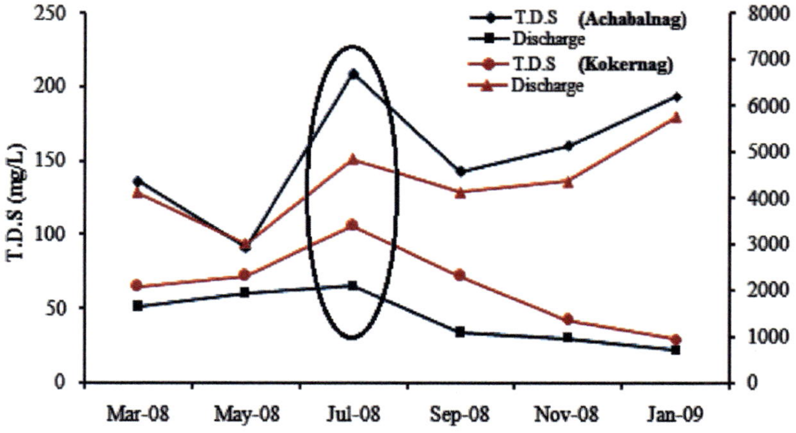

Fig. 20.12 Relationship of TDS with discharge of Achabalnag and Kokernag showing piston effect (*Source* Bhat et al. 2014)

The meteoric water line of study site is almost same as western Himalayas, $\delta D = 7.95 \times {}^{18}O + 11.51$ (Kumar et al. 2010). Shallower slope and low intercept than LMWL of western Himalayas and a shallow slope and high intercept to GMWL may be due to the different sources of moisture effect and/or effect of evaporation. Effect of temperature on the precipitation isotopic composition is observed in $\delta^{18}O$ versus δD plot (Fig. 20.13).

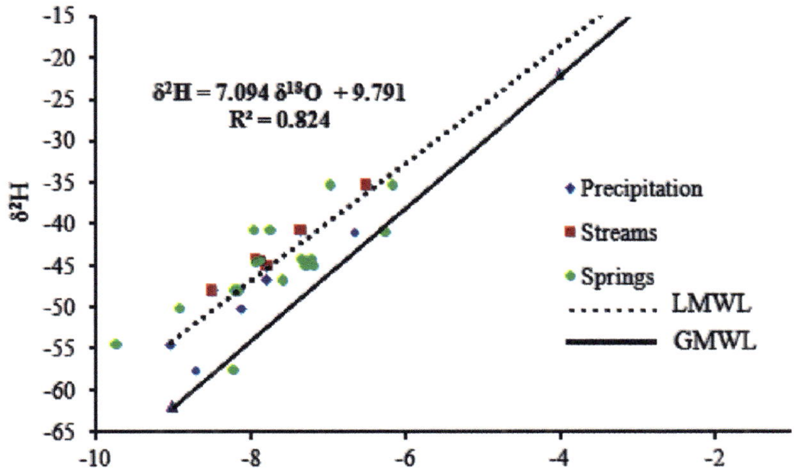

Fig. 20.13 $\delta^{18}O$ vs δD relationship of stream and spring water over GMWL and LMWL of Bringi watershed (*Source* Bhat and Jeelani 2015)

With increasing altitude the precipitation isotopic composition decreases, known as altitude effect (Ingraham and Taylor 1991), which is a significant tool to delineate the spring recharge areas (Jeelani et al. 2010). In the study area, altitude effect of −0.27‰ per 100 m was observed. The average elevation of area of spring recharge varies from 2500 to 2900 m amsl (Fig. 20.14).

There is a very good correlation ($r^2 = 0.97$) in seasonal $\delta^{18}O$ composition of streams and springs (Fig. 20.15), which indicates that the streams recharge these springs at different heights and share similar catchments.

20.7 Components and Mechanism of Groundwater Recharge

Various methods are present for defining the contribution of surface to groundwater. Chloride mass balance equation (CMBE) is used by a number of researchers to calculate the contribution of precipitation to groundwater (Bhat and Jeelani 2018).

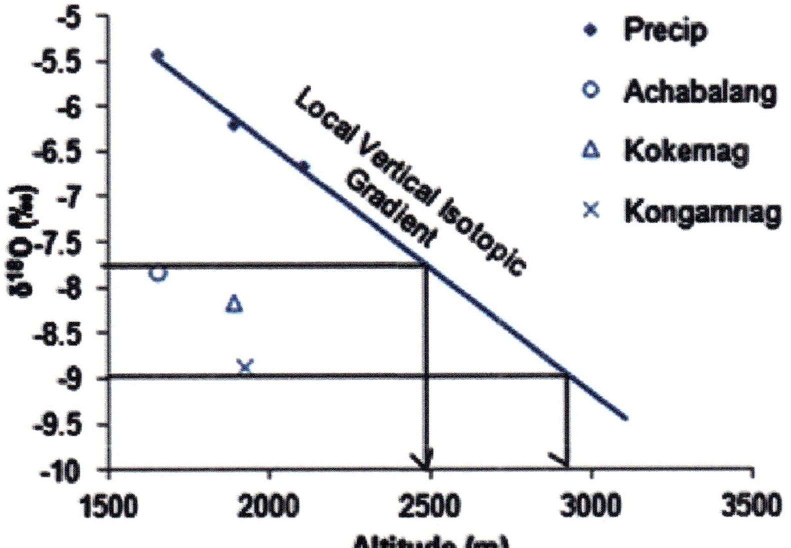

Fig. 20.14 The altitude versus $\delta^{18}O$, showing average elevation of spring recharge area (*Source* Bhat and Jeelani 2015)

$$\text{Recharge(mm)} = \text{Rainfall(mm)} * C_{\text{Rainfall}}/C_{\text{Spring water}} \quad (20.4)$$

where '*C*' is the concentration of chloride present in the precipitation and groundwater. The mean concentration of chloride of precipitation and springs is 0.57 mg/l and 3.55 mg/L. The estimated recharge through precipitation averages at 18.5%, with highest during July (about 22%) and lowest during November (<1%). Isotopic mass balance studies (IBME) defined that studies related to the isotopic mass balance indicate a mixture of two components (i.e. faction of groundwater (YG) and surface water (YS)

$$\text{YS} + \text{YG} = \text{YM} \quad (20.5)$$

$$\text{YS}\delta\text{S} + \text{YG}\delta\text{G} = \text{YM}\delta\text{M} \quad (20.6)$$

where YS and YG are the contribution of surface water and groundwater percentage to the mixture YM. δS, δG and δM are the isotopic composition of surface water, groundwater and admixture, respectively. Substituting Eqs. (20.5) in (20.6) for YG gives contribution of surface water component YS to the groundwater mixture.

$$\text{YS} = \text{YM}(\delta\text{M} - \delta\text{G})/(\delta\text{S} - \delta\text{G}) \quad (20.7)$$

The mean δ^{18}O composition of surface water was −8.077‰ in high flow time (May to July 2008) and for groundwater before the high flow period (March 2008) is −8.19‰. The mean δ^{18}O composition for mixture of groundwater and surface water in September was −7.3‰. Therefore, the components of surface recharge during high flow period, 'YS', average at 337.35 m^3/s, about 75% of total stream discharge.

20.8 Residence Time of Groundwater

As determined by (Clark and Fritz 1997), the groundwater residence time by decay equation (Eq. 20.8) is

$$a_t^3H = a_o^3He^{-\lambda t} \quad (20.8)$$

where a_o^3H is the concentration in precipitation or initial tritium activity (expressed in TU) and a_t^3H concentration of tritium in groundwater or residual activity remains after decay over time *t*. As mean residence time of ground water is quite short (<1 year) in the study, it is short for Achabalnag and longer for Kongamnag. During the present investigation, dye testing was carried out near Adigam (Fig. 20.16a) and Gadol (Fig. 20.16b), which confirmed connection between recharge sites and Karst springs.

Fig. 20.15 Positive temporal variations of δ^{18}O of springs and streams of Bringi watershed (*Source* Bhat and Jeelani 2018)

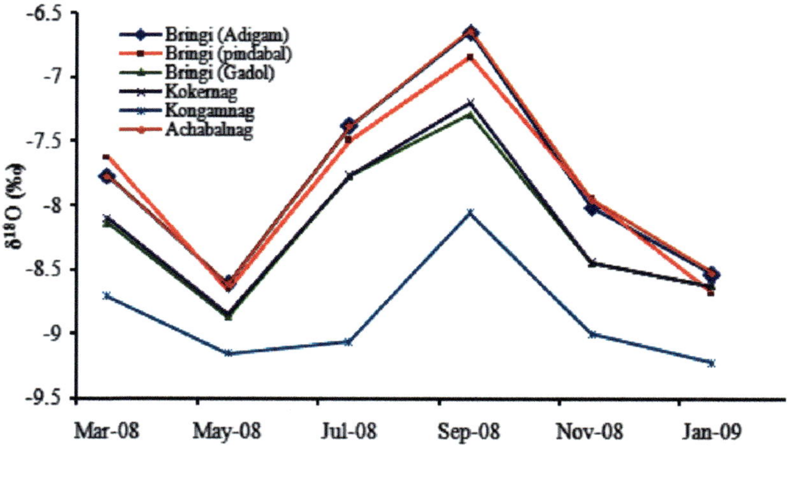

Fig. 20.16 Dye testing carried out near **a** Adigam and **b** Gadol confirming connection between recharge sites and Karst springs (*Source* Authors)

20.9 Quality of Water for Drinking, Agricultural and Livestock Purposes

The water quality was carried out as per Bureau of Indian standards (BIS 2012) and World Health Organization (WHO 2011) for drinking (Table 20.1). The TDS of water samples is within the prescribed limits for the livestocks (Hamill and Bell 1986; Ravindra and Garg 2007). Based on the classification of hardness (Sawyer and McCarthy 1967), 45% samples are categorized under soft, 41.6% under moderately hard and 20.8% under very hard.

A number of plots and formulas are available for determining the suitability of water for purpose of irrigation. Wilcox diagram with specific conductance plotted against percentage Na is used for evaluating water for irrigation purposes (Wilcox 1955). The diagram shows that water is good for irrigation (Fig. 20.17).

Appropriateness of water for the purpose of irrigation can also be determined by plotting electrical conductivity (EC) against sodium absorption ratio (SAR) (Fig. 20.18) on the US Salinity Laboratory (USSL) diagram (Richards 1954). About 62.5% (15 samples) fall in C1S1 field of the diagram indicates low sodium/low salinity-type water and 37.5% (9 samples) belong to C2S1 category indicating good condition of water for irrigation for most soils and crops.

20.10 Conclusions

The results concluded t with following points:

- Ca^{2+} was the dominant cation and HCO_3^- was the dominant anion whereas four hydrochemical types of water have been identified as Ca–Mg–HCO_3 > Ca–HCO_3 > Ca–Mg–Na–HCO_3 > Ca–Na–HCO_3.
- Carbonate weathering is mainly responsible for ions in groundwater as inferred from scatterplots and hydrogeochemical.
- Hydrographs and chemographs for both springs and streams showed high Ca, TDS, EC and HCO_3 during winter and low during summer. The positive correlation of

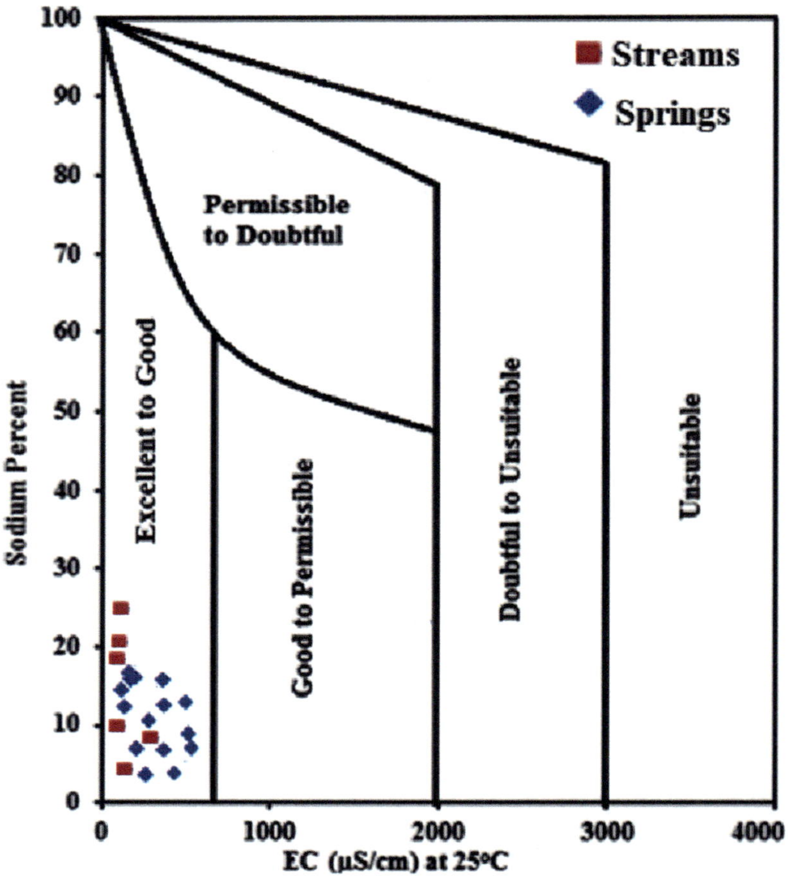

Fig. 20.17 Plot of electrical conductivity (EC) vs percent sodium (% Na) for water classification (*Source* Wilcox 1955)

chemographs of springs and streams indicates that Bringi Stream fed all the springs at various elevations.

- The LMWL for Bringi watershed is $\delta D = 7.094 \times \delta^{18}O + 9.791$ ($r^2 = 0.82$) whereas the springs are the major sources of recharge. The surface recharge component using IMBE averages at 337.35 m^3/s during high flow period, about 75% of total stream discharge and 7.5 m^3/s flow during low flow period, about 18.6% of total stream discharge.

20.11 Recommendations

Based on the present work, certain recommendations are made for preservation of the valuable water resource of the area.

- Fencing of the recharge sites near Adigam and Gadol is necessary to avoid contamination of water.
- Check dams may be built across Bringi Stream at Adigam Village and Gadol Stream near Gadol Village to maintain the flow of Karst springs during lean period.
- Continuous monitoring of water quality of streams and springs in terms of major and heavy metals. Continuous monitoring of stream and spring discharge to understand the response of springs to hydrological events.
- Public awareness programmes need to be conducted to create awareness among the people regarding the importance of preservation of the valuable resources of water of the area.

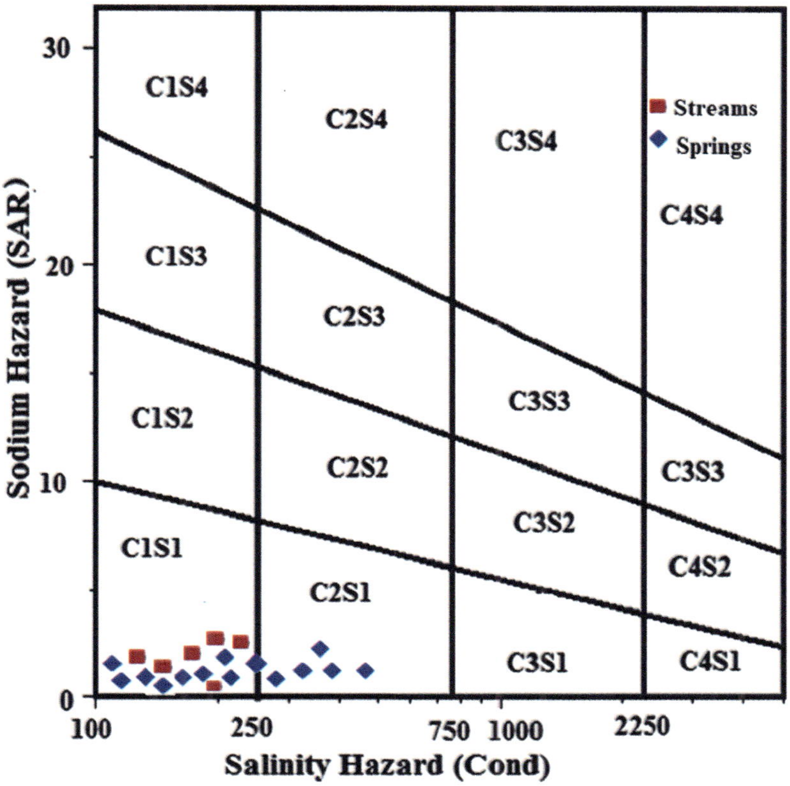

Fig. 20.18 Classification of water for irrigation purpose by using USSL Salinity Hazard Diagram (*Source* Richards 1954)

Acknowledgements The authors are grateful to the Editors of the Book for providing us an opportunity to publish this chapter. We are also extremely thankful to reviewers for constant support during the publication process.

References

Adimalla N, Dhakate R, Kasarla A, Taloor AK (2020) Appraisal of groundwater quality fordrinking and irrigation purposes in Central Telangana. India. Groundwater SustDevelop 10:P100334. https://doi.org/10.1016/j.gsd.2020.100334

Adimalla N, Taloor AK (2020a) Hydrogeochemical investigation of groundwater quality in the hard rock terrain of South India using Geographic Information System (GIS) and groundwaterquality index (GWQI) techniques. Groundwater Sust Develop 10:P100288. https://doi.org/10.1016/j.gsd.2019.100288

Adimalla N, Taloor AK (2020b) Introductory editorial for 'Applied Water Science'special issue: "Groundwater contamination and risk assessment with an application of GIS". Appl Water Sci 10:216. https://doi.org/10.1007/s13201-020-01291-3

Alam A, Bhat MS, Kotlia BS, Ahmad B, Ahmad S, Taloor AK, Ahmad HF (2017) Coexistent pre-existing extensional and subsequent compressional tectonic deformation in the Kashmir basin, NW Himalaya. Quat Int 444:201–208

Alam A, Bhat MS, Kotlia BS, Ahmad B, Ahmad S, Taloor AK, Ahmad HF (2018) Hybrid tectonic character of the Kashmir basin: Response to comment on "Coexistent pre-existing extensional and subsequent compressional tectonic deformation in the Kashmir basin, NW Himalaya (Alam et al. 2017)" by Shah (2017). Quat Int 468:284–289

APHA (2006) Standard methods for the examination of water and waste. American Public Health Association, Washington DC

Bhat MS, Alam A, Ahmad B, Kotlia BS, Farooq H, Taloor AK, Ahmad S (2019a) Flood frequency analysis of river Jhelum in Kashmir basin. Quat Int 507:288–294

Bhat NA, Ghosh P, Ahmed W, Naaz F, Priyadashinee A (2019b) Hydrochemical characteristics and quality assessment of stream water in parts of Gadag, Koppal and Ballery districts of Karnataka India. J Geol Soc Ind 94(6):635–640

Bhat NA, Bhat AA, Nath S, Singh BP, Guha DB (2016) Assessment of Drinking and irrigation water quality of surface water resources of South-West Kashmir India. J Civil Environ Eng 6(2):1000222. https://doi.org/10.4172/2165-784X.1000222

Bhat NA, Jeelani G (2015) Delineation of the recharge areas and distinguishing the sources of karst springs in

Bringi watershed, Kashmir Himalayas using hydrochemistry and environmental isotopes. J Ear Sys Sci 124(8):1667–1676

Bhat NA, Jeelani G (2018) Quantification of Groundwater —Surface water Interactions using Environmental Isotopes; A Case Study of Bringi Watershed, Kashmir Himalayas India. J Ear Sys Sci 127(5):63. https://doi.org/10.1007/s12040-018-0964-x

Bhat NA, Jeelani G, Bhat MY (2014) Hydrogeochemical assessment of groundwater in karst environments, Bringi watershed, Kashmir Himalayas India. Curr Sci 106(7):1000–1007

Bhat NA, Singh BP, Bhat AA, Nath S, Guha DB (2019c) Application of Geochemical Mapping in Unraveling Paleoweathering and Provenance of Karewa Sediments of South Kashmir, NW Himalayas India. J Geol Soc Ind 93(1):68–74

Bhat NA, Singh BP, Guha DB, Bhat AA, Nath S (2019d) Geochemistry of stream sediments and soil samples from Karewa deposits of South Kashmir, NW Himalaya, India. Environ Ear Sci 78(9):278. https://doi.org/10.1007/s12665-019-8212-5

Bisht H, Arya PC, Kumar K (2018) Hydro-chemical analysis and ionic flux of meltwater runoff from Khangri Glacier, West Kameng, Arunachal Himalaya, India. Environ Earth Sci. 77:1–16. https://doi.org/10.1007/s12665-018-7779-6

Clark ID, Fritz P (1997) Environmental isotopes in hydrogeology. Lewis Publishers, Boca Raton

Coward JMH, Waltham AC, Bowser RJ (1972) Karst springs in the Vale of Kashmir. J Hydrol 16:213–223

Craig H (1961) Isotopic variation in meteoric waters. Science 133:1702–1703

Epstein S, Mayeda T (1953) Variation of $\delta^{18}O$ content in waters from natural sources. Geochim Cosmochim Act 4:213–224

Eyankware MO, Nnajieze VS, Aleke CG (2018) Geochemical assessment of water quality for irrigation in abandoned limestone quarry pit at Nkalagu area, southern Benue Trough. Nigeria. Environ Earth Sci 77:66. https://doi.org/10.1007/s12665-018-7232-x

Fetter CW (1980) Applied hydrology, 2nd edn. Merrill, Columbus

Ford D, Williams P (1989) Karst Geomorphology and Hydrology. Chapman and Hall, London

Fredrickson GC, Criss RE (1999) Isotope hydrology and time constants of the unimpounded Meramec River basin, Missouri. Chem Geol 157:303–317

Freeze RA, Cherry JA (1979) Groundwater. Prentice Hall, N.J

Frondini F (2008) Geochemistry of regional aquifer systems hosted by carbonate–evaporite formations in Umbria and southern Tuscany (central Italy). App Geochem 23(8):2091–2104

Gat JR (1971) Comments on the stable isotope method in regional groundwater investigations. Wat Res Res 7:980–993

Gibbs RJ (1970) Mechanisms controlling world water chemistry. Science 170:1088–1090

Goldscheider N, Drew D (2007) Methods in Karst Hydrogeology. Taylor and Francis, London

Gonfiantini R, Gallo G, Payne BR, Taylor CB (1976) Environmental isotopes and hydrochemistry in groundwater of Gran Canaria. Interpretation of environmental isotope and hydrochemical data in groundwater hydrology. IAEA, Vienna, pp 159–170

Hamill L, Bell FG (1986) Groundwater resource development and management. The University Press, Cambridge, Great Britain, p 34

Ingraham NL, Taylor BE (1991) Light stable isotope systematic of large-scale hydrologic regimes in California and Nevada. Wat Res Res 27:77–90

Jasrotia AS, Kumar A (2014) Estimation of replenishable groundwater resources and their status of utilization in Jammu Himalaya, J&K, India. Eur Water 48:17–27

Jasrotia AS, Taloor AK, Andotra U, Bhagat BD (2018) Geoinformatics based groundwater quality assessment for domestic and irrigation uses of the Western Doon valley, Uttarakhand, India. Groundwater Sust Develop 6:200–212

Jasrotia AS, Taloor AK, Andotra U, Kumar R (2019) Monitoring and assessment of groundwaterquality and its suitability for domestic and agricultural use in the Cenozoic rocks ofJammu Himalaya, India: A geospatial technology based approach. Groundwater Sust Develop 8:554–566

Jeelani G, Bhat NA, Shivanna K (2010) Use of $\delta^{18}O$ tracer to identify stream and spring origins of a mountainous catchment: A case study from Liddar watershed, western Himalaya, India. J Hydrol 393:257–264

Jeelani G, Bhat NA, Shivanna K, Bhat MY (2011) Geochemical characterization of surface water and spring water in SE Kashmir valley, western Himalayas: Implication to water–rock interaction. J Ear Sys Sci 120(5):921–932

Jeelani G, Kumar SU, Bhat NA, Sharma S, Kumar B (2014) Variation of $\delta^{18}O$, δD and ^3H in karst springs of south Kashmir, western Himalayas (India). Hydrol Process 29:522–530. https://doi.org/10.1002/hyp.10162

Kattan Z (1997) Environmental isotope study of the major karst springs in Damuscus limestone aquifer system: case of the Figeh and Barada springs. J Hydrol 193:161–182

Kumar B, Rai SP, Kumar US, Verma SK, Garg P, Kumar SVV, Jaiswal R, Purendra BK, Kumar SR, Pande NG (2010) Isotopic characteristics of Indian precipitation. Wat Res Res 46:1–15

Kumar D, Singh AK, Taloor AK, Singh DS (2020) Recessional pattern of Thelu andSwetvarn glaciers between 1968 and 2019, Bhagirathi basin, Garhwal Himalaya, India. Quat Int. https://doi.org/10.1016/j.quaint.2020.05.017

Langelier WE, Ludwig HF (1942) Graphical method for indicating the mineral character of natural water. J Am Wat Wor Assoc 34:335–352

Lawrence W (1967) The Valley of Kashmir, Kesar Publishers

Lee KS, Wenner DB, Lee I (1999) Using H- and O-isotopic data for estimating the relative contributions of rainy and dry season precipitation to groundwater: Example from Cheju Island, Korea. J Hydrol 222:65–74
McConville C, Kalin RM, Johnston H, McNeil GW (2001) Evaluation of recharge in a small temperate catchment using natural and applied $\delta^{18}O$ profiles in the unsaturated zone. Groundwater 39:616–624
Middlemiss CS (1910) A revision of Silurian-Triassic sequence of Kashmir. Rec Geol Sur Ind 40(3):6–260
Piper AMA (1994) graphical procedure in the geochemical interpretation of water analysis. Trans Am Geophys Uni 25:914–928
Ravindra K, Garg VK (2007) Hydrochemical survey of ground water of Hisar city and assessment of defluoridation methods used in India. Environ Monit Assess 132:33–43
Richards LA (1954) Diagnosis Improvement Saline Alkali Soils. US Department of Agriculture Handbook. 60
Rozanski K, Arugu′as-Arugu′as L, Ganfiantini R (1993) Isotopic patterns in modern global precipitation. Geophys Mono 78:1–36
Sah S, Bisht H, Kumar K, Tiwari A, Tewari M, Joshi H (2017) Assessment of hydrochemical properties and annual variation in meltwater of Gangotri glacier system. ENVIS Bullet Himalayan Ecol 25:17–23
Sarkar T, Kannaujiya S, Taloor AK, Ray PKC, Chauhan P (2020) Integrated study of GRACE data derived interannual groundwater storage variability over water stressed Indian regions.Groundwater Sustain Develop 10:100376. https://doi.org/10.1016/j.gsd.2020.100376
Sawyer GN, McCarthy DL (1967) Chemistry of sanitary engineers, 2nd edn. McGraw Hill, New York, p 518
Siegenthaler U, Oeschger H (1980) Correlation of ^{18}O in precipitation with temperature and altitude. Nature 285:314–317
Singh AK, Jasrotia AS, Taloor AK, Kotlia BS, Kumar V, Roy S, Ray PKC, Singh KK, SinghAK SA (2017) Estimation of quantitative measures of total water storage variation from GRACE and GLDAS-NOAH satellites using geospatial technology. Quat Int 444:191–200
Taloor AK, Kotlia BS, Jasrotia AS, Kumar A, Alam A, Ali S, Kouser B, Garg PK, Kumar R, Singh AK, Singh B (2019) Tectono-climatic influence on landscape changes in the glaciatedDurung Drung basin, Zanskar Himalaya, India: A geospatial approach. Quat Int 507:262–273
Taloor AK, Kumar V, Singh VK, Singh AK, Kale RV, Sharma R, Khajuria V, Raina G, Kouser B, Chowdhary NH (2020a) Land Use Land Cover Dynamics Using Remote Sensing and GIS Techniques in Western Doon Valley, Uttarakhand, India. InGeoecology of Landscape Dynamics 2020 (pp. 37–51). Springer, Singapore
Taloor AK, Pir RA, Adimalla N, Ali S, Manhas DS, Roy S, Singh AK (2020b) Spring water quality and discharge assessment in the Basantar watershed of Jammu Himalaya using geographic information system (GIS) and water quality Index(WQI). Groundwater Sust Develop 10:P100364. https://doi.org/10.1016/j.gsd.2020.100364
Wilcox LV (1955) Classification and Use of Irrigation Waters. USDA. Circ 969, Washington, DC

Dr. Nadeem Ahmad Bhat is working as a Senior Geologist in Geological Survey of India. He has completed his M. Phil from University of Kashmir, Srinagar on Isotope Hydrology and Ph.D. from Banaras Hindu University, Varanasi, India on Sediment Geochemistry. He is honoured with high ranks in UPSC Geologist Examination. He has more than 15 years of professional experience in the fields of hydrology, geochemistry, field geology, geological mapping, mineral investigations, Himalayan geology and engineering geology. He has published 15 papers in reputed national and international journals, 7 technical reports and 13 abstracts in national and international conferences. He has served as a core faculty at GSI Field Training Centre, Aishmuqam, imparting training on Himalayan and Engineering Geology to newly recruited Geoscientists of Geological Survey of India

Prof. Ghulam Jeelani is presently serving as HOD, Department of Earth Science, University of Kashmir, Srinagar. He has carried out Master in Geology from Aligarh Muslim University, UP, India. Later, he has also carried out his Ph.D. from the Aligarh Muslim University in Hydrogeology. He has also carried out his Post-Doc from University of Kansas, Lawrence, USA. He has several awards to his credit like NET, CSIR/UGC, Excellent Grade Teacher Award by IQAC (University of Kashmir), INSA-Visiting Fellow and Fulbright-Nehru Senior Research Fellow and is a Visiting Scientist PRL, Ahmedabad, India. He has been Organizing Secretary/Convener of several workshops and is also acting as an Organizing Committee Member/Chairman of Committee/of several committees related to hydrogeology, climate change studies and sustainable development. He has completed several research projects and collaborations with many National Institutions. He has also supervised several candidates for their M.Phil. and Ph.D. degrees and has completed three Ph. D. and six M.Phils. He has authored about 50 articles published in International/national referred journals, about 40 abstracts presented in national/international, conferences and symposiums and six book chapters (Springer and Elsevier Publications). He has vast experience of more than 15 years in teaching and attended several national/international training programmes. He is a specialized worker in Hydrogeology, Hydrogeochemistry, Environmental Geology, Isotope Geology, Climate Change and Glaciology

Dr. Riyaz Ahmad Mir is working as a Sr. Geoscientist in Geological Survey of India for the last 8 years. He has been honored with high ranks in joint CSIR-UGC (NET+JRF) examination and UPSC Geologist/Geoscientists Examination. After receiving his Master's degree in Geology from Bundelkhand University, Jhansi, he completed his M.Phil. in Hydrogeochemistry from University of Kashmir, Srinagar and his Ph.D. in Geoinformatics from Indian Institute of Technology (IITR), Roorkee. He has also worked with the National Institute of Hydrology (NIH), Roorkee for his Doctorate degree. He has been awarded with best Paper Presentation (SiD-2018) at Cluster University, Srinagar. He has a good working experience in the areas of hydro-and-sediment geochemistry, remote sensing and GIS and its applications in glacier, Glacier Lake Outburst Flood (GLOF), landslide studies and its modelling, climate change studies and its effect of water resources. He has also a vast field expertise in Geochemical Mapping (GCM), Specialized Geological Thematic Mapping (STM) and landslide zonation and susceptibility mapping, etc. He has published about 20 research articles in refereed international/national journals, about 10 extended abstracts in national and international conferences/symposiums and 1 book chapter (Springer publication). He has also completed about 7 Geological Reports of different research projects at GSI. He has published two articles in Hindi Magazine 'Tawi' published by GSI, J&K and is also writing to daily locals regarding geoscientific issues. He has delivered several guest lectures at Degree College and Intermediate level pertaining to the role of Earth Sciences in society

Index

A. Taloor et al. (eds.), *Water, Cryosphere, and Climate Change in the Himalayas*,
Geography of the Physical Environment, https://doi.org/10.1007/978-3-030-67932-3

GPSR Compliance

The European Union's (EU) General Product Safety Regulation (GPSR) is a set of rules that requires consumer products to be safe and our obligations to ensure this.

If you have any concerns about our products, you can contact us on ProductSafety@springernature.com

In case Publisher is established outside the EU, the EU authorized representative is:

Springer Nature Customer Service Center GmbH
Europaplatz 3
69115 Heidelberg, Germany

Batch number: 10732807

Printed by Printforce, the Netherlands